The Editors

Dr. Ashok Aggarwal is currently working as Professor of Botany and teaches Plant Pathology, Mycology and Microbiology at the Department of Botany, Kurukshetra University, Kurukshetra, where he has been a faculty member for the last 30 years. He is also a member of several scientific organizations and has published more than 180 research papers and review articles in reputed journals of high impact factors. He has co-authored two monographs on the topic entitled, "Phytophthora Diseases in India" in 2001 and "Endomycorrhizal Diversity on ethnomedicinal plants of Himachal Pradesh" (VDM Verlag Pub. Germany) in 2013. He has also edited another book entitled, "Glimpses of Plant Sciences" in 2002. In addition, he has also written two textbooks entitled "Plant Pathology" and Fundamentals of Plant Pathology" published by McGraw Hill Education (India) Private Ltd., New Delhi with Professor Mehrotra. Dr. Aggarwal has guided 10 M.Phil and 20 Ph.D. students on diverse aspects of mycorrhizal fungi.

Dr. Kuldeep Yadav is working as Assistant Professor at Department of Botany, Kurukshetra University, Kurukshetra since 2014. He has done M.Sc., M.Phil and Ph.D. from Kurukshetra University, Kurukshetra. He has contributed extensively in the field of Plant Biotechnology, Molecular Biology and Mycology. He has been engaged in the micropropagation of several important medicinal plants like *Spilanthes acmella, Gloriosa superba, Acorus calamus, Aegle marmelos, Glycyrrhiza glabra, Stevia rebaudiana* etc. He has successfully developed protocol to acclimatize the tissue culture-raised plants using AM fungi as hardening media. He has published more than 50 research papers in both international/national peer reviewed journals and has also authored three book chapters. He is also a life member of Medicinal and Aromatic Plants Association of India.

Mycorrhizal Fungi

Mycorrhizal Fungi

— Editors —

Ashok Aggarwal

Kuldeep Yadav

Department of Botany,
Kurukshetra University,
Kurukshetra, Haryana

2017

Daya Publishing House®

A Division of

Astral International Pvt. Ltd.

New Delhi – 110 002

Cataloging in Publication Data--DK
  Courtesy: D.K. Agencies (P) Ltd. <docinfo@dkagencies.com>

    Mycorrhizal fungi / editors, Ashok Aggarwal, Kuldeep Yadav.
    pages cm
    Includes index.

    ISBN 978-93-86071-79-8 (International Edition)

    1. Mycorrhizal fungi.   I. Aggarwal, Ashok (Professor of botany),
  editor. II. Yadav, Kuldeep, editor.

  QK604.2.M92M93 2017   DDC 579.5   23

Published by   : **Daya Publishing House®**
                           A Division of
                           Astral International Pvt. Ltd.
                           – ISO 9001:2015 Certified Company –
                           4736/23, Ansari Road, Darya Ganj
                           New Delhi-110 002
                           Ph. 011-43549197, 23278134
                           E-mail: info@astralint.com
                           Website: www.astralint.com

Digitally Printed at   : **Replika Press Pvt. Ltd.**

— Dedicated to —

Dr. R.S. Mehrotra

for his inspiration as a teacher

Foreword

The increasing relevance of pesticides, fungicides and fertilizers in present agricultural system has created environmental pollution besides increasing crop productivity. Therefore, development of strategies and management of sustainable agriculture is the main concern in agriculture world wide. To increase agricultural production, one way is to utilize naturally occurring microbes in the ecosystem.

Arbuscular mycorrhizal (AM) fungi can form symbiosis with cultivated plants by infecting the roots. AM fungi can act as a "biofertilizer" by facilitating access of roots to nutrients. The demand of commercial application of AM fungal inoculum is increasing day by day. Reasons for mass multiplication of AM fungi include the fact that AM fungi are being considered as a "National Plant Health Insurance". Their positive impacts on land reclamation, phytoremediation and biocontrol are well recognized and there is higher awareness regarding the biodiversity issues. Probably, it is essential for the establishment of tree seedlings and their growth enhancement especially in low fertile soils. Furthermore, there is an increasing demand from the society regarding sustainable means of production with a feedback from growers and scientists

This book explores the various functions and potential applications of AM fungi including topics such as biodiversity and dynamics of root colonization, phosphorus acquisition by AM fungi, soil carbon sequestration and the functions of mycorrhizae under stress environments. Some authors focus on the use of AM fungi in the various crop production processes, including soil management practices, their use as biofertilizers in different crop systems. Other chapters elucidate the role of AM fungi in the alleviation of salt stress, water stress, heavy metal toxicity and in mining-site rehabilitation. In addition to their impact on ecosystem, the applications of AM fungi are also discussed.

It gives me a great pleasure to write a foreword of this edited book entitled, "Mycorrhizal Fungi" by Ashok Aggarwal and Kuldeep Yadav, Department of Botany, Kurukshetra University, Kurukshetra, which is one of the premier institutions in the field of Mycology and Plant Pathology.

The authors are to be congratulated on writing a book that gives so much information on various topics of AM fungi at single place in an uncomplicated way so as to be used and understood by teachers, students and scientists.

Dr. Kailash Chandra Sharma

Vice-Chancellor
Kurukshetra University,
Kurukshetra, Haryana

Preface

With an increase in human population and over exploitation of natural resources, there is an urgent need for quality food production and conservation of the environment. The unavoidable side effects of pesticides, fungicides and higher dozes of fertilizers have highlighted the use of microorganisms such as arbuscular mycorrhizal fungi (AMF). AMF are obligate biotrophs, living symbiotically in the roots of nearly eighty per cent of land plant species. Most of AMF spores germinate in soil but are unable to complete the life cycle without establishing a functional symbiosis with a host plant.

During the last 3-4 decades, substantial work has been done to understand the AMF symbiosis and its role in sustainable agriculture. Knowledge obtained from this host/fungal association underlines the interest of mycorrhizal research for understanding not only life cycle and symbiotic association but also the role of symbiotic association in sustainable agriculture.

There are several groups of mycorrhizal fungi of which the main types are the endomycorrhizal and ectomycorrhizal fungi and this book is written with the view that it will stimulate mutualism between mycorrhizal workers and ecologists.

Recent researches in the taxonomy including assessing the diversity of AMF offer opportunities for reinvigorating research on the management of mycorrhiza in various ecosystems. These advances further provide the incentive for promoting information on plant mycorrhizal association in debates about soil management, choice of plants, the application of minimum doze of fertilizers and environmental impacts of agricultural production. Understanding a particular phenomenon is also important for predicting the successful migration of plants and compatible mycorrhizal fungi during climate change. The recent PCR based molecular techniques have revolutionized our ability to identify the type of mycorrhizal association with plant and to derive phylogenetic relationship among mycorrhizal fungi. The presence of multiple host fungi versus host specific fungi has been tuned with the application of molecular techniques. Molecular studies will continue to provide a wealth of new information and approaches to determine the potential of mycorrhizal fungi.

Molecular studies have made a major breakthrough in the taxonomic organization of the phylum Glomeromycota. Fungi present in this group are known to form arbuscular mycorrhiza and this phylum has three classes, five orders and fourteen families with twenty-nine genera. The present classification based on molecular studies provides information to facilitate the classification of taxa from genus to class level. Based on morphological and molecular characteristics nineteen genera comprising more than 200 species are documented in AM fungi.

It may be mentioned that interest in this area has increased many folds in recent years due to their unique ability to increase the uptake of phosphorus by plants and for the utilization of AM as a substitute for phosphate fertilizers.

Even though each chapter is complete in every aspect but there has been a slight repetition of certain information but it was unavoidable as different aspects of mycorrhiza discussed are interrelated.

This book will throw light on diverse topics with latest updates like biodiversity of AM fungi, their mass multiplication and taxonomy, AM fungi as biofertilizers, as biocontrol agent, their role in alleviating various abiotic stresses and recent molecular approaches. AM fungi play an important role in enhancing soil fertility, crop yield and quality under different stress conditions leading towards sustainable agriculture. In the last two decades, interest in utilization of AM fungi in crop production has been practical due to commercialization. Nutrient management and disease control using plant- microbe interaction in agro-ecosystem would permit considerable reduction of chemical fertilizers and pesticide inputs, leading to sustainable agriculture.

The main purpose of writing this book is to provide a comprehensive source of current information and future prospects for further research. This book will be a rich source of inspiration for research and exploitation of the potential of mycorrhizal fungi. This book will ensure the relevant and latest comprehensive information useful to postgraduate students, researchers, scientists, teachers, different crop growers and industrialists who wish to adopt AM technology.

Ashok Aggarwal

Kuldeep Yadav

Contents

— Part I —
Mycorrhizal Fungal Biodiversity

1. Introduction – 2. Evolution of Mycorrhizae – 3. Types of Mycorrhizae – 4. Morphology and Development of Ectomycorrhizae – 5. Ecological Distribution – 6. Diversity of Ectomycorrhizal Fungi – 7. Ectomycorrhizal Host Plant Diversity – 8. Carbohydrate Physiology of EcM – 9. Applications of Ectomycorrhizal Symbioses – 10. Mineral Nutrient Cycling by Ectomycorrhizae – 11. EcM and Plant Root Disease Control – 12. Ectomycorrhizae in Forestry – 13. Ectomycorrhizal Fungi as a Source of Unconventional Food – 14. Ectomycorrhizae and Climate change – 15. Conclusions – References.

Yash Pal Sharma, Geeta Sumbali and R.S. Mehrotra

— Part III —
Arbuscular Mycorrhizal Fungi as Biofertilizers

— Part V —
Molecular Approaches and Biocontrol

The Contributors

R.S. Mehrotra

A Ph.D. from the University of Saugor, (now Dr. Hari Singh Gour University), Sagar, R.S. Mehrotra has a teaching and research experience of more than 35 years. He retired from Kurukshetra University as a Professor in 1997. He was a Post-Doctoral Fellow at the University of Western Ontario, London; Ontario Canada (1968-1970) and a Fulbright fellow at the University of California, Riverside Campus (1979). He is a fellow of National Academy of Sciences, Allahabad; and was elected President of the Indian Phytopathological Society (1990), President of Indian Society of Plant Pathologists (1992), President of the Mycological Society of India (1996), and President of the Botanical Section of Indian Science Congress (1997). He was the recipient of the Birbal Sahni Gold Medal of the Indian Botanical Society in 1998. Professor Mehrotra delivered the Dastur Memorial Lecture of the Indian Phytopathological Society in 2010. He has published around 200 research papers in national and international journals and has authored 5 books- *Plant Pathology* (McGraw Hill Education, India); *An introduction to Mycology*; *A Monograph on Phytophthora Diseases in India*; *Principles of Microbiology* (McGraw Hill Education, India) and *Fungal Diversity and Biotechnology*.

Geeta Sumbali

Professor Geeta Sumbali has been teaching for the last 32 years, various subjects like microbiology, mycology, plant pathology, economic botany, taxonomy and plant development in the Department of Botany, University of Jammu. She worked as Head, Department of Botany (Jan 2012- Jan 2015), Co-ordinator UGC-SAP Botany (2013-2016) and is presently the senior most faculty member working as

Convenor, Board of Studies in Botany, University of Jammu. She has been a Member Syndicate, University of Jammu (2012-2015) and Member Syndicate University of Kashmir (2012-2015). Her main focus of research work is post harvest pathology, mycotoxicology, mycodiversity, keratinophilic and onychomycotic fungi, cultivation of milky and oyster mushrooms. She has supervised 30 M.Phil and Ph.D. students and has published more than 150 research and review papers in National and International journals of repute. In addition she has written 2 books- The Fungi (Narosa Publishing House Private Limited) and Principles of Microbiology (Tata McGraw Hill Education private limited). She has handled 2 minor research projects (UGC funded) and 6 major research projects (CSIR, GBPIHED and UGC funded). She is also a Fellow of Indian Phytopathological Society (FIPS) and Fellow of Applied and Natural Science Foundation (FANSF).

Yash Pal Sharma

Dr. Yash Pal Sharma is presently working as a Professor and Head, Department of Botany, University of Jammu, Jammu. With a teaching carreer spanning over twenty years, Prof. Sharma has been teaching diverse courses of Botany including Mycology and Plant Pathology, Phycology, Gymnosperms, Plant Metabolism, Plant Anatomy, *etc.* He specializes in the area of Mycology and Plant Pathology with emphasis on Mushroom Diversity of North-West Himalaya and Ethnomycology. Having supervised one PDF, six Ph.D and twelve M.Phil students, he has published seventy-five research papers in national and international journals and has handled two research projects. He is a life member of many scientific organizations/societies. In recognition of his outstanding contributions to the field of Plant Sciences, he has been conferred with various awards such as Young Scientist Award (2003) of J and K State Council for Science and Technology, Jammu and Kashmir, UGC-Research award, Prof. P.N. Mehra Young Scientist Award (2005) and Prof. H. C. Dube outstanding Young Scientist Award (2010) by the Indian Society of Mycology and Plant Pathology, Udaipur.

Sharda R. Gupta

Professor S.R. Gupta taught Ecology for the last 34 years at the Department of Botany, and worked as Professor and Chairperson Department of Botany (1996-99, 2009- 2011), Dean Life Sciences (2009- 2011), Director Institute of Environmental Studies (2006-2007), and Professor Emeritus (UGC) from 2011-2013 at Kurukshetra University, Kurukshetra. She has been a Post Doctoral Fellow at the Rothamsted Experimental Station, Harpenden, Herts, U.K. under The Commission of European Communities and is Fellow of the National Institute of Ecology. The main focus of her research work is on biodiversity characterization and ecosystem functioning of grassland and forest systems, forest biomass assessment and soil carbon exchange, diversity of AM fungi in conservation agricutural systems and salt-affected soils, enhancing environmental services of salt –affected lands through agroforestry, carbon sequestration in forest and agroforestry systems for mitigating climate change, and the sustainability of agroforestry and agricultural systems. She has supervised twenty Ph.D. students and published more than one hundred research papers in national and international journals.

Solomon Das

Solomon Das has technical knowledge in microbiology, molecular biology and Biochemistry. He has analytical approach towards various aspects of microbial-ecology. He completed his Doctorate (2015) in Forest Biotechnology, entitled "Effect of *Pseudomonas fluorescens* and *Glomus* sp. on growth, nutrition and proliferation of *Dendrocalamus strictus*". During research, he has published 5 research papers in national and international journals, presented papers in 5 national and international conferences, authored one book and successfully completed three research projects.

Vipin Parkash

Dr. Vipin Parkash did his Ph.D. in the year 2004 from Botany Department Kurukshetra University, Kurukshetra, Haryana. He is presently working as Scientist - E in the Indian Council of Forestry Research and Education, an autonomous council of Ministry of Environment, Forest and Climate change, Govt. of India, at Rain Forest Research Institute, Jorhat, Assam, India. Dr. Parkash has significantly contributed in the field of Mycology, Plant Pathology and Soil Microbiology. He has received many prestigious awards like Young Achiever Award - 2014 and Achiever Award - 2015 instituted by SADHNA, Dr. Y. S. Parmar University of Horticulture and Forestry, Solan, Himachal Pradesh; George Bentham Research Award in Biodiversity for the year-2015 instituted by International Agency for Standards and Ratings; Scientist of the year- 2015 by International Foundation for Environment and Ecology *etc.* He has published more than 75 research articles and research papers in reputed International and National Journals having citations and impact factors. He has written one book/monograph on Endomycorrhizal Diversity on ethnomedicinal plants of Himachal Pradesh with Prof. Ashok Aggarwal (VDM Verlag Pub. Germany) and edited two books. He has also handled and completed research projects financed by Indian Council of Forestry Research and Education (India), CSIR, New Delhi. He has successfully guided 5 research scholars for Ph.D. Degree; 5 students for M. Phil. Degree and 15 students for M.Sc. dissertation. He has also acted as a member of board/committees in various Research Advisory Committees/Board/Ph.D. Board of National Universities and Institutions and in editorial and review board of more than fifteen International and National scientific journals related to Mycology, Plant pathology, Soil Microbiology and Environment and Ecology. He has also discovered and described two new fungal species *i.e. Lysurus habungianus* sp. nov. Gogoi and V. Parkash; MycoBank No. MB 812277; and *Gelatinomyces conus* sp. nov. V. Parkash; MycoBank No. MB 815447 (under review).

Tongmin Sa

Tongmin Sa is a Professor in the Department of Agricultural Chemistry, Chungbuk National University in South Korea, working on soil microbiology and fertilizers. His academic background includes bachelor's and master's degree from Seoul National University and doctoral degree in soil science from North Carolina State University. He is recognized for researches on development of biofertilizer and microbial diversity changes in agricultural soils under stress environmental conditions. He has written 18 book chapters, authored or co-authored over 280 peer reviewed publications and advised over 70 graduate students and postdoctoral researchers from many countries.

Sapana Sharma

Dr. Sapana Sharma is working as Post Doctoral Fellow at Regional Research Station (Punjab Agricultural University, Ludhiana), Ballowal Saunkhri in the Women Scientist Scheme funded by University Grants Commission, New Delhi. She did her B.Sc. (Medical) from H.P.U. Shimla in 2002, M.Sc. Botany from Kurukshetra University, Kurukshetra in 2005 and Ph.D. Botany from Kurukshetra University, Kurukshetra in 2009. She is recipient of Gold Medal in M.Sc. and University Research Scholarship in Ph.D. programme. She has worked on biodiversity, mass culture and inoculation techniques of AM fungi associated with medicinal plants. She has isolated 70 different species of AM fungi during Ph.D. research work. She has published 15 research papers in national and international journal of repute.

Chaya Pradeep Patil

Dr. (Mrs) Chaya P. Patil, is a Professor and Head at Department of Agricultural Microbiology, University of Horticultural Sciences, Bagalkot, Karnataka, India. She has been teaching Agricultural Microbiology, Applied Microbiology, Environmental Science, Food Microbiology and Biology of Mycorrhizae to under graduate and post graduate students for the last 26 years. She has specialized in VA Mycorrhizal fungi especially for horticultural and agricultural crops for Jamun, Mango, Papaya, Lime, Fig and Pomegranate for Package of Practice. The main focus of her research work is on applied aspects of VA mycorrhizae, mushroom production, organic farming, biofertilizers and bioformulations and microbial consortia. She has also focused on Homa Farming (Agnihotra). She has supervised more than 50 M.Sc. and Ph.D. students, and published more than one hundred research papers in national and international journals. She has also written three book chapters on biotechnology of VA Mycorrhiza.

B.N. Reddy

Prof. B.N. Reddy is presently heading the Department of Botany at Osmania University, Hyderabad. He has 30 years of teaching and 36 years of research experience at Osmania University. He published more than 70 research papers in peer reviewed National and International journals.

He is the author and Editor of few Books and contributed many chapters (25) for Books brought out by the reputed publishers in USA, UK, Germany, Hungary and Netherlands. He presented his work in several National and International Conferences (120) and delivered many Plenary and Key Note addresses in India and abroad. He successfully completed 4 major research projects and organized 15 National Conferences. He is the Chairman and Member of several academic bodies/ committees and Mentor/Resource Person of prestigious INSPIRE Programme of DST, Govt. of India. He is the Fellow of Andhra Pradesh Academy of Sciences and recipient of State Best Teacher Award for the year 2015.

A. Hindumathi

Dr. A. Hindumathi completed her graduation, post-graduation and Ph.D. from Osmania University, Hyderabad. She is working in field of mycorrhizae for the past 20 years and has contributed significantly in this subject. Her Areas of interest include mycorrhizal fungi, *Rhizobium*, plant growth promoting rhizobacteria, *Trichoderma* with reference to soil-plant-microbial interactions for improvement of plant growth, crop productivity, soil health and biocontrol of plant pathogens. She published her work in reputed National and International peer reviewed research journals and also presented her work in National and International conferences. She is the author of a Book- Systematics and Occurrence of Arbuscular Mycorrhizal Fungi, published by LAP LAMBERT Publishing, Germany. She was awarded the prestigious Women Scientist twice under Women Scientist Scheme – A, by DST, New Delhi, India. She has contributed several book chapters brought out by the reputed international publishers.

Richa Raghuwanshi

Dr. Richa Raghuwanshi is working as an Assistant Professor at the Department of Botany, Mahila Mahavidyalaya, Banaras Hindu University, Varanasi, since 2005. She has been teaching Botany, Applied microbiology and Bioinformatics at UG and PG levels. Three students have been awarded Ph. D degree under her supervision. Her research is focused on microbes as biofertilizers and biocontrol agents in cultivated crops and Indian medicinal plants, mycorrhizal technology and molecular basis of plant-microbe interaction. She has published 50 research papers, book chapters and also co-edited one book in Springer. She has received four awards in various national and international conferences. She has organized four conferences of national and international level in the capacity of Organizing Secretary. She has been working as member of editorial board of journals and books.

Neera Garg

Neera Garg is a Professor of Botany in Panjab University, Chandigarh, India. Her research work focuses on impact of abiotic stresses such as salinity, heavy metals, etc on biological nitrogen fixation in legumes and alleviation of these stresses using soil-microbes like Rhizobium, Arbuscular Mycorrhizal (AM) fungi. She has extended her work on the role of exogenous alleviants such as Silicon, flavonoids, polyamines, phytohormones etc in imparting tolerance to abiotic stresses. Her recent work is on physiological, biochemical and molecular aspects- such as nutrient uptake, activity and expression of antioxidant enzymes, osmolytes like proline and trehalose metabolism. She was awarded fellowship by ISPP, New Delhi and also received Shiksha Rattan Puraskar from India International Friendship Society, New Delhi in 2007. She has headed major UGC and DBT projects and has delivered lectures in European/International Nitrogen Fixation and Plant Physiology Conferences. She has published her work in more than 100 reputed journals of high impact factors and has guided a number of Ph.D/M.Phil students.

H.K. Kehri

Dr. (Ms.) H.K. Kehri, D.Phil., F.P.S.I., F.N.R.S., F.B.S., is a Professor in the Department of Botany, University of Allahabad. She has more than 35 years of research experience and 28 years of teaching experience. She is well known for her important contributions in the field of Rhizosphere microbiology with special reference to dry farming and application of AM fungi for the reclamation and revegetation of stressed sites. For more than a decade, she was associated with an International program of research namely Tropical Soil Biology and Fertility (TSBF) Kenya, devoted to development of sustainable agriculture and maintenance of soil fertility through biological means in tropical countries. She has handled a number of research projects funded by various agencies like DST, CSIR and UGC. She has published more than 100 research papers and review articles in journals of repute and is associated with more than a dozen scientific societies. She has edited three books and authored one. Ten students have been awarded D.Phil. degree under her guidance and more than seven are working at present. She is a recipient of a number of awards and honors including Silver TERI Medal award in 1992 sponsored by IDRC Canada for her research paper on VA Mycorrhizae.

V.K. Suri

Prof. V.K. Suri, Former Vice-Chancellor, Chandra Shekhar Azad University of Agriculture and Technology, Kanpur, Uttar Pradesh, obtained B.Sc. Agri. (Honours in Soil Sciences) and M.Sc. degree in Soil Science in 1976 and 1979 respectively from the Punjab Agricultural University (PAU), Ludhiana, Punjab. He did his Ph.D. in Soil Science from the Indian Institute of Technology (IIT), Kharagpur, West Bengal in 1986 where he also served as Senior Research Assistant/Scientist till 1986 (1982-1986). He has served as the Head, Department of Soil Science at the CSK Himachal Pradesh Agricultural University at Palampur (HP). Earlier, he has also served as the Head/OI, Agricultural Technology Information Centre at the above university. He worked as Assistant Soil Scientist/Assistant Professor during 1986-1994 (8 years) at the CSK H.P. Agricultural University, Palampur (HP), as Senior Soil Scientist/ Associate Professor during 1994- 2002 (8 years) and as Principal Scientist/Professor from 2002 to 2015 (12.5 years nearly). He has also served as the Chief Scientist (Water Management) of the university twice, total period being more than 4.5 years nearly. He has retired from the said position in Jan, 2015. Dr. Suri has also additionally served as the Associate Director (Research) of the university. His experience in agriculture (teaching, research, extension, development, administration, policy framework, *etc.*) spans more than 34 years out of which 24 years also involved administrative work. He has worked extensively in the areas of enhancing fertilizer use efficiency of rainfed lands using organic manures, green manures and biofertilizers. He has developed blue green algae technology for rice farmers and mycorrhizal biofertilizer (AM fungi) biotechnology for maize, wheat, soybean, okra and pea crops. Further, he has also developed integrated nutrient management technologies for various rainfed crops covering cereals, oilseeds, pulses and vegetables.

H.C. Lakshman

Prof. H. C. Lakshman did his M.Sc. degree in Botany at Banglore University in 1977. He joined to teaching profession in added private college (KLEs) from 1979-1996. He worked as Assistant Professor and Associate Professor in Botany for 20 years and joined Karnataka University as Professor in 1996. He is working on various aspects of mycorrhiza mainly on ecology, taxonomy, histochemistry, growth response physiology and interaction studies with other beneficial microorganisms. He served as Chairman (2009-2011), visited France and Italy to present his research paper in 2001. He was awarded F.B.S. (Fellow of Indian Botanical Society) in 2014, Award for Excellence in Research-2013 Rajasthan, Eminent Scientist of the year award -2011 New Delhi, F.I.S.E.C. 2009 Jharkhand, Dr. C. V. Raman Literature award 2006 Karnataka, F.S.E.Sc. (Fellow of Society of Environmental Sciences) 2000 Jharkhand. He successfully supervised 25 Ph.D. and 9 M.Phil students. He has completed three major and two minor research projects funded by UGC and DST. He has written/edited 17 books, published more than four hundred research papers in national and International journals and attended 98 National/International conferences.

— *Part I* —
Mycorrhizal Fungal Biodiversity

2017, Mycorrhizal Fungi
Editors: Ashok Aggarwal and Kuldeep Yadav
Published by: ASTRAL INTERNATIONAL PVT. LTD., NEW DELHI

Pages 3–32

1

Ectomycorrhizae in Natural Ecosystems: Structure, Development and Functions

Yash Pal Sharma¹*, Geeta Sumbali¹ and R.S. Mehrotra²

¹Department of Botany, University of Jammu, Jammu – 180 006, J&K
²Professor (Retd.), Department of Botany, Kurukshetra University,
Kurukshetra, Haryana
*Corresponding Author: yashdbm3@yahoo.co.in

ABSTRACT

Ectomycorrhizal symbiosis is a complex interaction between two eukaryotic species, the heterotrophic fungus and the autorophic host plant. The fungal hyphae form a sheath around all or some of the fine absorbing rootlets of the host tree roots and penetrate between the root cells and occasionally enter the cells. They, however, never penetrate beyond the cortex and any intracellular hyphae do not cause destruction of the host cell. The partners in this association are members of the fungal kingdom (Zygomycota, Ascomycota and Basidiomycota) and predominantly the vascular plants besides certain thallophytes. Based on the morphological relationship and anatomical variations, different types of mycorrhizal associations have been recognized. Ectomycorrhizae enhance the nutrient and water uptake of trees, protect roots from pathogens, have been implicated in interplant carbon and nutrient transfers besides being an important source of unconventional food and indicators of climate change.

Keywords: Ectomycorrhizae, Ecological distribution, Biodiversity, Application, Disease control.

1. Introduction

Ever since the origin of life around 3.5 billion years ago, fungi have remained crucial for life on earth because animal life depends on plant life for continued existence and plants depend on fungi (over 95 per cent of terrestrial plants require fungal infection of their roots in the form of mycorrhizae for adequate root function). The number of fungal species has been conservatively put at 1.5 million (Hawksworth, 1991). Among this number is included the largest organism on the earth; the mycelium of the larger basidiomyceteous fungus *Armillaria solidipes* Peck (formerly *A. ostoyoe*) has been spreading its black shoestring filaments, called rhizomorphs, through the forest for an estimated 2,400 years covering approximately 8.4 km^2 in the Malheur National Forest, Oregon (BBC-Earth, 2008). Since then, several workers have regarded fungi as 'Earth's natural internet'.

The term "mykorrhiza" was first coined in 1885 and used by the 19th-century German biologist Albert Bernard Frank to describe interdependent mutualistic relationships where the host plant receives mineral nutrients while the fungus obtains photosynthetically derived carbon compounds (Harley and Smith, 1983; Harley, 1989). The name mycorrhiza (pl. mycorrhizae or mycorrhizas), literally means "fungus root" and is given to structures formed as a beneficial symbiotic association between plant roots and specific soil fungi. The degree of colonization of plant host roots is influenced by the root physiology and morphology (Wilcox, 1968), by mycorrhizal fungal hyphal structure and phenology (Brundrett, 1991), and by edaphic and climatic factors (Halling, 2001).

The morphology and anatomy of the root is changed, and a new composite organ, "the mycorrhiza," is formed (Hacskaylo, 1972). Mycorrhizal organs can exist in many forms; their morphology is determined by the characteristics of each partner involved and by the specific plant host-fungus combination (Harley and Smith, 1983). During the infection process, the fungus invades the root epidermis and cortex, but does not enter the vasculature or the meristematic zone (the root tip) that is covered by the root cap. The partners in this association are members of the fungal kingdom (Zygomycota, Ascomycota and Basidiomycota) and predominantly the vascular plants besides certain thallophytes as well (Harley and Smith, 1983; Kendrick, 1985). According to different estimations, 80–94 per cent of the vascular plant species worldwide have mycorrhizal associations (Brundrett, 2009). Among these, around 90 per cent of the angiosperms, all known gymnosperms and more than 70 per cent of vascular cryptogams are in mutually-beneficial relationships with fungi (Harley and Harley, 1987).

2. Evolution of Mycorrhizae

The ancestors of modern land plants evolved in aquatic environments, where they existed and diversified over millions of years, from purely aquatic green algae to semi-aquatic algal forms and then became fully terrestrial during their evolution into the first land plants. However, the semi aquatic algal forms while invading the land between 490 and 409 million years ago, encountered a hostile soil environment, which was nutrient-deficient and organically poor and consequently, could not retain minerals and water which, would be quickly lost to the aquifer. However, the infertile and barren land did not offer any microbial competition and possessed plenty of carbon dioxide and sunlight. Similarly, the primitive fungi, the chytrids, were also aquatic, which possessed simple thalli with rhizoids for anchorage and to extract nutrients. Therefore, neither the algal nor the fungus was fully equipped to exploit the terrestrial environment while the algae lacked the ability to extract essential nutrients from the soil, the fungi could not manufacture their food *i.e.* carbohydrates. Desiccation on the land was another challenge for both these groups. Interestingly, out of the selection pressure, perhaps, the two organisms formed a mutualistic symbiosis. Subsequently, the autotrophic organism became morphologically superior and evolved rapidly than the heterotrophic component. Such mutualism led to enhanced performance vis-a vis uptake and assimilation of inorganic and organic nutrients. Thus, the selective advantage over non-symbiotic plant forms, not only provided the benefits to the symbiotic forms but also led to their dynamic evolution, since the extra energy generated using extra nutrients would allow for more differentiation and development of complex tissues.

The earliest known fungal partner facilitating this transition from water to land belonged to the order Endogonales, a group of saprobic, hypogeous and zygospore forming fungi (Morton and Benny, 1990). Later, the evolution of the more complex thalloid liverwort-Glomeromycota symbioses emerged. Many of the most convincing fungal fossils are associated with plant fossils, including glomeromycotan mycorrhizae, and ascomycete or chytridparasites. Most ancient of these fossil mycorrhizae in the form of entangled, occasionally branching, non-septate hyphae and globose spores are known from mid-Ordovician rocks of Wisconsin (460 million years old) and the Devonian Rhynie Chert of Aberdeenshire in the north of Scotland (fossil spores identifiable as *Scutellospora*, about 400 million years old), in which mycorrhizal fungi and several other fungi have been found associated with the preserved tissues of early vascular plants (Taylor *et al.*, 2004). The age of these fossil glomeromycotan fungi indicates that such fungi were present before the first vascular plants arose, when the land flora most likely only consisted of bryophytes, lichens and cyanobacteria. Presently, the arbuscular mycorrhizal (AM) symbiosis has diversity of fungi from Glomeromycota and is ubiquitous in modern vascular plants besides, their occasional association with modern liverworts and hornworts. Therefore, it can now be reasonably appropriate to believe that arbuscular mycorrhizae played an important role in the success of early terrestrial plants (Redecker *et al.*, 2000a,b). Ectomycorrhizal fungi, on the other hand, concomitantly diversified in the Jurassic with the establishment and dominance of gymnosperms, probably the family Pinaceae. The subsequent diversification of angiosperms catalyzed a boost in the diversity of ectomycorrhizal fungi soon after these spermatophytes appeared in the Cretaceous (130mya) (Raven and Axelrod, 1974; Truswell *et al.*, 1987; Berbee and Taylor, 1993). Owing to their microscopic size, soft tissues, and ephemeral nature, Ectomycorrhizae (EcM) did not form well preserved fossils. But the permineralized fossils of The Princeton Chert display Hartig net, pseudoparenchymatous mantle, and extramatrical hyphae. These features combined with *Pinus*-like root morphology including the coralloid root clusters, dichotomized roots, and lack of root hairs, clearly demonstrate that ectomycorrhizal associations evolved at least 50 million years ago (LePage *et al.*, 1997)

3. Types of Mycorrhizae

Ectomycorrhizae symbiosis is a complex interaction between two eukaryotic species. Based on the morphological relationship of the fungus to the root cells, and the fungal and plant species involved, following types of mycorrhizal associations have been recognized (Harley and Smith, 1983; Read, 1983; Harley, 1989; Finlay, 2008):

(i) *Ectomycorrhizae* (*EcM*): In these associations fungi form a mantle around the roots and a Hartig net whose hyphae grow around cells in the epidermis or cortex of short swollen lateral roots (Figures 1.1 a-b).

(ii) *Arbuscular mycorrhizae*: These are the most common associations in which mycorrhizal fungi produce arbuscules, hyphae and vesicles within root cortex cells. These are the most abundant and commonest endotrophic mycorrhizae. In these fungal-root associations, the fungi belong to the Glomeromycota that are obligate biotrophs. The fungal ramification inside the roots may occur by the linear or coiled hyphae. They are associated with roots of about 80 per cent of plant species, including many crop plants (Figure 1.1c).

(iii) *Ectendomycorrhizae*: Ectendomycorrhizae exhibit anatomical characteristics of both ecto and endomycorrhizae. The mycosymbiont form a hyphal mantle and Hartig net but also shows extensive intracellular penetration. The EcM association sometimes has hyphae entering a few dead cells, but in ectendomycorrhizae infection can occur in living plant cells. Ectendomycorrhizas are restricted mostly to the *Pinus, Picea* and to a lesser extent *Larix*. Analysis of nuclear and mitochondrial genomes of the fungi confirmed that most belong to the genus *Wilcoxina* (*W. mikaloe* and *W. rehmii*) of order Pezizales under the phylum Ascomycota.

(iv) *Ericoid endomycorrhizae*: These are formed in three plant families, the Ericaceae, Empetraceae, and Epacridaceae (now in Ericaceae), all belonging to the order Ericales. These plants are mostly found in upland areas in temperate grasslands, savannas, and shrublands characterised by low-growing vegetation on acidic soils (moorlands) and similar challenging environments. Around 3400 plant species form this type of mycorrhizal association with various fungi from the Ascomycota but only one species, *Rhizoscyphus ericae* (syn. *Hymenoscyphus ericae*) has been studied in detail. This fungus penetrates the cell walls of roots and forms coiled structures within each cell without penetrating the host plasmalemma. The plant's rootlets are covered with a sparse network of hyphae; the fungus digests polypeptides saprotrophically and passes absorbed nitrogen to the plant host. In extremely harsh conditions, the mycorrhiza may even provide the host with carbon sources (by metabolising polysaccharides and proteins for their carbon content).

(v) *Orchidaceous mycorrhizae*: The family Orchidaceae is the largest in the plant kingdom and is estimated to contain 30,000 species. Although most orchids have green leaves and are autotrophic when fully established yet

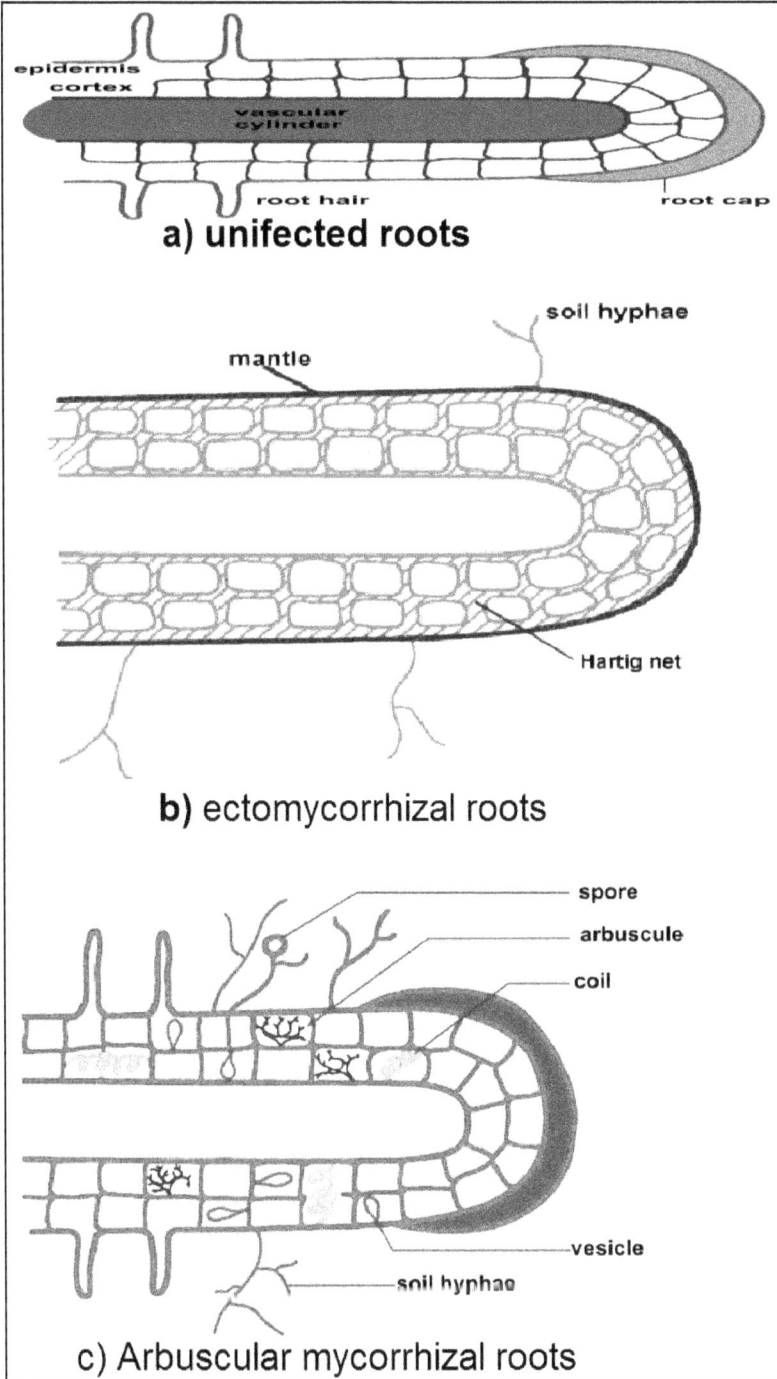

a) unifected roots

b) ectomycorrhizal roots

c) Arbuscular mycorrhizal roots

Figure 1.1 a-c: Diagrammatic Representation of the Relationship of the Symbionts in Mycorrhizal Associations.
(a) Uninfected roots; (b) ectomycorrhizal roots; (c) Arbuscular mycorrhizal roots.

about 100 species are achlorophyllous as adults and they pass through a germination and early development phase when they are dependent on an external supply of nutrients and organic carbon. There numerous minute seeds have no reserves, and are initially entirely dependent upon the supply of carbon and nitrogen from fungi and hence are described as myco-heterotrophic orchid mycorrhizae which consist of coils of hyphae called peltons within roots or stems of orchidaceous plants. Their carbon nutrition is more dedicated to supporting the host plant as the young orchid seedling is non-photosynthetic and depends on the fungal partner utilising complex carbon sources in the soil, and making carbohydrates available to the young orchid. Since all orchids are non chlorophyllous in the early seedling stages, but usually chlorophyllous as adults, at the seedling stage orchid can be interpreted as parasitizing the fungus. A characteristic example of the fungus supplying carbon resources is the basidiomycetous genus *Rhizoctonia*. The myco-heterotrophic plants, such as the phantom orchid (*Cephalanthera austiniae*), have no chlorophyll and are extremely dependent on symbiotic association and get the carbon via the mycelia of fungi from nearby trees that both are connected to. Other orchids only steal when it suits them. These 'mixotrophs' can carry out photosynthesis, but they also extract carbon from other plants using the fungal network that links them.

(vi) *Arbutoid mycorrhizae*: Typical arbutoid mycorrhizal associations such as these formed by *Arbustus*, *Pyrola* and *Arctostaphylos* (Ericales) show extensive intracellular penetration, with coils of hyphae filling large volumes in many cells. Similar to ectomycorrhizas, the arbutoid associations produce an intercellular Hartig net, usually restricted to the outer layer of root cells. This 'paraepidermal' net is also seen in ectomycorrhizas in the majority of angiosperms. A hypodermis, formed by deposits of suberin and a casparian strip in the outer layer of cortical cells, prevents deeper penetration of the Hartig net. However, a major difference between the arbutoid and ectomycorrhizal association is that the hyphae of the former do actually penetrate the outer cortical cells, and fill them with coils, which the hyphae of EcM do not. The intracellular coils, along with the mantle sheath and Hartig net are the diagnostic features of arbutoid mycorrhizas. These mycorrhizal roots have been described by Largent *et al.* (1980), Molina and Trappe (1982) and Massicotte *et al.* (1998, 2005a,b).

(vii) *Monotropoid mycorrhizas:* These are the EcM associations of a few genera of myco-heterotrophic plants in the Ericaceae. These associations are characterised by limited hyphal penetration into the epidermal cells. Fungi colonising achlorophylous plants such as *Monotropa hypopitys* (in Europe) and *M. uniflora* (North America) never actually penetrate the plant cell walls. Since all the 10 genera of Monotropaceae (now in Ericaceae) are entirely achlorophyllous, they are unable to photosynthesise and produce carbohydrates. Therefore, the mycorrhizae play a dual role of obtaining minerals and nutrients, as well as tapping the carbon supplies of nearby plants such beech, oak, cedar, pine, spruce and fir via their roots. The *Monotropa* symbiosis, is therefore a three tier system, involving chlorophyllous (tree) and non-chlorophyllous (*Monotropa*) and the fungal channel (EcM fungus *e.g. Boletus edulis*).

(viii) *Subepidermal association of Thysanotus:* The Australian lilies in the genus *Thysanotus* (Laxmaniaceae) have unique mycorrhizae where fungal hyphae grow in a cavity under epidermal cells. Other members of this family have AM or non-mycorrhizal roots.

4. Morphology and Development of Ectomycorrhizae

Mycorrhizae are said to be the most important symbioses on earth (Bucking *et al.*, 2012). Geologically, the evolution of EcM is relatively a recent phenomenon with a fossil record dating back to the Mesozoic. This is consistent with the origin of its non-flowering gymnospermous plant families such as Pinaceae originating in Triassic to Cretaceous followed by angiospermic families including Belutaceae, Fagaceae, Salicaceae, *etc.* in the Cretaceous of Mesozoic (Malloch *et al.*, 1980). Likewise, the fungal associates including septate Ascomycotina and Basidiomycotina were already present during the Cretaceous of the Mesozoic or even before this period.

EcM are morphologically and anatomically distinguished by the presence of a dense mass of fungal hyphae upto 40 µm thick known as the mantle forming a pseudoparenchymatous tissue ensheathing the root surface. This hyphal network called as Hartig net is characterized by labyrinthine branching and an outward network of hyphae that aids in water and nutrient uptake (Kottke and Oberwinkler, 1987). Development of a mature mantle proceeds through a programmed series of events (Horan *et al.*, 1988; Martin *et al.*, 1997). Due to the encapsulating nature of the mantle, the normal development of host root is often affected as the fungal hyphae originating from a germinating basidiospore penetrates into the root cap cells and grow through them. This is followed by the ramification of mycelium towards root epidermal cells. Consequently EcM fungal partners characteristically suppress root hair development of the host plant (Chilvers, 1968; Feugey *et al.*, 1999). The fungal invasion further disorganizes the root cap, which becomes an inner part of the mantle. The hyphae in these structures are encased in an extracellular polysaccharide and proteinaceous matrix. Air and water channels that allow the flow of nutrients into the symbiosis innervate these structures (Ashford *et al.*, 1989). There is a structural (Paris *et al.*, 1993) and physiological (Cairney and Burke, 1994) heterogeneity within the ectomycorrhizal mantle, and between the mantle and the inward and outward fungal networks. These morphological analyses of the infection process have shown a fairly complex influence of the root on the fungus, including a general growth stimulus, a trophic response directing hyphal growth inwards towards the plant tissues and a morphogenetic effect leading to compact hyphal mantle development. In addition to putative morphogens, the supply of nutrients, the presence of a physical support and the supply of water likely play a role in mantle formation (Read and Armstrong, 1972; Martin *et al.*, 1999). The mantles of different EcM partners may vary with respect to colour, extent of branching, and degree of thickness and complexity. On the other hand, fungal hyphae stimulate lateral root formation, dichotomy of the apical meristem in conifer species, and cyto-differentiation (radial elongation, root hair decay) of root cells (Horan *et al.*, 1988).

Hormonal relationships in mycorrhizal development are also significant. It is evident that the benefits to the host plant provided by the fungus are not confined to supplying water, mineral and flavonoids. Plant hormones, including auxins, cytokinins, abscisic acid and ethylene, produced by ectomycorrhizal fungi (Gogala, 1991) may have a possible role to play in EcM development. It seems likely, however, that auxin and cytokinin are essential for organogenesis and maintenance of EcM (Rupp *et al.*, 1989; Gay *et al.*, 1994; Karabaghli- Degron *et al.*, 1998; Kaska *et al.*, 1999). Enhanced proliferation of short roots (Carnero Diaz *et al.*, 1996) and proliferation and dichotomous branching of lateral roots (Smith and Read, 1997) are controlled by fungal indole acetic acid (IAA). They probably are produced in small quantities as secondary metabolites, synthesized from root-originated substrates and present at fairly constant, low concentrations. Morphological differentiation is accompanied by the onset of novel metabolic organizations in fungal and plant cells leading to the distinct organ, the mycorrhiza (Cairney *et al.*, 1989; Martin and Botton, 1993; Rygiewicz and Andersen, 1994; Hampp *et al.*, 1995) and its extramatrical hyphae, the mantle and the intra-radicular hyphal network ensure active metabolite exchange and nutrient transfer (*e.g.* phosphate, nitrogen) to the host plant. The host in turn compensates by providing stable carbohydrate-rich niche in the roots for the fungal partner, making the relationship a mutualistic symbiosis.

Plants have uniquely co-evolved physiological connectivity between them using underground fungus as an information channel. This natural biological internet allows the plant to readily and effectively communicate with each other (Babikova *et al.*, 2013). The communication taking place in EcM occurs at the cellular and molecular level involving signally molecules, genes and proteins. Mutual signal exchange triggers the developmental process (Figure 1.2).

By contrast to some other plant-microbe interactions (Denarie *et al.*, 1996; Moller and Chua, 1999), the nature of the signalling molecules and the molecular basis of signal perception and transduction in mycorrhizae are unknown or ill-defined. Identifying the processes that regulate the information flow between mycorrhizal fungi and host root is an active research area. The physiology and morphology of infection and EcM development essentially involves a) the release of exudates by the host plant into the rhizosphere that are able to trigger basidiospore germination (Fries *et al.*, 1987); b) chemotactic growth of hyphae towards the root in response to rhizospheric signals (Horan and Chilvers, 1990); c) attachment and invasion of host tissues by hyphae (adhesins, hydrolases); d) Induction of organogenetic signals in both fungal and root cells (hormones and secondary signals); e) facilitating survival of the mycobiont despite plant defense responses; f) Coordinating strategies for exchanging carbon and other metabolites (*e.g.* vitamins) for plant colonization (mycorrhiza formation) and for balancing growth of the soil fungal web with its role in gathering minerals from the soil (Beguiristain and Lapeyrie, 1997; Salzer *et al.*, 1997a,b,c; Ditengou and Lapeyrie, 2000). Interestingly, the nature and the

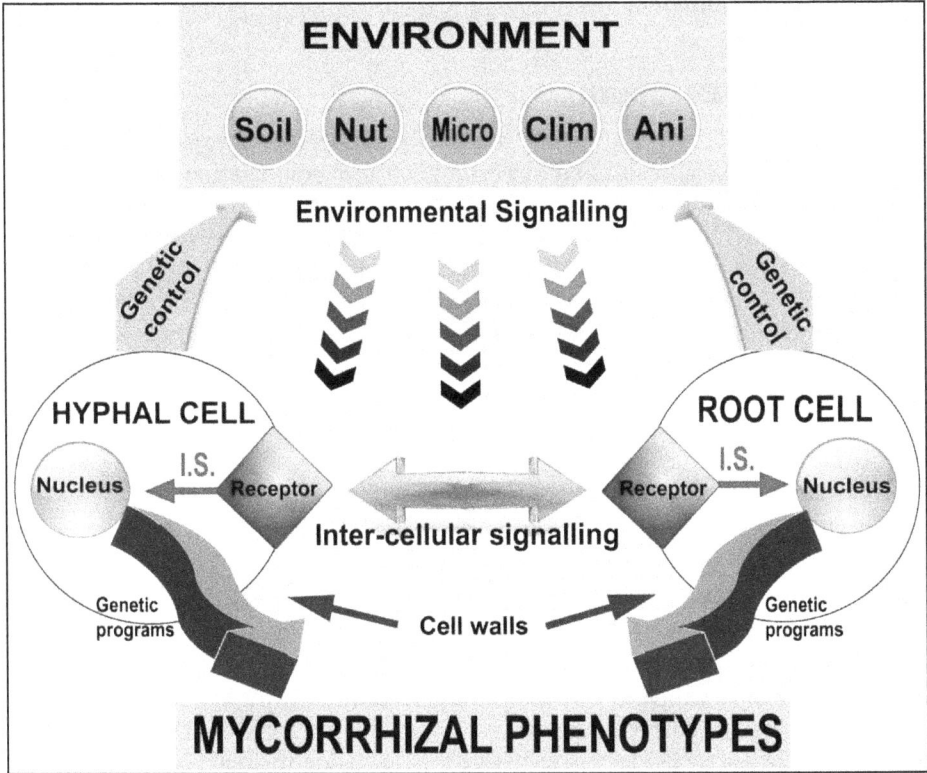

Figure 1.2: Inter- and Intracellular Communications between Fungal Hyphae and Root Cells in the Ectomycorrhizal Symbiosis.

Changes in environmental conditions may produce signals sensed by both partner cells which probably transduce this information to their nuclei leading to gene expression and consequently change in phenotypes. (abbreviations: I.S. - Intracellular signalling. Nut - nutrients. Micro - microorganisms. Clim - climatic factors. Ani - animals (humans included) (Modified and redrawn from Tagu *et al.*, 2002).

transduction of the signals in the EcM interactions within the partners, and the mechanisms underlying the triggering the expression of symbiosis-regulated genes that assist in partner recognition and the formation of symbiotic tissues are being unravelled gradually. About 100 SR-genes have been identified in various EcM associations, the products of which may play a role in recognition and attachment of the mycobiont onto root surfaces, formation of the symbiotic interface, signalling networks, protein turnover, organogenesis, and novel symbiotic metabolism (Martin *et al.*, 2001) However, many questions concerning the differentiation of plant and fungal symbiotic structures still remain unanswered. It is likely that new technical approaches such as application of reverse genetics, whole genome sequencing (Tagu *et al.*, 2002), transcriptomics (Velculescu *et al.*, 1999), proteomics (Pandey and Mann, 2000) or metabolomics (Raamsdonk *et al.*, 2001) would unravel the putative signalling networks and early gene regulation processes involved in EcM development.

5. Ecological Distribution

More than half of the afforested surface of earth is on poor soil. For these 20 million km², this symbiosis is an important factor in colonization and stability of forest ecosystems. The best form of humus for EM-fungi is a slightly acidic leaf litter with a low P and N content but good aeration and reasonable moisture content. Any increase in soil moisture above a certain level results in reduced mycorrhizal formation (Boucher and Malajczuk, 1990). Ectomycorrhizae are abundant mainly under temperate (deciduous and evergreen forests) and boreal climates. Besides their abundance in Northern Hemisphere, the mountain regions of Africa and South America and the *Eucalyptus* and *Nothofagus* forests of Australia and New Zealand, EcM are also abundant in the forests of Mediterranean lands. Obligate EM-species mainly belong to the genera *Pinus, Picea, Abies, Larix, Quercus, Fagus* and *Carpinus*, whereas facultative EM-species belong especially to the genera *Acer, Alnus, Betula, Corylus, Cupressus, Eucalyptus, Juniperus, Salix* and *Ulmus*.

The distribution of EM-fungi reflects the incidence of their host trees (Dighton *et al.,* 1986). In pure stands of one tree species, a succession of fruiting bodies can be observed. Thus, in a new *Betula* plantation, species of *Hebeloma* are seen in the first few years. After 6 years, *Leccinum* species dominate, to be succeeded after 10 years by *Russula* species. Chilvers *et al.* (1987) reported that during the development of *Pinus contorta* trees, a characteristic succession of fungi was observed by their fruiting bodies: *Paxillus-Laccaria- Cortinarius/Inocybe* to *Russula* spp. In mixed forests where young trees continually spring up, it is more difficult to observe such successions and a large spectrum of species is stably produced. In *Betula*, it has been observed how mycorrhizal fungi infecting side roots change in accordance to the distance from the tree trunk. At a distance of 25 cm from the trunk 10 per cent of roots were infected with *Hebeloma* species. This infection rate increased to over 50 per cent at a distance of 100cm. Mycelium of *Lactarius* reached an optimum at 50cm distance whereas *Leccinum* mycelium could not be found 75-100cm away. Increasing soil depth also affected the percentage of mycorrhizal root tips in different tree species. 70 per cent and 30 per cent of possible infection sites were occupied even at a depth of 2m in *Fagus sylvatica* and *Pinus sylvestris* respectively (Werner, 1992). However, the ectomycorrhizal fungal richness on plant roots declined with distance from the center of the host species range (Lankau and Keymar, 2016).

Hypogeous macrofungi, such as *Mesophelia* sp. and *Castorium* sp., develop their sporopcarps in close connection with the host plant root (*e.g. Eucalyptus diversicolor*). A single sporocarp of *Mesophellia trabalis* can entrap upto 5m of roots with a surface of 45cm² (Dell *et al.,* 1990).

6. Diversity of Ectomycorrhizal Fungi

EcM are true "myco-indicators" of forest ecosystems, where an enormous diversity of ectomycorrhizal fungi can be found. A wide range of fungi, including thousands of species, belonging to Basidiomycota, Ascomycota and Zygomycota can form EcM associations (Miller, 1982) Many of these include common forest mushrooms, puffballs and truffles. Typical EM fungi include *Amanita, Boletus, Russula, Lactarius, Laccaria, Hebeloma, Pisolithus, Rhizopogon, Tricholoma, Scleroderma* and *Suillus* species (Table 1.1). Only 3-5 per cent of plant species form EcM associations, but in contrast to AM fungi, there are at least 7750 documented EcM fungal species and some estimates of EcM species richness range as high as 20,000-25,000 species (Rinaldi *et al.*, 2008; Tedersoo *et al.*, 2010; Long *et al.*, 2016). In natural forest ecosystem, the roots of single tree are almost invariably associated with several different EcM species (Guidot *et al.*, 2003). Some hosts form associations with many fungi while others are more specific (Trappe, 1962; Duddridge, 1987). Host preferences, metabolic diversity and responses to habitat conditions provide evidence that considerable diversity occurs within this group of fungi. Debaud *et al.* (1988) found that homokaryotic mycelia of the fungus *Hebeloma cylindrosporum* produced EcM associations that were similar to those produced by hyphae of the dikaryotic parent. Unfortunately, little is known about the genetics (cytology, nuclear behaviour, mating systems, *etc.*) of EcM fungi (Harley and Smith, 1983; Trappe and Molina, 1986). The ericoid mycorrhizal fungi that have been identified include ascomycetous forms such as *Hymenoscyphus* (*Pezizella*), *Myxotrichum* and *Gymnascella* (Read, 1983; Dalpe, 1989). Fungi forming mycorrhizal associations with orchids include many *Rhizoctonia* anamorphs (some of which have known telomorphs), as well as other fungi that may form specific or non-specific associations with their hosts (Warcup, 1981, 1985; Currah *et al.*, 1987; Ramsay *et al.*, 1987).

Propagules of EcM fungi include hyphae, mycelial strands and rhizomorphs (Ogawa, 1985; Read *et al.*, 1985). basidiospores (Bowen and Theodorou, 1973; Fox, 1983, 1986a), sclerotia (Fox, 1986b; Gibson *et al.*, 1988) and probably also mycorrhizal roots, but these fungi typically do not produce asexual (conidial) spores (Hutchinson, 1989). Boreal forest soil and leaf litter contain basidiospores, which can initiate mycorrhizas (Parke *et al.*, 1983, 1984; Perry *et al.*, 1987). Localized patterns of EcM fungus proliferation depend on the production of hyphae, mycelial strands, or rhizomorphs by a particular endophyte (Ogawa, 1985). Hyphal strands of some EcM fungi will only initiate new mycorrhizas if attached to living host roots (Fleming, 1984). Roots with EcM usually live for one or more years and are protected by mantle hyphae (Harley and Smith, 1983), suggesting that they may be important perennating structures.

Many fungi forming EcM associations have large fruiting structures (mushrooms) that produce abundant quantities of wind-borne spores, but survival and dispersal of these spores may be limited (Bowen and Theodorou, 1973; Harley and Smith, 1983). Sclerotia, including those produced by EcM fungi, can be moved in spring run-off water by floating or adhering to organic material (Malloch *et al.*, 1987). Some EcM fungi produce hypogeous fruiting bodies that are excavated and consumed by small mammals or marsupials and thus spread to new locations. Spores of EcM fungi contained in animal faeces can be a viable source of inoculum (Kotter and Farentinos, 1984; Lament *et al.*, 1985). In western North America, hypogeous fungi form a major part of the diet of squirrels, which in turn are beneficial to the community by transporting nitrogen-fixing bacteria and fungal spores, which can

Table 1.1: Fungal Genera Involved in the Formation of Ectomycorrhiza (After Harley and Smith 1983; Singer *et al.,* 1983; Smith and Read 1997; Peterson *et al.,* 2004)

Fungal Group/Family	*Genera Involved in Ectomycorrhizal Associations*
Zygomycota	*Densospora, Endogone, Peridiospora,*
Endogonaceae	
Ascomycota	
Balsamiaceae	*Balsamia*
Elaphomycetaceae	*Elaphomyces, Pseudotulostoma*
Geneaceae	*Genea*
Geoglossaceae	*Cudonia, Spathularia*
Helvellaceae	*Helvella, Leucangium*
Hydnotryaceae	*Bassia, Choiromyces, Hydnotrya*
Otidiaceae	*Otidia*
Pyronemaceae	*Geopora, Lachnea, Sepultaria*
Rhiziniaceae	*Gyromitra*
Terfeziaceae	*Mukomyces, Picoa, Terfezia, Tirmania*
Tuberaceae	*Tuber, Kalaharituber, Labyrinthomyces*
Basidiomycota	
Agaricaceae	*Lepiota*
Amanitaceae	*Amanita, Amanitopsis, Limacella*
Boletaceae	*Boletinus, Boletus, Callostoma, Pulveroboletus, Fistulinella, Gyrodon, Gyroporus, Krombholzia, Leccinum, Suillus, Tylopilus, Xerocomus*
Cortinariaceae	*Alnicola, Cortinarius, Dermocybe, Hebeloma, Inocybe, Rozites, Thaxterogaster*
Gomphidiaceae	*Gomphidius, Gomphus, Phellodon*
Hygrophoraceae	*Hygrophorus, Humidicutis, Gliophorous*
Paxillaceae	*Paxillus*
Rhodophyllaceae	*Clitopilus, Rhodophyllus*
Russulaceae	*Lactarius, Russula, Cystangium, Gymnomyces, Leucogaster, Stephanospora, Zelleromyces*
Strobilomycetaceae	*Boletellus, Strobilomyces*
Tricholomataceae	*Laccaria, Leucopaxillus, Lyophyllum, Tricholoma*
Entolomataceae	*Entoloma*
Cantharellaceae	*Cantharellus, Craterellus, Hydnangium*
Hydnaceae	*Hydnum*
Thelephoraceae	*Bankera, Boletopsis, Corticium, Thelephora*
Hymenogastraceae	*Alpova, Rhizopogon*
Geastraceae	*Geastrum, Astraeus*
Lycoperdaceae	*Calvatia, Lycoperdon*
Phallaceae	*Clathrus, Phallus*
Pisolithaceae	*Pisolithus*
Sclerodermataceae	*Scleroderma*
Clavariaceae	*Aphelaria, Clavaria, Clavariadelphus, Clavicorona, Clavulina, Clavulinopsis, Ramariopsis*
Mitosporic Fungi	*Cenococcum*

establish new mycorrhizal fungus colonies or transfer genetic material to existing colonies (Maser and Maser, 1988). Similar tree-mycorrhizal fungus-dispersing animal inter-relationships also occur in Europe, Australia and New Zealand (Malajczuk *et al.*, 1987; Blaschke and Baumler, 1989; Cowan, 1989). The supply of mycorrhizal inoculum could be limited in some recently created or disturbed habitats if these fungi were less readily dispersed than their host plants.

7. Ectomycorrhizal Host Plant Diversity

Approximately 95 per cent of the world's living species of vascular plants belong to the families that are characteristically mycorrhizal (Trappe, 1977). Globally, approximately 8000 plant species enter into EcM symbiosis (Taylor and Alexander, 2005). The plant species involved are usually trees or shrubs from cool, temperate boreal or montane forests, but also include arctic-alpine dwarf shrub communities, and Mediterranean/chaparral vegetation (Table 1.2). EcM are the main type of association for the Araucariaceae, Pinaceae, Betulaceae, Casuarinaceae, Fagaceae and Salicaeae, and are also common for the Caesalpiniaceae, Cupressaceae, Dipterocarpaceae, Juglandaceae, and Myrtaceae (Wang and Qiu, 2006; Smith and Read, 2008) The composition and type of plant species in a terrestrial ecosystem is a primary determinant of ecosystem productivity and sustainability (Tilman *et al.*, 1996). It has now been realized that the plant diversity may primarily be regulated by the diversity of mycorrhizal fungi.

Table 1.2: Genera of Seed Plants with Ectomycorrhiza (Modified from Marks and Kozlowski, 1973 and Harley and Smith, 1983)

Family	Plant Genera Involved in Ectomycorrhizal Associations
Aceraceae	*Acer*
Asteraceae	*Lactuca*
Betulaceae	*Alnus, Benda, Carpinus, Corylus, Ostrya, Ostryopsis*
Bignoniaceae	*Jacaranda, Phyllarthron*
Caesalpiniaceae	*Afzelia, Aldina, Anthonota, Bauhinia, Brachystegia, Cassia, Eperua, Gilbertiondendron, Julbernardia, Monopetalanthus, Paramacrolobium, Swartzia*
Caprifoliaceae	*Sambucus*
Casuarinaceae	*Casuarina*
Cistaceae	*Helianthencum, Cistus*
Combretaceae	*Terminalia*
Cupressaceae	*Cupressus, Juniperus*
Cyperaceae	*Kobresia*
Dipterocarpaceae	*Anisoptera, Balanocarpus, Cotylelobium, Dipterocarpus, Thyobalanops, Hopea, Monotes, Shorea, Valica*
Elaeaganaceae	*Shepherdia*

Contd...

Table 1.2–*Contd...*

Family	Plant Genera Involved in Ectomycorrhizal Associations
Ericaceae	*Arbutus, Arctostaphylos, Chimaphila, Gaultheria, Kalmia, Ledum, Leucothoe, Rhododendron, Vaccinium*
Euphorbiaceae	*Poranthera, Uapaca*
Fabaceae	*Bartonia, Brachysema, Chorizema, Daviesia, Dillwynia, Eutaxia, Gompholobium, Hardenbergia, Jacksonia, Kennedya, Mirbelia, Oxylobium, Platylobium, Pultenaea, Robinia, Vicia, Viminaria*
Fagaceae	*C Castanea, Castanopsis, Fagus, Lithocarpus, Nothofagus, Pasania, Quercus, Trigonobalanus*
Globulariaceae	*Globularia*
Gnetaceae	*Gnetum*
Goodenaceae	*Brunonia, Goodenia*
Hammamelidaceae	*Parrotia*
Juglandaceae	*Catya, Juglans, Pterocarya*
Lauraceae	*Sassafras*
Mimosaceae	*Acacia*
Myricaceae	*Comptonia, Myrica*
Myrtaceae	*Angophora, Callistemon, Camponesia, Eucalyptus, Leptospermum, Melaleuca, Tristania*
Nyctinaginaceae	*Neea, Pisonia, Torrubia*
Oleaceae	*Fraxinus*
Pinaceae	*Abies, Cathaya, Cedrus, Keteleeria, Larix, Picea, Pinus, Pseudolarix, Pseudotsuga, Tsuga*
Platanaceae	*Platanus*
Polygonaceae	*Coccoloba, Polygonum*
Pyrolaceae	*Pyrola*
Rhamnaceae	*Cryptandra, Pomaderris, Rhamnus, Spyridium, Trygmalium*
Rosaceae	*Chamaebatia, Circocarpus, Crataegus, Dryas, Malus, Prunus, Pyrus, Rosa, Sorbus*
Rubiaceae	*Galium, Opercularia, Rubia, Psychotria*
Salicaceae	*Populus, Salix*
Sapindaceae	*Allophylus, Nephelium*
Sapotaceae	*Glycoxylon*
Saxifragaceae	*Ribes*
Sterculiaceae	*Lasiopetalum, Thontasia*
Stylidiaceae	*Stylidium*
Thymeliaceae	*Pimelia*
Tiliaceae	*Tilia*
Ulmaceae	*Ulmus, Celtis*
Vitaceae	*Vitis*
Zygophyllaceae	*Peganum*

8. Carbohydrate Physiology of EcM

Carbohydrates are synthesized by photosynthesis in green plants and are translocated predominantly as sucrose, via the phloem, to organs and tissues which have a demand for carbon and form a 'sink'. Ectomycorrhizal fungi, which live in intimate symbiosis with trees, receive up to 30 per cent of the total carbon fixed by the plant host (Finlay and Soderstrom, 1992) and thus function as an additional sink. In exchange, the tree receives mineral nutrients from the fungus. Although sucrose has been detected as the main translocated form of photoassimilate in plant root cells, it has not been detected in symbiotic fungal tissues, where mannitol, trehalose, and glycogen are the main labelled carbohydrates (Soderström *et al.*, 1988; Hampp and Schaeffer, 1995; Smith and Read, 1997). This is due to enhanced invertase activity in mycorrhizal roots. Glucose resulting from sucrose catabolism is thought to be the primary source of carbon for the generation of ATP, reducing power, and carbon skeletons for biosynthetic pathways in ectomycorrhizae (Hampp and Schaeffer, 1995). The metabolic pathways leading to the synthesis of major fungal carbohydrates, such as mannitol and trehalose have been characterized in several free-living ectomycorrhizal fungi (Martin *et al.*, 1985, 1988; Ramstedt *et al.*, 1989). Glucose labeling to study carbohydrate and amino acid metabolism in *Eucalyptus globulus* subsp. *bicostata* and in *Pisolithus tinctorius*, growing separately and in mycorrhizal association has demonstrated significant mutual effects on fungal and host-plant metabolism (Martin *et al.*, 1998).

9. Applications of Ectomycorrhizal Symbioses

Over the last century, there has been considerable progress in our understanding of the occurrence and multifaceted role of mycorrhizae in ecosystems. The application of mycorrhizae technology has a major impact on world agriculture by aiding the establishment of plants in poor soils, nutrient and water absorption, root health and longevity, tolerance to drought, high soil temperatures, toxic heavy metals, soil salinity, extremes in pH *etc*. In addition, the fungal symbiont in this relationship can transport metabolites from one plant to another, produce plant growth regulators and antibiotics and may protect roots from invasion by pathogens (Maronek *et al.*, 1981; Schenck, 1982; Harley and Smith, 1983; Sikes 2010).

Ectomycorrhizal fungi during symbioses with higher plants play a role of natures' bioprotectors. Due to their ability to access both mobile and immobile forms of nitrogen and phosphorus, plant-fungus symbioses are important for the nutrient acquisition of plants (Smith *et al.*, 2011; Pena and Polle, 2014). Besides nutrient acquisition, mycorrhizal fungi facilitate plant water relations (Auge, 2001), protect plants against pathogens (Newsham *et al.*, 1995; Borowicz, 2001), and reduce uptake of toxic metals (Gadd, 1993; Mcharg, 2003). Communities of ectomycorrhizal fungi are central to many aspects of bio-geochemical cycles of interest for ecosystem services: carbon fixation, water flow, preservation of soil fertility, emission of greenhouse gases, *etc*. Ectomycorrhizal communities are very diverse, taxonomically as well as functionally. Thread-like fungi colonise tree roots and spread out like a huge net through the soil, helping trees gather water and nutrients in exchange for sugars made by photosynthesis. In addition to providing a certain degree of protection to seedlings in harsh circumstances, such as increased salinity or heavy metal pollution, the EcM are also instrumental in improving soil quality by preventing soil erosion and thus establishing vegetation and restoring habitats. New plantations may benefit considerably from EcM fungal inoculation.

10. Mineral Nutrient Cycling by Ectomycorrhizae

Mycorrhizae are considered bio-fertilizers as they extend roots' nutrient depletion zone especially in phosphorus and nitrogen-deficient soils. In mycorrhizal associations, plants provide fungi with food in the form of carbohydrates. In exchange, the fungi help the plants to take up water, and provide nutrients like phosphorus and nitrogen, via their mycelia. Increased plant nutrient supply by acquiring nutrient forms that would not normally be available to plants help individual plants to grow (Tarafdar and Marschner, 1994; Schweiger *et al.*, 1995; Kahiluoto and Vestberg, 1998; Hawkins *et al.*, 2015; Shah *et al.*, 2016). Smith and Read (1997) have demonstrated the potential of extracellular proteases and phosphatase secreted by EcM fungi in releasing nitrogen and phosphorus from simple organic substrates. Some EcM and ericoid fungi have the capacity to breakdown phenolic compounds in soils which can interfere with nutrient uptake (Bending and Read, 1997). Mycorrhizal benefits can include greater yield, nutrient accumulation, and/ or reproductive success (Lewis and Koide, 1990; Stanley *et al.*, 1993). Suppression of competing non-host plants by mycorrhizal fungi could reduce competition between plants and contribute to the stability and diversity of ecosystems (Allen *et al.*, 1989).

11. EcM and Plant Root Disease Control

Root colonisation by EcM and AM fungi can provide protection from parasitic fungi and nematodes and is more effective in EcM roots than in AM-roots (Duchesne *et al.*, 1989; Grandmaison *et al.*, 1993; Cordier *et al.*, 1998; Morin *et al.*, 1999). *Pisolithus tinctorius* could significantly enhance the survival rate of *Pinus taeda* plants after infection with *Rhizoctonia solani*. Infection by pathogens can be reduced by the following mechanisms (Werner, 1992):

a) The excretion of antifungal and antibiotic substances by the ectomycorrhizal mycobiont for inhibition of other microorganisms.

b) The production of antibiotics by the host under the control of symbiont;

c) The stimulation of other micro-organisms, which themselves inhibit pathogens;

d) The reduction of carbon and energy sources on the root surface that inhibit the spore germination of pathogens;

e) The structural reinforcement of the root surface by the thick fungal covering.

Furthermore, EcM fungi influence soil microbial populations and exudates in the mycorrhizosphere and hyphosphere (Ames *et al.*, 1984; Bansal and Mukerji, 1994; Andrade *et al.*, 1998). Fungal networks also boost the resistance of host plants by triggering the production of defense-related chemicals at the time of root colonization. Simply plugging in to mycelial networks makes plants more resistant to disease. Researchers have just documented how plants use underground fungal networks to warn neighbouring plants of impending insect attack, uniquely illustrating the complex and highly designed interconnected cooperation found in nature. Studies have shown that tomato and broad beans use fungal networks to pick up on impending threats such as aphids (Babikova *et al.*, 2013) activated their anti-aphid chemical defences. Some form of signalling was going on between these plants about herbivory by aphids, and those signals were being transported through mycorrhizal mycelial networks.

12. Ectomycorrhizae in Forestry

Forest communities cover approximately 33 per cent world's land surface (Rumney, 1968) with EcM as the most frequent and widespread mycorrhizal type in forest and woodlands of cool temperate and boreal latitudes. In forest soils, EcM fungi can contribute upto one-third of microbial biomass (Hogberg and Hogberg, 2002). They are associated with almost all feeder roots of woody plants in temperate, boreal and some subtropical forests (Smith and Read, 1997). Recently, the United States' National Aeronautics and Space Administration (NASA) spots an 'underground network of fungi' that can help track the spread of forests suggesting the role of EcM in establishment and enhancing the range of forests across the globe (Fisher *et al.*, 2016). For successful afforestation in barren lands, transplanting of crop trees, such as *Eucalyptus* and *Pinus* species in new locales often requires an accompanying ectomycorrhizal partner. Different isolates of *Pisolithus tinctorius* as EcM inoculum have been tried for growth of *Acacia* while *Boletus luridus* and *Rhizopogon* species have significantly increased the EcM colonization of *Pinus* species. Promising results have been achieved in reducing transplanting period and successful establishment of seedlings of *Cedrus deodara, Pinus wallichiana, P. gerardiana, P. roxburghii, Grevillea robusta* etc. in India (Bakshi *et al.*, 1972; Lakhanpal and Kumar, 1984).

13. Ectomycorrhizal Fungi as a Source of Unconventional Food

EcM fungi are economically and nutritionally important as human food resources (Arora, 1991; Kalotas, 1996) as some of them produce edible sporocarps, *i.e.*, fruiting bodies, which are important for the food industry. Cultivation of associated fungi, particularly those of high commercial value is highly demanding. Ectomycorrhizal fungi also produce various metal chelating molecules and some unique secondary metabolites, which are of remarkable biotechnological significance. These mushrooms have also have been used as medicines and as natural dyes (Arora; 1991, Morgan, 1995). Epigeous and hypogeous sporocarps of EcM and VAM fungi are important food sources for placental and marsupial mammals (Reddell and Spain, 1991; McGee and Baczocha, 1994; Janos *et al.*, 1995; Mcllwee and Johnson, 1998; Claridge, 2002). Mycorrhizal roots, fungal fruit bodies and hyphae are important food sources and habitats for soil invertebrates (Fogel and Peck, 1975; Rabatin and Stinner, 1989; Setala, 1995; Lawrence and Milner, 1996).

14. Ectomycorrhizae and Climate change

Mycorrhizal fungi contribute to carbon storage in soil by altering the quality and quantity of soil organic matter (Ryglewicz and Andersen, 1994). Climate change can induce a number of changes in the environment, and subsequently, ectomycorrhizal communities. Many of these studies are in their infancy, but it is clear that they often exhibit some effect. In some studies, elevated CO_2 levels increased fungal mycelium growth due to increased carbon allocation (Fransson *et al.*, 2005) and increased EcM root colonization by 14 per cent. (Garcia *et al.*, 2008). Increased temperature also appears to affect EcM communities, though the results cover a range of responses. Some studies have shown that respiration is reduced in certain species in response to warming (Malcolm *et al.*, 2008), whereas others demonstrated increased total colonization of host plants (Swaty *et al.*, 1998). However, many species provide protection against root desiccation and improve water uptake ability of the roots. In this sense, they provide a general benefit to plants during times of drought. Regardless of how variably these EcM symbioses may change in response to the changing environmental conditions, it is clear that they do at least change. Therefore, as more research is accomplished, and as patterns and general effects emerge, we will have a better understanding of the critical consequences that climate change has on EcM communities.

15. Conclusions

It is now well established that the most plants in ecosystems have mycorrhizae, so studies of nutrient uptake, soil resource competition, nutrient cycling, *etc.* in ecosystems may be of little value if they do not consider the role of these associations. Significantly, the mycorrhizal fungi connect the primary producers of ecosystems, plants, to the heterogeneously distributed nutrients required for their growth, enabling the flow of energy-rich compounds required for nutrient mobilization whilst simultaneously providing conduits for the translocation of mobilized products back to their hosts. The value of mycorrhizal associations to plants, which occur in natural ecosystems has been demonstrated for a limited number of species by growth experiments at realistic nutrient levels. Therefore, understanding the diversity of pathways involved in cross talk, the range of plant-microbe interactions, and the ways in which these are regulated remains the ultimate challenge in understanding the role of these fungi in ecosystem functioning. Evolutionists are hard-pressed to explain how complex, cooperative networks between completely different types of organisms such as these could have come about through Darwinian evolution— particularly when they involve dynamic biochemical networks of interaction in two separate types of organisms. The more we learn about these underground networks, the more our ideas about plants have to change. They aren't just sitting there quietly growing. By linking to the fungal network they can help out their neighbours by sharing nutrients and information – or sabotage unwelcome plants by spreading toxic chemicals through the network. It appears that this "wood wide web" has its own adaptation of operational strategy and sustainability.

References

Allen, M.F., Allen, E.B. and Friese, C.F. 1989. Responses of the non-mycotrophic plant *Salsola kail* to invasion by vesicular arbuscular mycorrhizal fungi. *New Phytol.,* 111: 45-49.

Ames, R.N., Reid, C.P.P. and Ingham, E.R. 1984. Rhizosphere bacterial populations responses to root colonization by a vesicular-arbuscular mycorrhizal fungus. *New Phytol.,* 96: 555-563.

Andrade, G., Mihara, K.L., Linderman, R.G. and Bethlenfalvay, G.J. 1998. Soil aggregation status and rhizobacteria in the mycorrhizosphere. *Plant Soil,* 202: 89-96.

Arora, D. 1991. All that the rain promises and more. A hip pocket guide to Western Mushroom. Ten Speed Press, New York.

Ashford, A.E., Allaway, W.G., Peterson, C.A. and Cairney, J.W.G. 1989. Nutrient transfer and the fungus root interface. *Aust. J. Plant Physiol.,* 16: 85-97.

Auge, R.M. 2001. Water relations, drought and vesicular-arbuscular mycorrhizal symbiosis. *Mycorrhiza,* 11: 3-42.

Babikova, Z., Gilbert, L., Toby. J. Bruce, A., Birkett, M., Caulfield, J.C., Woodcock, C., Pickett, J.A. and Johnson D. 2013. Underground signals carried through common mycelial networks warn neighbouring plants of aphid attack. *Ecol. Lett.,* 16: 835-843.

Bakshi, B.K., Reddy, M.A.R., Thapar, H.S. and Khan, S.N. 1972. Studies of silver fir regeneration. *Indian Forester,* 88: 135-144.

Bansal, M. and Mukerji, K.G. 1994. Positive correlation between AM induced changes in root exudation and mycorrhizosphere mycoflora. *Mycorrhiza,* 5: 39-44.

B.B.C. Earth. 2008. Strange but true: The largest organism on earth is a fungus. www.bbc.com/earth/story/20141114-the-biggest-organism-in-the-world.

Beguiristain, T. and Lapeyrie, F. 1997. Host plant stimulates hypaphorine accumulation in *Pisolithus tinctorius* hyphae during ectomycorrhizal infection while excreted fungal hypaphorine controls root hair development. *New Phytol.,* 136: 525-532.

Bending, G.D. and Read, D.J. 1997. Lignin and soluble-phenolic degradation by ectomycorrhizal and ericoid mycorrhizal fungi. *Mycol. Res.,* 101: 1348-1354.

Berbee, M.L. and Taylor, J.W. 1993. Dating the evolutionary radiations of the true fungi. *Can. J. Bot.,* 71: 1114- 1127.

Blaschke, H. and Baumler, W. 1989. Mycophagy and spore dispersal by small mammals in Bavarian forests. *Forest Ecol.,* 26: 237-245.

Borowicz, V.A. 2001. Do arbuscular mycorrhizal fungi alter plant-pathogen relations? *Ecol.,* 82: 3057-3068.

Boucher, N.L. and Malajczlk, N. 1990. Effects of high soil moisture on formation of ectomycorrhizas and growth of karri (*Eucalyptus diversicolor*) seedlings inoculated with *Descolea maculata*, *Pisolithus tinctorius* and *Laccaria laccata*. *New Phytol.*, 114: 87-91.

Bowen, G.D. and Theodorou, C. 1973. Growth of ectomycorrhizal fungi around seeds and roots. In: Marks, G.C. and Kozlowski, T.T. (Eds.). Ectomycorrhizae - Their Ecology and Physiology, Academic Press, New York, pp. 107-150.

Brundrett, M.C. 1991. Mycorrhizas in natural ecosystems. *Adv. Ecol. Res.*, 21: 171–311.

Brundrett, M.C. 2009. Mycorrhizal associations and other means of nutrition of vascular plants: understanding the global diversity of host plants by resolving conflicting information and developing reliable means of diagnosis. *Plant Soil.*, 320: 37-77.

Bucking, H., Liepold, E. and Ambilwade, P. 2012. The role of the mycorrhizal symbiosis in nutrient uptake of plants and the regulatory mechanisms underlying these transport processes. In: Dhal, N.K. and Sahu, S.C. (Eds.) *Plant Science Intech*, Rijeka, pp. 107-138; DOI: 10.5772/52570.

Cairney, J.W.G., Ashford, A.E. and Allaway, W.G. 1989. Distribution of photosynthetically fixed carbon within root systems of *Eucalyptus pilularis* plants ectomycorrhizal with *Pisolithus tinctorius*. *New Phytol.*, 112: 495-500.

Cariney, J.W.G. and Burke, R.M. 1994. Fungal enzymes degrading plant cell walls: their possible significance in the ectomycorrhzal symbiosis. *Mycol. Res.*, 98: 1345-1346.

Carnero Diaz, E., Martin, F. and Tagu, D. 1996. Eucalypt α -tubulin: cDNA cloning and increased level of transcripts in ectomycorrhizal root system. *Plant Mol. Biol.*, 31: 905-910.

Castellano, M.A. and Trappe, J.M. 1985. Mycorrhizal associations of five species of Monotropoideae in Oregon. *Mycologia*, 77: 499-502.

Chilvers, G.A. 1968. Some distinctive type of eucalypt mycorrhiza. *Aust. J. Bot.*, 16: 49.

Chilvers, G.A., Lapeyrie, F.F. and Horan, D.P. 1987. Ectomycorrhizal vs endomycorrhizal fungi within the same root system. *New Phytol.*, 107: 441-448.

Claridge, A.W. 2002. Ecological role of hypogeous ectomycorrhizal fungi in Australian forests and woodlands. *Plant Soil*, 244: 291-305.

Cordier, C., Pozo., M.J., Barea., J.M., Gianinazzi, S. and Gianinazzi-Pearson, V. 1998. Cell defense responses associated with localized and systemic resistance to *Phytophthora parasitica* induced in tomato by an arbuscular mycorrhizal fungus. *Mol. Plant Microbe Interact.*, 11: 1017-1028.

Cowan, P.E. 1989. A vesicular-arbuscular fungus in the diet of brushtail possums. *New Zeal. J. Bot.*, 27: 129-131.

Currah, R.S., Sigler, L. and Hambleton, S. 1987. New records and new taxa of fungi from the mycorrhizae of terrestrial orchids of Alberta. *Can. J. Bot.*, 65: 2473-2482.

Dalpe, Y. 1989. Ericoid mycorrhizal fungi in the Myxotrichaceae and Gymnoascaceae. *New Phytol.*, 113: 523-527.

Debaud, J.C., Gay, C., Prevost, A., Lei, J. and Dexheimer, J. 1988. Ectomycorrhizal ability of genetically different homokaryotic and dikaryotic mycelia of *Hebeloma cylindrosporum*. *New Phytol.*, 108: 322-328.

Dell, B., Malajczuk, N. and Thompson, G.T. 1990. Ectomycorrhiza formation in *Eucalyptus*. V. A tuberculate ectomycorrhiza of *Eucalyptus pilularis*. *New Phytol.*, 114: 633-640.

Denarie, J., Debelle, F. and Prome, J.C. 1996. Rhizobium lipochitooligosaccharide nodulation factors: signaling molecules mediating recognition and morphogenesis. *Ann. Rev. Biochem.*, 65: 503-535.

Dighton, J., Poskitt, J.M. and Howard, D.M. 1986. Changes in occurrence of basidiomycete fruit bodies during forest stand development: with specific reference to mycorrhizal species. *Trans. Brit. Mycol. Soc.*, 87: 163-171.

Ditengou, F.A. and Lapeyrie, F. 2000. Hypaphorine from the ectomycorrhizal fungus *Pisolithus tinctorius* counteracts activities of indole-3-acetic acid and ethylene but not synthetic auxins in eucalypt seedlings. *Mol. Plant Microbe Interact.*, 13: 151-158.

Duchesne, L.C., Peterson, R.L. and Ellis, B.E. 1989. The time course of disease suppression and antibiosis by the ectomycorrhizal fungus *Paxillus involutus*. *New Phytol.*, 111: 693-698.

Duddridge, J. A. 1987. Specificity and recognition in ectomycorrhizal associations. In: Pegg, G.F. and Abres, P.G. (Eds.). Fungal Infection of Plants, Cambridge University Press, Cambridge, pp. 25-44.

Feugey, L., Strullu, D.G., Poupard, O. and Simoneau, P. 1999. Induced defense limit Hartig net formation in ectomycorrizal birch roots. *New Phytol.*, 144: 541-547.

Finlay, R.D. 2008. Ecological aspects of mycorrhizal symbiosis with special emphasis on the functional diversity of interactions involving the extraradical mycelium. *J. Exp. Bot.*, 59: 1115-1126.

Finlay, R.D. and Sonderstrom, B. 1992. Mycorrhiza and carbon flow to the soil. In: Alien, M.F. (Ed). Mycorrhizal Functioning, Chapman and Hall, New York, pp. 134-160.

Fisher, J.B., Sweeny, S., Brzostek, E.R., Evans, T.P., Johnson, D.J., Myers, J.A., Bourg, N.A., Wolf, A.T., Howe, R.W. and Phillips, R.P. 2016. Tree-mycorrhizal associations detected remotely from canopy spectral properties. *Global Change Biol.*, 22: 2596-2607.

Fleming, L.V. 1984. Effect of soil trenching and coring on the formation of octomycorrhizas on birch seedlings grown around mature trees. *New Phytol.*, 98: 143-153.

Fogel, R. and Peck, S.B. 1975. Ecological studies of hypogeous fungi. I. *Coleoptera* associated with sporocarps. *Mycologia*, 67: 741 -747.

Fox, F.M. 1983. Role of basidiospores as inocula of mycorrhizal fungi of birch. *Plant Soil*, 71: 269-273.

Fox, F.M. 1986a. Groupings of ectomycorrhizal fungi of birch and pine, based on establishment of mycorrhizas on seedlings from spores in unsterile soils. *Trans. Brit. Mycol. Soc.*, 87: 371-380.

Fox, F.M. 1986b. Ultrastructure and infectivity of sclerotium-like bodies of the ectomycorrhizal fungus *Hebeloma sacchariolens*, on birch (*Betula* spp.). *Trans. Brit. Mycol. Soc.*, 87: 359-369.

Fransson, P.M.A., Taylor, A.F.S., Finlay, R.D. 2005. Mycelial production, spread and root colonisation by the ectomycorrhizal fungi *Hebeloma crustuliniforme* and *Paxillus involutus* under elevated atmospheric CO_2. *Mycorrhiza*, 15: 25-31.

Fries, N., Serck-Hanssen, K., Dimberg, L.H. and Theandek, O. 1987. Abietic acid, an activator of basidiospore germination in ectomycorrhizal species of the genus *Suillus* (Boletaceae). *Exp. Mycol.*, 11: 360-363.

Gadd, G.M. 1993. Interactions of fungi with toxic metals. *New Phytol.*, 124: 25-60.

Garcia, M.O., Ovaspyan, T., Greas, M. and Treseder, K.K. 2008. Mycorrhizal dynamics under elevated CO^2 and nitrogen fertilization in a warm temperate forest. *Plant Soil.*, 303: 301-310.

Gay, G., Normand, L., Marmeisse, R., Sotta, B. and Debaud, J.C. 1994. Auxin overproducer mutants of *Hebeloma cylindrosporum* Romagnesi have increased mycorrhizal activity. *New Phytol.*, 128: 645-657.

Gibson, F., Fox, F. M. and Deacon, J.W. 1988. Effects of microwave treatment of soil on growth of birch (*Betula pendula*) seedlings and infection of them by ectomycorrhizal fungi. *New Phytol.*, 108: 189-204.

Gogala, N. 1991. Regulation of mycorrhizal infection by hormonal factors produced by hosts and fungi. *Experientia*, 47: 331-339.

Grandmaison, J., Olah, G.M., VanCalsteren, M.R., and Furlan, V. 1993. Characterization and localization of plant phenolics likely involved in the pathogen resistance. *Mycorrhiza*, 3: 155-164.

Guidot, A., Debaud, J.C., Effosse, A., and Marmeisse, R. 2003. Below-ground distribution and persistence of an ectomycorrhizal fungus. *New Phytol.*, 161: 539-547.

Hacskaylo, E. 1972. Mycorrhiza: The ultimate in reciprocal parasitism. *BioSci.*, 22: 577-583.

Halling, R.E. 2001. Ectomycorrhizae: Co-evolution, significance, and biogeography. *Ann. Missouri Bot. Garden*, 88: 5-13.

Hampp, R. and Schaeffer, C. 1995. Mycorrhiza- carbohydrate and energy metabolism. In: Varma A and Hock, B. (Eds.). Mycorrhiza structure, function, molecular biology and biotechnology. Springer Verlag, Berlin, Germany, pp. 267-296.

Hampp, R., Schaeffer, C., Wallenda, T., Stulten, C., Johann, P. and Einig, W. 1995. Changes in carbon partitioning or allocation due to ectomycorrhiza formation: biochemical evidence. *Can. J. Bot.*, 73: 548-556.

Harley, J.L. 1989. The significance of mycorrhiza. *Mycological Research* 92: 129-139.

Harley, J.L. and Harley, E.L. 1987. A check-list of mycorrhiza in the British flora. *New Phytol.*, 105: 1-102.

Harley, J.L. and Smith, S.E. 1983. Mycorrhizal symbiosis. Academic Press, Toronto, London.

Hawkins, B.J., Jones, M.D. and Kranabetter, J.M. 2015. Ectomycorrhizae and tree seedling nitrogen nutrition in forest restoration. *New Forests,* 46: 747-771.

Hawksworth, D.L. 1991. The fungal dimension of biodiversity: magnitude, significance, and conservation. *Mycol. Res.,* 95: 641-655.

Hogberg, M.N. and Hogberg, P. 2002. Extramatrical ectomycorrhizal mycelium contributes one-third of microbial biomass and produces, together with associated roots, half the dissolved organic carbon in a forest soil. *New Phytol.,* 154: 791-795.

Horan, D.P. and Chilvers, G.A. 1990. Chemotropism: the key to ectomycorrhizal formation? *New Phytol.,t* 116: 297-301.

Horan, D.P., Chilvers, G.A. and Lapeyrie, F. 1988. Time sequenee of the infection process in *Eucalyptus* ectomycorrhizas. *New Phytol.,* 109: 451-458.

Hutchinson, L.J. 1989. Absence of conidia as a morphological character in ectomycorrhizal fungi. *Mycologia,* 81: 587-594.

Janos, D. P., Sahley, C.T. and Emmons, L.H. 1995. Rodent dispersal of vesicular-arbuscular mycorrhizal fungi in Amazonian Peru. *Ecol.,* 76: 1852-1858.

Kahiluoto, H. and Vestberg, M. 1998. The effect of arbuscular mycorrhiza on biomass production and phosphorus uptake from sparingly soluble sources by leek (*Allium porrum* L.) in Finnish field soils. *Biol. Agric. Hort.,* 16: 65-85.

Kalotas, A.C. 1996. Aboriginal knowledge and use of fungi. *Fungi of Australia* 1: 268-295.

Karabaghli-Degron, C., Sotta, B., Bonnet, M., Gay, G. and Le Tacon, F. 1998. The auxin transport inhibitor 2,3,5-triiodobenzoic acid (TIBA) inhibits the stimulation of *in vitro* lateral root formation and the colonization of the tap-root cortex of Norway spruce (*Picea abies*) seedlings by the ectomycorrhizal fungus *Laccaria bicolor*. *New Phytol.,* 140: 723-733.

Kaska, D.D., Myllyla, R. and Cooper, J.B. 1999. Auxin transport inhibitors act through ethylene to regulate dichotomous branching of lateral root meristems in pine. *New Phytol.,* 142: 49-58.

Kendrick, B. 1985. The Fifth Kingdom. Mycologue Publications. Waterloo, Ontario.

Kotter, M.M. and Farentinos, R.C. 1904. Formation of ponderosa pine ectomycorrhizae after inoculation with feces of tassel-eared squirrels. *Mycologia,* 76: 758-760.

Kottke, I. and Oberwinkler 1987. The cellular structure of Hartig net. The coenocytic and transfer cell like organization. *Nordic J. Bot.,* 7: 75-85.

Lakhanpal, T.N. and Kumar, S. 1984. Influence of ectomycorrhiza on growth and elemental position of *Picea smithiana* seedlings. *J. Tree Sci.*, 2: 38-41.

Lament, B., Ralph, C. and Chrisiensen, P.E.S. 1985. Mycophagous marsupials as dispersal agents for ectomycorrhizal fungi on *Eucalyptus calophylla* and *Gastrolobium bilobum*. *New Phytol.*, 101: 651-656.

Lankau, R.A. and Keymer, D.P. 2016. Ectomycorrhizal fungal richness declines toward the host species' range edge. *Mol. Ecol.*, 25: 3224-3241.

Largent, D.L., Sugihara, N. and WIshner, C. 1980. Occurrence of mycorrhizas in ericaceous and pyrolaceous plants in northern California. *Can. J. Bot.*, 58: 2274-2279.

Lawrence, J.F. and Milner, R.J. 1996. Associations between arthropods and fungi. In: Mallett, K. and Grgurinovic, C.A (Eds.). Fungi of Australia 1B - Introduction - Fungi in the Environment. ABRS Australia, Canberra, pp. 137-202.

Lepage, B.A., Randolph, S. Currah, S., Ruth A. Stockey, R.A. and Rothwell, G.W. 1997. Fossil ectomycorrhizae from the Middle Eocene. *Am. J. Bot.*, 84: 410-412.

Lewis, J.D. and Koide, R.T. 1990. Phosphorus supply, mycorrhizal infection and plant offspring vigour. *Funct. Ecol.*, 4: 695-702.

Long, D., Liu, J., Han, Q., Wang, X. and Huang, J. 2016. Ectomycorrhizal fungal communities associated with *Populus simonii* and *Pinus tabuliformis* in the hilly-gully region of the Loess Plateau, China. *Sci. Rep.* 6: 24336 doi: 10.1038/srep24336

Malajczuk, N., Trappe, J.M. and Molina, R. 1987. Interrelationships among some ectomycorrhizal trees, hypogeous fungi and small mammals: Western Australian and northwestern American parallels. *Aust. J. Ecol.*, 12: 53-55.

Malcolm, G.M., Lopez-Gutierrez, J.C., Koide, R.T and Eissenstat, D.M. 2008. Acclimation to temperature and temperature sensitivity of metabolism by ectomycorrhizal fungi. *Global Change Biol.*, 14: 1169-1180.

Malloch, D.W., Pirozynski, K.A. and Raven, P.H. 1980. Ecological and evolutionary significance of mycorrhizal symbioses in vascular plants. *Proc. of Nat. Acad. of Sci. USA.*, 77: 2113–2118.

Malloch, D., Grenville, D. and Hubart, J.M. 1987. An unusual subterranean occurrence of fossil fungal sclerotia. *Can. J. Bot.*, 65: 1281-1283.

Marks, G.C. and Kozlowski, T.T. 1973. Ectomycorrhizae - Their ecology and physiology. Academic Press, Inc., New York and London.

Maroneck, D.M., Hendrix, J.W. and Kiernan, J. 1981. Mycorrhizal fungi and their importance in horticultural crop production. *Hort. Rev.*, 3: 172-212.

Martin, F., Boiffin, V. and Pfeffer, P.E. 1998. Carbohydrate and amino acid metabolism in the Eucalyptus globulus *Pisolithus tinctorius* ectomycorrhiza during glucose utilization. *Plant Physiol.*, 118: 627–635.

Martin, F. and Botton, B. 1993. Nitrogen metabolism of ectomycorrhizal fungi and ectomycorrhiza. *Adv. Plant Pathol.*, 9: 83-102.

Martin, F., Canet, D. and Marchal, P. (1985). 13^C NMR study of the mannitol cycle and trehalose synthesis during glucose utilization by ectomycorrhizal ascomycete *Cenococcum graniforme*. *Plant Physiol.*, 77: 449-502.

Martin, F., Duplessis, S., Ditengou, F., Lagrange, H., Voiblet, C. and Lapeyrie, F. 2001. Developmental cross talking in the ectomycorrhizal symbiosis: signals and communication genes. *New Phytol.*, 151: 145-154.

Martin, F., Lapeyria, F. and Tagu, D. 1997. Altered gene expression using ectomycorrhizal development. In: Tudzynski, C. (ed.). Mycota V Part A Plant Relationships, Springer Verlag, pp. 223-242.

Martin, F., Ramstedt, M., Solerhall, K. and Canet, D. 1988. Carbohydrate and amino acid metabolism in the ectomycorrhizal ascomycete *Sphaerosporella brunnea* during glucose utilization. *Plant Physiol.*, 86: 935-940.

Martin, F., Laurent, P., de Carvalho, D., Voiblet, C., Balestrini, R., Bonfante, P. and Tagu, D. 1999. Cell wall proteins of the ectomycorrhizal basidiomycete *Pisolithus tinctorius*: Identification, function, and expression in symbiosis. *Fungal Genetics Biol.*, 27: 161–174.

Maser, C. and Maser, Z. 1988. Interactions among squirrels, mycorrhizal fungi, and coniferous forests in Oregon. *Great Basin Naturalist*, 48: 358-369.

Massicotte, H.B., Melville, L.H. and Peterson, R.L. 2005a. Structural features of mycorrhizal associations in two members of the Monotropoideae, *Monotropa uniflora* and *Pterospora andromedea*. *Mycorrhiza*, 15:101-110

Massicotte, H.B., Melville, L.H. and Peterson, R.L. 2005b. Structural characteristics of root-fungal interactions for five ericaceous species in eastern Canada. *Can. J. Bot.*, 83: 1057-1064.

Massicotte, H.B., Melville, L.H., Peterson, R.L. and Luoma, D.L. 1998. Anatomical aspects of field mycorrhizas on *Polygonum viviparum* (Polygonaceae) and *Kobresia bellardi* (Cyperaceae). *Mycorrhiza*, 7: 287-292.

McGee, P.A. and Baczocha, N. 1994. Sporocarpic Endogonales and Glomales in the scats of *Rattus* and *Perameles*. *Mycol. Res.*, 98: 246-249.

McIlwee, AP. and Johnson, C.N. 1998. The contribution of fungus to the diets of three mycophagous marsupials in *Eucalyptus* forests, revealed by stable isotope analysis. *Funct. Ecol.*, 12: 223-231.

Meharg, A.A. 2003.The mechanistic basis of interactions between mycorrhizal associations and toxic metal cations. *Mycol. Res.*, 107: 1253-1265.

Miller, O.K. Jr. 1982. Taxonomy of ecto- and ecto-endomyccorhzal fungi. In: Schenek, N.C. (ed.). Methods and Principles of Mycorrhizal Research. American Phytopathological Society. Minncosta, pp.91-101.

Molina, R. and Trappe, J.M. (1982). Lack of mycorrhizal specificity in the ericaceous hosts *Arbutus menziesii* and *Arctostaphylos uvaursi*. *New Phytol.*, 90: 495-509.

Moller, S.G. and Chua, N.H. 1999. Interactions and intersections of plant signaling pathways. *J. Mol. Biol.*, 293: 219–234.

Morgan, A. 1995.Toads and toadstools. Celestial Arts Publishing, Berkeley.

Morin, C., Samson J., and Dessureault, M. 1999. Protection of black spruce seedlings against *Cylindrocladium* root rot with ectomycorrhizal fungi. *Can. J. Bot.*, 77: 169-174.

Morton, J.B and Benny, G.L. 1990. Revised classification of arbuscular mycorrhizal fungi (Zygomycetes): A new order, Glomales, two new sub-orders Glomineae and Gigasporineae and new families, Acaulosporaceae and Gigasporineae, with an emendation of Glomaceae. *Mycotaxon*, 37: 471-491.

Newsham, K.K., Watkinson, A.R. and Fitter, A.H. 1995. Rhizosphere and root-infecting fungi and the design of ecological field experiments. *Oecologia*, 102: 230-237.

Ogawa, M. 1985. Ecological characters of ectomycorrhizal fungi and their mycorrhizae - an introduction to the ecology of higher fungi. *Jpn. Agric. Res. Q.*, 18: 305-314.

Pandey, A. and Mann, M. 2000. Proteomics to study genes and genomes. *Nature*, 405: 837-846.

Paris, F., Dexheimer, J. and Lapeyrie, F. 1993. Cytochemicl evidence of *Eucalyptus* roots by the ectomycorrhizal fungus *Conococcum geophilum*. *Arc. Microbiol.*,159: 526-529.

Parke, J. L., Linderman, R. G. and Trappe, J.M. 1983. Effects of forest litter of mycorrhiza development and growth of douglas-fir and western red cedar seedlings. *Can. J. Forest Res.*, 13: 666-671.

Parke, J. L., Linderman, R.G. and Trappe, J. M. 1984. Inoculum potential of ectomycorrhizal fungi in forest soils of southwest Oregon and northern California. *Forest Sci.*, 30: 300-304.

Pena, R and Polle, A. 2014. Attributing functions to ectomycorrhizal fungal identities in assemblages for nitrogen acquisition under stress. *The ISME J.*, 8: 321–330.

Perry, D. A., Molina, R. and Amaranthus, M.P. 1987. Mycorrhizae, mycorrhizospheres, and reforestation: current knowledge and research needs. *Can. J. Forest Res.*, 17: 929-940.

Peterson RL, Massicotte HB, and Melville LH. 2004. Mycorrhizas: Anatomy and Cell Biology. NRC Research Press, Ottawa.

Rabatin, S.C. and Stinner, B.R. 1989. The significance of vesicular-arbuscular mycorrhiza1 fungi-soil macroinvertebrate interactions in agroecosystems. *Agric. Ecosyst. Environ.*, 27: 195-204.

Ramsay, R.R., Sivasithamparam, K. and Dixon, K.W. 1987. Anastomosis groups among rhizoctonia-like endophytic fungi in southwestern Australian *Pterostylis* species (Orchidaceae). *Lindlyana*, 2: 161- 166.

Raamsdonk L.M, Teusink, B., Broadhurst, D., Zhang, N., Hayes, A., Walsh, M.C., Berden, J.A. Brindle, K.M., Kell, D.B., Rowland, J.J., Westerhoff, H.V., Van Dam,

K and Oliver, S.G. 2001. A functional genomics strategy that uses metabolome data to reveal the phenotype of silent mutations. *Nature Biotechnol.,* 19: 45-50.

Ramstedt, M., Martin, F. and Soderhall, K. 1989. Mannitol metabolism in the ectomycorrhizal basidiomycete *Piloderma croceum* during glucose utilization. A 13C NMR study. *Agric. Ecosyst. Environ.,* 28: 409-414.

Raven, P. and Axelrod, D.I. 1974. Angiosperm biogeography and past continental movements. *Ann. Missouri Bot. Gard.,* 61: 539-673.

Read, D.J. 1983. The biology of mycorrhiza in the Ericales. *Can. J. Bot.,* 61: 985-1004.

Read, D.J. 1991. Mycorrhizas in ecosystems. *Experimentia,* 47: 376-391.

Read, D.J. and Armstrong, W. 1972. A relationship between oxygen transport and the formation of the ectotrophic mycorrhizal sheath in conifer seedlings. *New Phytol.,* 71: 49-53.

Read, D.J., Francis, R. and Finlay, R. 1985. Mycorrhizal mycelia and nutrient cycling in plant communities. In: Fitter, A.H., Atkinson, D., Read, D.J. and Usher, M.B. (Eds.). Ecological Interactions in Soil: Plants, Microbes and Animals. Blackwell, Oxford, pp. 193-217.

Reddell, P. and Spain, A.V. 1991. Earthworms as vectors of viable propagules of mycorrhizal fungi. *Soil Biol. Biochem.,* 23: 767-774.

Redecker, D., Kodner, R. and Graham, L.E. 2000a. Glomalean fungi from the Ordovician. *Sci.,* 289: 1920-1921.

Redecker, D., Morton, J.B. and Bruns, T.D. 2000b. Ancestral lineages of arbuscular mycorrhizal fungi. *Mol. Phylogenet. Evol.,* 14: 276-284.

Rinaldi, A.C, Comandini, O. and Kuyper, T.W. 2008. Ectomycorrhizal fungal diversity: separating the wheat from the chaff. *Fungal Divers.,* 33: 1-45.

Rumney, G.R. 1968. Climatology and the world's climate. The Macmillan Company, New York, pp. 643.

Rupp, L., Devries, H., and Mudge, K. 1989. Effect of aminocyclopropane carboxylic acid and aminoethoxyvinylglycine on ethylene production by ectomycorrhizal fungi. *Can. J. Bot.,* 67: 483-485.

Rygiewicz, P.T. and Andersen, C.P. 1994. Mycorrhizae alter quality and quantity of carbon allocated below ground. *Nature,* 369: 58-60.

Salzer, P., Hebe, G. and Hager, A. 1997a. Cleavage of chitinous elicitors from the ectomycorrhizal fungus *Hebeloma crustuliniforme* by host chitinases prevents induction of K^+ and Cl^- release, extracellular alkalinization and H_2O_2 synthesis of *Picea abies* cells. *Planta,* 203: 470-479.

Salzer, P., Hubner, B., Sirrenberg, A. and Hager, A. 1997b. Differential effect of purified spruce chitinases and β-1,3-glucanases on the activity of elicitors from ectomycorrhizal fungi. *Plant Physiol.,* 114: 957-968.

Salzer, P., Munzenberger, B., Schwacke, R., Kottke, I. and Hager, A. 1997c. Signalling in ectomycorrhizal fungus-root interactions. In: Rennenberg, H., Eschrich,

W. and Ziegler, H. (Eds.). Trees- Ccontributions to Modern Tree Physiology. Leiden, Backhuys Publishers, The Netherlands, pp. 339-356.

Schenck, N.C. 1982. Methods and principles of mycorrhizal research. The American Phytopathological Society, St. Paul, Minnesota.

Schweiger, P., Robson, A. and Barrow, N. 1995. Root hair length determines beneficial effect of a *Glomus* species on shoot growth of some pasture species. *New Phytol.*, 131: 247-254.

Setala, H. 1995. Growth of birch and pine-seedlings in relation to grazing by soil fauna on ectomycorrhizal fungi. *Ecol.*, 76: 1844-1851.

Shah, F., Nicolas, C., Bentzer, J., Ellstrom, M., Smits, M., Rineau, F., Canback, B., Floudas, D., Carleer, R., Lackner, G., Braesel, J., Hoffmeister, D., Henrissat, B., Ahren, D., Johansson, T., Hibbett, D.S., Martin, F., Persson, P. and Tunlid, A. 2016. Ectomycorrhizal fungi decompose soil organic matter using oxidative mechanisms adapted from saprotrophic ancestors. *New Phytol.*, 209: 1705-1719.

Sikes B.A. 2010. When do arbuscular mycorrhizal fungi protect plant roots from pathogens? *Plant Signal Behav.*, 5: 763-765.

Singer, R., Araujo, I. and Ivory, M.H. 1983. The ectotropically mycorrhizal fungi of the neotropical lowlands, especially Central Amazonia. *Nova Hedwigia Beihefte*, 77: 1-352

Smith, S.E. and Read, D.J. 1997. Mineral nutrition, heavy metal accumulation and water relations in VA mycorrhizas. In: Smith, S.E. and Read, D.J. (Eds.). Mycorrhizal Symbiosis. Academic Press, San Diego, pp. 126-160.

Smith, S.E and Read, D.J. 2008. Structure and development of ectomycorrhizal roots. In: Smith, S.E. and Read, D.J. (Eds.). Mycorrhizal Symbiosis (3rd Ed), pp. 191-268.

Smith, S.E., Jakobsen, I., Gronlund, M. and Smith, F.A. 2011. Roles of arbuscular mycorrhizas in plant phosphorus nutrition: interactions between pathways of phosphorus uptake in arbuscular mycorrhizal roots have important implications for understanding and manipulating plant phosphorus acquisition. *Plant Physiol.*, 156: 1050-1057.

Soderstrom, B., Finlay, R.D. and Read, D.J. 1988. The structure and function of the vegetative mycelium of ectomycorrhizal plants IV. Qualitative analysis of carbohydrate contents of mycelium interconnecting host plants. *New Phytol.*, 109, 163-166.

Stanley, M.R., Koide, R.T., Shumway, D.L. 1993. Mycorrhizal symbiosis increases growth, reproduction and recruitment of *Abutihn theophrasti* Medic, in the field. *Oecohgia* 94: 1-35

Swaty, R. L., Gehring, C, VanErt, M., Theimer, T. C., Keim, P. and Whitham, T.G. 1998. Temporal variation in temperature and rainfall differentially affects ectomycorrhizal colonization at two contrasting sites. *New Phytol.*, 139: 733-739. doi:10.1046/j.1469-8137.1998.00234.x.

Tagu, D., Lapeyrie, F., and Martin, F. 2002. The ectomycorrhizal symbiosis: Genetics and developments. *Plant Soil*, 244: 97-105.

Tarafdar, J.C. and Marschner, H. 1994b. Phosphatase activity in the rhizosphere and hyposphere of VA-Mycorrhizal wheat supplied with inorganic and organic phosphorus. *Soil Biol. Biochem.*, 26: 385-395.

Taylor, A.F.S. and Alexander I. (2005). The ectomycorrhizal symbiosis: Life in the real world. *Mycologist*, 19: 102-112.

Taylor, D.L., Bruns, T.D. and Hodges, S.A. 2004. Evidence for mycorrhizal races in a cheating orchid. *Proc. R. Soc. Lond. B Biol. Sci.*, 271: 35-43.

Tedersoo, L., May, T. and Smith, M. 2010. Ectomycorrhizal lifestyle in fungi: global diversity, distribution, and evolution of phylogenetic lineages. *Mycorrhiza*, 20: 217-263.

Tilman, D., Wedin, D. and Knops, J. 1996. Productivity and sustainability influenced by biodiversity in grassland ecosystems. *Nature*, 379: 718-720.

Trappe, J.M. 1962. The fungus associates of ectotrophic mycorrhizae. *Bot. Rev.*, 28: 538-606.

Trappe, J.M. 1977. Selection of fungi for ectomycorrhizal inoculation in nurseries. *Ann. Rev. Phytopathol.*, 15: 203-222.

Trappe, J. M. and Molina, R. 1986. Taxonomy and genetics of mycorrhizal fungi: their interactions and relevance. In: Gianinazzi-Pearson, V. and Gianinazzi, S. (Eds.). Physiological and genetical aspects of mycorrhiza, INRA. Paris, pp. 133-146.

Truswell, E.M., Kershaw, A.P.J. and Sluiter, I.R. 1987. The Australian-southeast Asian connection: Evidence from the paleobotanical record. In: Whitmore, T.C (Ed.). Biogeographical Evolution of the Malay Archipelago. Oxford Univ. Press, Oxford, pp. 32-49.

Velculescu, V.E., Madden, S.L., Zhang, L., Lash, A.E., Yu, J., Rago, C., Lai, A., Wang, C.J., Beaudry, G.A., Ciriello, K.M., Cook, B.P., Dufault, M.R., Ferguson, A.T., Gao, Y.H., He, T.C., Hermeking, H., Hiraldo, S.K., Hwang, P.M., Lopez, M.A., Luderer, H.F., Mathews, B., Petroziello, J.M., Polyak, K., Zawel, L., Zhang, W., Zhang, X.M., Zhou, W., Haluska, F.G., Jen, J., Sukumar, S., Landes, G.M., Riggins, G.J., Vogelstein, B. and Kinzler, K. W. 1999. Analysis of human transcriptomes. *Nat. Genet.*, 23: 387-388.

Wang, B. and Qiu, Y.L. 2006. Phylogenetic distribution and evolution of mycorrhizas in land plants. *Mycorrhiza*, 16: 299-363.

Warcup, J.H. 1981. Mycorrhizal relationships of Australian orchids. *New Phytol.*, 87: 371-381.

Warcup, J.H. 1985. Pathogonic *Rhizoctonia* and orchids. In: Parker, C.A., Rovira, A. D. Moore, K.J., Wong, P.T.W. and Kollmorgen, J.F. (Eds.). Ecology and Management of Soilborne Plant Pathogens. The American Phytopathological Society, St Paul. Minnesota, pp. 69-70.

Werner, D. 1992. Symbiosis of plants and microbes. Chapman and Hall London.

Wilcox, H.E. 1968. Morphological studies of the roots of red pine, *Pinus resinosa*. II. Fungal colonization of roots and the development of mycorrhizae. *Am. J. Bot.*, 55: 686-700.

2017, Mycorrhizal Fungi   *Pages 33–73*
Editors: **Ashok Aggarwal and Kuldeep Yadav**
Published by: **ASTRAL INTERNATIONAL PVT. LTD., NEW DELHI**

2

Diversity of Arbuscular Mycorrhizal Fungi and their Role in Conservation Agriculture

S.R. Gupta* and Pushpa Devi

*Department of Botany, Kurukshetra University,
Kurukshetra – 136 119, Haryana
Corresponding Author: sgupta2002158@gmail.com

ABSTRACT

Conventional agriculture, involving deep ploughing and soil inversion, often leads to disruption of soil structure, loss of soil biodiversity, and decreased soil organic matter. Conservation agriculture is system of agronomic practices including no-tillage, zero-tillage, direct seeding, reduced tillage, ridge tillage, crop rotation, and cover crops with minimal soil disturbance. A diversity of conservation agriculture are practiced in diverse agro-ecologies and farming systems worldwide. In recent years, the zero-tillage with wheat succeeding rice is now the most widely adopted resource conserving technology in the Indo-Gangetic plains of India. Conservation agriculture systems are reported to have a favourable effect on the delivery of regulatory ecosystem services including climate regulation through carbon sequestration and reducing greenhouse gas emissions. The arbuscular mycorrhizal (AM) fungi are an ancient group of obligate root symbionts which are present in about 80 per cent of land plant species, including most agricultural crops. The no-tillage, reduced tillage, crop rotations, and cover crops promote the diversity AM fungi and AMF root colonization in diverse types of agricultural systems. AM fungi are important for crop productivity, nutrient cycling, soil aggregation, and as potential drivers of biodiversity and soil regulatory processes for sustainable agriculture.

AM fungi play a key role in soil conservation, nutrient cycling, soil carbon sequestration. AMF produce a glycoprotein called glomalin which is important to bind soil aggregates together for protecting soil carbon, and enhancing soil carbon sequestration. Under no-tillage and reduced tillage, the enhanced diversity of AMF in soil has the potential to enhance the sustainability. More diverse AMF communities would be able to impart greater adaptability and resilience of agricultural systems due to changing environmental conditions over space and time. By favouring diversity of AMF, conservation tillage has the greater possibility of encouraging mycorrhiza as the possible combinations of host plant and AMF functional symbiosis, depending on biotic or abiotic conditions.

Keywords: *Conventional Tillage, Zero Tillage, AM fungi, Carbon cycling, N and P cycling, Soil carbon sequestration*

1. Introduction

Production of food is one of the important provisioning services of agriculture. Intensification in agriculture and various human activities are disturbing many regulating and supporting ecosystem services of the agricultural systems (Power, 2010). For agriculture to be sustainable in future, it is important to increase cropping efficiency (Foley *et al.*, 2011), adopting conservation agriculture practices (Hobbs *et al.*, 2008; Palm *et al.*, 2014), and exploiting arbuscular mycorrhizal fungal symbiosis for increased plant productivity (Pellegrino *et al.*, 2015; van der Heijden *et al.*, 2015). Conservation agriculture is system of agronomic practices through the application of minimal soil disturbance that includes no-tillage, zero-tillage, direct seeding, reduced tillage, ridge tillage and cover crops, so as to improve livelihoods of farmers (Baker *et al.*, 2002; Derpsch *et al.*, 2010, 2011; Palm *et al.*, 2014). Conservation agriculture provides multiple ecosystem services such as climate regulation, soil carbon sequestration, soil conservation, and regulation of water and nutrients by influencing several soil properties and processes (Palm *et al.*, 2014). Increasing food production at the cost of ecosystem services is going to affect crop production and the agricultural sustainability in the long-term. The arbuscular mycorrhizal fungi regulate plant–soil interactions and nutrient fluxes, which can play a key role achieving agricultural sustainability (Verbruggen and Kiers, 2010).

The arbuscular mycorrhizal (AM) fungi constitute an ancient group of obligate root symbionts whose origin is considered to be coinciding with the origin of land plants (Bonfante and Genere, 2008). AM characterize a delicate balance between plant, fungus and the soil (Mosse, 1986) and are present in about 80 per cent of land plant species, including most agricultural crops (Smith and Read, 2008). In this symbiotic relationship, the fungus receives the plant synthesized carbon, while the plant is benefited with an increased capacity for nutrient uptake as well as enhanced tolerance to drought and pathogens (Smith and Read, 2008). This symbiosis is known to influence interactions among the plants and the structure of plant communities, thereby affecting the agricultural production, and the conservation of ecosystems (van der Heijden *et al.*, 2008). There are direct effect of AM fungi on the ecosystem by regulating the structure of plant communities and productivity (van der Heijden *et al.*, 1998), and improving the soil structure and aggregation (Rillig and Mummey,

2006; Rillig *et al.*, 2015). The ecosystem services provided by AM fungi include increased plant productivity (van der Heijden *et al.*, 1998; Lekberg and Koide, 2005), improved soil aggregate stability and high soil organic matter level (Rillig and Mummey, 2006), and regulating carbon, phosphorus, and nitrogen cycling (Fitter *et al.*, 2011; van der Heijden *et al.*, 2015), and improving phosphorus uptake. The influence of AM symbiosis on greenhouse gas (GHG) emissions from agricultural systems has recently been investigated (Bender *et al.*, 2014; Lazcano *et al.*, 2014).

AM fungi are important for crop productivity, nutrient cycling, soil aggregation, and as potential drivers of biodiversity and soil regulatory processes for sustainable agriculture. This paper gives an overview of conservation agriculture and ecosystem services, the diversity of AM fungi in soil, and the role of AM fungi in nutrient cycling, soil carbon sequestration and reducing greenhouse gas emission from soil.

2. Conservation Agriculture and Ecosystem Services

Conservation agriculture was originally applied for controlling wind and water erosion (Baveye *et al.*, 2011). The term "Conservation Agriculture" (CA) was coined in the late 1990s, now there is marked diversity in approaches and understanding different issues concerning CA (Drepsch *et al.*, 2014). CA is being adopted on large scale and on mechanized farms in different regions of the world (Derpsch and Friedrich, 2009). Conservation tillage is defined as any tillage system that leaves sufficient crop residue in the agricultural field to cover at least 30 per cent of the soil surface after planting (CTIC, 1995; FAO, 2001; Lal, 2003). The information on various aspects of conservation agriculture has been recently discussed by various workers (Stevenson *et al.*, 2014; Erenstein *et al.*, 2015; Jat *et al.*, 2014; Giller *et al.*, 2015).

2.1. Types of Conservation Agriculture

The diversity of conservation agriculture as practiced in the world's diverse agro-ecologies and farming systems are summarized (Table 2.1).

In the Great Plains of US and Canada, switching to no-till/stubble retention allowed replacement of fallow with cropping of cash crops like oilseeds and legumes (Kirkegaard *et al.*, 2008). In Brazil's Cerrados, land productivity benefits of CA have occurred through the possibility of growing two crops sequentially in the same growing season (Bolliger *et al.*, 2006). Rice-wheat systems are among the most widespread cropping system in India. The conventional systems of growing rice and wheat are not very efficient in terms of water economy and resource use. In recent years, zero-tillage with wheat crop succeeding rice crop is now the most widely adopted resource conserving technology in the Indo-Gangetic plains of India, including regions of Haryana and Punjab (RWC, 2005). The soils of the indo-Gangetic plains are poor in the organic carbon and hold a great potential for soil carbon sequestration in agriculture systems. Conservation tillage, along with some complimentary practices such as soil cover and crop diversity (Corsi *et al.*, 2012) has emerged as a viable option to ensure sustainable food production and maintain environmental sustainability (Busari *et al.*, 2015).

Table 2.1: Different Types of Conservation Agriculture Systems and their Main Characteristics in different Regions of the World (Modified from Giller et al., 2015)

Type	Region (Examples)	Tillage	Mulch	Rotation	Farm Type/Input	References
Animal driven no-tillage	Subtropical southern Brazil	Use of direct seeder for planting directly through mulch into soil	Mulch of crop residues and cover crops	Maize, soybean and beans followed by winter wheat, black oats, rye, or leguminous cover crop	Medium-sized mixed crop-livestock farms (20–50 ha) Medium level of fertilization and use of herbicides	Bolliger et al. (2006)
Tractor-operated no/reduced tillage (small-medium scale)	NW Indo-Gangetic Plains (India/Pakistan)	Use of no-till tractor-mounted direct seeder (locally manufactured)	Partial mulch of crop residues	Wheat crop only (in irrigated wheat-based double crop systems, e.g., rice-wheat)	Mixed crop-livestock farms (<20ha) Irrigation, fertilization and use of herbicides	Erenstein and Laxmi (2008)
Tractor-operated no/reduced tillage (small-medium scale)	West Asia-North Africa (dry Mediterranean climate)	Use of no-till tractor-mounted direct seeder (locally manufactured)	Limited mulch of crop residues	Wheat, barley, legumes (lentil, chickpea)	Mechanized mixed crop-livestock (sheep) farms (<200ha) Medium use of fertilizer and herbicides	Kassam et al. (2012)
Tractor operated reduced tillage (medium scale)	North-west Europe (cool temperate climate)	Some superficial soil tillage before direct seeding	Mulch of crop residues	Fodder and grain maize, wheat, barley, and cruciferous cover crops, ryegrass	Mechanized medium-scale (arable) farms (30–300 ha Intensive use of fertilizer and herbicides	Cannell (1985); Soane et al. (2012)
Tractor operated direct seeding (large scale)	Australian wheat belt (subtropical and Mediterranean climate)	Use of no-till tractor-mounted director seeder (large tractor implements)	Mulch of crop residues	Cereal-legumes (oilseed)	Mechanized large scale farms and enterprises (1000–10,000 ha) Reliance on herbicides and fertilizer	Llewellyn et al. (2012); Kirkegaard et al. (2014)

Contd...

Table 2.1–*Contd...*

Type	Region (Examples)	Tillage	Mulch	Rotation	Farm Type/Input	References
Tractor operated direct seeding (large scale)	North-America (Canada and the mid-west)	Use of no-till tractor-mounted director seeder (large tractor implements)	Mulch of crop residues	Maize-soybean	Soybean Mechanized large scale farms (<500 ha)	Hansen *et al.* (2015)
Tractor operated direct seeding (large scale)	Cerrado region, Brazil (tropical sub-humid climate)	Use of no-till tractor-mounted director seeder (large tractor implements)	Mulch of crop residues		Mechanized large scale farms and enterprise (500–5000 ha)	Bolliger *et al.* (2006)

The area under conservation tillage is estimated to be over 111 million-hectares worldwide (Derpsch *et al.*, 2011), which accounts for 8 per cent of global cropland including diverse farming systems under temperate, subtropical and tropical conditions. Recent estimates in Table 2.2, shows some 66.4 M ha (42 per cent) of the total global area under CA is in South America, which accounts for ~ 60 per cent of the cropland in the region, 54 M ha (34 per cent) is in the USA and Canada accounting for 24 per cent of the cropland of the region (Kassam *et al.*, 2015). 17.9 M ha (11 per cent) is in Australia and New Zealand, corresponding to 36 per cent of the cropland. In Asia, conservation agriculture is being practiced on some 10.6 M ha (7 per cent) corresponding to 3 per cent of the cropland in the region. About 8.4 M ha (5 per cent) of the total global CA area is in the rest of the world (Table 2.2).

Table 2.2: Cropland Area under CA (M ha), CA Area as per cent of Total Cropland, and CA Area as per cent of Cropland by Continent, in 2013 (from Kassam *et al.*, 2015).

Continent	Cropland under CA (M ha)	Per cent of Global CA Area	Per cent of Cropland
South America	66.2	42.3	60.0
North America	54.0	34.4	24.0
Australia and NZ	17.9	11.4	35.9
Asia	10.3	6.6	3.0
Russia and Ukraine	5.2	3.3	3.3
Europe	2.0	1.3	2.8
Africa	1.2	0.8	09
Global Total	157.0	100	10.9

The recent comprehensive meta-analysis of 5463 paired yield observations from 610 studies suggests that no-till could result in a yield reduction of around 10 per cent overall (Pittelkow *et al.*, 2015a). The reduction in yield was greatest for cereal crops; however oilseeds, cotton, and legumes gave comparable yields under no-till and conventional tillage (Pittelkow *et al.*, 2015b). The negative effects of no-till could be minimized when combined with mulching and crop rotation, characterizing the systems to be truly CA (Pittelkow *et al.*, 2015a). Yield benefits of CA under such conditions are mostly due to timelier sowing, early crop establishment, and summer fallow weed control.

2.2. Ecosystem Services of Conservation Agriculture

The ecosystem services are the benefits that the natural environment provides to humanity. The Millennium Ecosystem Assessment (MEA) highlighted the condition of ecosystem and ecosystem services, and distinguished four broad categories of ecosystem services, *i.e.*, provisioning, regulating, cultural, and supporting services. The provisioning services describe the processes that yield foods, fibers, fuels, water, biochemical, medicinal plants, pharmaceuticals, and genetic resources. The regulating services are the benefits obtained from regulation of ecosystem processes; include erosion control or soil stabilization, waste treatment; climate regulation, and hydrological flows. The Supporting services are those that are necessary for the production of all other ecosystem services. Achieving many of the Millennium Development Goals depends directly or indirectly on the ecosystem services of the soil (MEA, 2005). The delivery of different ecosystem services is determined to a large extent by soils, vegetation and climate, and resulting ecosystem services.

Agricultural intensification can adversely affect many of the regulating and supporting services including nutrient cycling, climate regulation, regulation of water quantity and quality, pollination services and pest control (Figure 2.1). The conventional agricultural practices have significant impact on plant and soil biodiversity underpinning many of these ecosystem services. Conservation agriculture practices have favourable effect on soil properties and processes, which in turn affect the delivery of ecosystem services including climate regulation through carbon sequestration and greenhouse gas emissions, and provision of water through regulation of physical, chemical and biological properties of the soil (Palm *et al.*, 2014).

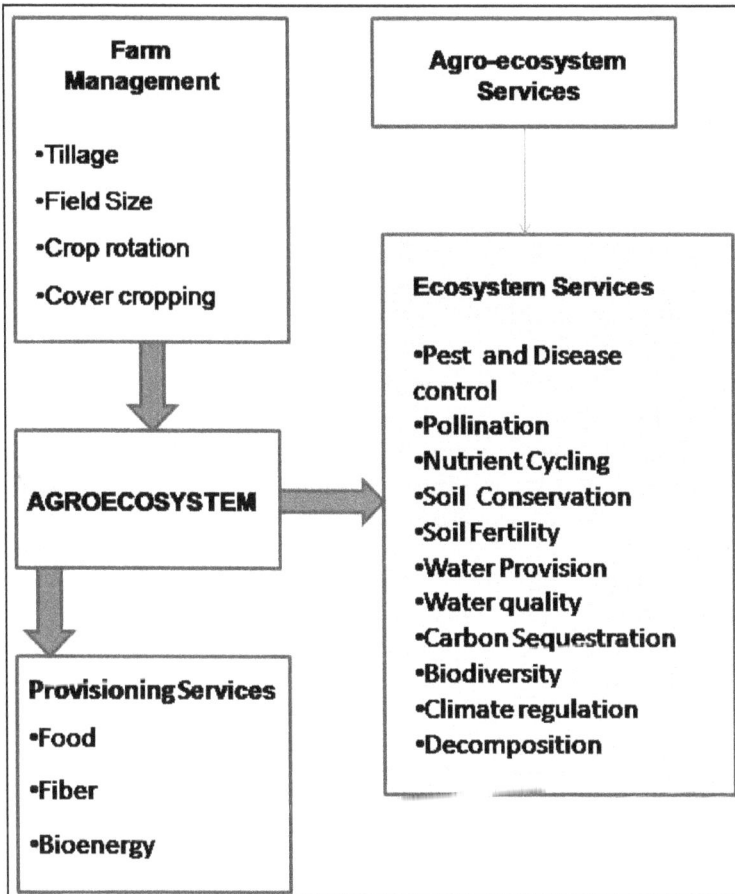

Figure 2.1: The Potential Agroecosystem Services as Influenced by different Agricultural Management Practices (from Power, 2010).

Soil biodiversity is responsible for supplying the environment with a number of critically important ecosystem goods and services (Gupta *et al.*, 2009). The ability of soils to deliver the ecosystem services directly depends on soils regulatory services of soil biodiversity, decomposition of organic materials, regulation of fluxes of greenhouse gases to and from the atmosphere, and plant-soil nutrient cycles (Palm *et al.*, 2007). A number of processes mediated by the soil biota such as waste recycling, soil formation, nitrogen fixation, bioremediation of chemicals, biotechnology, biocontrol of pests, and pollination by organisms having edaphic phase in their life cycle, provide the ecosystem services(Palm *et al.*, 2007; Gupta *et al.*, 2009). The ecosystem services can provide a framework for analyzing the differences in specific ecosystem services among soil types, and interconnections between the ecosystem services and key soil processes (Palm *et al.*, 2007). The maintenance of fertile soil is one of the most vital ecological services the soil biota performs.

3. Biodiversity of AM Fungi

Frank (1885) was probably the first to recognize the widespread nature of associations between plant roots and mycorrhizal fungi (Frank and Trappe, 2005). Arbuscular mycorrhiza is the most ancient and widespread form of fungal symbiosis (Smith and Read, 2008). On the basis of paleobotanical and molecular sequence data, it has been suggested that the first land plants formed associations with Glomalean fungi from the Glomeromycota about 460 million years ago (Redecker *et al.*, 2000). These are soil borne fungi belonging to six fungal genera (*Glomus, Sclerocystis, Acaulospora, Entrophospora, Gigaspora* and *Scutellospora*) of the single order Glomales within Zygomycetes (Morton and Benny 1990). Recently, two new genera, *i.e.*, *Archaeospora* and *Paraglomus* have been added to the existing six genera of AM fungi (Redecker *et al.*, 2000). These fungi are keystone organisms that form an interface between soils and plant roots, and are indicator of changes in soil and plant conditions (Power and Mills, 1995).

On basis of morphological characteristics of the spores, only 244 species of Glomeromycota have been described (Schussler, 2013, 2014; Oehl *et al.*, 2011). Estimates of AM fungal richness based on environmental ribosomal DNA sequences (operational taxonomic units, OTUs) range from 341 (Opik *et al.*, 2013) to 1600 worldwide (Koljalg *et al.*, 2013).These 300–1600 AM fungal taxa associate with approximately 200,000 plant species (Brundrett, 2009), indicating host specificity to be very low. The host preferences and host selectivity of AM fungi have been widely reported (Helgason *et al.*, 1998). Richness and composition of AM fungal communities depend on host plant, climate, and soil conditions (Opik *et al.*, 2006). The natural communities of AM fungi are mostly composed of uncultured taxa and it remains a challenge to elucidate if uncultivated fungi differ functionally from cultured taxa (van der Heijden *et al.*, 2015).

The populations of AM fungi are greatest in plant communities with high diversity such as tropical rainforests and temperate grasslands where they have many potential host plants and can take advantage of their ability to colonize a broad host range (Smith and Read, 2008). The biogeography of Glomeromycota is influenced by environmental factors such as climate (Leckberg *et al.*, 2007) soil series and soil pH, soil nutrients and plant community (Allen *et al.*, 1995). Davidson *et al.* (2015) have analyzed the global distribution of AM fungi on the basis of 1014 plant root samples collected from different ecosystems worldwide. These workers have identified DNA-based AM fungal taxa (virtual taxa (VT) after Optik *et al.*, 2010) by employing 454 sequencing. This continent wide comparison of AM fungal diversity showed that 93 per cent of recorded VT were present on more than one continent, and one-third (34 per cent) were present across all six sampled continents (Davidson *et al.*, 2015). This study further revealed that more than 90 per cent of virtual taxa were found in more than one climate zone, and 79 per cent were present in both grassland and forests.

4. Diversity of AM Fungi in Conservation Agriculture

There is a marked diversity among AM fungal communities belowground, depending on plant species diversity, soil type, and season, or a combination of these factors (Smith and Smith, 2012). Many modern agronomic practices have adverse effects on mycorrhizal symbiosis. There is great potential for low-input agriculture to manage the system in a way that promotes mycorrhizal symbiosis. Conventional agriculture practices, such as tillage, heavy fertilizers and fungicides, poor crop rotations interfere with the ability of plants to form symbiosis with AM fungi. Proper management of AM Fungi in the agroecosystems can improve the quality of the soil and the productivity of the land. The tillage practices, crop rotations, and cover crops, have been found to promote functional mycorrhizal symbiosis in terms of Size of AMF community, AMF root colonization, nutrient cycling and soil properties and processes (Figure 2.2). The composition and diversity of AMF spore communities were affected by tillage in a number of studies (Kabir, 2005; Douds *et al.*, 2005).

4.1. Tillage Practices Influence on AM Diversity

The negative impact of tillage on AM fungi may reduce mycorrhizal development in plant root (Kabir *et al.*, 1997a; McGonigle and Miller, 1996; Roldan *et al.*, 2007) and have a selective effect on AM fungal communities (Jansa *et al.*, 2002; Jansa *et al.*, 2003). Soil tillage can modify the physical, chemical and biological properties of a soil (Beauregard *et al.*, 2010). Some researchers pointed out that tillage could modify AM fungal spore density indirectly by altering soil properties, such as soil moisture and soil organic matter content (Anderson *et al.*, 1984) or directly through the dilution of AM fungal spores in a greater volume of soil (Kabir, 2005). Tillage has been found to reduce AMF spore and hyphal length densities, and reduction in glomalin concentrations in both temperate and tropical soils (Wright *et al.*, 1999). Conventional tillage is detrimental to AMF, reducing the abundance of soil mycelium in spring, and delaying the mycorrhizal development of the following crop (Kabir, 2005). Reduced tillage had much less severe negative impact on the abundance of soil hyphae and mycorrhizal colonization of corn by indigenous AMF (Kabir *et al.*, 1997b). For example, spores of *Glomus etunicatum* and *G. caledonium* were predominant in tilled temperate soils, whereas other species like *Glomus occultum*, *Scutellospora pellucida*, *Acaulospora paulinae* and *Entrophospora infrequens* were more abundant in non-tilled soils (Galvez *et al.*, 2001; Jansa *et al.*, 2002).

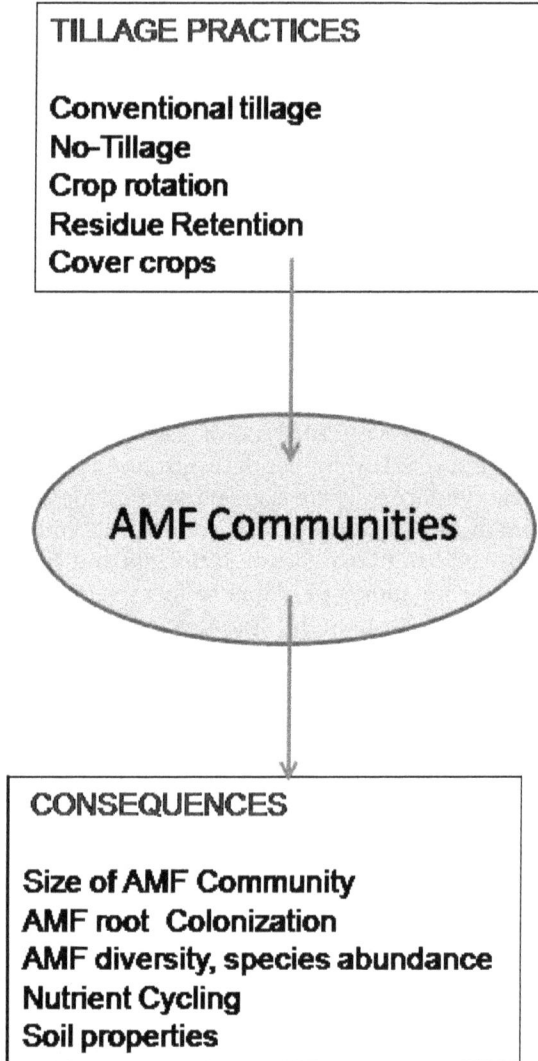

Figure 2.2: Effect of Tillage Practices on the Size and Diversity of AM Fungi, Nutrient Cycling and Soil Properties (Modified from Jansa *et al.,* 2002).

In a 13 yr-old no-tillage soil, the shifts in species composition based on spores have been reported, non *Glomus* AM fungal spores including species of *Gigaspora, Scutellaspora, Entrophospora* were found to be more frequent in no-tilled soils, whereas in conventional tilled soils, AM fungal spores belonged to *Glomus* spp (Jansa *et al.,* 2002). The spores from *Glomus occulutum* group were more numerous in no-tilled soils, whereas *Glomus etunicatum*, other *Glomus* species were predominant in tilled soils (Douds *et al.,* 1995). *Glomus* spp. are believed to survive disturbance well and are found generally prevalent in highly disturbed agricultural systems (Douds *et al.,* 1995; Dodd *et al.,* 2000). The changes in the community structure (abundance and diversity) of AMF, which colonised maize roots, could result from differences between species of AM fungi in their tolerance to the tillage-induced disruption of the hyphae, changes in the nutrient content of the soil, soil microbial activity, and changes in weed populations (Jansa *et al.*, 2003).

Several workers have reported that species richness of AM fungi could be reduced by intensive tillage (Jansa *et al.*, 2003; Alguacil *et al.*, 2009; Schnoor *et al.*, 2011). A reduction in the diversity of AM fungal communities in distinct agro-ecosystems under conventional tillage has been reported (Alguacil *et al.*, 2008; Brito *et al.*, 2012). On the basis of AM fungal sequences, obtained by nested PCR, Brito *et al.* (2012) found decrease of AM fungal diversity by 40 per cent in conventional tillage as compared to no-till systems of wheat-triticale-sunflower rotation under Mediterranean conditions (Brito *et al.*, 2012).

The diversity of AM fungi, spore density and relative spore density has been found to be affected by tillage systems in a rice- wheat system in northern India (Neelam *et al.*, 2010). In the zero-tillage system, the number of mycorrhizal fungi was higher compared to that of the conventional tillage and the furrow irrigated raised bed system. A total of 42 species of arbuscular mycorrhizal (AM) fungal species belonging to six genera, *i.e. Acaulospora, Entrophospora, Glomus, Gigaspora, Sclerocystis and Scutellospora* were recorded in the wheat cropping system under different tillage practices (Neelam *et al.*, 2010). The relative density of four groups of mycorrhizal spores was: 64.45 to 73.08 per cent (*Glomus* spp.), 14.14 to 20.38 per cent (*Acaulospora* spp.); 2.70 to 13.68 per cent (*Gigaspora* spp.) and 3.88 to 6.71 per cent (other species). The AM fungal root colonization of wheat roots ranged from 78.98 to 93.96 per cent in zero-tillage, furrow irrigated raised bed tillage and the conventional tillage system at crop maturity in 7.5 to 7.5 to 15 cm soil depths.

AM fungal spore diversity in rice-wheat crop rotation under zero-tillage, reduced tillage, and conventional tillage in Ambala district of northern India has been analyzed (Devi, 2015; Devi *et al.*, 2014). In the wheat systems, total of 55 species of mycorrhizal fungi belonging to six genera, *i.e. Acaulospora, Entrophospora, Glomus, Gigaspora, Sclerocystis* and *Scutellospora* were identified. In the zero-tillage system, the number of mycorrhizal fungi was higher comprised of 55 species belonging to all the six genera (Table 2.3). About 22 species belonging to four genera, *i.e. Acaulospora, Entrophospora, Gigaspora* and *Glomus* were recorded in soil of the rice system (Devi, 2015). In this study, twenty two species of *Glomus*, twenty species of *Acaulospora*, nine species of *Gigaspora*, four species of *Entrophospora,* and two species of *Scutellospora* and four species of *Sclerocystis* were recorded in the case of both rice and wheat system.

Devi (2015) in a study in rice-wheat crop rotation in northern India showed that there was significant effect of tillage systems, soil depth and crop growth stage on the total number of AM fungal spores (Table 2.4). The number of AM fungal spores tended to increase with crop growth, the number of AM fungal spores was highest at crop maturity. The number of AM fungal spores (per 10g soil) across soil depths was: 48.6 to 95.08 per l0g soil at 0 to 15 cm soil depth and 17.25 to 25.18 per l0g soil at 15 to 30 cm soil depth. During the wheat growing period, the spore number was generally higher in zero-tillage (87.6 to 96.3 per l0g soil) and reduced

Table 2.3: Distribution of VAM Fungi Species in under Conventional Tillage (CT), Zero-Tillage (ZT) and Reduced-Tillage (RT) Systems in Wheat Crop (Devi, 2015)

AM Fungal Species	RT	CT	ZT
Acaulospora bireticulata (Roth. and Trappe)	–	+	+
Acaulospora gerdemannii Schenck & Nicoloson	+	+	+
Acaulospora lacunosa (Morton)		+	+
Acaulospora laevis (Gerdemann and Trappe)	+	+	+
Acaulospora mellea (Spain and Schenck)	+	+	+
Acaulospora nocolsonii (Walk. Read Sand.)	+	+	+
Acaulospora rehmii (Sieverding and Toro)	+	+	+
Acaulospora scrobiculata (Trappe)	+	+	+
Acaulospora trappei Ames & Linderman	+	+	+
Acaulospora capsicula Błaszk	+	+	+
Acaulospora denticulata Sieverd. & S. Toro	+	+	–
Acaulospora spinosa Walker and Trappe	+	+	+
Acaulospora tuberculata Janos and Trappe	–	+	–
Entrophospora nevadensis Oehl, Palenz., Silva & Sieverd	+	+	+
Entrophospora sp.1	+	+	+
Entrophospora sp.11	–	+	–
Entrophospora sp.111	+	+	–
Gigaspora albida (Walker and Rhodes)	+	+	+
Gigaspora calospora (Nicolson & Gerd.) Gerd. & Trappe	–	+	–
Gigaspora gigantea (Nicol. and Gerde.) (Gerdemann and Trappe)	–	+	–
Gigaspora rosea Nicolson & Schenck	+	+	–
Gigaspora sp.1	–	–	–
Gigaspora sp.11	–	–	–
Gigaspora sp.111	–	+	+
Gigaspora margirata Becker & Hall	–	+	–
Gigaspora albida Schenck and Sm	–	+	+
Gigaspora albida Schenck & Sm.	+	+	+
Glomus claroideum (Schenk and Smith)	+	+	+
Glomus clarum Nicolson & Schenck			
Glomus constrictum (Trappe)	–	+	–
Glomus deserticola (Trappe, Bloss and Menge)	–	+	+
Glomus diaphanum (Morton and Walker)	+	+	+
Glomus fasciculatum (Gerd. and Trappe)	+	+	+
Glomus fulvum (Berk and Broome)	+	+	+
Glomus geosporum (Nicol. and Gerde.) Walker	+	+	+
Glomus hoi Berch & Trappe	–	+	+

Contd...

Table 2.3–*Contd...*

AM Fungal Species	RT	CT	ZT
Glomus sp1	+	+	+
Glomus indica (Mano. Sharat. and Adho.)	+	+	+
Glomus intraradices Schenck & Smith	+	+	+
Glomus microaggregatum Koske, Gemma & Olexia	+	+	+
Glomus macrocarpum (Tulosne and Tulosne)	+	+	–
Glomus manihotis Howeler, Sieverd. &. Schenck			
Glomus monosporum (Gerd and Trappe)	+	+	+
Glomus mosseae (Gerd and Trappe)	+	+	+
Glomus pallidum (Hall)	–	+	+
Glomus reticulatam (Bhattacharjee and Mukerjee)	+	+	+
Glomus sp II	+	+	+
Glomus sp III	–	+	–
Glomus sinuosum (Gerd. & B.K. Bakshi) Almeida & Schenck	+	+	+
Sclerocystis coremioides (Berk. & Broome) Redecker & Morton	–	+	+
Sclerocystis sp. 1	–	+	+
Scutellospora sp.	–	+	+
Scutellospora calospora Nicolson & Gerd.	–	+	–
Scutellospora cerradensis Spain & Miranda	+	+	+
Scutellospora pellucida (Nicolson and Schenck)	+	+	+

tillage systems (72.83 to 94.83 per 10g soil) as compared to the conventional system (48.6 to 82.78 per l0g soil).

Sale *et al.* (2015) studied the impact of reduced tillage under organic farming and conventional farming on the diversity of arbuscular mycorrhizal fungi in the Sissle valley (Canton Aargau, Switzerland). In total, >50,000 AMF spores were identified on the species level, and 53 AMF species were found in different systems in the order as 38 species in the permanent grassland, 33 each in the two reduced till organic farming systems, 28-33 in the regularly plowed organic farming systems, 28-33 in the non-organic conventional farming systems (Sale *et al.*, 2015). The results of the study revealed that AMF spore density and species richness, and the Shannon- Weaver AMF diversity index increased in the top-soils under reduced tillage as compared to the conventional farming. AMF communities in clay soils were affected by land use type, farming system, tillage as well as fertilization strategy across soil depth (Sale *et al.*, 2015).

4.2. Crop Rotation Effects on AMF

Crop rotation is important to maintain and improve soil quality, nutrient and water availability; enhance N inputs through biological N fixation; prevent soil erosion, and to control weeds (Watson *et al.*, 2002). Crop rotation strongly affects both the diversity and composition of AMF spore communities in the soil, with higher

Table 2.4: Seasonal Variation in Arbuscular Mycorrhizal Spores in Wheat growing Soil under Conventional Tillage (CT), Zero-Tillage (ZT) and Reduced-Tillage (RT) System during 2012 at Mirzapur in Ambala District, Northern India (from Devi, 2015)

Systems/Soil Depth (cm)	AM Fungal Spore Density (10g⁻¹ soil)			
	January 2012	*February 2012*	*March 2012*	*April 2012*
CT				
0-15 cm	48.6±3.36	68.65±7.08	82.23±2.63	82.78±2.98
15-30 cm	20.65±3.88	20.3±2.37	24.05±2.70	25.18±1.56
LSD (p<0.05)	9.61	17.98	6.83	9.86
F-ratio	37.11	62.86	327.49	317.99
ZT				
0-15 cm	87.6±4.28	84.5±2.28	96.3±3.62	95.08±4.67
15-30 cm	18.4±3.75	24.7±0.91	21.75±1.70	17.8±2.53
LSD (p<0.05)	18.13	6.59	7.99	9.43
F-ratio	123.98	649.98	746.72	447.89
RT				
0-15 cm	72.83±2.05	80.88±2.34	92.5±2.60	94.83±2.30
15-30 cm	20.65±3.88	17.25±1.96	23.23±2.59	22.65±2.83
LSD (p<0.05)	9.00	6.00	5.85	3.74
F-ratio	135.04	497.43	1.088E3	1.439E3

AMF diversity usually found under crop rotations than under monocultures (Oehl *et al.*, 2003). In various studies carried out in the USA, *Gigaspora gigantea*, *Glomus albidum*, *G. mosseae* and *G. etunicatum* dominated AMF communities under maize; *G. caledonium* and *G. microcarpum* were abundant under soybeans; and *Glomus occultum* was most abundant under wheat and barley (Johnson *et al.*, 1991; Troeh and Loynachan, 2003). The relative species abundances of AMF spore communities were significantly affected by crop rotation; the abundance of *Acaulospora scrobiculata* and *Scutellospora verrucosa* spores was significantly higher in soil under maize–crotalaria rotation than under continuous maize (Mathimaran *et al.*, 2007).

4.3. Cover Crops Effects on AMF Diversity

Cover crops are widely recognized as an important management practice their contributions to soil conservation and quality, crop performance, and maintaining or increasing mycorrhizal potential of soils, and providing nourishment during winter periods to AMF (Kabir and Koide, 2002). Cover crops are grown in the fall, winter, and spring, covering the soil during periods when it would commonly be left without a cover of growing plants. Since AM fungi are biotrophic, they are dependent on plants for the growth of their hyphal networks. Growing a cover crop extends the time for AM growth into the autumn, winter, and spring. The mycorrhizal colonization increase found in cover crops systems may be largely attributed to an increase in the extraradical hyphal network that can colonize the roots of the new crop (Boswell *et al.*, 1998). Mycorrhizal cover crops can be used to improve the mycorrhizal inoculum potential and hyphal network (Kabir and Koide 2000; Boswell *et al.*, 1998; Sorensen *et al.*, 2005).

The effect of cover crop diversity on mycorrhizal colonization in subsequent organic maize cultivars differing in the level of genetic diversity has been studied by Njeru *et al.* (2015). The crop rotations included maize (*Zea mays*), common wheat (*Triticum aestivum*), sunflower (*Helianthus annuus* L.), pigeon bean (*Vicia faba*), and durum wheat (*Triticum durum* Desf.). These workers showed a higher mycorrhizal colonization in maize plants grown after hairy vetch, of 35.0 per cent, and Mix cover crops, of 29.4 per cent, compared to Indian mustard, of 20.9 per cent, and Control, of 21.3 per cent.

Bowles *et al.* (2015) carried out a meta-analysis of the effect of cover crops and reduced tillage on AM colonization of roots and AM communities in annual cropping systems. For cover crops, they compiled information on 93 comparisons of winter cover crops vs. fallow management from 17 field studies, comprising a variety of cover crop species and main crops. Cover crops increased mycorrhizal root colonization of subsequent main crops by 33 per cent; AM root colonization increased by 35 per cent vs. 20 per cent for AM and non-AM cover crop hosts, respectively, relative to a winter fallow.

5. AM Fungal Root Colonization

Distribution of AM fungi in agricultural soils can vary with the season of the year, plant growth, edaphic factors, fertilization, and cropland management practices (Kabir, 2005). Agricultural practices including tillage, crop rotation and crop management systems have a marked effect on the development, activity and diversity of AM fungi (Douds *et al.*, 1995; Kabir *et al.*, 1997a, b; Castillo *et al.*, 2006). The densities of total and metabolically active soil hyphae and AM fungal root colonization were significantly lower in conventional tillage than in zero-tillage (Kabir *et al.*, 1997a, b). AM fungi exist in two different phase: inside the root and in the soil. The intra-radical mycelium consists of hyphae and other fungal structures such as arbuscules and vesicles. This phase is connected to the soil mycelium. The extra-cellular mycelium forms spores; explore soil and new area for colonization is assessed by measuring per cent root colonization (Rillig, 2004).

McGonigle and Miller (1993) reported higher levels of arbuscular mycorrhizal colonization in ridge tillage and no-tillage systems as compared to moldboard tillage systems. Kabir *et al.* (1998) showed differences in vertical distributions of arbuscular mycorrhizal fungi under corn in no-till and conventional tillage systems. S p o r e populations and colonization of winter wheat roots by AM fungi were higher under low-input than conventional agriculture (Galvez *et al.*, 2001).

The effect of tillage practices on AM fungal root colonization in subtropical region in northern Tamaulipas, Mexico showed that the lowest degree of root colonization was found in bean plants and greatest in the sorghum species (Alguacil *et al.*, 2008). The AM fungal colonization percentage of bean and sorghum roots under subsoil-bedding and no-tillage were significantly higher than under moldboard plowing (Figure 2.3). The maize roots under no tillage showed significantly higher colonization rates than both under shred-bedding and moldboard plowing.

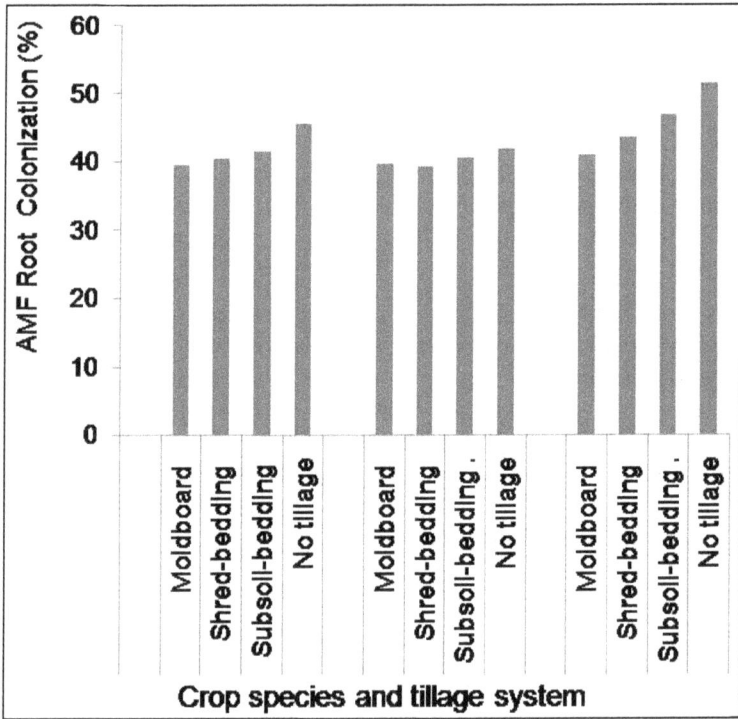

Figure 2.3: AMF Colonization of the Roots of Maize, Bean, and Sorghum under different Management Tillage Systems in Northern Tamaulipas, Mexico (from Alguacil *et al.*, 2008).

Neelam *et al.* (2010) reported that maximum roots colonization by AM fungi was recorded at flowering stage and at the time of crop maturity in the case both rice and wheat under conservation tillage in northern India. In the 7.5-15 cm depth range, AM fungal root colonization was greatest under ZT and FIRB than that of the corresponding depth in CT (Figure 2.4). AM fungal root colonization ranged between 17.60-87.78 per cent in CT, ZT and FIRB tillage systems, respectively. In rice root, the value of AM fungal root colonization was 13.22-68.16 per cent in CT and ZT systems. In different tillage systems, AM fungal colonization of root decreased significantly at 15-30 cm depth.

AM fungal root colonization in rice and wheat crop rotation under reduced tillage (RT) and conventional tillage (CT) at Mirzapur in Ambala district in northern India was studied by Devi *et al.* (2014). The AM fungal infection consisted of both fine and coarse hyphae with distinct vesicles and arbuscules. In the 0-15 cm soil depth, AM fungal root colonization was greatest under RT than that of the corresponding depth in CT. AM fungal root colonization ranged between 34.74 – 93.36 per cent in CT

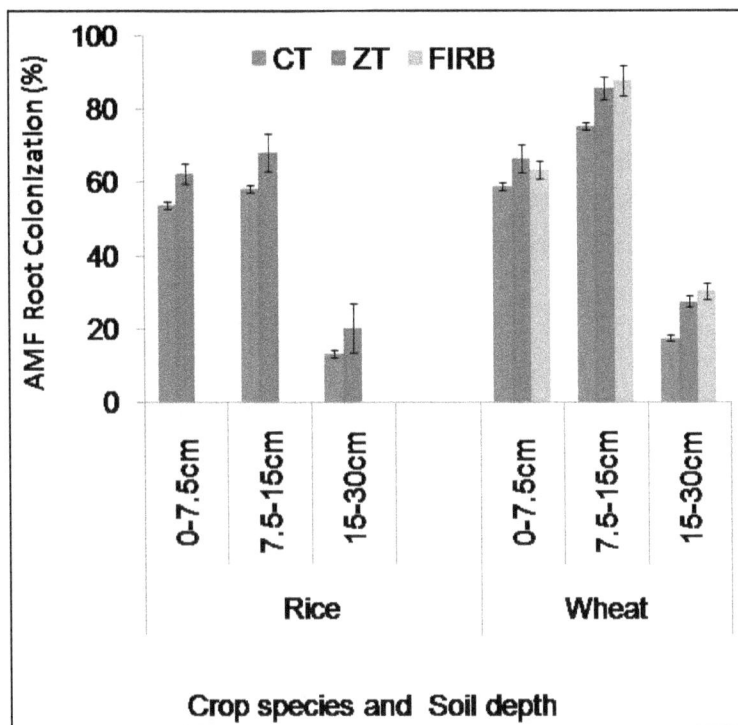

Figure 2.4: AMF Colonization of the Roots of Rice and Wheat at different Soil Depths under Conventional Tillage (CT), Zero-Tillage (ZT) and Furrow Irrigated Raised Bed (FIRB) Tillage in a Rice-wheat Cropping System at Uchana, Karnal, Northern India (Neelam *et al.,* 2010).

and RT systems, respectively. In rice roots, the values of AM fungal root colonization ranged from 34.74 to 66.85 per cent in CT and RT systems, respectively. In different tillage systems, AM fungal colonization of root decreased significantly at 15-30 cm depth (Table 2.5). Zero tillage and reduced –tillage in wheat had a favorable effect on mycorrhizal root colonization (Figures 2.5 and 2.6).

Table 2.5. Effect of tillage on soil organic carbon storage by soil depth in continuous corn under no-tillage and moldboard plowing after 14 yr in each treatment at Minnesota in USA (adapted from Huggins *et al.*, 2007).

Soil Depth	No-tillage (Mg C ha⁻¹cm⁻¹)	Moldboard Plowing (Mg C ha⁻¹cm⁻¹)
0–7.5 cm	4.49	3.89
7.5–15 cm	4.94	4.66
5–30 cm	3.51	3.17
30–45 cm	2.45	1.48
0-45 cm (Mg C ha⁻¹)	160	133

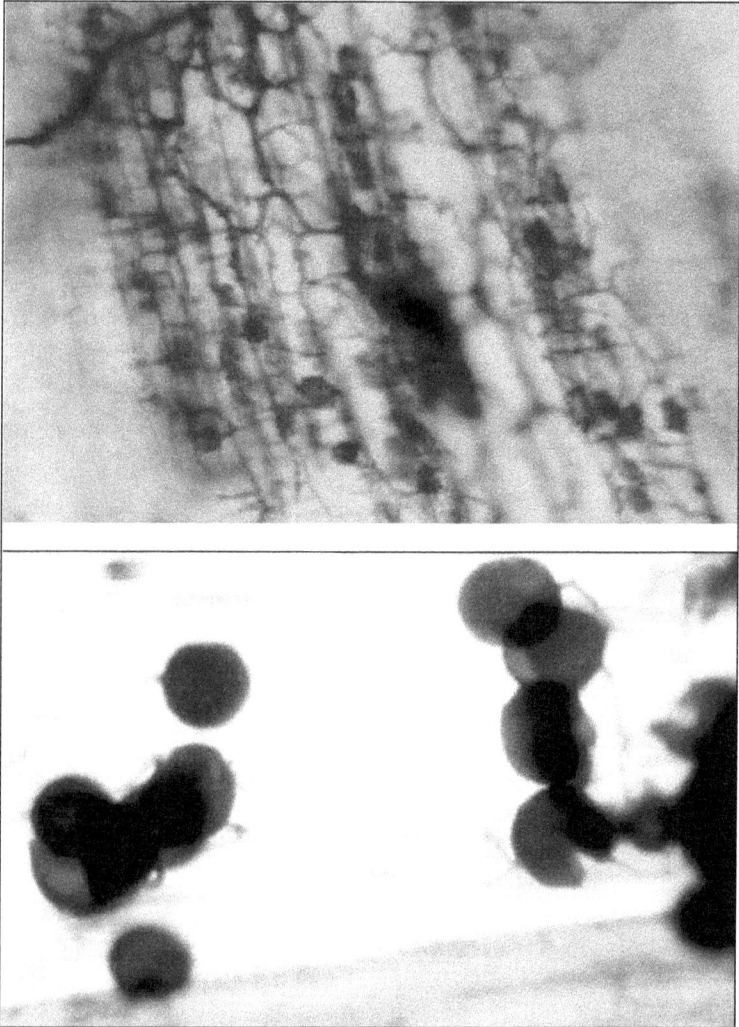

Figure 2.5: Arbuscular Mycorrhizal Infection in Roots Showing the Formation of (A) Arbuscules in cortical cells with H and Y type hyphe, and (B) Formation of globose vesicles with attached hyphae in wheat roots under zero tillage (Devi, 2015).

6. AM Fungi and Nutrient Cycling

AMF are important microbial symbioses for plants and under conditions of P-limitation and are significant in the maintenance of soil health and fertility, plant community development, nutrient uptake and above-ground productivity (Smith and Read, 1997). van der Heijden *et al.* (1998) found that plant biodiversity, nutrient capture, and productivity in macrocosms increase significantly with increasing AM hyphal length and AMF species richness.

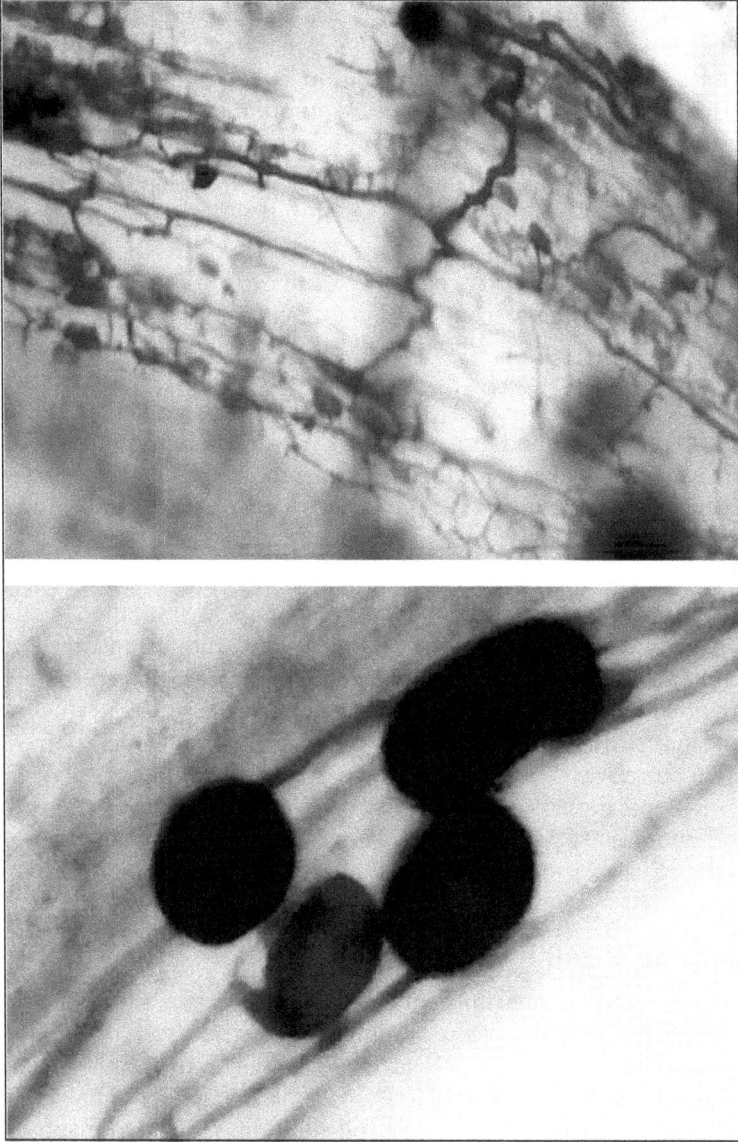

Figure 2.6: Arbuscular Mycorrhizal Infection in Rice Roots Showing the Formation of (A) Arbuscules in cortical cells with H and Y type hyphe, and (B) formation of globose vesicles with attached hyphae in rice roots under reduced tillage (Devi, 2015).

6.1. Effects on C Cycling

AM fungi control critical processes within the C cycle, and can mediate soil C storage (Rillig, 2004a, b; Verbruggen *et al.*, 2012; Mohan *et al.*, 2014; Verbruggen *et al.*, 2016). In fact, AM fungi may receive about 37–47 per cent of belowground NPP in ecosystems dominated by AM host plants (Johnson *et al.*, 2002; Treseder and Cross,

2006). AM fungi can enhance the removal of CO_2 from the atmosphere by plants, and then deposit a portion of that additional C in the soil. Fine roots and AM fungi contribute the largest input of carbon (C) into soils and are therefore of primary relevance to the soil C balance. AM fungi can stimulate litter decomposition as well as stabilize C during litter decomposition, which persist through time.

Glasshouse experiments and field studies suggest that plants allocate between 10 and 20 per cent of their photosynthates to AM fungi (Johnson *et al.*, 2002; Nottingham *et al.*, 2010). In intensively managed agricultural ecosystems the abundance of AM fungi is often reduced as a result of heavy fertilization, soil disturbance. The majority of crops (*e.g.* maize, cereals, soybean, potato, rice) are colonized by AM fungi in the field and thus allocate C to the fungal compartment below ground The recent observations that mycorrhizal fungi are important regulators of C dynamics because of impaired degradation of fungal residues (Clemmensen *et al.*, 2013). AM fungi associate with plant roots and form a network of extra-radicle hyphae (ERH). Thus, AMF-ERH act as an extension of the host plant root system and increases surface area for plant nutrient acquisition beyond the rhizosphere (Johnson and Gehring, 2007; Camenzind *et al.*, 2013).

Soil organic carbon associated with aggregates is an important reservoir of carbon, protected from mineralization and enzymatic degradation. Conservation tillage systems increase the storage of soil organic matter and the stability of macroaggregates as compared to conventional tillage in various types of soils and climatic regions. The soil microaggregates can play an important role to stabilize soil organic matter in tropical agriculture systems. Six *et al.* (2002) showed that soil organic matter turnover was slower in conservation agriculture systems as compared to the conventional agriculture. In tropical soils, an increase in soil aggregation has been found under no-tillage practices (Six *et al.*, 2002). Long-term minimum tillage enhanced the physical protection of organic carbon and nitrogen in Haplic Luvisols (Jacobs *et al.*, 2009).

6.2. Effects on N and P Cycles

Phosphorus is an essential nutrient for all organisms, including plants and fungi, which is not readily available as the dominant form absorbed is orthophosphate. It is known that orthophosphate reacts strongly with calcium in soils of high pH and with iron and aluminium at low pH. Hence available P in soil is normally much lower than total P and Pi concentrations in the soil solution. When AM plants are grown on soils with poorly available P their responsiveness to colonization is higher than when grown on soils with equivalent amounts of readily available P. This effect may be due to better access to adsorbed P by AM fungal hyphae compared with roots (Bolan *et al.*, 1984) (Figure 2.7). In Figure 2.7, the various pathways are: 1) Extensive physical exploration of soil by fungal hyphae; 2) higher substrate affinity of P uptake into hyphae than directly into roots; 3) combined extensive physical exploration and chemical modification of adsorbed P that speeds up soluble P release; 4) possible fungal hydrolysis of organic P. The possibility that AM fungal hyphae have strong capability to hydrolyze organic P is still unresolved (Joner *et al.*, 2000).

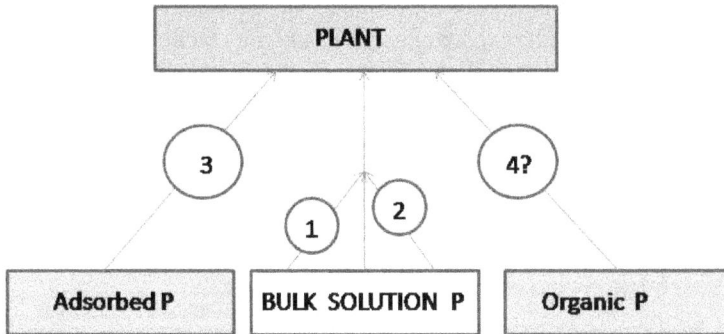

Figure 2.7: Possible Mechanisms for Increased Uptake of P by Arbuscular Mycorrhizal Plants, Modified from Bolan *et al.* (1984).

There is increasing evidence that AM hyphae can acquire substantial amounts of nitrogen from decomposing litter and that they can transport the nitrogen to the host plant (Herman *et al.*, 2012; Hodge and Fitter, 2010; Koller *et al.*, 2013). AMF probably stimulate the decomposition of organic material, and thus liberate mineral nutrients, via the supply of carbon containing exudates to the surrounding decomposer microbial community (Herman *et al.*, 2012). The composition of hyphal exudates ranges from low-molecular-weight sugars and organic acids to high-molecular-weight polymeric compounds, substances that can both enhance and reduce bacterial growth in the soil (Toljander *et al.*, 2007).

Phosphorus (P) is an essential macronutrient for plant growth and its uptake from soil is effective in the form of soluble phosphate anions (Schachtman *et al.*, 1998). The establishment of AMF colonization in roots under no-tillage is faster resulting in early uptake of P and Zn by plants as compared to conventional tillage (Miller, 2000),which could translate to yield-increases in crops such as maize (Grant *et al.*, 2001). Many studies have reported the significance of AMF for growth of crop plants by regulating nutrient supply and influencing the changes in plant physiology and morphology (Pellegrino and Bedini, 2014). In particular, it has been shown that the AM symbiosis promotes the inflow of slowly mobile nutrients to plant roots, predominantly P (Antunes *et al.*, 2007). Koide and Kabir (2000) showed that extraradical hyphae of the AM fungus, *Glomus intraradices*, can hydrolize organic P (*i.e.* phytate) and that the resultant inorganic P can be taken up and transferred to plant root. It has also been reported that mycorrhized plants respond positively to the soil amendment with insoluble forms of inorganic phosphorus such as rock phosphates (RPs) (Cabello *et al.*, 2005; Duponnois *et al.*, 2005). Antunes *et al.* (2007) showed that the mechanisms underlying increased P uptake by the AM symbiosis establishment did not result from the fungal release of H+ ions alone or in combination with organic acid anions. It is well known that extraradical hyphae of AMF provide an important area for interactions with soil microbes and a large pathway for the translocation of energy-rich plant assimilates to the soil (Johansson *et al.*, 2004).

Several studies have shown that some phosphate-solubilizing bacteria can interact synergistically with mycorrhizal fungi that may facilitate phosphorus uptake by the plants (Caravaca *et al.*, 2004; Cabello *et al.*, 2005). Under field conditions, it has been reported that the abundance of phosphate-solubilizing bacteria belonging to the fluorescent pseudomonad group was correlated to the level of plant mycorrhizal colonization (Duponnois *et al.*, 2011). AMF communities from different tillage systems can change plant productivity, and AMF communities of non-tilled soils enhanced plant P uptake (Kohl *et al.*, 2014).

7. Soil Carbon Sequestration

Soil is a vital component in the functioning of terrestrial ecosystems provides a habitat for diverse and interacting populations of soil organisms, accounts for decomposition processes, and a critical link in carbon sequestration to mitigate climate change. Soils deliver provisioning, regulating, cultural and supporting ecosystem services, which are regulated by the physical, chemical and biological properties of the soil (Palm *et al.*, 2007). Soil organic matter plays a key role in regulating climate, soil water and soil biodiversity. In the context of mitigating global climate change, it is important to create a well quantified carbon sink in agricultural soils worldwide. Conservation tillage has been found to reduce the carbon footprint of agriculture and could also reduce environmental impact (Ilan and Lal, 2013).

Soil carbon sequestration involves transferring atmospheric carbon into the soil via plant photosynthesis and keeping those soil-based carbon pools protected as effectively as possible from microbial activity that will release the carbon back to the air. There are agricultural management practices that show promise for restoring soils and sequestering a very significant portion of atmospheric carbon. Soil microorganisms play a crucial role in the carbon sequestration process by transforming plant residues into smaller carbon molecules that are more likely to be protected and sequestered (Six *et al.*, 2006).

The effect of tillage on soil organic carbon storage by soil depth in continuous corn under no-tillage and moldboard plowing after 14 yr in each treatment at Minnesota in USA was reported by Huggins *et al.* (2007). The soil carbon storage was greatest at 7.5–15 cm soil depth (Table 2.5). The soil organic carbon was 20 per cent greater in continuous corn NT as compared to moldboard plowing.

On the basis of paired treatment comparisons of conventional and conservation agriculture, the global rate of SOC sequestration with conversion to no-tillage farming have been reported to 0.57 t C ha^{-1}yr^{-1} (West and Post, 2002). Govaerts *et al.* (2009) reviewed 78 studies of the potential impact of conservation agriculture on carbon sequestration; the soil carbon stock being higher in 40 cases of no-tillage as compared to conventional agriculture. According to Lal (2011), the technical potential of carbon sequestration in soils of the agroecosystems of the world is 1.2–3.1 billion tons C yr^{-1}. Soil organic carbon (SOC) storage in conventional and conservation agriculture as compiled from various sources are given in Table 2.6.

In rice-wheat cropping system in northern India, soil organic carbon storage in 0-30 cm soil layer was greater in zero-tillage (14.42 Mg C ha^{-1}) as compared to conventional tillage (10.52 Mg C ha^{-1}) (Neelam and Gupta 2009). The total soil organic carbon (SOC) pool up to 100 cm soil depth ranged from 19.665 Mg C ha^{-1} to 25.452 Mg C ha^{-1} in conventional tillage and from 22.574 Mg C ha^{-1} to 30.294 Mg C ha^{-1} in zero-tillage/reduced tillage systems in rice – wheat crop rotation in northern India (Devi, 2015), Table 2.6. The soil organic carbon pool at 0-15 cm soil depth accounted for 54.73 per cent to 64.52 per cent of the total organic carbon pool up to 100 cm soil depth. In this study, it was found that water stable aggregates were influenced by the tillage systems. In conservation agriculture of maize-wheat rotation in Central highlands of Mexico, the soil organic carbon content in 0-60 cm soil layer was higher in conservation agriculture, 117.7 Mg C ha^{-1} as compared to conventional tillage, 69.7 Mg C ha^{-1} (Dendooven *et al.*, 2011).

No-tillage systems are the most widely studied for their effects on soil carbon. Reduced-tillage or NT as a component of conservation agriculture may increase soil carbon compared to that of the conventional tillage, which are generally less than 10 cm soil depth, at deeper depths, soil C in CA maybe equal or even lower compared with CT (Palm *et al.*, 2014). A meta-analysis found increased soil C in the topsoil (0-10 cm) due to conversion of CT to NT, but no significant difference over the soil profile to 40 cm because of redistribution of C in the profile (Luo *et al.*, 2010). A majority of studies on no-till and conservation tillage show differences in carbon concentrations at the soil surface, while ignoring lower depths retention of crop residues is an essential component of CA for increasing or maintaining soil C (Powlson *et al.*, 2014 and Baker *et al.*, 2002).

In tropical and subtropical conditions, soil aggregation has been found to increase during the early years of no-tillage adoption (Six *et al.*, 2002). In minimum tillage systems, new aggregates were formed due to incorporation of crop residues in the soil and storage of excess organic matter in biochemically degraded fraction, especially in the surface soil (Jacobs *et al.*, 2010). Conservation or no tillage reduces soil mixing and soil disturbance, which improves SOM accumulation.

In a recent meta-analysis, researchers found that more diverse crop rotations consistently have higher soil carbon and soil microbial biomass than less diverse systems, especially when cover crops were included in the rotation (McDaniel *et al.*, 2014). The rotational diversity has important impacts on soil carbon accrual by improving the ability of soil microbial communities to rapidly process plant residues and protect them in aggregates Tiemann *et al.* (2015) The inclusion of several different crops in a rotation also introduces a greater diversity of carbon compounds into the soil, some of which may be more resistant to decomposition. A recent laboratory experiment found that the initial chemistry of the plant residues and the microbial community had a strong influence on which carbon compounds are present in the soil (Wickings *et al.*, 2012).

Jat *et al.* (2009) studied the effect of conservation agriculture (CA)-based resource-conserving technologies in the rice–wheat rotation on soil physical properties in Uttar Pradesh, India. The Permanent beds (PBDSR-PBW) and double no-till (ZTDSR-ZTW) had significantly higher soil aggregates (>0.25 mm) than conventional- tillage (PTR-CTW). Further, under conventional-tillage, soil aggregation was static across the seasons, whereas it improved over time under double no-till and permanent beds (Table 2.7).

Table 2.6: Soil Organic Carbon (SOC) Storage in Conventional and Conservation Agriculture [Adapted from Lal (2015); Devi (2015)].

Location	Study Duration (yr)	Crop	Soil Depth	SOC (Mg C ha⁻¹)	Reference
Huan Province, China	4	Rice	80	NT (129.4), PT (126.3), RT (122.5)	Xu *et al.* (2013)
Georgia, United States	41	Corn, sorghum	200	NT (60), CR (52),	Devine *et al.* (2011)
Central Brazil (Cerrado)	1 to 13	Soybean	40	+400 to 1,700	Blanchart *et al.* (2007)
Southeastern Australia	20	Grain crops	–	+312 to 544	West and Post (2002)
Uchana, KarnalIndia	15	Rice-wheat rotation	100	20.661(CT) 27.328 (ZT)	Devi (2105)
Tharwa, AmbalaIndia	12	Rice-wheat rotation	100	25.452(CT) 30.294(ZT)	Devi (2105)
Global (24 studies)	>5	Grain crops	>30	NT (100.3), PT (95.4)	Angers and Eriksen-Hamel (2008)
Global (52 Studies)	–	Grain crops	–	+47 to 620	Puget and Lal (2005)

NT: Not-till; PT: Plow tillage; RT: Rotary tillage; Conventional tillage (CT); ZT (Zero-tillage).

Table 2.7: Effect of Tillage and Crop Establishment Techniques on Soil Aggregates (>0.25 mm) in the Rice-Wheat Cropping System (from Jat *et al.*, 2009)

Tillage and Crop Etablishment Methods	Soil Aggregates (>0.25 mm) Rice	Soil Aggregates (>0.25 mm) Wheat
PTR-CTW	51.0	52.0
ZTDSR-ZTW	66.0	67.5
CTDSR-CTW	59.5	59.5
BDSR-PBW	64.0	68.0

PTR-CTW: Conventional puddled-transplanted rice and conventional-tillage wheat.

ZTDSR-ZTW: Zero-till direct drill-seeded rice and wheat after no-tillage.

CTDSR-CTW: Conventional-till direct drill-seeded rice and wheat after conventional- tillage.

BDSR-PBW: Direct drill-seeded rice and wheat on permanent raised beds.

In a tropical rice-wheat cropping system in northern India, microaggregates (<250µm to 53µm) formed 48.47 to 61.64 per cent of total soil aggregates fractions and protected most of organic carbon in soils of the zero-tillage (ZT) and furrow irrigated raised bed (FIRB) tillage systems (Neelam *et al.*, 2010), Figure 2.8. The microbial biomass by acting as a labile pool of carbon plays a crucial role in soil carbon dynamics; the soil microbial biomass was higher in zero-tillage systems in rice-wheat crop rotation. The water stable soil aggregates >0.25 mm were 28 per cent higher in zero-tillage direct seeding than in conventional tillage under 7 year rice-wheat rotation (Gathala *et al.*, 2011).

Devi *et al.* (2015) reported that the carbon concentration was higher in macroaggregates at 0 to 7.5 cm and 7.5 to 15 cm soil depth in the case of both rice (0.585-0.354 per cent) and wheat (0.67-0.354 per cent) cropping as compared to that of microaggregates (0.458-0.246 per cent) under no-tillage (Figure 2.8). In clay and silt-associated microaggregates, the carbon concentration ranged from 0.315 to 0.255 per cent in soils of the rice crop, and from 0.416-0.274 per cent in case of the wheat crop.

During the process of decomposition, different types of carbon compounds of differing size and chemical complexity that are incorporated into soil aggregates (Six *et al.*, 2006; Grandy and Neff, 2008; Grandy and Wickings, 2010). Soil fungi play an additionally important role in soil carbon sequestration by maximizing the amount of carbon allocated to the soil and producing compounds that improve aggregate stability. As AMF fungi derive carbon from plants, their biomass increases effectively increasing the amount of carbon in soil (Rillig *et al.*, 2001). In a long-term manipulation of field experiments to produce a gradient of AMF abundance, Wilson *et al.* (2009) found that AMF was strongly positively correlated to soil aggregation and carbon levels.

One of the most important ways carbon is sequestered in soils is through the process of soil aggregation. AM fungi also produce glomalin, a recalcitrant C-N rich glycoprotein (Rilling *et al.*, 2005). Glomalin is a major component of organic matter in soil and contributes to better soil aggregate formation, which is important for soil

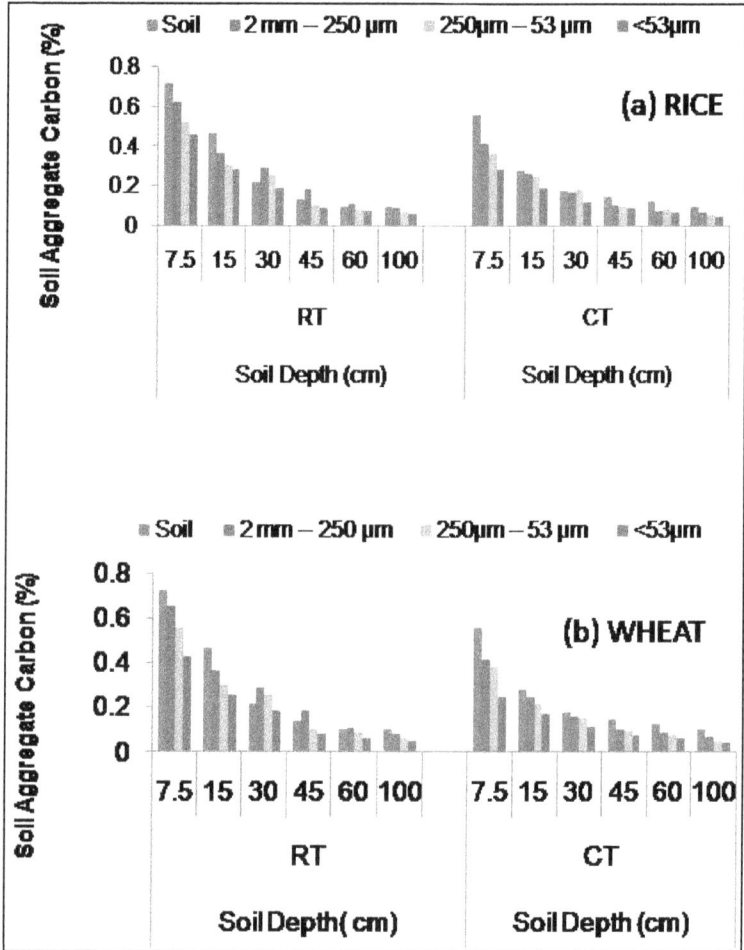

Figure 2.8: Per cent Organic Carbon in Soil and Aggregate Fractions in Conventional Tillage (CT), Zero-Tillage (ZT) in (A) Rice and (B) Wheat (from Devi, 2015).

structure and stability against erosion (Wright *et al.*, 1996; Wright and Upadhyaya, 1998). Glomalin has been shown to make up 3-8 per cent of total organic carbon in some ecosystems, including agriculturally managed land (Treseder and Turner, 2007; Wilson *et al.*, 2009). Together, AMF-ERH and glomalin contribute to water stable aggregate formation, and increase overall soil fertility (Rillig, 2004b).

Soil aggregates are formed by combining of smaller soil particles to form larger, more stable groups, bound together by clay particles present in the soil and by glomalin produced by AM fungi (Oades, 1984; Six *et al.*, 2004; Wilson *et al.*, 2009). As these aggregates form, small particles of carbon are captured in the center of the aggregates, which are physically protected from microbial attack. When aggregates remain stable and undisturbed, they can protect soil carbon for an extended period of time. However, tillage in conventional agriculture can lead to breakdown of aggregates, exposing soil carbon to microbial attack (Grandy and Robertson, 2006; 2007).

8. Reducing Greenhouse Gas Emission

Direct-seeded rice is a feasible alternative to conventional puddled transplanted rice having good potential to mitigate and adapt to climate change (Pathak *et al.*, 2010). The average global warming potential (GWP) due to three GHGs (CO_2, CH_4 and N_2O) in transplanted rice was 2.91 Mg ha^{-1} whereas in Direct-seeded rice the GWP was 1.9 Mg ha^{-1} (Pathak *et al.*, 2011).

In conservation agriculture of maize in Central Highlands of Mexico, the cumulative greenhouse gas emissions (CO_2, CH_4, N_2O) have been found to be similar in both conventional and conservation agriculture (Dendooven *et al.*, 2012b). The global warming potential of agricultural systems was calculated on the basis of soil C sequestration, greenhouse gas emissions, fuel use, fertilizer use, and seed production (Dendooven *et al.*, 2012); the GWP was found lower in conservation agriculture (-7729 to -7892 kg CO_2 ha^{-1}yr^{-1}) than in conventional tillage (1327 to 1156 kg CO_2 ha^{-1}yr^{-1}).

The soils in rice systems are a major source of nitrous oxide, a greenhouse gas 298 times more effective than CO_2. Soil contributes about 65 per cent of the total nitrous oxide emission. The major sources are soil cultivation, fertilizer and manure application, and burning organic material and fossil fuels (Pathak *et al.*, 2011). Appropriate crop-management practices favoring increased N-use efficiency seem promising to reduce nitrous oxide emission in conservation agriculture (Pathak *et al.*, 2011). Direct-seeded rice has been shown to be a viable technology to save water, to mitigate green-house gas emission and adapt to climatic change (Singh and Ladha, 2011; Pathak *et al.*, 2011).

AM fungi could have an indirect influence on GHG emissions, and also change the physical conditions of soil, *i.e.*, moisture, aggregation, and aeration, all of which influence the production and transport of GHG in soil. Lazcano *et al.* (2014) have reported that AM symbiosis helps to regulate N_2O emissions at high soil moisture levels and suggested that the control of N_2O emissions by AM plants could be driven by a higher use of soil water rather than by increased N uptake.

9. Managing Groundwater Use in CA

There are significant zero tillage-induced resource-saving effects on farmers' fields in terms of diesel, tractor time, and cost savings for wheat cultivation (Erenstein *et al.*, 2008; Erenstein and Laxmi, 2008). Zero tillage has been shown to save irrigation water through the combined effect of pre-irrigation saving of water and modest water use during the crop growing season (Erenstein *et al.*, 2008). Recently, the laser leveling of soil has been taken up by farmers, which has helped to save water due to leveling off the field. The farmers participatory trials have shown that laser assisted precision land leveling saved a minimum of 15 cm water in rice-wheat system and improved yield up to 25 per cent (Pathak *et al.*, 2011). Laser leveling of fields with bed planting and zero tillage also helps to improve water use efficiency through uniformity in water application, better crop stand, improved nutrient- water interaction (Pathak *et al.*, 2011).

10. Conclusion

There is large potential for managing soil carbon in agriculture ecosystems for climate change mitigation and enhancing the sinks of greenhouse gases (Lal, 2011). However, for effective soil organic carbon sequestration, there is need to develop conservation tillage systems over a long-term, to reduce emission of nitrous oxide from soil, to measure soil carbon in deep soil layer, and to improve efficiency of water and fertilizer use in cropping systems. The additional requirements of nutrients and water for soil carbon sequestration across sites, cropping systems and climatic conditions must be taken into consideration. The inclusion of a diversity of crops might ensure that a diversity of carbon compounds is present in the soil, improving soil carbon sequestration potential. The permanence of soil carbon sequestration in view of climate change and anthropogenic activities also need attention. The AM fungal symbioses play a key role in shaping plant communities and agricultural production as these fungi connect the plants to the heterogeneously distributed nutrients in soil that are required for their growth. The AM fungi are of high value for the ecosystem functioning and sustainability. Studies on AMF species diversity and their functions in conservation tillage are crucial to understand the impact of land use changes on ecosystem services. There are enormous possibilities to explore the role of mycorrhizal fungal diversity in relation to functional processes and delivery of ecosystem services. Comparative analysis of conventional and conservation tillage would improve our understanding of responses to changing environmental and climatic conditions. Thus, new knowledge is an important prerequisite for sustainable agricultural systems. It is crucial to gain a better understanding of functional variation among AM fungal species to formulate appropriate nutrient conserving strategies and soil carbon sequestration.

References

Alguacil, M.M., Diaz-Pereira, E., Caravaca, F., Fernandez, D.A. and Roldan, A. 2009. Increased diversity of arbuscular mucorrhizal fungi in a long-term field experiment via application for organic amendments to a semiarid degraded soil. *Appl. Environ. Microbiol.*, 75: 4254–4263.

Alguacil, M.M., Lumini, E., Roldan, A., Salinas-Garcia, J.R., Bonfante, P. and Bianciotto, V. 2008. The impact of tillage practices on arbuscular mycorrhizal fungal diversity in subtropical crops. *Ecol. Appl.*, 18: 527–536.

Allen, E.B., Allen, M.F., Helm, D.J., Trappe, J.M., Molina, R. and Rincon, E. 1995. Patterns and regulation of mycorrhizal plant and fungal diversity. *Plant Soil*, 170: 47–62.

Anderson, R.C., Liberta, A.E. and Dickman, L.A. 1984. Interaction of vascular plants and vesicular-arbuscular mycorrhizal fungi across a soil moisture nutrient gradient. *Oecologia*, 64: 111-117.

Angers, D.A., and Eriksen-Hamel, N.S. 2008. Full-inversion tillage and organic carbon distribution in soil profiles: A meta-analysis. *Soil Sci. Soc. Am. J.*, 72: 1370-1374.

Antunes, P.M., Lehmann, A., Hart, M.M., Baumecker, M. and Rillig, M.C., 2012. Long-term effects of soil nutrient deficiency on arbuscular mycorrhizal communities. *Funct. Ecol.*, 26: 532-540.

Antunes, P.M., Schneider, K., Hillis, D. and Kiloronomos, J.N., 2007. Can the arbuscular mycorrhizal fungus *Glomus intraradices* actively mobilize P from rock phosphates? *Pedobiologia*, 51: 281-286.

Baker, C.J, Saxton, K.E. and Ritchie, W.R. 2002. No-tillage Seeding: Science and Practice. Second edition. CAB International, Oxford, U.K. p. 352.

Baveye, P.C., Rangel, D., Jacobson, A.R., Laba, M., Darnault, C., Otten, W. 2011. From Dust Bowl to Dust Bowl: soils are still very much a frontier of science. *Soil Sci. Soc. Am. J.*, 75: 2037–2048.

Beauregard, M.S., Hamel, C., Atul, N. and St-Arnaud, M. 2010. Long-term phosphorus fertilization impacts soil fungal and bacterial diversity but not AM fungal community in alfalfa. *Microb. Ecol.*, 59:379-389.

Bender, S.F., Plantenga, F., Neftel, A., Jocher, M., Oberholzer H.R. and Köhl, L. 2014. Symbiotic relationships between soil fungi and plants reduce N_2O emissions from soil. *ISME J*, 8: 1336–1345.

Blanchart, E., M. Bernoux, X. Sarda, M. Siqueira Neto, C.C. Cerri, M. Piccolo, J. Douzet, E. Scopel, and C. Feller. 2007. Effect of direct seeding mulch-based systems on soil carbon storage and macrofauna in Central Brazil. *Agric. Conspec. Sci.*, 72: 81-87.

Bolan, N.S., Robson, A., Barrow, N.J. and Aylemore, L.A.G. 1984. Specific activity of phosphorus in mycorrhizal and nonmycorrhizal plants in relation to the availability of phosphorus to plants. *Soil Biol. Biochem.*, 16: 299–304.

Bolliger, A., Magid, J., Amado, T.J.C. Skora Neto, F., Ribeiro, M.F.S. and Calegari, A. 2006. Taking stock of the Brazilian "zero-till revolution": a review of landmark research and farmer's practice. *Adv. Agron.*, 91: 49-110.

Bonfante, P. and Genre, A. 2008. Plants and arbuscular mycorrhizal fungi: an evolutionary-developmental perspective. *Tren. Plant Sci.*, 13: 492-498.

Boswell, E.P., Koide, R.T., Shumway, D.L. and Addy H.D. 1998. Winer wheat cover cropping, VA mycorrhizal fungi and maize growth and yield. *Agric. Ecosyst. Environ.*, 67: 55–65.

Bowles, T., Loeher, M., Jackson, L. and Cavagnaro, T. 2015. Effects of cover crops and reduced tillage on arbuscular mycorrhizal abundance and diversity: A meta-analysis SSSA, Nov, 17 2015: Strategies for Managing Microbial Communities and Soil Health.

Brito, I., Goss, M.J., de Carvalho, M., Chatagnier, O. and van Tuinen, D. 2012. Impact of tillage system on arbuscular mycorrhiza fungal communities in the soil under Mediterranean conditions. *Soil Till. Res.*, 121: 63-67.

Brundrett, M. 2009. Mycorrhizal associations and other means of nutrition of vascular plants: understanding the global diversity of host plants by resolving conflicting information and developing reliable means of diagnosis. *Plant Soil.,* 320: 37–77.

Busari, M.A., Kukal, S.S., Kaur, A., Bhatt, R. and Dulazi, A.A. 2015. Conservation tillage impacts on soil, crop and the environment. *International Soil and Water Conservation Research,* 3: 119–129.

Cabello, M., Irrazabal, G., Bucsinszky, A.M., Saparrat, M. and Schalamuk, S. 2005. Effect of an arbuscular mycorrhizal fungus, *Glomus mosseae*, and a rock-phosphate-solubilizing fungus, *Penicillium thomii*, on *Mentha piperita* growth in a soilless medium. *J. Basic Microbiol.,* 45: 182- 189.

Camenzind, T. and Matthias, C. and Rillig, M.C. 2013. Extraradical arbuscular mycorrhizal fungal hyphae in an organic tropical montane forest soil. *Soil Biol. Biochem.,* 64: 96-102.

Cannell, R.Q. 1985. Reduced tillage in north-west Europe – a review. *Soil Till. Res.,* 5: 129-177.

Caravaca, F., Figueroa, D., Barea, J. M., Azcon-Aguilar, C. and Roldan, A. 2004. Effect of mycorrhizal inoculation on nutrient acquisition, gas exchange, and nitrate reductase activity of two Mediterranean-autochthonous shrub species under drought stress. *J. Plant Nutr.,* 27: 57–74.

Castillo, C.G., Rubio, R., Rouanet, J.L. and Borie, F. 2006. Early effects of tillage and crop rotation on arbuscular mycorrhizal fungal propagules in an ultisol. *Biol. Fert. Soils,* 43: 83-92.

Clemmensen, K.E., Bahr, A., Ovaskainen, O., Dahlberg, A., Ekblad, A., Wallander, H., Stenlid, J., Finlay, R.D., Wardle, D.A. and Lindahl, B.D. 2013. Roots and associated fungi drive long-term carbon sequestration in boreal forest. *Science,* 339: 1615–1618.

Corsi, S., Friedrich, T., Kassam, A., Pisante, M. and de Moraes Sà, J. 2012. Soil Organic Carbon Accumulation and Greenhouse Gas Emission Reductions from Conservation Agriculture: A literature review. *Integrated Crop Management,* 16: 89.

CTIC. 1995. Conservation Technology Information Center, 12: 1-6.

Davison, J., Moora, M., Öpik, M. and Adholeya, A. 2015. Global assessment of arbuscular mycorrhizal fungus diversity reveals very low endemism. *Science,* 349: 970-973.

Dendooven, L., Gutiérrez-Oliva, V.F., Patiño-Zúñiga, L., Ramírez-Villanueva, D.A., Verhulst, N., Luna-Guido, M., Marsch, R., Montes-Molina, J., Gutiérrez-Miceli, F.A., Vásquez-Murrieta, S. and Govaerts, B. 2012. Greenhouse gas emissions under conservation agriculture compared to traditional cultivation of maize in the central highlands of Mexico. *Sci. Total Environ.,* 431:237–244.

Dendooven, L., Patino-Zuniga, L., Verhulst N., Luna-Guido M., Marsch, R. and Govaerts, B. 2011. Global warming potential of agricultural systems with contrasting tillage and residue management in the central highlands of Mexico. *J. Farming Systems Res. Extension*, 4: 35-66.

Derpsch, R. and Friedrich, T. 2009. Global Overview of Conservation Agriculture Adoption. In: Proceedings of 4th World Congress on Conservation Agriculture, New Delhi, India, pp. 429-438.

Derpsch, R., Friedrich, T., Kassam, A. and Li, H.W. 2010. Current status of adoption of no-till farming in the world and some of its main benefits. *Int. J. Agric. Biol. Eng.*, 3: 1-26.

Derpsch, R., Friedrich, T., Landers, J.N., Rainbow, R., Reicosky, D.C., Sa, J.C.M., Sturny, W.G., Wall, P., Ward, R.C. and Weiss K. 2011. No-tillage – a discussion paper. In: Proc. 5th World Congress of Conservation Agriculture, Brisbane, Australia.

Derpsch, R., Franzluebbers, A.J., Duiker, S.W., Reicosky, D.C., Koeller, K., Sturny, W.G., Sá, J.C.M. and Weiss, K. 2014. Why do we need to standardize no-tillage research? *Soil Till. Res.*, 137: 16–22.

Devi, P., Aggarwal, A. and Gupta, S.R. 2014. Carbon Accumulation, Nitrogen Uptake and Mycorrhizal Root Colonization in a Tropical Rice-wheat System in Northern India. *Ind. J. Sci.*, 11: 21-31.

Devi, P., Aggarwal, A. and Gupta, S.R. 2015. Effect of Zero Tillage on Soil Carbon Storage and Nitrogen Uptake of Rice-Wheat Systems of Northern India. *American-Eurasian J. Agri. and Environ. Sci.*, 15: 923- 931.

Devi, P. 2015. Soil Microbial Diversity, Residue Decomposition and Carbon Storage in Conservation Agriculture Systems. Ph.D. Thesis, Department of Botany, Kurukshetra University, Kurukshetra, India.

Devine, S., Markewitz, D., Hendrix, P. and Coleman, D. 2011. Soil carbon change through 2 m during forest succession alongside a 30-year agroecosystem experiment. *For. Sci.*, 57: 36-50.

Dodd, J.C., Boddington, C.L., Rodriguez, A., Gonzalez-Chavez, C. and Mansur, I. 2000. Mycelium of arbuscular mycorrhizal fungi (AMF) from different genera: form, function and detection. *Plant Soil.*, 226: 131–151.

Douds Jr. D.D., Nagahashi, G., Pfeffer, P.E., Kayser, W.M., Reider, C. 2005. On-farm production and utilization of arbuscular mycorrhizal fungus inoculum. *Can. J. Plant Sci.*, 85: 15–21.

Douds Jr. D.D., Janke, R.R. and Peters, S.E. 1993. VAM fungus spore populations and colonization of roots of maize and soybean under conventional and low-input sustainable agriculture. *Agric. Ecosyst. Environ.*, 43: 325–335.

Douds, D.D., Galvez, L., Janke, R.R. and Wagoner, P. 1995. Effect of tillage and farming system upon populations and distribution of vesicular-arbuscular mycorrhizal fungi. *Agric. Ecosyst. Environ.*, 52: 111-118.

Duponnois, R., Ouahmane, L., Kane, A., Thioulouse, J., Hafidi, M., Boumezzough, A., Prin, Y., Baudoin, E., Galiana, A. and Dreyfus, B. 2011. Nurse shrubs increased the early growth of Cupressus seedlings by enhancing belowground mutualism and soil microbial activity. *Soil Biol. Biochem.*, 43: 2160-2168.

Duponnois, R., Colombet, A., Hien, V. and Thioulouse, J. 2005b. The mycorrhizal fungus Glomus intraradices and rock phosphate amendment influence plant growth and microbial activity in the rhizosphere of *Acacia holosericea*. *Soil Biol. Biochem.*, 37: 1460-468.

Erenstein, O. and Laxmi, V. 2008. Zero tillage impacts in India's rice-wheat systems: a review. *Soil Till. Res.*, 100: 1-14.

Erenstein, O., Farooq, U., Malik, R.K. and Sharif, M. 2008. On-farm impacts of zero tillage wheat in South Asia's rice–wheat systems. *Field Crop Res.,* 105: 240–252.

Erenstein, O., Gérard, B. and Tittonell, P. 2015. Biomass use trade-offs in cereal cropping systems in the developing world: overview. *Agric. Syst.*, 134: 1–5.

FAO. 2001. World Soil Resources Reports 96: Soil carbon sequestration for improved land management. Food and Agriculture Organization of the United Nations, Rome, Italy.

Fitter, A.H., Helgason, T. and Hodge, A. 2011. Nutritional exchanges in the arbuscular mycorrhizal symbiosis: implications for sustainable agriculture. *Fungal Biol. Rev.*, 25: 68-72.

Foley, J.A., Ramankutty, N., Brauman, K.A., Cassidy, E.S., Gerber, J.S., Johnston, M., *et al.,* 2011. Solutions for a cultivated planet. *Nature,* 478: 337–342.

Frank, A.B. 1885. Ueber die auf Wurzelsymbiose beruhende Ern hrung gewisser Ba_me durch unterirdische Pilze. *Ber. Dtsch. Bot. Ges.,* 3: 128–145.

Frank, A.B. and Trappe, J.M. 2005. On the nutritional dependence of certain trees on root symbiosis with belowground fungi (an English translation of A.B. Frank's classic paper of (1885). *Mycorrhiza*, 15: 267–275.

Galvez, L., Douds, D.D., Drinkwater, L.E. and Wagoner, P. 2001. Effect of tillage and farming system upon VAM fungus populations and mycorrhizas and nutrient uptake of maize. *Plant Soil*, 228: 299–308.

Gathala, M.K., Ladha, J.K., Saharawat, Y.S. Kumar, V., Kumar, V. and P.K. Sharma. 2011. Effect of tillage and crop establishment methods on physical properties of a medium-textured soil under 7-year rice-wheat rotation. *Soil Sci. Soc. Am. J.*, 75: 1851-1862.

Giller, K.E., Andersson, J.A., Corbeels, M., Kirkegaard, J., Mortensen, D., Erenstein, O. and Vanlauwe, B. 2015. Beyond conservation agriculture. *Front. Plant Sci.*, 6: 870.

Govaerts, B., Verhulst, N., Castellanos-Navarrete, A., Sayre, K., Dixon, J. and Dendooven, L. 2009. Conservation agriculture and soil carbon sequestration: between myth and farmer reality. *Cr. Rev. Plant Sci.,* 28: 97–122.

Grandy, A.S. and Neff, J.C. 2008. Molecular C dynamics downstream: the biochemical decomposition sequence and its impact on soil organic matter structure and function. *Sci. Total Environ.*, 404: 297–307.

Grandy, A.S. and Robertson, G.P. 2006. Aggregation and organic matter protection following tillage of a previously uncultivated soil. *Soil Sci. Soc. Am. J.*, 70: 1398–1406.

Grandy, A.S. and Robertson, G.P. 2007. Land-use intensity effects on soil organic carbon accumulation rates and mechanisms. *Ecosystems*, 10: 58–73.

Grandy, A.S. and Wickings, K. 2010. Biological and biochemical pathways of litter decomposition and soil carbon stabilization. *Geochim. Cosmochim. Acta.*, 74: A351–A351.

Grant, C.A., Flaten, D.N., Tomasiewicz, D.J. and Sheppard, S.C. 2001. The importance of early season phorphorus nutrition. *Can. J. Plant Sci.*, 81: 211-224.

Gupta, S.R., Neelam and Kumar, R. 2009. Soil Ecology, Biodiversity and carbon mangement. *Int. J. Ecol. Environ. Sci.*, 35: 129-161.

Hansen, N.C,. Tubbs, S., Fernandex, F., Green, S., Hansen, N.E. and Stevens, W.B., 2015. Conservation agriculture in North America. In: *Conservation Agriculture,* (Eds.) M. Farooq and K.H.M. Siddique. Dordrecht: Springer International Publishing, pp. 417-441.

Helgason, T., Daniell, T.J., Husband, R., Fitter, A.H. and Young, J.P.W. 1998. Ploughing up the wood-wide web? *Nature*, 394: 431.

Herman, D.J., Firestone, M.K., Nuccio, E. and Hodge, A. 2012. Interactions between an arbuscular mycorrhizal fungus and a soil microbial community mediating litter decomposition. *FEMS Microbiol. Ecol.*, 80: 236-47.

Hobbs, P.R., Sayre, K. and Gupta, R. 2008. The role of conservation agriculture in sustainable agriculture. *Philos. Trans. R. Soc. B.*, 363: 543–555.

Hodge, A. and Fitter, A.H. 2010. Substantial nitrogen acquisition by arbuscular mycorrhizal fungi from organic material has implications for N cycling. *Proc. Natl. Acad. Sci. USA.*, 107: 13754–13759.

Huggins, D.R., Allmaras, R.R., Clapp, C.E., Lamb, J.A. and Randall, G.W. 2007. Corn-Soybean Sequence and Tillage Effects on Soil Carbon Dynamics and Storage. *J. Soil Sci. Soc. Am. J.*, 71: 145-154.

Ilan, S. and Lal, R. 2013. Agriculture and greenhouse gases, a common tragedy. A review. *Agron. Sustain. Dev.*, 33: 275-289.

Jacobs, A., Rauber, R. and Ludwig, B. 2009, Impact of reduced tillage on carbon and nitrogen storage of two Haplic Luvisols after 10 years. *Soil Till. Res.*, 102: 158-164.

Jansa, J., Mozafar, A., Anken, T., Ruh, R., Sanders, I.R. and Frossard, E. 2002. Diversity and structure of AMF communities as affected by tillage in a temperate soil. *Mycorrhiza*, 12: 225–234.

Jansa, J., Mozafar, A., Kuhn, G., Anken, T., Ruh, R., Sanders, I.R. and Frossard, E. 2003. Soil tillage affects the community structure of mycorrhizal fungi in maize roots. *Ecol. Appl.*, 13: 1164-1176.

Jat, M.L., Gathala, M.K., Ladha, J.K., Saharawat, Y.S., Jat, A.S., Kumar, V., Sharma, S.K., Kumar V. and Gupta, R., 2009. Evaluation of precision land leveling and double zero-till system in the rice-wheat rotation: Water use, productivity, profitability and soil physical properties. *Soil Till. Res.*, 105: 112-121.

Jat, R.A., Sahrawat, K.L. and Kassam, A.H. 2014. Conservation Agriculture: Global Prospects and Challenges. CABI, Wallingford. pp. 393.

Johansson, J.F., Paul, L.R. and Finlay, R.D. 2004. Microbial interactions in the mycorrhizosphere and their significance for sustainable agriculture. *FEMS Microbiol. Ecol.*, 48: 1–13.

Johnson, D., Leake, J.R., Ostle, N., Ineson, P. and Read, D.J. 2002. *In situ* CO_2-13C pulse labelling of upland grassland demonstrates a rapid pathway of carbon flux from arbuscular mycorrhizal mycelia to the soil. *New Phytol.*, 153: 327–334.

Johnson, N.C. and Gehring C.A. 2007. "Mycorrhizas: symbiotic mediators of rhizosphere and ecosystem processes," In: Cardon, Z. and Whitbeck, J. (Eds.) The Rhizosphere: An Ecological Perspective, New York: Academic Press, pp. 73–100.

Johnson, N.C., Zak, D.R., and Tillman, D. 1991. Dynamics of vesicular-arbuscular mycorrhizae during old field succession. *Oecologia*, 86: 349-358.

Joner, E.J., van Aarle, I.M. and Vosatka, M. 2000. Phosphatase activity of extra-radical arbuscular mycorrhizal hyphae: a review. *Plant Soil.*, 226:199–210.

Kabir, Z., O'Halloran, I.P., Widden, P. and Hamel, C. 1998. Vertical distribution of arbuscular mycorrhizal fungi under corn (*Zea mays* L.) in no-till and conventional tillage systems. Mycorrhiza, 8: 53–55.

Kabir, Z. 2005. Tillage or no-tillage: impacts on mycorrhizae. *Can. J. Plant Sci.*, 85: 23–29.

Kabir, Z. and Koide, K.T. 2002. Mixed cover crops, mycorrhizal fungi, soil properties and sweet corn yield. *Plant Soil*, 238: 205–215.

Kabir, Z., O'Halloran, I.P. and Hamel, C. 1997a. Overwinter survival of arbuscular mycorrhizal hyphae is favored by attachment to roots but diminished by disturbance. *Mycorrhiza*, 7: 197-200.

Kabir, Z., O'Halloran, I.P., Fyles, J.W. and Hamel, C. 1997b. Seasonal changes of arbuscular mycorrhizal fungi as affected by tillage practices and fertilization: Hyphal density and mycorrhizal root colonization. *Plant Soil*, 192: 285-293

Kirkegaard, J.A., Conyers, M. K., Hunt, J.R., Kirkby, C.A., Watt, M. and Rebetzke, G.J. 2014. Sense and nonsense in conservation agriculture principles, pragmatism and productivity in Australina mixed farming systems. *Agric. Ecosyst. Environ.*, 187:133-145.

Kassam, A., Friedrich, T., Derpsch, R., Lahmar, R., Mrabet, R. and Basch, G., *et al.*, 2012. Conservation agriculture in the dry Mediterranean climate. *Field Crop Res.*, 132: 7-17.

Kassam, A., Friedrich, T., Derpsch R. and Kienzle, J. 2015. Overview of the Worldwide Spread of Conservation Agriculture. *Field Actions Sci. Rep.*, 8: 1-11.

Kassam, A., Friedich, T., Shaxson, F., Bartz, H., Mello, I., Kienzle, J., *et al.*, 2014. The spread of conservation agriculture. Policy and institutional support for adoption and uptake. *Field Actions Sci. Rep.*, 7: 1-12.

Kirkegaard, J., Christen, O., Krupinsky, J. and Layzell, D., 2008. Break crop benefits in temperate wheat production. *Field Crop Res.*, 107: 185-195.

Kohl, L., Oehl, F. and van der Heijden, M.G.A., 2014. Agricultural practices indirectly influence plant productivity and ecosystem services through effects on soil biota. *Ecol. Appl.*, 24: 1842-1853.

Koide, R.T. and Kabir, Z. 2000. Extraradical hyphae of the mycorrhizal fungus Glomus intraradices can hydrolyse organic phosphate. *New Phytol.*, 148: 511–517.

Kõljalg, U., Nilsson, R.H., Abarenkov, K., Tedersoo, L., Taylor, A.F.S., Bahram, M., Bates, S.T., Bruns, T.D., Bengtsson-Palme, J., Callaghan, T.M., Douglas, B., Drenkhan, T., Eberhardt, U., Dueñas, M., Grebenc, T., Griffith, G.W., Hartmann, M., Kirk, P.M., Kohout, P., Larsson, E., Lindahl, B.D., Lücking, R., Martín, M.P., Matheny, P.B., Nguyen, N.H., Niskanen, T., Oja, J., Peay, K.G., Peintner, U., Peterson, M., Põldmaa, K., Saag, L., Saar, I., Schüßler, A., Scott, J.A., Senés, C., Smith, M.E., Suija, A., Taylor, D.L., Telleria, M.T., Weiss, M. and Larsson, K.H. 2013. Towards a unified paradigm for sequence-based identification of fungi. *Mol. Ecol.*, 22: 5271–5277.

Koller, R., Rodriguez, A., Robin, C., Scheu, S. and Bonkowski, M. 2013. Protozoa enhance foraging efficiency of arbuscular mycorrhizal fungi for mineral nitrogen from organic matter in soil to the benefit of host plants. *New Phytol.*, 199: 203-211.

Lal, R. 2003. Global potential of soil carbon sequestration to mitigate the greenhouse effect. *Cr. Rev. Plant Sci.*, 22: 151-184.

Lal, R. 2011. Sequestering carbon in soils of agro-ecosystems. Food Policy 36: S33–S39.

Lazcano, C., Barrios-Masias, F.H. and Jackson, L.E. 2014. Arbuscular mycorrhizal effects on plant water relations and soil greenhouse gas emissions under changing moisture regimes. *Soil Biol. Biochem.*, 74: 184–192.

Lekberg, Y. and Koide, R.T. 2005. Is plant performance limited by abundance of arbuscular mycorrhizal fungi? A meta-analysis of studies published between 1988 and 2003. *New Phytol.*, 168: 189-204.

Lekberg, Y., Koide, R.T., Rohr, J.R., Aldrich-Wolf, L. and Morton, J.B., 2007. Role of niche restrictions and dispersal in the composition of arbuscular mycorrhizal fungal communities. *J. Ecol.*, 95: 95-105.

Llewellyn, R.S., D'Emden, F.H., and Kuehne, G., 2012. Extensive Use of no-tillage in grain-growing regions of Australia. *Field Crop Res.*, 132: 204-212.

Mathimaran, N., Ruh, R., Jama, B., Verchot, L., Frossard, E. and Jansa, J. 2007. Impact of agricultural management on arbuscular mycorrhizal fungal communities in Kenyan ferralsol. *Agric Ecosyst. Environ.*, 119: 22-32.

McDaniel, M.D., Tiemann, L.K. and Grandy, A.S. 2014. Does agricultural crop diversity enhance soil microbial biomass and organic matter dynamics? A meta-analysis. *Ecol. Appl.*, 24: 560–570.

McGonigle, T.P. and Miller, M.H. 1993. Mycorrhizal development and phosphorus absorption in maize under conventional and reduced tillage. *Soil Sci. Soc. Am. J.*, 57: 1002–1006.

McGonigle, T.P. and Miller, M.H. 1996. Development of fungi below ground in association with plants growing in disturbed and undisturbed soils. *Soil Biol. Biochem.*, 28: 263–269.

MEA. 2005. Millenium Ecosystem Assessment, Our human planet: summary for decision makers. Island Press, Washington, DC, USA.

Miller, M.H., 2000. Arbuscular mycorrhizae and the phosphorus nutrition of maize: A review of Guelph studies. *Can. J. Plant Sci.*, 80: 47-52.

Mohan, J.E., Cowden, C.C., Baas, P., Dawadi, A., Frankson, P.T., Helmick, K., Hughes, E., Khan, S., Lang, A., Machmuller, M., Taylor, M. and Witt, C.A., 2014. Mycorrhizal fungi mediation of terrestrial ecosystem responses to global change: Mini review. *Fungal Ecol.*, 10: 3-19.

Morton, J.B. and Benny, G.L., 1990. Revised classification of arbuscular mycorrhizal fungi (Zygomycetes): A new order, Glomales, two new suborders, Glomineae and Gigasporineae, and two families, Acaulosporaceae and Gigasporaceae, with an emendation of Glomaceae. *Mycotaxon*, 37: 471-491.

Mosse, B. 1986. Mycorrhiza in a sustainable agriculture. In: Lopez-Real J.M. and Hodges R.H. (Eds.) Role of Microorganisms in a Sustainable Agriculture, Academic publishers, London. pp. 105-123.

Neelam, Aggarwal, A., Asha, Ekta and Gupta, S.R. 2010. Soil aggregate carbon and diversity of mycorrhiza as affected by tillage practices in a rice-wheat cropping system in northern India. *Int. J. Ecol. Environ. Sci.*, 36: 20-26.

Neelam and Gupta, S.R. 2009. Conservation tillage effects on crop yield, carbon accumulation and soil nitrogen transformation in rice-wheat systems in northern India. *Int. J. Ecol. Environ. Sci.*, 35: 199-209.

Njeru, E., Avio, L., Bocci, G., Sbrana, C., Turrini, A., B_arberi, P., Giovannetti, M. and Oehl, F. 2015. Contrasting effects of cover crops on 'hot spot' arbuscular mycorrhizal fungal communities in organic tomato. *Biol. Fert. Soils*, 51: 151-166.

Nottingham, A.T., Turner, B.L., Winter, K., van der Heijden, M.G.A. and Tanner, E.V.J. 2010. Arbuscular mycorrhizal mycelial respiration in a moist tropical forest. *New Phytol.*, 186: 957–967.

Oades, J.M. 1984. Soil organic matter and structural stability: mechanisms and implications for management. *Plant Soil,* 76: 319–337.

Oehl, F., Sieverding, E., Palenzuela, J., Ineichen, K. and Silva, G.A. 2011. Advances in Glomeromycota taxonomy and classification. *IMA Fungus,* 2: 191–199.

Oehl, F., Sieverding, E., Ineichen, K., Mäder, P., Boller, T. and Wiemken, A. 2003. Impact of land use intensity on the species diversity of arbuscular mycorrhizal fungi in agroecosystems of Central Europe. *Appl. Environ. Microbiol.,* 69: 2816-2824.

Opik, M., Moora, M., Liira, J. and Zobel, M. 2006. Composition of root-colonizing arbuscular mycorrhizal fungal communities in different ecosystems around the globe. *J. Ecol.,* 94: 778–790.

Opik, M., Zobel, M., Cantero, J.J., Davison, J., Facelli, J.M., Hiiesalu, I., Jairus, T., Kalwij, J.M., Koorem, K., Leal, M.E. *et al.,* 2013. Global sampling of plant roots expands the described molecular diversity of arbuscular mycorrhizal fungi. *Mycorrhiza,* 23: 411–430.

Opik, M., Vanatoa, A., Vanatoa, E., Moora, M., Davison, J., Kalwij, J.M., Reier, U. and Zobel, M. 2010. The online database MaarjAM reveals global and ecosystemic distribution patterns in arbuscular mycorrhizal fungi (Glomeromycota). *New Phytol.,* 188: 223-241.

Palm, C., Blanco-Canqui, H., Declerck, F., Gatere, L. and Grace, P. 2014. Conservation agriculture and ecosystem services: an overview. *Agric. Ecosyst. Environ.,* 187: 87–105.

Palm, C., Sanchez, P., Ahamed, S. and Awiti, A. 2007. Soils: A contemporary perspective. *Annu. Rev. Environ. Resour.,* 32: 99–129.

Pathak, H., Bhatia, A., Jain, N. and Aggarwal, P.K. 2010. Greenhouse Gas Emission and Mitigation in Indian Agriculture – A Review, In: Singh, B. (Ed.) ING Bulletins on Regional Assessment of Reactive Nitrogen, Bulletin No. 19, SCON-ING, New Delhi, pp. i-iv and 1-34.

Pathak, H., Tewari, A.N., Sankhyan, S., Dubey, D.S., Mina, U., Singh, V.K., Jain, N. and Bhatia, A. 2011. Direct seeded rice: Potential, performance and problems-A review. *Curr. Adv. Agric. Sci.,* 3: 77-88.

Pellegrino, E. and Bedini, S. 2014. Enhancing ecosystem services in sustainable agriculture: biofertilization of chickpea (*Cicer arietinum* L.) by arbuscular mycorrhizal fungi. *Soil Biol. Biochem.,* 68: 429–439.

Pellegrino, E., Opik, M., Bonari, E. and Laura Ercoli. 2015. Responses of wheat to arbuscular mycorrhizal fungi: A meta-analysis of field studies from 1975 to 2013. *Soil Biol. Biochem.,* 84: 210-217.

Pittelkow, C.M., Liang, X., Linquist, B.A., van Groenigen, K.J., Lee, J., Lundy, M.E., *et al.,* 2015a. Productivity limits and potentials of the principles of conservation agriculture. *Nature,* 51: 365-368.

Pittelkow, C.M., Linquist, B.A., Lundy, M.E., Liang, X., van Groenigen, K.J., Lee, J., *et al.* 2015b. When does no-till yield more? A global meta-analysis. *Field Crop Res.* 183: 156-168.

Power, A.G. 2010. Ecosystem services and agriculture: tradeoffs and synergies. *Phil. Trans. R. Soc. B.*, 365: 2959–2971.

Power, M.E. and Mills, L.S. 1995. The keystone cops meet in Hilo. *Trend Ecol. Evol.*, 10: 182-184.

Powlson, D.S., Stirling, C.M., Jat, M.L., Gerard, B.G., Palm, C.A., Sanchez, P.A. and Cassman, K.G. 2014. Limited potential of no-till agriculture for climate change mitigation. *Nat. Clim. Change*, 4: 678–683.

Puget, P. and Lal, R. 2005. Soil organic carbon and nitrogen in a Mollisol in central Ohio as affected by tillage and land use. *Soil Till. Res.*, 80: 201-213.

Redecker, D., Kodner, R. and Graham, L.E. 2000. Glomalean fungi from the Ordovician. *Science*, 289: 1920–1921.

Rillig, M.C., Aguilar-Trigueros, C.A., Bergmann, J., Verbruggen, E., Veresoglou, S.D. and Lehmann, A. 2015. Plant root and mycorrhizal fungal traits for understanding soil aggregation. *New Phytol.*, 205: 1385–1388.

Rillig, M.C. and Mummey, D.L. 2006. Mycorrhizas and soil structure. *New Phytol.*, 171: 41–53.

Rillig, M.C. 2004a. Arbuscular mycorrhizae and terrestrial ecosystem processes. *Ecol. Lett.*, 7: 740–754.

Rillig, M.C. 2004b. Arbuscular mycorrhizae, glomalin, and soil aggregation. *Can. J. Soil Sci.*, 84: 355–363.

Rillig, M.C., Lutgen, E.R., Ramsey, P.W., Klironomos, J.N. and Gannon, J.E. 2005. Microbiota accompanying different arbuscular mycorrhizal fungal isolates influence soil aggregation. *Pedobiologia*, 49: 251-259.

Rillig, M.C., S.F. Wright, K.A. Nichols, W.F. Schmidt, and M.S. Torn. 2001. Large contribution of arbuscular mycorrhizal fungi to soil carbon pools in tropical forest soils. *Plant Soil*, 233: 167–177.

Rillig, M.C., Wright, S.F. and Eviner, V.T. 2002. The role of arbuscular mycorrhizal fungi and glomalin in soil aggregation: comparing effects of five plant species. *Plant Soil*, 238: 325-333.

Roldan, A., Salinas-Garcia, J.R., Alguacil, M.M. and Caravaca, F. 2007. Changes in soil sustainability indicators following conservation tillage practices under subtropical maize and bean crops. *Soil Till. Res.*, 93: 273–282.

RWC (Rice Wheat Consortium for the Indo-Gangetic Plains), Research Highlights 2005. In: Gupta, R.K. (Regional Facilitator) Rice-Wheat Consortium for the Indo-Gangetic Plains, CG Centre, New Delhi. pp. 1-6 (http://www. rwc. cgiar. org.).

Säle, V., Aguilera, P., Laczko, E., Mäder, P., Berner, A., Zihlmann, U., van der Heijden, M.G.A. and Oehl, F. 2015. Impact of conservation tillage and organic farming on the diversity of arbuscular mycorrhizal fungi. *Soil Biol. Biochem.*, 84: 38-52.

Schachtman, D.P., Reid, R.J. and Ayling, S.M. 1998. Phosphorus uptake by plants: from soil to cell. *Plant Physiol.*, 116: 447–453.

Schnoor, T.K., Lekberg, Y., Rosendahl, S. and Olsson, P.A. 2011. Mechanical soil disturbance as a determinant of arbuscular mycorrhizal fungal communities in semi-natural grassland. *Mycorrhiza*, 21: 211-220.

Schuessler, A. 2013. Phylogeny and taxonomy of Glomeromycota ('arbuscular mycorrhizal (AM) and related fungi'). Available at: <http://schuessler.userweb.mwn.de/amphylo/>.

Schussler, A. 2014. Glomeromycota: species list. [WWW document]. *http://schuessler.userweb.mwn.de/amphylo*

Singh, V. and Ladha, J.K. 2011. Direct seeding of rice: Recent developments and future research needs. *Adv. Agron.*, 11: 297-413.

Six, J., Conant, R.T., Paul, E.A. and Paustian, K. 2002. Stabilization mechanisms of soil organic matter: Implication for C- saturation of soils. *Plant Soil*, 241: 155- 176.

Six, J., Frey, S.D., Thiet, R.K. and Batten, K.M. 2006. Bacterial and fungal contributions to carbon sequestration in agroecosystems. *Soil Sci. Soc. Am. J.*, 70: 555–569.

Six, J., Ogle, S.M., Breidt, F.J., Conant, R.T., Mosier, A.R. and Paustian, K. 2004. The potential to mitigate global warming with no-tillage management is only realized when practised in the long term. *Global Change Biol.*, 10: 155–160.

Smith, S.E. and Smith, F.A. 2012. Fresh perspectives on the roles of arbuscular mycorrhizal fungi in plant nutrition and growth. *Mycologia*, 104: 1–13.

Smith, S.E. and Read, D.J. 2008. Mycorrhizal symbiosis, 3rd Edn. Academic Press, London, UK.

Smith, S.E. and Smith, F.A. 2011. Roles of arbuscular mycorrhizas in plant nutrition and growth: new paradigms from cellular to ecosystem scales. *Ann. Rev. Plant Biol.*, 62: 227–250.

Smith, S.E. and Read, D.J. 1997. Mycorrhizal symbiosis. Academic Press, London, UK.

Soane, B.D., Ball, B.C., Arvidsson, J., Basch, G., Moreno, F. and Roger-Estrade, J. 2012. No-till in northern, western and souther-western Europe: a review of problems and opportunities for crop production and the environment. *Soil Till. Res.*, 118: 66-87.

Sorensen, J.N., Larsen, J. and Jakobsen, I. 2005. Mycorrhizae formation and nutrient concentration in leeks (*Allium porrum*) in relation to previous crop and cover crop management on high P soils. *Plant Soil*, 273: 101-114.

Stevenson, J.R., Serraj, R. and Cassman, K.G., 2014. Evaluating conservation agriculture for small-scale farmers in Sub-Saharan Africa and South Asia. *Agric. Ecosyst. Environ.*, 187: 1-10.

Tiemann, L.K., Grandy, A.S., Atkinson, E.E., Marin-Spiotta, E., and Mcdaniel, M.D. 2015. Crop rotational diversity enhances belowground communities and functions in an agroecosystem. *Ecol. Lett.*, 18: 761–771.

Toljander, J.F., Lindahl, B.D., Paul, L.R., Elfstrand, M. and Finlay, R.D., 2007. Influence of arbuscular mycorrhizal mycelial exudates on soil bacterial growth and community structure. *FEMS Microbiol. Ecol.*, 61: 295-304.

Treseder, K.K. and Cross, A. 2006. Global distributions of arbuscular mycorrhizal fungi. *Ecosystems*, 9: 305–316.

Treseder, K.K. and Turner, K.M. 2007. Glomalin in Ecosystems. *Soil Sci. Soc. Am. J.*, 71:1257-1266.

Troeh, Z.I. and Loynachan, T.E. 2003. Endomycorrhizal fungal survival in continuous corn, soybean, and fallow. *Agron. J.*, 95: 224–230.

Van der Heijden, M.G.A., Bardgett, R.D. and Van Straalen, N.M. 2008. The unseen majority: Soil microbes as drivers of plant diversity and productivity in terrestrial ecosystems. *Ecol. Lett.*, 11: 296 – 310.

van der Heijden, M.G.A., Martin, F.M., Selosse, M.A. and Sanders, I.R. 2015. Mycorrhizal ecology and evolution: the past, the present, and the future. *New Phytol.*, 205: 1406-23.

van der Heijden, M.G.A., Klironomos, J.N., Ursic, M., Moutoglis, P., Streitwolf-Engel, R. and Boller, T. 1998. Mycorrhizal fungal diversity determines plant biodiversity, ecosystem variability and productivity. *Nature*, 396: 69–72.

Verbruggen, E. and Kiers, E.T. 2010. Evolutionary ecology of mycorrhizal functional diversity in agricultural systems. *Evol. Appl.*, 3: 547–560.

Verbruggen, E., Jansa, J., Hammer, E.C. and Rillig, M.C. 2016. Do arbuscular mycorrhizal fungi stabilize litter-derived carbon in soil? *J. Ecol.*, 104: 261–269.

Verbruggen, E., van der Heijden, M.G.A., Weedon, J.T., Kowalchuk, G.A. and Roling, W.F.M. 2012. Community assembly, species richness and nestedness of arbuscular mycorrhizal fungi in agricultural soils. *Mol. Ecol.*, 21: 2341-2353.

Wang, Y., Tu, C., Cheng, L., Li, C., Gentry, L.F., Hoyt, G.D., Zhang, X. and Hu, S. 2011. Long-term impact of farming practices on soil organic carbon and nitrogen pools and microbial biomass and activity. *Soil Till. Res.*, 117: 8–16.

Watson, C.A., Bengtsson, H., Ebbesvik, M., Loes, A.-K., Myrbeck, A., Salomon, E., Schroder, J. and Stockadale, E.A., 2002. A review of farm-scale nutrient budgets for organic farms as a tool for management of soil fertility. *Soil Use and Manage.*, 18: 264-273.

West, T.O. and Post, W.M. 2002. Soil organic carbon sequestration rates by tillage and crop rotation: a global data analysis. *Soil Sci. Soc. Am. J.*, 66: 1930-1946.

Wickings, K., Grandy, A.S., Reed, S.C. and Cleveland, C.C. 2012. The origin of litter chemical complexity during decomposition. *Ecol. Lett.*, 15: 1180–1188.

Wilson, G.W.T., Rice, C.W., Rillig, M.C., Springer, A. and Hartnett, D.C. 2009. Soil aggregation and carbon sequestration are tightly correlated with the abundance of arbuscular mycorrhizal fungi: results from long-term field experiments. *Ecol. Lett.*, 12: 452–461.

Wright, S.F., Franke-Snyder, M., Morton, J.B. and Upadhyaya, A. 1996. Time-course study and partial characterization of a protein on hyphae of arbuscular mycorrhizal fungi during active colonization of roots. *Plant Soil*, 181: 193–203.

Wright, S.F., Starr, J.L. and Paltineau, I.C. 1999. Changes in aggregate stability and concentration of glomain during tillage management transition. *Soil Sci. Soc. Am. J.*, 63: 1825–1829.

Wright, S.F. and Upadhyaya, A. 1998. A survey of soils for aggregate stability and glomalin, a glycoprotein produced by hyphae of arbuscular mycorrhizal fungi. *Plant Soil*, 198: 97-107.

Xu, S.Q., Zhang, M.Y., Zhang, H.L., Chen, F., Yang, G.L. and Xiao, X.P. 2013. Soil organic carbon stocks as affected by tillage systems in a double-cropped rice field. *Pedosphere*, 23: 696-704.

2017, Mycorrhizal Fungi
Editors: Ashok Aggarwal and Kuldeep Yadav
Published by: ASTRAL INTERNATIONAL PVT. LTD., NEW DELHI

Pages 75–85

3

Bamboo-Mycorrhizal Association

Solomon Das[1]*, Abhaya Garg[2] and Y.P. Singh[3]

[1]*Vimix Solutions Private Limited, Dehradun*
[2]*INDUS Environmental Services Private Limited, New Delhi*
[3]*Forest Research Institute, Dehradun*
*Corresponding Author: solo77das@gmail.com

ABSTRACT

Bamboo in many ways is the mainstay of the rural Indian economy, sparking considerable social and ecological spin-offs. Bamboos cover 12.8 per cent of the total forest area of India; out of this, Dendrocalamus strictus covers 53 per cent. One of the important problems in tropical biomass production is the low soil phosphate (P) availability. The consumption of chemical fertilizers is significantly increased in the last three decades. Bamboo having the shallow root system and fast growing capacity, it widely depended upon the microbial interaction for their nutrient availability. Arbuscular mycorrhizae (AM) played a significant role in the sustainability of these Bamboos, also known as Green Gold or poor man's timber. The interaction of AM fungi with other beneficial microbes around bamboo rhizospheres cannot be ignored. Such interactions are much needed to be explored for developing efficient biofertilizers in mitigating the phosphate deficient condition of Indian soil.

Keywords: Bamboo, Mycorrhizae, Microbial interactions.

1. Introduction

India is one of the 12 mega diversity countries of the world which has 7 per cent of world's biodiversity and total forest cover of 21.05 per cent (692,027 km^2)

of the geographical area of the country. The forest cover of the country has been classified on the basis of tree canopy density into pre-defined classes, *viz.*, very dense forest (VDF), moderately dense forest (MDF), and open forest (OF). The area under VDF is 2.54 (83,471 km^2), MDF 9.76 (320,736 km^2), and of 8.75 per cent (287, 820 km^2) (Forest Survey of India, 2011). Forestry is the second largest land use after agriculture. Having population over 1.21 Bn (www.censusindia.gov.in; based on 2011 census), more than 300 M rural people depend on forests, 100 M people are having forests as main source of income, nearly 35 M tribal people subsist on forest resources and about 500 M livestock use forests as grazing lands.

Bamboo has traditionally being used for paper manufacturing, scaffolding, construction materials and handicrafts. Now, there is more diversification such as bamboo flooring and panelling, though only a few people are doing this work. Bamboo is also used for making incense sticks, foot rulers and matchsticks. The biggest growth is expected to come from use of bamboo as a replacement for wood. Bamboo housing is slowly growing in concept. The bamboo is also called 'green gold' in India; the country that has the second-largest reserve of bamboo in the world (Nayar, 2009). Bamboo in many ways is the mainstay of the rural Indian economy, sparking considerable social and ecological spin-offs.

Of all the commonly occurring genera of bamboos, the genus *Bambusa* is the widely distributed in India followed by genus *Dendrocalamus*. It occurs in the plains of south and central India and dry hills of north India. From east to west, it occurs from Punjab to Assam and other north-eastern states (Thomas *et al.*, 1985; Biswas, 1988). It is extensively used as raw material in paper mills and also for a variety of purposes such as construction, agricultural implements, musical instruments and furniture. Young shoots are commonly used as food and a decoction of leaves, nodes and silicaceous matter are used in traditional medicine (Muthukumar and Udaiyan, 2006). Deogun (1937) has differentiated three major growth forms of *D. strictus* on the basis of edaphic conditions, isolation aspect, temperature and humidity. These growth forms growing in different geographical regions show variations in their appearances, arrangement of clumps and cross-sectional thicknesses of culms. These variations, evolved due to the environmental and other factors, are a matter of study.

2. Field Status of Arbuscular Mycorrhizae

Although AM fungi are widespread and are distributed in different parts of the world especially in the tropics, little functional information was known about them, until the mid 1950s (Smith and Read, 1997). They are reported to be found in diverse land areas such as calcareous grasslands, arid/semi arid grasslands, several temperate forests, tropical rain forests and shrub lands in diverse parts of the world (Muthukumar and Udaiyan, 2002; Oehl *et al.*, 2003; Renker *et al.*, 2005). Recently, AM fungi have received more attention especially in African countries such as Namibia, Cameroon, Kenya, Morocco, Nigeria, Senegal, Zambia and South Africa. These studies have concentrated on AM fungal diversity in various regions and soil types or the mycorrhizal status of indigenous crop and plant species (Diop *et al.*, 1994; Bâ *et al.*, 2000; Dalpé *et al.*, 2000; Stutz *et al.*, 2000; Hawley and Dames, 2004; Bouamri *et al.*, 2006). Results from these studies revealed that different species of AM fungi are obtained depending on plant species and geographic location. Amongst AM fungal species, *Glomus* sp. were consistently isolated while, others species belonging to the genera, *Acaulospora, Gigaspora* and *Scutellospora* were either absent or found in few numbers (Dames, 1991; Stutz *et al.,* 2000; Bouamri *et al.,* 2006; Uhlmann *et al.,* 2006).

The essence of mycorrhizzal relation has perhaps been best captured in the surveys of AMF for grassland and shrub-lands (Miller, 1987) and for humid tropical ecosystem (Janos, 1987). The arid and semi-arid regions of the world, where soil organic matter is low, typically have AMF (Read 1993; Allen *et al.,* 1995; Muthukumar and Udaiyan, 2002). The tundra region with high organic matter has virtually no AMF (Bledsoe *et al.,* 1990). A study conducted in South India showed that endomycorrhizal fungi were found to be widespread in and around the Nagarjuna University campus contains red sandy loam soil. Roots of all the 21 grass species formed multiple mycorrhizzal associations. Most of them were colonised by the VAM fungi showing infection of 50 to 80 per cent of root length. The amount of external and internal colonisation of the roots varied with species and 14 mycorrhizal fungi were identified. The development pattern of mycorrhiza was studied on two selected grass species raised in autoclaved soil, supplemented with root inoculum derived from the same species. These results clearly show that VAM fungi are wide spread in the grassland and they colonise the roots of all grasses forming multiple mycorrhizzal associations (Ammani *et al.,* 1994).

A preliminary survey on the association of AM in bamboo species in Western Ghats, Kerala was taken up by Appaswamy and Ganapati (1992) with an aim to find out its possible role in plant growth and biomass production. Rhizosphere soils and root samples from areas were screened for VAM spore and infection, respectively. Out of nine species of *Glomus* recorded, *G. Fasiculatum* and *G. mosseae* were found to be widespread in distribution. Cent per cent infection was observed in *B. bambos* in all the places surveyed except in Kasargod district.

Muthukumar *et al.* (1996) examined mycorrhizal colonisation and spore numbers of *Cyprus*, growing in the semi-arid tropical grassland during the 1993 and 1994 monsoons. The soil nutrient exhibited seasonal variations, but were highly variable between years. Intercellular hyphae and vesicles with occasional intra-radical spores characterise mycorrhizzal association in the sedge plants. Temporal variations in mycorrhizzal colonisation and spore number occurred, indicating seasonality. However, the patterns of mycorrhizzal colonisation were similar during both the years. The VAM fungal structures observed were intercellular hyphae and vesicles. Changes in the proportion of root length with VAM structure, total colonisation levels and spore numbers were related to climate and edaphic factors. However, the intensity of influence of climatic and soil factors on VAM tended to vary with sedge species.

Mohanan and Sebastian (1999) reported the field mycorrhizzal status in 19 species of bamboos in Kerala mostly belonging to nine genera, *viz., Bamboo, Dendrocalamus, Melocanna, Oxytenanthera, Phyllostachys, Thyrsostachus, Arundinaria* and *Ochlandra* which are grown in natural forests and plantations in the state. They observed high per cent of AM root colonisation (81-99 per cent) with spores ranged from 65 to 205/10g of soil. Majority of the spores were belonging to *Glomus* sp., *Gigaspora* sp., *Acaulopsora* sp. and *Scutellospora* sp.

Muthukumar and Udayan (2002) examined the plants of Western Ghats regions of southern India for their mycorrhizal associations. Root and soil samples of 329 species representing 61 families (in which 54 species of family Poaceae) were examined. Out of them, 174 were mycorrhizal. The mycorrhizal colonization characterized by arbuscules, intaradical hyphae, intracellular hyphal coils with or without vesicles. Vesicles and hyphae (but no arbuscules) were observed in 135 plant species. Among the plant families examined, 26 per cent contained only mycorrhizal species, 36 per cent had mycorrhizal and non-mycorrhizal species and 38.5 per cent had only non-mycorrhizal species. Mycorrhizal incidence in dicots (57 per cent) was higher than in monocots (39 per cent).

Rawat (2005) surveyed the field populations of indigenous mycorrhizae in the roots of different growth forms of *D. strictus* growing in three locations. She also studied the root colonization and spore populations in different seasons. She reported that Chiriapur (one of the growth forms) having the lowest soil P supported highest root colonization, while, Shivpuri had maximum spore population. *D. strictus* supported highest root colonization during July (monsoon) and lowest in January (winter).

There is evidence that local ecotypes of plants and mycorrhizal fungi co-adapt to each other and to their local soil environment. A comparison of *Andropogon gerardii* ecotypes from phosphorus-rich and phosphorus-poor prairies (ecosystem consist of temperate grasslands, savannas and shrublands biome) show that each ecotype grew best in the soil of its origin. Furthermore, the *A. gerardii* ecotype from the phosphorus-poor soil was three times more responsive to mycorrhizal colonization and had a significantly coarser root system than the ecotype from the phosphorus-rich soil (Schultz *et al.*, 2001). These results suggest that the genetic composition of plant populations evolve, so that, mycorrhizal costs are minimized and benefits are maximized within the local soil fertility conditions (Cordon and Whitbeck, 2007).

3. Interaction between *D. strictus* and AM Fungus

Bhattacharya *et al.* (1999) studied the relevance of mycorrhiza for bamboo cultivation in laterite wastelands along with mycorrhizal dependency and phosphorous utilization efficiency. One of the major problems with commercially important bamboo species is their shallow root systems. Natural root colonization by arbuscular mycorrhizae and inoculum availability in the rhizosphere of most species of bamboo in wasteland soils are reported to be low. *D. strictus*, an important bamboo species, raised in a phosphorous deficient laterite soil showed above 40 per cent arbuscular dependency for shoot dry matter production and phosphorus acquisition at 360 days. He also reported that mycorrhiza inoculated seedlings acquired more phosphorus as compared to non-inoculated seedlings. This phosphorous acquisition efficiency decreased with the increase in soil P. The author concluded that the root morphological traits and the reported mycorrhizal relation of bamboo species such as *D. strictus* may be considered suitable for application of AM technology in wasteland plantations.

The proliferation method is potentially universally applicable and can be used for mass scale production of field plantable saplings of sympodial bamboos. Studies on the macroproliferation, a standardised technique of *D. strictus* multiplication, showed that invariably, mycorrhiza (either as indigenous or inoculated) supported better plant growth and accumulated more nutrients than its uninoculated counterparts. The effect of phosphorus and mycorrhiza on plant growth was mutually synergistic at lower P doses (1/2 or 1/4th P). Likewise, the rhizome (1.5) as well as tiller number (1.8) had significantly more count at 1/2 and 1/4th P and mycorrhiza (indigenous or inoculated) in comparison to uninoculated plants with similar P doses. Phosphorus content of *D. strictus* shoots of inoculated plants (0.86 of indigenous mycorrhizae and 1.00 mg/g tissue of *G. etunicatum*) was significantly more than the control plants (0.64 mg/g; Rawat, 2005). Better performance of mycorrhiza inoculated *D. strictus* plants at lower P doses affirms the old observations regarding better expression of the symbiont at low P regimes. It may help in formulating a strategy for afforestation of vast tracts of Indian wastelands (with low P) with bamboos that too with seedlings fortified with endomycorrhizae so as to improve their field survival as well as performance.

Anuradha *et al.* (2012) identified suitable arbuscular mycorrhizal (AM) fungi for inoculation of *B. bamboos* and *D. strictus* at nursery stage for increasing growth and productivity. In *B. bamboos*, total dry weight and P uptake were significantly increased by all tested fungi and shoot length was increased by eight AM inoculants. In *D. strictus*, all endophytes significantly increased shoot length, dry shoot weight and P uptake. Dry root weight was significantly increased by only two inoculants namely, *G. cerebriforme* and *G. etunicatum*. The results supported that utilisation of effective AM fungi can enhance the productivity of bamboo in Bundelkhand region.

4. Interactions between *Pseudomonas*–AM Fungi

The solubilisation of P is reported to be the common mode of action for PGPR and studies by Singh and Kapoor (1998). They showed that PSB such as *B. circulans* together with AM fungi increased plant yield and P uptake of wheat. Phosphate solubilising bacteria (PSB) have great prospects to improve plant growth under P deficient soils when used in conjunction with AM fungi (Gryndler, 2000). They are known to mobilise phosphate ions from sparingly soluble organic and inorganic P sources. However, the released P does not reach the root surface as a result of inadequate diffusion (Azcün-Aguilar and Barea, 1992; Barea *et al.*, 2005). It was proposed that AM fungi could improve the uptake of the solubilised P, hence, this combined interaction should improve P nutrition and supply to plants (Barea *et al.*, 2002).

The interactive effects of AM fungi and PSB on plant use of soil P in the form of either endogenous or added rock P was studied using a soil microcosm system integrated with ^{32}P isotopic dilution. Results revealed that the PSB (*Enterobacter* sp. and *B. subtilis*) promoted mycorrhizal establishment of *G. intraradices* and their combined inoculation increased biomass, N and P accumulation in the onion plant tissues (Toro *et al.*, 1997). Thus, the inoculation of organisms may result in utilisation of P fertilisers that quickly become unavailable in soils (Picini and Azcon, 1987). Multi-microbial interactions between AM fungi, PSB and *Azospirillum* have been reported to be synergistic when inoculated together (Belimov *et al.*, 1995;

Muthukumar *et al.*, 2001). Muthukumar *et al.* (2001) confirmed it by inoculating the neem seedlings with *G. intraradices, G. geosporum, Azospirillum brasilense* and isolated PSB individually or in various combinations under nursery conditions. They also reported that mycorrhizal colonisation, leaf area and number, plant height and biomass, nutrient content (N, P and K) and seedling quality were found to be significantly increased because of combined microbial inoculations.

Bacteria which can promote mycorrhizal development are collectively called as mycorrhiza helper bacteria (MHB) (Duponnois and Garbaye, 1991; Garbaye, 1994). But there are some inconsistencies in reports of the effects of PGPR on AM fungi as well as in their mode of action. Sometimes, these bacteria and mainly *Pseudomonas* strains that produce non-volatile diffusible compounds such as methane, acetaldehyde, acetoin and diacetyl may or may not reduce the inoculum load of mycorrhiza (Linderman, 1992; Gryndler, 2000; Aspray *et al.*, 2006). While, studies by Walley and Germida (1997) using different *Pseudomonas* strains with the co-inoculation of AM fungi observed varying effects, *i.e.*, some strains of *Pseudomonas* hindered AM fungal germination. Hence, it can be argued that not all PGPR are mycorrhizal helper bacteria (MHB) or MHB as PGPR. Only when PGPRs are found to stimulate mycorrhizal formation, then, they can be regarded as MHB (Fitter and Garbaye, 1994), this interchangeable characteristic brings about the overlap that exists between these two groups (MHB and PGPR). It is also reported that *P. fluorescens* and *P. putida* both as PGPR and biocontrol strains did not reduce the root colonisation by AM fungi (Meyer and Linderman, 1986; Paulitz and Linderman, 1991; Staley *et al.*, 1992; Burla *et al.*, 1996).

5. Interaction among *D. strictus–P. fluorescens–Glomus* sp.

Kim *et al.* (1997) studied, in non-sterile soil containing hydroxyapatite and glucose, the interaction of vesicular-arbuscular mycorrhizae (VAM) and phosphate-solubilising bacteria (PSB) on plant growth, soil microbial activities and the production of organic acids. *G. etunicatum*, a fungus and *Enterobacter agglomerans*, a bacterium able to solubilise insoluble phosphate, were used as inocula. Inoculation with E, G, or E+G had increased plant growth by days 35, 55, and 75 compared with the control. A significantly higher soluble phosphorus (P) concentration was observed in treatments E and E+G on day 55 compared with control. However, there was no significant difference in soluble P concentration in the rhizosphere between treatments with time. The P concentration was greatest in all treatments on day 55.

Gamalero *et al.* (2004) reported that two strains of *P. fluorescens* and *G. mosseae* when co-inoculated had a synergistic effect on root fresh weight of tomato plants. Moreover, co-inoculation of the three microorganisms synergistically increased plant growth compared with singly inoculated plants. Both the fluorescent pseudomonads and the myco-symbiont, depending on the inoculum combination, strongly affected the root architecture. *P. fluorescens* 92rk increased mycorrhizal colonisation, suggesting that this strain is a mycorrhization helper bacterium. Finally, the bacterial strains and the AMF, alone or in combination, improved plant mineral nutrition by increasing leaf P content. These results support the potential use of fluorescent pseudomonads and AMF as mixed inoculants for tomato and suggest that improved tomato growth could be related to the increase in P acquisition.

Muthukumar and Udaiyan (2006) studied the effect of inoculation of arbuscular mycorrhizzal fungi and PGPR on the growth of bamboo (*D. strictus*) in different tropical soils. They observed that combined inoculation of AM fungi, P solubilising bacteria (PSB) and *A. brasilense* resulted in maximum growth response under both fertilised and unfertilised conditions and soil types (alfisol and vertisol). Fertiliser application enhanced N, P and K uptake whereas, reduced their usage efficiencies. Though, soil type did not eect microbial inoculation response, but fertilizer application significantly effected the plant response to microbial inoculation.

6. Conclusion

India is a growing economy that demands sustainable resources. However, the burgeoning population, rising urbanization, static forests are some of the major challenges that have to be addressed by a nation that aspires to be sustainable as way of life. Green India Mission is one such approach. It has species-specific targets for different agro-climatic zones. Bamboos have a major share of about 13 per cent in India's forest. Moreover, by its multifarious uses, it is poor man's timber. To achieve the magic figure of 33 per cent forest cover, greening of the countryside has to be largely depending on the vast 175Mha of wastelands. The soil types and nutrient status of wastelands are critical impediment in greening/afforestation. The inorganic fertilizers cannot be solely relied upon owing to global energy crisis, their costs, limited role in soil amelioration, their fixation and loss in various ways, *etc.* A model needs to be envisaged where nutrient solubilisation and mobilization may be achieved with a effective cost cutting. Based on past information and experience, the *D. strictus-* AM model was extended to a PGPR, fluorescent pseudomonad.

References

Allen, E.B., Allen, M.F. Helm, D.J., Trappe, J.M., Molina, R. and Ricon, E. 1995. Patterns and regulation of mycorrhizal plants and fungal diversity. *Plant Soil*, 170: 47–62.

Ammani, K., Venkatswarlu, K. and Rao, A. S. 1994. Vesicular-arbuscular mycorrhiza in grasses: Their occurrence, identity and development. *Phytomorphol.*, 44: 159–168.

Anuradha J., Kumar, A., Saxena, R.K., Kamalvanshi, M. and Chakravarty, N. 2012. Effect of arbuscular mycorrhizal inoculations on seedling growth and biomass productivity of two bamboo species. *Ind. J. Microbiol.*, 52: 281–285.

Appasamy, T. and Ganapati, A. 1992. Preliminary survey of vesicular-arbuscular mycorrhiza (VAM) association with bamboos in Western Ghats. *BIC-India Bull.*, 2: 13–16.

Aspray, T.J., Eirian, J.E., Whipps, I.M. and Bending, C.D. 2006. Importance of mycorrhization helper bacteria cell density and metabolite localization for the *Pinus sylvestris-lactarius rufus* symbiosis. *FEMS Microbiol. Res.*, 56: 25–33.

Azcón-Aguilar, C. and Barea, J. 1992. Interactions between mycorrhizal fungi and other rhizosphere micororganisms. In: Allen, M.F. (Ed.) Mycorrhizal functioning. New York, Chapman and Hall Inc. pp. 163–198.

Bâ, M.A., Plenchette, C., Danthu, P., Duponnois, R. and Guissou, T. 2000. Functional compatibility of two arbuscular mycorrhizae with thirteen fruit trees in Senegal. *Agrofor. Syst.*, 50: 95–105.

Barea, J., Azcón, R. and Azcón-Aguilar, C. 2002. Mycorrhizosphere interactions to improve plant fitness and soil quality. *Antonie Van Leeuwenhoek,* 81: 343–351.

Barea, J., Pozo, M. J. and Azcón-Aguilar, C. 2005. Microbial co-operation in the rhizosphere. *J. Exp. Bot.* 56 (417):1761–1778.

Belimov, A.A., Kojemiakov, A.P. and Chuvarliyeva, C.V. 1995. Interaction between barley and mixed cultures of nitrogen fixing and phosphate-solubilizing bacteria. *Plant Soil*, 173: 29–37.

Bhattacharya, P.M., Misra, D., Saha, J. and Chaudhuri, S. 1999. Mycorrhizal dependency, phosphorus utilization efficiency and relevance of mycorrhiza for bamboo cultivation in laterite wasteland. In: Chandra, S. and Kehri, H.K. (Eds.). Biotechnology of VA Mycorrhiza: Indian Scenario. New India Publishing, New Delhi, 413p.

Biswas, S. 1988. Studies on bamboo distribution in North-eastern region of India. *Ind. For.*, 114: 514–517.

Bledsoe, C., Klier, P. and Bliss, L.C. 1990. A survey of plants on Truelove Lowlands, Devon Island, N.W.J. Canada. *Can. J. Bot.*, 68: 1848–1856.

Bouamri, R., Dalpé, Y., Serrhini, M.N. and Bennani, A. 2006. Arbuscular mycorrhizal fungi species associated with rhizosphere of *Phoenix dactylifera* L. in Morocco. *Afr. J. Biotechnol.*, 5: 510–516.

Burla, M., Goverde, M., Schwinn, F.J. and Wiemken. A. 1996. Influence of biocontrol organisms on root pathogenic fungi and on plant symbiotic micro-organisms *Rhizobium phaseoli* and *Glomus mosseae. Z. Pflanzenkrankh. Pflanzenschutz*, 103: 156-163.

Cardon, Z.G. and Whitbeck, J.L. 2007, The Rhizosphere: An ecological perspective. San Diego, Academic Press, 201p.

Dalpé, Y., Diop, T.A., Plenchette, C. and Gueye, M. 2000. *Glomales* species associated with surface and deep rhizosphere of *Faidherbia albida* in Senegal. *Mycorrhiza,* 10: 125–129.

Dames, J.F. 1991. The distribution of vesicular arbuscular mycorrhizal fungi in the savanna region of Nylsvley nature reserve in relation to soil fertility factors. Master's Thesis, University of the Witswaterand, Johannesburg, 105p.

Deogun, P.N. 1937. The silviculture and management of the bamboo *Dendrocalamus strictus* Nees, *Ind. For.*, 2: 173.

Diop, T.A., Gueye, M., Dreyfus, B.L., Plenchette, C. and Strullu, D.G. 1994. Indigenous arbuscular mycorrhizal fungi associated with *Acacia albida* Del. in different areas of Senegal. *Appl. Environ. Microbiol.*, 60: 3433–3436.

Duponnois, R. and Garbaye, J. 1991. Mycorrhization helper bacteria associated with the Douglas fir *Laccaria laccata* symbiosis: Effects *in vitro* and in glasshouse conditions. *Ann. Sci. For.*, 48: 239–251.

Fitter, A.H. and Garbaye, J. 1994. Interaction between mycorrhizal fungi and other soil organisms. In: Robson, A.D.; Abbot, L.L. and Malajckzuk, N. (Eds.). Management of mycorrhizas in agriculture, horticulture and forestry. Netherlands, Kluwer Academic Publishers, pp.47–68.

Forest Survey of India. 2011. State of forest report. Dehra Dun, FSI (www.fsi.nic. in). 210p.

Gamalero, E., Trotta, A., Massa, N., Copetta, A., Martinotti, M.G. and Berta, G. 2004. Impact of two fluorescent pseudomonads and an arbuscular mycorrhizal fungus on tomato plant growth, root architecture and P acquisition. *Mycorrhiza*, 14: 185-192.

Garbaye, J. 1994. Mycorrhiza helper bacteria: A new dimension to the mycorrhizal symbiosis. *New Phytol.*, 128: 197–210.

Gryndler, M. 2000. Interactions of arbuscular mycorrhizal fungi with other soil organism. In: Kapulnik, Y. and Douds, J.D.D. (Eds.). Arbuscular mycorrhizas: Physiology and function. Netherlands, Kluwer Academic Publishers, pp. 239–262.

Hawley, G.L. and Dames, J.F. 2004. Mycorrhizal status of indigenous tree species in forest biome of the Eastern Cape, South Africa. *South Afr. J. Sci.*, 100: 633–637.

Janos, D.P. 1987. VA mycorrhizas in humid tropical ecosystem. In: Safir, G.R. (Ed.). Ecophysiology of VA mycorrhizal plants. Boca Raton, Florida, CRC Press, pp. 107–134.

Kim, D., Jordan, G.A. and McDonald, G.A. 1997. Effect of phosphate-solubilizing bacteria and vesicular-arbuscular mycorrhizae on tomato growth and soil microbial activity. *Biol. Fert. Soils*, 26: 79–87.

Linderman, R.G. 1992. Vesicular-arbuscular mycorrhizae and soil microbial interactions. In: Bethlenfalvay, G.J. and Linderman, R.G. (Eds.). Mycorrhizae in sustainable agriculture. Madison, ASA Special Publication, pp.65–77.

Meyer, J.R. and Linderman, R.G. 1986. Response of subterranean clover to dual inoculation with vesicular-arbuscular mycorrhizal fungi and a plant growth-promoting bacterium, *Pseudomonas putida*. *Soil Biol. Biochem.*, 18: 185–190.

Miller, R.M. 1987. The ecology of vesicular arbuscular mycorrhizae in grass and shrub lands. In: Safir, G.R. (Eds.). Ecophysiology of VA mycorrhizal plants. Boca Raton, FA, CRC Press, pp. 135–170.

Mohanan, C. and Sebastian, M. 1999. Mycorrhizal status of nineteen species of bamboos in Kerala, India. In: Proceedings of the National Conference on Mycorrhiza, Institute of Microbiology and Biotechnology, Barkatullah University, Bhopal, 5-7 March, pp. 4–9.

Muthukumar, T. and Udaiyan, K. 2002. Arbuscular mycorrhizal fungal composition in semi-arid soils of Western ghats, Southern India. *Curr. Sci.*, 82: 624–628.

Muthukumar, T. and Udaiyan, K. 2006. Growth of nursery-grown bamboo inoculated with abruscular mycorrhizal fungi and plant growth promoting rhizobacteria in two tropical soil types with and without fertilizer application. *New Forests*, 31: 469–485.

Muthukumar, T., Udaiyan, K. and Manians, S. 1996. Vesicular-arbuscular mycorrhizae in tropical sedges of southern India. *Bio. Fertil. Soil.*, 22: 96–100.

Muthukumar, T., Udaiyan, K. and Rajeshkannan, V. 2001. Response of neem (*Azadirachta indica* A. juss) to indigenous arbuscular mycorrhizal fungi, phosphate-solubilizing and asymbiotic nitrogenfixing bacteria under tropical nursery conditions. *Biol. Fertil. Soil.*, 34: 417–426.

Nayar. L. 2009. Bamboo is India's 'Green Gold'. In: Indo-Asian News Service, (accessed on: 20 January, 2009).

Oehl, F., Sieverding, E., Ineichen, K., Mader, P., Boller, T. and Wiemken, A. 2003. Impact of land use intensity on the species diversity of arbuscular mycorrhizal fungi in agro-ecosystems of central Europe. *Appl. Environ. Microbiol.*, 69: 2816–2824.

Paulitz, T.C. and Linderman, R.G. 1991. Lack of antagonism between the biocontrol agent *Gliocladium vixens* and vesicular arbuscular mycorrhizal fungi. *New Phytol.*, 117: 303–308.

Picini, D. and Azcon, R. 1987. Effect of phosphate-solubilising bacteria and vesicular-arbuscular mycorrhizal fungi on the utilization of Bayovar rock phosphate by alfalfa plants using a sand vermiculite medium. *Plant Soil*, 50: 45–50.

Rawat, S. 2005. Bamboo-endomycorrhiza: Ecology, growth and macroproliferation. Ph. D. thesis, Forest Research Institute (Deemed) University, Dehradun, 200p.

Read, D.J. 1993. Mycorrhizas in plant communities. *Adv. Plant Pathol.*, 9: 1–31.

Renker, C., Blanke, V. and Buscot, F. 2005. Diversity of arbuscular mycorrhizal fungi in grassland spontaneously developed on area polluted by a fertilizer plant. *Environ. Pollut.*, 135: 255–266.

Schultz, P.A., Miller, R.M., Jastrow, J.D., Rivetta, C.V. and Bever, J.D. 2001. Evidence of a mycorrhizal mechanism for the adaptation of *Andropogon gerardii* (Poaceae) to high- and low-nutrient prairies. *Am. J. Bot.*, 88: 1650-1656.

Singh, S. and Kapoor, K.K. 1998. Effects of inoculation of phosphate-solubilizing microorganisms and an arbuscular mycorrhizal fungus on mungbean grown under natural soil conditions. *Mycorrhiza*, 7: 249–253.

Smith, S.E. and Read, D.J. 1997. Mycorrhizal symbiosis, 2nd edition, San Diego, Academic Press. 212p.

Staley, T.E., Lawrence, E.G. and Nance, E.L. 1992. Influence of a plant growth-promoting pseudomonad and vesicular-arbuscular mycorrhizal fungus on alfalfa and birds foot trefoil growth and nodulation. *Biol. Fertil. Soils*, 14: 175–180.

Stutz, J.C., Copeman, R., Martin, C.A. and Morton, J.B. 2000. Patterns of species composition and distribution of arbuscular mycorrhizal fungi in arid regions of southwestern north America, Namibia and Africa. *Can. J. Bot.*, 28: 237–245.

Thomas, T.A., Arora, R.K. and Singh, R. 1985. Genetic resources of bamboo in India: Their diversity, utilization and socio-economic role. In: Proceedings of the International Conference, Honghzou, China, October 6-14, 1985.

Toro, M., Azcon, R. and Barea, J. 1997. Improvement of arbuscular mycorrhiza development by inoculation of soil with phosphate-solubilizing rhizobacteria to improve rock phosphate bioavailability (P) and nutrient cycling. *Appl. Environ. Microbiol.*, 63: 4408–4412.

Uhlmann, E., Görke, C., Petersen, A. and Oberwinkler, F. 2006. Arbuscular mycorrhizae from arid parts of Namibia. *J. Arid Environ.*, 64: 221–237.

Walley, F.L. and Germida, J.J. 1997. Response of spring wheat (*Triticum aestivum*) to interactions between *Pseudomonas* species and *Glomus clarum* NT 4. *Biol. Fert. Soils*, 24: 365–371.

www.censusindia.gov.in/2011-Common/IntroductionToNpr.html (accessed on 24-09-2014).

2017, Mycorrhizal Fungi   Pages 87–115
Editors: Ashok Aggarwal and Kuldeep Yadav
Published by: ASTRAL INTERNATIONAL PVT. LTD., NEW DELHI

4

Distribution and Diversity of Arbuscular Mycorrhizal Fungi in Nongkhyllem Reserve Forest, Meghalaya

Vipin Parkash* and Ankur Jyoti Saikia

*Mycology and Soil Microbiology Research and Technology Laboratory,
Rain Forest Research Institute, Jorhat – 785 001, Assam
Corresponding Author: bhardwajvpnpark@rediffmail.com

ABSTRACT

Nongkhyllem Reserve Forest, Nongpoh, Meghalaya were surveyed during 2009-2012 and 117 forest plant species and their rhizospheric soil samples were collected and screened for endomycorrhizal qualitative and quantitative analysis. All the three types of root infection/colonization namely Hyphal, Vesicular and Arbuscular were observed in varying degree of occurrence. Hyphal infection/colonization was present in all the collected rhizospheric soil samples, whereas Vesicular infection was found scarce and absent in most of the root samples. The range of percentage Hyphal infection/ colonization was from 10 per cent (lowest) to 100 per cent (highest). The percentage Vesicular infection/colonization was ranging from 5 per cent (lowest) to 60 per cent (highest). The percentage of Arbuscular infection was ranging from 5 per cent (lowest) to 90 per cent (highest). The percentage colonization intensity in roots showed the range from 25 per cent (lowest) to 94.74 per cent (highest) in the roots of forest plant species. The quantification of AM spore numbers showed the range from 7/50g soil^{-1} (lowest) in to 903/50g soil^{-1} (highest). There are 14 Acaulospora spp., 5 Gigaspora spp. and only one Entrophospora, Scutellospora, Sclerocystis species each had been identified while 45 Glomus species were isolated and identified. Natural occurrence of AM fungi showed *highest frequency in case of Glomus species (64 per cent) followed by Acaulospora species (33 per cent), Entrophospora species (2 per cent), Gigaspora species (1 per cent) and Scutellospora and Sclerocystis species approximately negligible (almost 0 per cent) each. The Species Richness (unique) in Umtasoar range was ranging from 3 (lowest in Kyrdemkulia compartment) to 249.89 (highest in Umsaw compartment), while the Species Richness (unique) in Nongpoh range was ranging from 14 (lowest in Umsiling compartment) to 133.93 (highest in Leprosy Colony compartment). The Diversity Index (compartment wise) of Kyrdemkulia and Pen-Point was found with lowest value (0) while maximum was found in Ben-Point and Zero-Point (0.039) each respectively of the Umtasoar Forest Range. The Diversity Index of Umsiling compartment was found negligible (0) while maximum diversity index was found in Towar-Point (0.059) of the Nongpoh Forest Range. The Diversity Index of Umtasor range was low (0.036) than that of Nongpoh range (0.049). The arbuscular mycorrhizal status in important forest plant species occurring in NRF is discussed in detail in this article.*

Keywords: *Arbuscular mycorrhizae, Colonization, Diversity, Vesicles, Wild medicinal plants.*

1. Biodiversity of AM Fungi

Soil consists mainly of mineral, organic matter, water and gaseous phases besides hosting a dynamic population of microorganisms. One of the important members of functional soil microbial community is the *'mycorrhiza'* that forms symbiotic relationships with the roots of the higher plants. Natural ecosystems are undisturbed habitats, where the diversity of Arbuscular Mycorrhizal (AM=VAM) species is maintained on the basis of natural phenomena of ecosystem *i.e.* survival of the fittest. Mycorrhizae are believed to protect the environmental quality by enhancing beneficial biological interactions (Bhatia *et al.*, 1996). Mycorrhizae also regulate the composition and functioning of plant communities by regulating the resource allocation and growth characteristics of interacting plants (Allen, 1991). These VAM fungi are therefore, responsible for forming a potential factor in determining diversity in ecosystem (Giovannetti and Gianinazzi-Pearson, 1994).

Mycorrhizal fungal biomass can account for as much as 15 per cent of the net primary production and a majority of it enters the soil organic matter annually (Bagyaraj and Menge, 1978). Any reduction in the richness of population of Arbuscular Mycorrhizal fungi or their functional diversity could have important consequence for the equilibrium of natural plant community structure (Martins, 1993). The principal way of safeguarding biodiversity of VAM should be through protection of the ecosystem and habitat (Hawksworth, 1991). Although, these fungi show no specific variations in their range of host plants, a large variability can be seen in their population biology, ecological specificity and symbiotic activity (McGraw and Hendrix, 1984). The difference in edaphic, climatic factors and host plants among sample studies of distribution of VAM have been so different that it is difficult to explain the factors responsible for distribution and abundance of a particular species (Walker *et al.*, 1982). Diversity in Arbuscular Mycorrhizal fungi can be explored by studying spore characteristics, ultra structural features and colonization patterns (Giovannetti and Gianinazzi-Pearson, 1994). Still the great interest in VAM in recent years has promoted the numerous surveys at enumerating the species of VAM fungi in a particular region. The studies on the distribution and mycorrhizal status of plants would enable us to understand the influence of these mycobionts on plant species diversity, distribution and growth effects. The study of mycorrhizal association on tree species is important from the angle of biomass enhancement so that tree species inoculated with VAM will have better vigour and would release the pressure on reserved forests for meeting the increasing demand for fuel, fodder and timber.

Tree species differ in their dependency on mycorrhizae. There is hardly any tree species that does not form mycorrhizal association in nature. Studies revealed that most woody plants require mycorrhizal association to survive and most herbaceous plants need them to thrive (Malloch *et al.*, 1980). Most of the economically important plants and trees form endomycorrhizal associations. The subject is currently attracting much attention in agricultural, horticultural and forestry research. Desirable strains of VAM fungi are being introduced to new areas in different plants. The possible interaction between introduced and indigenous VAM species and the persistence of the introduced species is therefore of great concern. If the indigenous mycorrhizal fungi are to be managed in field conditions, the keystone protocol is to determine the colonization patterns and the functional efficiency of these fungi and to determine what interactions are occurring between the native endophyte in a natural situation before the introduction of other species. The selection of the most appropriate plant-VAM fungus association for each specific environmental and ecological situation is one of the main challenges in current research on VAM fungi. Moreover, prediction of degree of colonization of VAM fungi prior to plant establishment is essential, if inoculation with functionally efficient VAM fungi is to be considered (Singh and Adholeya, 2002). Screening of VAM fungal species is also essential, in the areas not studied to prepare a calendar of existing VAM fungi with their distribution, which will be helpful to prepare a mycorrhizal database on a National level including different geographic and agro- climatic zones.

Although, VAM fungi are known to exhibit little or no host specificity but ecological specificity may exist (McGonigle and Fitter, 1990). Therefore, a large number of surveys have been conducted by several workers in their chosen areas to find the diversity of the mycorrhizal fungi in particular selected areas. Jagpal and Mukerji (1987) surveyed semi arid areas around Delhi and reported VAM association in almost all the plants scanned. Tarafdar and Rao (1990) investigated the occurrence of VAM colonization in different tree species growing in arid zone of Rajasthan. Shanker *et al.* (1990) studied mycorrhizal association with twenty-four desert plants from different regions of Rajasthan and isolated two species of *Endogone*, two species of *Gigaspora*, six species of *Glomus*, two species of *Sclerocystis*, one species each of *Scutellospora* and *Acaulospora*. They also reported that *Glomus macrocarpum* was dominant among the isolated VAM spores. Ganesan *et al.* (1990) screened 203 soil samples for studying VAM association in coastal region and plains of Tamil Nadu and found that VAM root colonization ranged from 90 per cent to 100 per cent and also isolated twenty-three species of VAM fungi belonging to *Glomus*, *Gigaspora*, *Sclerocystis* and *Acaulospora*. Thapar *et al.* (1991) conducted survey of sodic soils of Haryana under barren conditions and identified thirteen species of *Glomus*, two species of *Sclerocystis* and one *Scutellospora* species.

Thapar *et al.* (1992) also studied the frequency and level of VAM colonization in fourteen forest tree species growing at New Forest, Dehradun and reported *Glomus macrocarpum* and *Sclerocystis coremioides* as dominant VAM spores in the rhizospheric soil of the selected tree species. Muthukumar *et al.* (1994) evaluated the mycorrhizal status of thirty-eight wild legumes in the Western Ghats region and observed that all the sampled legume species were heavily colonized by VA mycorrhizal fungi with per cent root colonization ranging from 66.65 per cent to 100 per cent and also isolated two species of *Gigaspora, Glomus, Sclerocystis* and one species each of *Acaulospora* and *Scutellospora*. Bhat *et al.* (1994) studied occurrence of VA mycorrhizae in eleven tree species of Nilgiri hill area, Tamil Nadu and noticed VAM colonization in all the ten trees except *Pinus patula* and also reported maximum mycorrhizal colonization of 88 per cent in *Tectona grandis* and minimum VAM colonization of 4 per cent in *Acacia mearnsii*. Santhaguru *et al.* (1995) surveyed VAM association with tree legumes of Alagar hills in the Eastern Ghats and noticed that mycorrhizal colonization ranged from 16 per cent to 100 per cent. In this region twenty-one species of VAM fungi were recorded of which nine belonged to *Glomus*, four to *Acaulospora*, three to *Sclerocystis*, two each of *Gigaspora* and *Scutellospora* and one belonged to *Entrophospora*. Vijaya *et al.* (1995) did survey of important tree species of Rayalaseema region of Andhra Pradesh to ascertain the status of VAM fungi and found *Acaulospora* and *Glomus* spp. predominating in Rayalaseema region.

Ravi *et al.* (1995) carried out investigations to study the nature of VAM existence in agro-forestry trees in Vamban district of Tamil Nadu and observed maximum mycorrhizal colonization in *Azadirachta indica* and *Tectona grandis* whereas moderate root colonization was observed in *Mangifera indica* and *Cassia* species and minimum root colonization was noticed in *Cordia* species. They also observed that *Glomus* and *Gigaspora* species dominated in this region. Rani *et al.* (1995) studied the VAM distribution associated with trees in Nagaon district of Assam. Mohan and Singh (1997) studied endomycorrhizal association of different *Acacia* species from Jodhpur district of Rajasthan and observed root colonization ranging from 32 per cent to 92 per cent in studied species of *Acacia*. VAM fungi belonging to genera *Glomus, Gigaspora* and *Sclerocystis* were isolated from this region. Among these genera *Glomus* was reported to be the dominant one. Mohan and Babbar (1997) carried out intensive survey of important arid and semi-aid zone tree species of Rajasthan to investigate the mycorrhizal status of these tree species. VAM root colonization in different tree species ranged from 47 per cent to 68.4 per cent and also twelve different fungal species of *Glomus* and *Gigaspora* were isolated. Among the isolated VAM species *Glomus fasciculatum* was found frequently followed by *Glomus microcarpum*. Nagabhushanam *et al.* (1999) surveyed forty forest tree species belonging to twenty genera growing in different places of Godavari belt to study their VAM association. Of all the trees with varying incidence of mycorrhizal colonization, twelve species of *Acaulospora,* sixteen species of *Glomus,* three species each of *Gigaspora* and *Scutellospora,* two species of *Entrophospora* and one species of *Sclerocystis* were isolated. Rahangdale and Gupta (1999) carried out investigations to find mycorrhizal association of forty-six tree species of Raipur, Madhya Pradesh and reported 10 per cent to 94 per cent VAM colonization varying from species to species, while the VAM fungi associated with these trees mainly belonged to genera *Glomus*, *Acaulospora* and *Gigaspora*.

Kumar *et al.* (2000) did survey in Warangal, Andhra Pradesh to study the VAM association with twenty-nine agro-forestry tree species and reported highest mycorrhizal association in *Azadirachta indica* A. Juss. and *Albizia lebbeck* (L.) Benth. Santhaguru and Sadhana (2000) studied mycorrhizal status of *Acacia* spp. from Madurai district in Tamil Nadu and reported that the extent of VAM colonization ranged from 56 per cent to 80 per cent; maximum colonization was observed in *Acacia leucophloea* (Roxb.) Willd. and nineteen species of VAM fungi were recorded, of which seven belonged to *Acaulospora* and *Glomus*, two to *Entrophospora* and one each of *Gigaspora*, *Sclerocystis* and *Scutellospora*. The VAM spore number varied from 69 to 165 spores per 50g of soil. Senapati *et al.* (2000) listed predominating VAM fungi associated with twenty-one forest tree species in Orissa. Mehrotra and Mehrotra (2000) conducted investigations to find the composition of VAM fungi in the rhizosphere of *Acacia* spp. and other forest trees and reported that in all the forest trees *Glomus* predominated followed by *Acaulospora* although the genus *Gigaspora* were frequently recorded from *Acacia* species. Ravikumar *et al.* (2001) carried out investigations to find the VAM fungal association in the rhizosphere of bamboo growing areas of Tamil Nadu and isolated a total of nineteen VAM species of which eleven were of *Glomus*, three were of *Gigaspora* and *Scutellospora* and two were of *Sclerocystis*; while AMycorrhizal colonization ranged from 51 per cent to 81 per cent and spore count ranged from 306 to 481 spores per 500g of soil samples. Panwar and Vyas (2002) undertook investigations to study the distribution of VAM fungal association with *Tamarix aphylla* growing in different areas of Rajasthan and isolated ten VAM fungal species, of which four species were of *Glomus*, two of *Gigaspora* and *Scutellospora* and one species belonged to *Sclerocystis*.

Similarly, Aggarwal *et al.* (2005a, 2005b) and Parkash *et al.* (2009) had done survey of ethno-medicinal plants of Himachal Pradesh and Haryana. They had listed VAM fungi associated with them and reported that the rhizospheres were dominated with *Glomus mosseae* only. Raj and Chandrashekar (2003) studied the association of mycorrhizal fungi with endemic dipterocarps of Sharavathi valley forest, Karnataka region. They isolated thirty-six species of mycorrhizal fungi. Among these, *Glomus* was represented by twenty species, *Acaulospora* by eight, *Scutellospora* by four, *Gigaspora* and *Entrophospora* by only two species each. The root colonization ranged from 84 per cent to 92 per cent. Uniyal (2003) isolated ten species of *Glomus*, three species of *Acaulospora*, *Scutellospora* and *Sclerocystis* and two species of *Gigaspora* from the rhizosphere of *Populus deltoides* plantations in New Forest, Dehradun.

Considering the importance of forest plant species, the present chapter has highlighted the mycorrhizal association and AM fungal population from the rhizospheres of the selected forest plant species of North- East India. These selected plant species were of multipurpose uses, economically important and frequently which are used for timber, fuel wood, fodder, sources of tannins, gums, dying agents and agro-forestry species and traditionally medicinal purposes. Information on the status of AM fungal association with the forest plant species in Nongkhyllem Reserve Forest, Nongpoh, Meghalaya is very little. Nongkhyllem Reserve Forest, with an area of 9,691 ha is located in the Ri- Bhoi district (erstwhile East Khasi Hills district) of Meghalaya, which are part of the Archaean Meghalaya Plateau (BirdLife International, 2016). The major part of the habitat is Tropical Moist Deciduous forest with patches of Tropical Semi-evergreen forest, especially in the river valleys and stream. The deciduous forests can be classified as '*Khasi hill sal*' and '*Kamrup sal*' (Champion and Seth, 1968).

Surveys of different remote sites and areas of study region *i.e.* Nongkhyllem Reserve Forest, Nongpoh, Meghalaya, India (Figure 4.1) were done for the collection of rhizospheric soil samples and plant specimens of some economically important tree/s. Rhizospheric soil samples (at least three samples) were taken by digging out a small amount of soil (500g) close to plant roots up to the depth of 15-30 cm and these samples were kept in sterilized polythene bags at 10°C for physiochemical analysis of soil, mycorrhizal colonization and quantification to see the dominance status and diversity parameters of AM fungi using standard methods and protocols (Adholeya and Gaur, 1994; Blaszkowski, 2003; Curtis and McIntosh, 1950; Gerdemann and Nicolson 1963; Phillips and Hayman, 1970; Miller *et al.,* 1987; Morton and Benny, 1990; Morton and Redecker, 2001; Morton, 2014; Mukerji, 1996; Parkash, 2012; Schenck and Perez, 1990; Sharma *et al.,* 2008, 2009; Trappe, 1982; Walker, 1992).

2. Distribution and Diversity Analysis of AM Fungal Populace

2.1. AM Colonization Status

The list of collected plant species according to different compartments of Nongkhyllem Reserve Forest, Nongpoh is shown in Table 4.1. All 117 collected rhizospheric soil samples were screened for endomycorrhizal qualitative and quantitative analysis. Different types of root colonizations in collected rhizospheric soil samples are shown in the same table. All the three types of root infection/ colonization namely Hyphal, vesicular and Arbuscular were observed in varying degree of occurrence. Hyphal infection/colonization was present in all the collected rhizospheric soil samples, whereas Vesicular infection was found scarce and absent in most of the samples. The range of percentage hyphal infection/colonization was from 10 per cent (lowest in *Cinnamomum zeylanicum*) to 100 per cent (highest in 16 samples namely *Dioscorea* sp., *Asplenium* sp., *Polypodium* sp., *Thysanoleana maxima, Hyptis suaveolens, Bidens pilosa, Dicliptera* sp., *Houttuynia cordata, Abroma augusta, Melastoma malabathricum, Solanum torvum, Alpinia allughas, Euphorbia* sp., *Grewia, Callicarpa arborea, Ficus glomerata*). The percentage Vesicular infection/colonization was ranging from 5 per cent (lowest in 12 samples namely *Smilax aspera, Pteridium* sp., *Richardia scabra, Bidesh pilosa, Hypoespes, Dicliptera* sp., *Manihot esculenta, Leea indica, Elaeocarpus floribundus, Ehretia* sp., *Cassia occidentalis, Artocarpus chaplasha, Caryota urens*) to 60 per cent (highest in *Mussaenda frondosa*). The percentage of Arbuscular infection was ranging from 5 per cent (lowest in 8 samples namely *Dictiptera* sp., *Crassocephalum crepidioides, Carissa* sp., *Ehretia* sp., *Cassia occidentalis,*

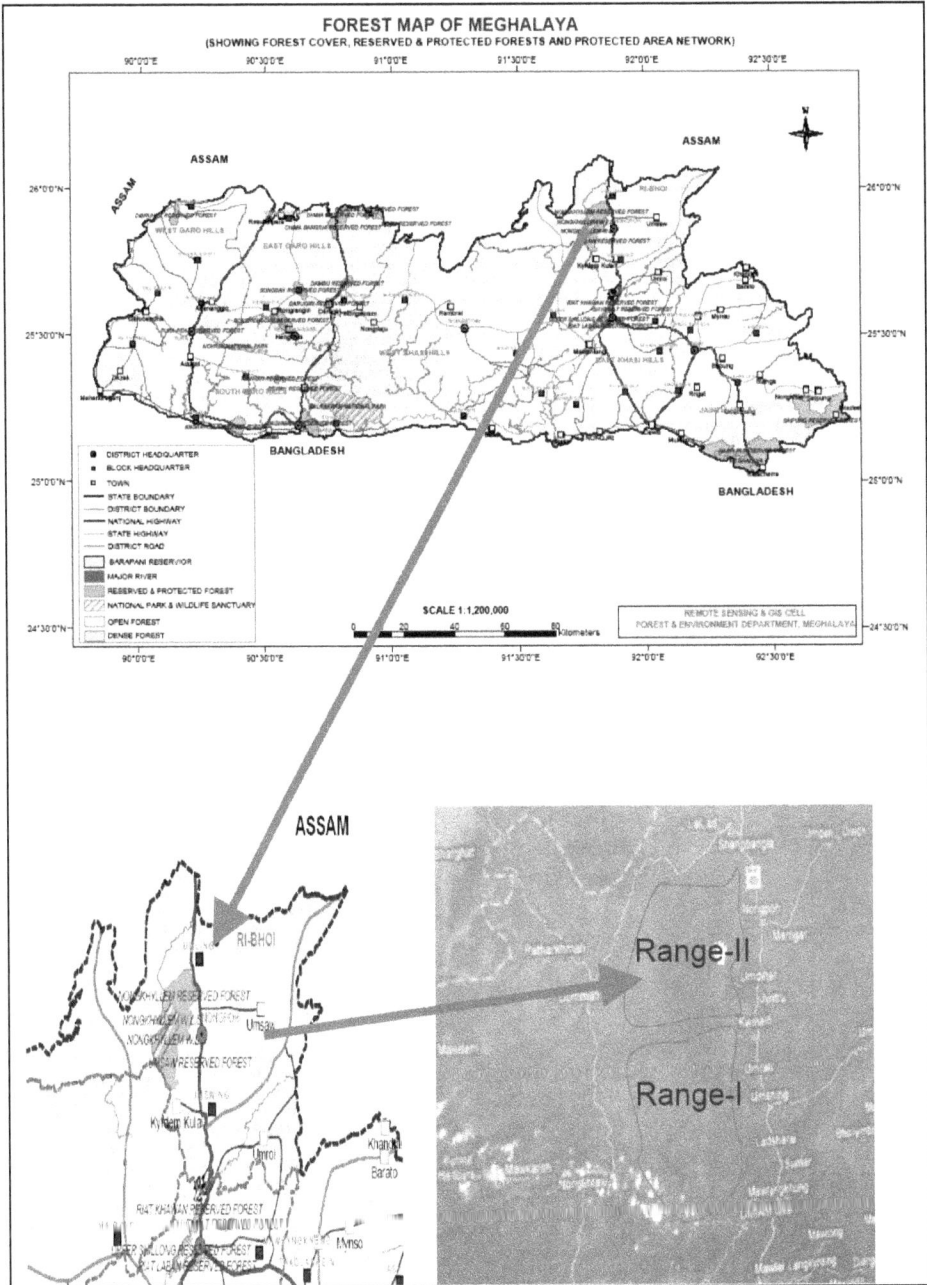

Figure 4.1: Study Site, Nongkhyllem Reserve Forest, Nongpoh, Meghalaya, India.

Table 4.1: Different Types of Root Colonization in some Useful Plants of Nongkhyllem Reserve Forest, Nongpoh, Meghalaya

Sl.No.	Botanical Names	Local Names	Habit	Family	Per cent Hypal Infection	Per cent Arbuscular Infection	Per cent Vesicular Infection	Per cent VAM Root Colonization	Per cent Colonization Intensity	VAM Spore (/50 gmsoil)
1.	Castanopsis indica (Roxb.)DC.	Khasi badam	Tree	Fagaceae	45±1.41	0	35±0.72	60±0.94	48±0.45	255±4.08
2.	Sterculia villosa Roxb.	Dieng-star	Tree	Sterculiaceae	57.14±1.19	7.14±0.27	28.6±0.72	64.28±1.24	45.45±0.07	673±4.89
3.	Melastoma malabathricum Linn.	Dieng-soh-khing	Shrub	Melastomaceae	100±0.27	1±0.27	0	100±0.27	68.75±0.53	605±8.16
4.	Curcuma pseudomontana Graham	*	Herb	Zingiberaceae	28.6±0.72	0	42.9±0.82	50±0.98	50±0.31	155±22.04
5.	Aquilaria agallocha Roxb.	Makhi sal	Tree	Thymelaeaceae	20±0	5±0.27	0	25±0.27	40±0.34	194±2.45
6.	Artocarpus chaplasha Roxb.	Dieng-soh-ram	Tree	Artocarpaceae	70±0.27	30±0.47	5±0.27	75±0.47	51.6±0.2	476±21.23
7.	Mussaenda frondosa Sensu G. Forst., non L.	*	Shrub	Rubiaceae	55±0.82	25±0.54	60±0.72	85±1.09	57.35±0.96	651±20.55
8.	Garcinia sp. Linn.	*	Tree	Clusiaceae	85±0.27	35±0.72	20±0.72	90±0.47	91.67±0.24	378±12.25
9.	Shorea robusta C.F. Gaertn.	Dieng-blei (K), Bolsal, Borsal (G)	Tree	Dipterocarpaceae	70±0.72	0	0	70±0.72	71.23±0.54	130±4.08
10.	Smilax sp. Linn.	*	Shrub	Smilacacea	30±0.54	0	0	30±0.54	29±0.65	560±4.08
11.	Emblica officinalis Gaertn.	Dieng-sohmylleng (K)	Tree	Euphorbiaceae	25±0.27	5±0.27	0	25±0.27	25±0.14	372±1.63
12.	Lagerstroemia parviflora Roxb.	Dieng-lang-sing (K)	Tree	Lythraceae	50±0.27	10±0.27	0	20±0.27	31.8±0.18	630±26.13
13.	Solanum khasianum C.B. Clarke	*	Shrub	Solanaceae	60±0	40±0.27	0	70±1.19	46.43±0.56	903±1.63
14.	Sida cordifolia Linn.	*	Shrub	Malvaceae	20±0.27	0	0	20±0.27	31.25±0.18	566±10.61
15.	Solanum torvum Sw.	Katahi Bengelr	Shrub	Solanaceae	100±0.27	0	0	100±0.27	68.75±0.53	743±7.34
16.	Asplenium sp.	*	Herb	Aspleniaceae	100±0.27	70±0.98	30±0.47	100±0.27	80±0.36	191±4.08
17.	Mesua ferrea Linn.	Dieng-ngai (K), Khimdi (G)	Tree	Clusiaceae	20±0.27	0	0	20±0.27	30±0.34	301±5.71

Contd...

Table 4.1—Contd...

Sl.No.	Botanical Names	Local Names	Habit	Family	Per cent Hypal Infection	Per cent Arbuscular Infection	Per cent Vesicular Infection	Per cent VAM Root Colonization	Per cent Colonization Intensity	VAM Spore (/50 gmsoil)
18.	*Spilanthes paniculata* Linn.	*	Herb	Asteraceae	45±0	45±0.47	0	80±0.27	66.7±0.54	542±5.71
19.	*Eryngium foetidum* Linn.	Dhania-khlaw	Herb	Apiaceae	70±0.27	0	0	80±0.27	43.75±0.18	476±4.08
20.	*Polystichium braunii*	*	Herb	Dryopteridaceae	80±0.27	0	30±0	80±0.27	66.7±0.54	236±3.56
21.	*Polypodium* sp.	*	Herb	Polypodiaceae	100±0.27	70±0.72	15±0.82	100±0.27	80±0.36	439±5.71
22.	*Pteridium* sp. Gleditsch ex Scop.	*	Herb	Dennstaedtiaceae	60±0.47	10±0.27	5±0.27	60±0.47	62.5±0.66	366±10.61
23.	*Alpinia allughas* (Retz.) Rose	Tara plant	Shrub	Zingiberaceae	100±0.27	20±0.54	0	100±0.27	68.75±0.53	175±7.35
24.	*Callicarpa arborea* Roxb.	Lakhiat	Tree	Verbenaceae	100±0.27	90±0.47	0	100±0.27	70±0.61	471±11.89
25.	*Thysanoleana maxima* (Roxb.) O. Ktze	Thadu plant	Shrub	Poaceae	100±0.27	50±1.09	10±0.27	100±0.27	72.5±0.6	464±3.26
26.	*Manihot esculenta* Crantz.	*	Scrub	Euphorbiacae	90±0.47	10±0.54	5±0.27	90±0.54	35.52±0.36	504±2.45
27.	*Ficus glomerata* Linn.	Tejmuri (Assamese)	Tree	Moraceae	100±0.27	30±0.82	20±0.72	100±0.27	70±0.61	582±4.89
28.	*Pueraria tuberosa* (Roxb. ex Willd.) DC.	*	Climber	Papilionaceae	80±0.27	40±0.72	10±0.54	85±0.27	35.29±0.24	530±4.08
29.	*Pteris cretica* Linn.	*	Herb	Adiantaceae	40±0.72	0	0	40±0.72	25±0.18	497±1.63
30.	*Cocculus* sp.	*	Climber	Menispermaceae	95±0.54	55±1.09	10±0.54	95±0.54	65.79±1.24	297±4.9
31.	*Nephrodium* sp.	*	Herb	Dryopteridaceae	85±0.27	35±0.54	0	85±0.27	41.18±0.38	406±6.53
32.	*Adiantum tenerum* Sw.	*	Herb	Pteridaceae	90±0.47	55±0.98	30±1.25	90±0.47	63.89±0.99	276±8.18
33.	*Bridelia retusa* Spreng.	Jati	Scrub	Euphorbiaceae	15±0.47	0	0	20±0.27	25±0.07	121±0.82
34.	*Mimosa himalayensis* Gamble	Sohshih	Shrub	Mimosaceae	90±0.27	45±0.47	10±0.27	85±0.27	52.63±0.53	112±2.05
35.	*Crotolaria alata* Buch.-Ham. ex D.Don	Turin	Herb	Fabaceae	50±0.54	0	0	50±0.27	45±0.42	200±14.29

Contd...

Table 4.1–Contd...

Sl.No.	Botanical Names	Local Names	Habit	Family	Per cent Hypal Infection	Per cent Arbuscular Infection	Per cent Vesicular Infection	Per cent VAM Root Colonization	Per cent Colonization Intensity	VAM Spore (/50 gmsoil)
36.	Castonopsis concinna (Champ. ex Benth.) A.DC.	Sohot	Tree	Fagaceae	95±0.54	30±1.25	25±0.27	95±0.54	78.95±0.62	59±1.25
37.	Boehmeria macrophylla Hornem	Sohbyrthid	Shrub	Urticaceae	93.75±0.27	25±0.72	25±0.72	100±0.27	40.62±0.38	181±4.9
38.	Uncaria sp. Schreb.	Jermi	Climber	Rubiaceae	50±0.98	10±0.27	10±0.27	60±0.94	58.33±0.53	277±1.63
39.	Festuca pratensis Huds.	*	Herb	Poaceae	95±0.54	30±0.54	45±0.47	95±0.54	55.26±0.12	394±3.26
40.	Leea indica (Burm.)Merr	*	Shrub	Lauraceae	95±0.27	35±0.72	5±0.27	95±0.27	52.63±0.36	444±3.26
41.	Duabanga grandiflora Roxb. ex DC	Diengbai	Tree	Sonneratiaceae	80±1.36	35±1.19	25±0.27	80±1.36	65.62±1.16	226±4.08
42.	Boehmeria nivea (L.) Gaudich	*	Shrub	Urticaceae	90±0	45±0.47	10±0.27	90±0.27	60.53±0.38	130±4.08
43.	Begonia sp. Linn.	Jajew	Herb	Begoniaceaea	80±0.27	10±0.27	0	80±0.27	75±0.36	322±4.89
44.	Dioscorea sp. Linn.	Sohksiew	Climber	Dioscoriaceae	100±0.27	80±0.54	25±0.54	100±0.27	85±0.54	143±2.45
45.	Costus speciosus (J. Koenig) Sm.	*	Shrub	Costaceae	50±0.27	30±0	0	50±0.27	30±0.12	257±6.53
46.	Crotalaria juncea L.	*	Shrub	Leguminosae	95±0.27	75±0.47	30±0.47	95±0.27	94.74±0.24	254±3.26
47.	Elephantopus scaber Linn.	*	Herb	Asteraceae	85.75±0.82	42.85±0.94	7.14±0.27	85.75±0.82	58.3±0.76	127±3.26
48.	Richardia scabra Linn.	*	Herb	Rubiaceae	60±0.47	25±0.27	5±0.27	75±0	50±0.24	228±4.08
49.	Zanthoxylum armatum DC	Dieng-sohkhlam (K)	Scrub	Rutaceae	20±0.54	0	0	20±0.54	25±0.14	49±2.45
50.	Abrus fruticulosus Wight & Arn.	*	Shrub	Fabaceae	68.42±0.54	31.58±0.47	15.79±0.47	84.21±0.27	56.25±0.2	230±4.9
51.	Holmskioldia sanguinea Retz	*	Shrub	Verbenaceae	95±0.54	25±0.72	50±0.27	95±0.54	65.79±0.58	56±4.9
52.	Semecarpus anacardium Linn. f.	*	Tree	Anacardiaceae	50±1.19	40±0.72	20±0.54	55±0.72	71.23±0.54	347±4.89
53.	Hyptis suaveolens (Linn.) Poit.	Bontulsi	Herb	Lamiaceae	100±0.27	90±0.47	40±0.27	100±0.27	80±0.36	184±3.26
54.	Schima wallichii (DC.) Korth.	Dieng-nganbuit	Tree	Theaceae	90±0	75±0.47	40±0.72	90±0	91.67±0.24	448±7.34

Contd...

Table 4.1–Contd...

Sl.No.	Botanical Names	Local Names	Habit	Family	Per cent Hypal Infection	Per cent Arbuscular Infection	Per cent Vesicular Infection	Per cent VAM Root Colonization	Per cent Colonization Intensity	VAM Spore (/50 gmsoil)
55.	Leucas cephalotes (Roth) Spreng.	*	Herb	Lamiaceae	95±0.54	80±0.54	10±0.27	100±0.27	72.5±0.6	106±4.08
56.	Bidens pilosa Linn.	*	Herb	Asteraceae	100±0.27	50±1.36	5±0.27	100±0.27	65±0.59	168±3.26
57.	Phyllanthus glaucus Wall. Ex Mull.Arg.	*	Shrub	Phyllanthaceae	100±0.27	75±0.94	10±0.27	100±0.27	68.75±0.53	237±4.89
58.	Dicliptera sp Juss.	*	Herb	Acanthaceae	100±0.27	85±0.72	0	100±0.27	80±0.59	202±5.71
59.	Lepidagathis hyalina Nees	*	Herb	Acanthaceae	85±0.27	55±0.27	0	85±0.27	73.53±0.92	112±2.45
60.	Achyranthes sp Linn.	*	Herb	Amaranthaceae	25±0.27	0	0	25±0.27	40±0.18	286±3.26
61.	Urena picta Linn	*	Herb	Malvaceae	44.44±0.72	55.55±0.27	44.44±0.98	88.88±0.54	59.375±0.34	170±4.89
62.	Selaginella exalta (Kunze) Spring.	*	Herb	Selaginellaceae	15±0.47	0	0	20±0.54	31.25±0.18	130±2.45
63.	Carissa sp. Linn.	*	Shrub	Apocynaceae	45±0.82	5±0.27	0	70±0.54	46.43±0.56	112±4.08
64.	Pteris umbrosa R.Br.	*	Herb	Adiantaceae	45±0.47	20±0.72	0	60±0.47	62.5±0.66	120±2.45
65.	Crepis japonica (L.) Benth.	*	Herb	Asteraceae	85±0.72	40±0.27	20±0.72	95±0.54	84.21±0.68	122±2.05
66.	Hypoestes sp. Sol. ex R.Br.	*	Herb	Labiatea	75±0.94	30±1.25	5±0.27	85±0.72	67.65±0.92	61±3.26
67.	Salix sp. Linn.	*	Tree	Salicaceae	90±0.47	25±0.72	30±0.82	100±0.27	70±0.6	72±2.45
68.	Uncaria sessilifructus Wall. Ex Roxb.	*	Climber	Rubiaceae	65±0.54	15±0	0	80±0.72	59.36±0.41	159±2.45
69.	Macaranga denticulata Muell.Arg.	Dieng-lakhor	Tree	Euphorbiaceae	85±0.27	10±0.27	35±0.27	85±0.27	61.76±0.47	108±3.26
70.	Phlogacanthus thyrsiflorus (Roxb.) Nees	Tadang-kakseu, Dieng-soh-kajut	Shrub	Acanthaceae	25±0.54	0	0	25±0.54	25±0.14	63±1.63
71.	Ficus elastica Roxb. ex Hornem	Dieng-jiri (K), Phrap-ramkhet (G)	Tree	Moracaea	25±0.54	0	0	25±0.54	40±0.34	123±4.08
72.	Salix daphnoides Vill.	Padem-dieng	Tree	Salicaceae	15±0.47	0	0	15±0.47	25±0.12	45±1.63

Contd...

Table 4.1–Contd...

Sl.No.	Botanical Names	Local Names	Habit	Family	Per cent Hypal Infection	Per cent Arbuscular Infection	Per cent Vesicular Infection	Per cent VAM Root Colonization	Per cent Colonization Intensity	VAM Spore (/50 gmsoil)
73.	Elaeocarpus floribundus Bl.	Lamsu	Shrub	Elaeocarpaceae	60±0.94	20±0.72	5±0.27	70±0.54	60.71±0.65	45±3.26
74.	Alangium platanifolium Linn.	Lobong-kakseu	Scrub	Alangiaceae	15±0.47	10±0.27	0	40±0.72	31.25±0.25	66±0.82
75.	Toona ciliata Roem.	Dieng-Sali	Tree	Meliacaea	21.42±0.47	0	0	21.42±0.47	25±0.12	49±0.82
76.	Litsea sp. Lam.	*	Scrub	Lauraceae	20±0.47	0	20±0.47	33.33±0.27	30±0.12	345±3.26
77.	Paspalidium sp. Stapf	Tuilaasu bon	Herb	Poaceae	95±0.54	50±0.27	30±0.47	95±0.54	65.79±0.85	417±2.45
78.	Dendrocalamus hamiltonii Nees and Arn. Ex Murno	Seij-lai	Herb	Poaceae	53.33±0.27	0	0	73.33±0.27	40.91±0.12	499±4.9
79.	Cinnamomum tamala (Buch.-Ham.) T.Nees and Nees	Dieng-syiem	Tree	Lauraceae	10±0.27	0	0	20±0.27	25±0.07	229±3.26
80.	Eheretia sp. Linn.	*	Shrub	Boraginaceae	70±0.98	5±0.27	5±0.27	70±0.98	57.14±0.98	211±4.9
81.	Cyperus involculatus Linn.	*	Herb	Cyperaceae	70±0.27	25±0.27	0	70±0.27	67.86±0.54	202±4.08
82.	Terminalia bellirica (Gaertn.) Roxb.	Dieng-sohkhoru (K)	Tree	Combretaceae	56.25±0.82	31.25±0.27	0	75±0.82	47.91±0.45	247±4.9
83.	Terminalia arjuna (Roxb.) Wight and Arn.	*	Tree	Combretaceae	70±0.27	10±0.27	0	80±0.27	50±0.18	113±5.71
84.	Smilex aspera Linn.	*	Climber	Smilacaceae	85±0.72	55±0.27	5±0.27	85±0.72	79.41±0.51	101±1.63
85.	Michelia champaca Linn.	Shap (K)	Tree	Magnoliaceae	52.94±0.47	0	5.89±0	64.7±0.54	45.45±0.07	107±4.08
86.	Tectona grandis Linn.f.	Dieng-rang	Tree	Verbenaceae	85.71±0	0	0	92.85±0.27	46.15±0.24	132±2.6
87.	Chlorophytum sp. Ker Gawl.	*	Herb	Liliaceae	81.25±0.27	0	0	87.5±0.27	35.71±0.14	105±3.26
88.	Amorphophallus Blume ex Decne.	*	Herb	Araceae	35±0.54	0	0	35±0.54	25±0.14	88±2.86
89.	Impatiens balsamina Linn.	*	Herb	Balsaminaceae	60±1.25	0	0	65±0.98	50±0.82	75±3.26
90.	Boechmeria sp. Jacq.	*	Herb	Urticaceae	25±0.27	0	0	33.33±0.27	30±0.12	82±4.08

Contd...

Table 4.1–Contd...

Sl.No.	Botanical Names	Local Names	Habit	Family	Per cent Hypal Infection	Per cent Arbuscular Infection	Per cent Vesicular Infection	Per cent VAM Root Colonization	Per cent Colonization Intensity	VAM Spore (/50 gmsoil)
91.	Peperomia pellucida (Linn.) Kunth	*	Herb	Piperaceae	95±0.35	60±0.72	30±0.65	85±0.52	82±0.2	43±1.63
92.	Centella asiatica (L.) Urban	Badmaina	Herb	Apiaceae	90±0.27	70±0.27	20±0.27	80±0.72	85±0.36	26±4.08
93.	Abroma augusta (Linn.) Linn. f.	Dieng-tyrkhum	Scrub	Sterculiaceae	100±0.27	35±0.47	10±0.54	100±0.27	68.75±0.53	163±2.45
94.	Houttuynia cordata Thunb.	Ja-myrdoh	Herb	Saururaceae	100±0.27	70±0.54	15±0	100±0.27	80±0.41	37±4.9
95.	Bauhinia purpurea Linn.	Dieng-long (K) Bol-Megong(G)	Shrub	Ceasalpineaceae	50±0.47	25±0.27	0	66.66±0.27	56.25±0.62	35±1.63
96.	Ficus sp. Linn.	*	Tree	Moraceae	66.66±0.27	13.33±0.54	0	66.66±0.27	37.5±0.31	53±2.86
97.	Phyllanthus fraternus G.L. Webster	*	Herb	Phyllantheceae	50±1.09	6.25±0.27	0	62.5±0.98	40±0.45	41±3.26
98.	Clerodendrum sp. Linn	*	Shrub	Verbenaceae	35±0.72	0	0	40±0.98	25±0.25	228±5.71
99.	Cassia obtusifolia (L.) Irwin and Barneby	Amoora	Shrub	Caesalpiniaceae	40±0.54	0	0	60±0	33.33±0.14	104±4.08
100.	Roylea cineria (D. Don) Baillon	*	Shrub	Labiatae	85±0.27	0	0	95±0.27	55.26±0.47	149±6.53
101.	Grewia sp. Linn	*	Shrub	Tiliaceae	100±0.27	0	0	100±0.27	71.4±0.67	142±4.08
102.	Cassia occidentalis Linn	*	Shrub	Caesalpiniaceae	50±0.72	5±0.27	5±0.27	60±0.47	41.66±0.54	49±1.63
103.	Selaginella selaginoides (L.) P. Beauv. ex Mart. and Frank	*	Herb	Selaginellaceae	55±0.54	25±0.27	0	75±0.47	66.67±0.54	21±4.08
104.	Trevesia sp.	Phunlut	Tree	Araliaceae	18.75±0.47	0	0	18.75±0.47	33.33±0.18	36±4.49
105.	Phoebe attenuata (Nees) Nees	Bonsum	Tree	Lauraceae	40±0.47	6.66±0.27	0	60±0.47	50±0.42	24±4.08
106.	Celtis orientalis Linn.	*	Tree	Ulmaceae	46.66±0.72	20±0.47	0	60±0.47	38.8±0.29	09±0.81
107.	Caryota urens Linn.	Dieng-klai	Tree	Arecaceae	45±0.47	50±0.27	5±0.27	80±0.54	51.56±0.2	07±1.63
108.	Smilax Linn.	*	Climber	Smilacaceae	87.5±0.72	37.5±0.47	6.25±0.27	87.5±0.72	53.6±0.66	20±4.89
109.	Dicliptera sp. Juss.	*	Herb	Acanthaceae	65±0.98	5±0.27	5±0.27	90±0.47	55.55±0.68	41±2.45

Contd...

Table 4.1—Contd...

Sl.No.	Botanical Names	Local Names	Habit	Family	Per cent Hypal Infection	Per cent Arbuscular Infection	Per cent Vesicular Infection	Per cent VAM Root Colonization	Per cent Colonization Intensity	VAM Spore (/50 gmsoil)
110.	*Celastrus paniculatus* Willd.	*	Herb	Celastraceae	65±0.54	20±0.27	0	65±0.54	69.23±0.54	18±3.26
111.	*Albizia chinensis* (Osb) Merr.	*	Tree	Leguminosae	55±0.54	5±0.27	0	55±0.54	31.8±0.18	19±3.26
112.	*Rubia* sp. Linn.	*	Shrub	Rubiaceae	95±0.54	0	0	100±0.27	73.75±0.29	35±2.44
113.	*Commelina* sp. Linn.	*	Herb	Commelinaceae	33.3±0.72	0	0	50±0.47	50±0.31	41±1.63
114.	*Michelia* sp. Linn.	*	Tree	Magnoliaceae	50±0.72	0	0	90±0.47	72.2±0.36	30±2.44
115.	*Indigofera hebepetala* (Benth)	*	Shrub	Fabaceae	85±0.72	20±0.72	10±0.27	85±0.72	57.35±0.96	09±2.44
116.	*Rubus* sp. Linn.	*	Shrub	Rosaceae	61.11±0.27	5.55±0.27	0	77.77±0.27	41.07±0.56	179±2.45
117.	*Crassocephalum crepidioides* (Benth.) S. Moore	*	Herb	Asteraceae	55±1.44	5±0.27	10±0.27	70±1.09	55.35±1.07	174±4.08

*: Local names could not be traced out or retrieved; ±Standard error of mean (three replications).

Aquilaria agallocha, Emblica officinalis, Albizia chinensis) to 90 per cent (highest) in 2 samples namely *Hyptis suaveolens, Callicarpa arborea* respectively.

Endomycorrhizal quantification and per cent colonization in roots of collected useful plants of Nongkhyllem Reserve Forest, Nongpoh is also shown (Table 4.1). The percentage endomycorrhizal colonization in roots was found lowest (10 per cent) in *Cinnamomum zeomumylanicum* and highest (100 per cent) in 20 samples namely *Dioscorea* sp., *Sida cordifolia, Polystichum* sp., *Pteridium* sp., *Richardia scabra, Hyptis suaveolens, Callicarpa arborea, Leucas cephalotus, Biden pilosa, Centella asiatica, Litsea* sp., *Abroma augusta, Solanum torvum, Alpinia allughas, Boechmeria malabarica, Euphorbia* sp., *Grewia, Rubia* sp., *Callicarpa arborea, Ficus glomerata, Smilax* sp. The study of percentage colonization intensity in roots showed the results ranging from 25 per cent (lowest) in *Pteris cretica, Amorphophallus* sp., *Bridelia retusa, Zanthoxylum armatum, Carissa* sp., *Bauhinia purpurea, Emblica officinalis, Salix daphnoides, Toona ciliata, Cinnamomum zeylanicum* to 94.74 per cent (highest) in *Costos speciossus*. The quantification of AM spore numbers showed the range from 7 (lowest) in *Caryota urens* to 903 (highest) in *Solanum khasianum*.

Per cent colonization in roots was also low in scrubs and trees than herbs and shrubs where as climbers were having the high total colonization in roots. Shrubs and trees were having more AM spore numbers than scrubs and trees. Per cent colonization intensity and per cent Hyphal infection in roots in case of herbs, shrubs and climbers were high than trees and scrubs. Per cent Vesicular infection was also low in trees and shrubs than herbs. Per cent Arbuscular infection was again low in case of scrubs and trees than shrubs, herbs and climbers but trees were having more per cent organic carbon than shrubs, herbs and climbers. The soil temperature in scrubs and trees was high than shrubs, climbers and herbs.

The variation in AM fungal colonization pattern within the same species in different location was earlier noticed by (Lakshmipathy *et al.*, 2003) in their study with medicinal plants in Karnataka. The variation of AM fungal root colonization in different locations could be due to the change in the habitat, environmental factors, soil fertility and acclimatization of a particular AM fungal genus/species to a particular location (Brundrett, 1991).This study also reveals that AM Fungal colonization pattern is related to soil pH and available phosphorous in the soil. AM Fungal colonization was high in slightly acidic soils compared to the neutral and alkaline soils. Koide (1991) had observed a hindrance in colonization of AM fungi at very low (<3 ppm) and high (>9 ppm) phosphorus content of soil will hinder colonization of AM fungi. According to Sumana (1998), the acid phosphatase activity increases with increased colonization by AM fungi. AMF colonization is known to alter the inherent phosphorous supply by increasing the phosphatase activity in the rhizosphere (Allen *et al*, 1995)

The variation in colonization could be related to the conclusion by other workers (Brundrett and Kendrick, 1988; Sander and Fitter, 1992) when they found a great range in VAM infection density of the roots of coexisting species. The variation in VAM infection could be a result of environmental pressures that determine the growth of plant roots or the fungus, the survival strategies that adaptive by the different plant species or the fungus distribution in space and time. At the other, the variation in VAM infection could be a result of highly controlled phenomenon determined by plant, fungus or their interaction in response either to the flow of materials across the plant-fungus interface or to some other benefit or cost being exchanged. These events are different by the different plant and fungus species that reflect the symbiotic association.

Competition for a limited resource is one of the strongest selective pressures operative in native plant communities (Read, 1991). The observations of Fitter (1977) that mycorrhizal colonization increases under competitive conditions and those of Fitter and Merryweather (1992) that a clear benefit - an increase in phosphorus inflow – is correlated with colonization under field conditions. Mycorrhizal associations can be beneficial because colonized hosts have improved access to limited nutrients and water when compared to nonmycorrhizal plants. However, it must be noted that increased mycorrhizal colonization also may be due to the increase in the number of roots present in soils supporting several species.

The range of *p*H in collected rhizospheric soil samples was varying from 4.89 (lowest in *Amorphophallus* sp.) to 8.89 (highest in *Centella asiatica*). In general, it is observed that scrubs, herbs and shrubs had high *p*H values than trees with low *p*H values. The *p*H value also affects the AM colonization. Survival of VA mycorrhizal fungi and subsequent spore germination may depend on a species' adaptation and on the influence of physical parameters of the soil such as *p*H (Green *et al.*, 1976). Powell and Bagyaraj (1984) concluded that *p*H can influence spore germination in VAM fungal species, and that spore germination occurs within a range that is acceptable for plant growth. Friese and Koske (1991) found no significant correlation between VA mycorrhizal fungal spore clumping and soil *p*H. Bagyaraj (1992) points out that the interpretation of a *p*H effect on VAM fungal spore germination is difficult because many chemical properties of soil vary with changes in *p*H. Soil *p*H within a range of 4.8-8.0 significantly influenced spores germination of *Glomus epigaeum*; optimum germination occurred at *p*H 7.0 (Daniels and Trappe, 1980).

3. Natural Occurrence Range of AM Fungal Spore, including their Ecological Indices

The natural occurrence and diversity of *Acaulospora, Gigaspora, Entrophospora, Scutellospora, Sclerocystis* sp. and *Glomus* species has been discussed. There are 14 *Acaulospora* sp., 5 *Gigaspora* sp. and only one *Entrophospora, Scutellospora, Sclerocystis* species each had been identified from the collected rhizospheric soil samples of Nongkhyllem Reserve Forest (NRF), Nongpoh. While 45 *Glomus* species were isolated and identified with the help of available literatures, flora and manuals in the laboratory. Natural occurrence of VAM fungi in NRF showed highest frequency in case of *Glomus* species (64 per cent) followed by *Acaulospora* species (33 per cent), *Entrophospora* species (2 per cent), *Gigaspora* species (1 per cent) and *Scutellospora* and *Sclerocystis* species approximately negligible (almost 0 per cent) each. Density, frequency (per cent) and abundance was studied sample and compartment wise. Among the *Acaulospora* species, *Acaulospora laevis* had the highest frequency of occurrence (27 per cent) and *Acaulospora trappei* and *Acaulospora capsicula* had negligible frequency of occurrence (almost 0 per cent) each. In case of natural occurrence of different species of *Gigaspora*, *Gigaspora rosea* were having maximum natural occurrence (43 per cent) followed by *Gigaspora nigra* (15 per cent) and *Gigaspora gregaria* and *Gigaspora margarita* (14 per cent) each respectively, whereas *Entrophospora* species had 14 per cent natural occurrence. The natural occurrence of different *Glomus* species showed that *Glomus clavisporum* (27 per cent), *Glomus mosseae* (19 per cent) and *Glomus macrocarpum* (19 per cent) had maximum natural occurrence whereas *Glomus callosum*, *Glomus diaphanum*, *G. formasanum*, *G. verruculosum*, *G. maculosum*, *G. pansihalos*, *G. microcarpum* had lowest/negligible natural occurrence (0.1 to 2.45 per cent) respectively each.

A. laevis (7.36, 90.91, 8.10), *G. mosseae* and *G. macrocarpum* (6.0, 81.82, 7.33), *G. clavisporum* (5.64, 72.73, 60.75), *G. geosporum* (3.91, 81.82, 4.73), *G. maculosum* (2.00, 62.0, 3.14) and *G. multicaule* and *G. albidum* (0.73, 49.45, 1.60) were having high density, frequency (per cent) and abundance in the rhizosphere of NRF respectively.

The Species Richness (unique) in Umtasoar range was ranging from 3 (lowest in Kyrdemkulia compartment) to 249.89 (highest in Umsaw compartment), while the Species Richness (unique) in Nongpoh range was ranging from 14 (lowest in Umsiling compartment) to 133.93 (highest in Leprosy Colony compartment). The Diversity Index (compartment wise) of Kyrdemkulia and Pen-Point was found with lowest value (0) while maximum was found in Ben-Point and Zero-Point (0.039) each respectively of the Umtasoar Forest Range. The Diversity Index of Umsiling compartment was found lowest (0) while maximum diversity index was found in Towar-Point (0.059) of the Nongpoh Forest Range. The Diversity Index of Umtasor range was 0.036 and that of Nongpoh range was 0.049. The correlation coefficient (r) between Species Richness and Diversity Index was 0.54 (positive correlation). It suggests that Species Richness is positively correlated with Diversity Index. The similarity index of Nongkhyllem Reserve Forest, Nongpoh was found 0.77 while dissimilarity index was 0.23.

There are many factors that could affect spore density and species richness in a given host rhizosphere. Values for arbuscular mycorrhizal fungal spore density associated with different plants at different sites have varied greatly in previous reports (Walker *et al.*, 1982; Sylvia, 1986; Koske, 1987). Seasonality, edaphic factors, host-dependence, age of the host plants, the sporulation abilities of arbuscular mycorrhizal fungi, and the dormancy and the distribution patterns of arbuscular mycorrhizal fungal spores in the soils, have been reported previously (Walker *et al.*, 1982; Sylvia, 1986; Koske, 1987; Gemma and Koske, 1988; Bever *et al.*, 1996; Zhao *et al.*, 2003; Mangan *et al.*, 2004). Guadarrama and Alvarez-Sanchez (1999) reported that disturbance, but not seasonality, affects the abundance and richness of mycorrhizal spores in a tropical wet forest in Mexico. The uneven spatial distribution (clumped distribution) of arbuscular mycorrhizal fungal spores and the complex below ground structure of tropical rainforests are major factors that affect the spore density (Zhao *et al.*, 2001). Similar results were obtained by Basumatary *et al.* (2015) along with Parkash and Saikia (2015). However NRF Meghalaya, disturbance from anthropogenic factors resulted in scantiness with respect to AM fungal composition as well as diversity.

Similarly, Koske (1987) found 8 to 14 different VAM species in the roots of *Ammophila reiligulata, Solidago sempervirens* and *Uniola paniculata* along transect on the barrier dunes of the United States. Also, Rosendahl *et al.* (1989) were able to identify four different mycorrhizal types in the roots of several coexisting plants, while Sanders and Fitter (1992) found that different VAM fungal species sporulated when a number of coexisting grassland plant species were grown either as seedlings in sterilized soil or from seeds in field soil. These were evidences of no host specificity, either in fungal distribution or in the ability of one fungal species to infect a restricted range of hosts. The un-specificity in VAM fungi also has been proofed (Read *et al.,* 1976; Johnson *et al.,* 1997; Hirrel *et al.,* 1979).

It was observed that there were variations in AM fungal spore density at NRF, Meghalaya. Picone (2000) recorded more sporulation by AM fungi in pastures compared to forests. Further, spore densities in pastures were found to be more compared to secondary forests (Fischer *et al.,* 1994). Sieverding and Leihner, (1984) reported that combination of graminaceous and leguminous crops generally increase mycorrhizal population. The other plants associated with *Sida cordifolia* in the natural habitats may influence the variations in spore density. The genus, *Glomus* was more abundant compared to *Acaulospora* and *Scutellospora* among these three genera. The variation in spore abundance of different AM fungi in different locations was observed in earlier studies also. Schenck and Kinloch (1980) observed the dominance of species in *Gigaspora margarita, Gigaspora gigantea* and *Gigaspora gregaria* in soyabean fields while *Glomus fasciculatum* and *Glomus clarum* in bahia grass and *Acaulospora* in cotton and peanut fields. The same results were obtained by Lakshmipathy *et al.* (2004) in cashew plantations where *Glomus etunicatum* was most abundant. Vijayalakshmi and Rao (1988) while working with mycorrhizal association in ten members of Asteraceace and seven Amaranthaceace growing in different locations having different soil types observed variation in mycorrhizal association and they found that *Glomus mosseae* and *Glomus macrocarpum* were dominant.

The plant community composition is likely to affect mycorrhizal fungi (Janos, 1980; Kormanik *et al.,* 1980). Bayils (1962) observed that Arbuscular mycorrhizal populations readily respond to mycotrophic plants in a community. This kind of diversity was noticed earlier by Oehl *et al.* (2003) in different land use types and found a maximum diversity index being in grasslands compared to cultivated fields. However, Johnson and Wedin (1997) did not find any significant variation in diversity due to change in land use types. The variation in AM fungal diversity in rhizosphere soil of a particular plant in different locations may be due to the influence of adjoining plants (Krucklemann, 1975; Bagyaraj and Manjunath, 1980). However, the qualitative and quantitative differences in diversity and other parameters like pH, available phosphorous and phosphatase activity with regard to AM fungal association with *Sida cardifolia* was observed (Kumar *et al.,* 2008).

Generally, it is seen during analysis in this study that AM spore numbers increased with elevation. Total root colonization (per cent), Hyphal infection (per cent), Arbuscular infection (per cent) and Colonization intensity (per cent) gradually decreased with elevation. However, vesicular infection (per cent) abruptly decreased to zero at elevation 569 to 570 amsl and it started increasing from 576 to 616 amsl elevation, but again it decreased at 628amsl. There was no any significant change in *p*H value, but little fluctuation was reported with elevation and it occurred in acidic range. Temperature at lowest elevation was 27.3° C. At elevation 536 amsl, the temperature of the soil abruptly increased to 29.2° C and then gradually decreased with elevation. However, at highest elevation the temperature reached 28.3° C. At lowest elevation the organic Carbon (per cent) was low. As the organic Carbon (per cent) increased in middle range elevation but at higher elevation it again decreased.

In ecosystem surveys, the degree of mycorrhizal colonization should be expressed as the proportion of susceptible roots that were mycorrhizal, by excluding woody roots. This requires an understanding of root structure and phenology (Brundrett *et al.*, 1996). The taxonomic classification of most mycorrhizal fungi have not been fully resolved. Consequently, it is vitally important to submit voucher specimens of any fungi used in experiments to a registered herbarium to allow their names to be confirmed and updated in the future (Agerer *et al.*, 2000). Mycorrhizas are three-way interactions of plants, fungi, and soils (Brundrett, 1991). Consequently, descriptions of mycorrhiza types should include information about the soils and habitats where they occur which can be as valuable as information about the taxonomic identity of fungi (Brundrett, 1991). Studies of mycorrhizal synthesis under artificial conditions should include comparisons with the same host and fungus in natural habitats to identify artifacts due to cultural conditions. Combinations of host plants, fungi and soils that do not occur in nature may provide inaccurate knowledge of structure and physiology (Brundrett, 2004).

The taxonomy of arbuscular mycorrhizal fungi is based on the morphology of spores obtained from pot-grown symbioses and precise identification of spores from the field is difficult, as their morphology might vary. Since sporulation depends on environmental conditions and both host and fungal genotype, it is likely to be a very unreliable indication of root colonization in the field; methods that can reveal directly the colonizing fungal taxa within a root will be of great importance in the elucidation of mycorrhizal biodiversity and ecology (Clapp, 1995).

Glomalean fungi consist of large asexual spores and coenocytic hyphae distributed throughout the soil. But there is little idea of the size of individuals of these fungi or how much genetic diversity occurs within their multinucleate organs (Tommerup, 1988). Many species of Glomalean fungi have worldwide distribution patterns and have apparently adapted to diverse habitats. However, it is known that soil factors such as *p*H restrict the distribution of some taxa (Abbott and Robson, 1991) and some of these widespread taxa are now known to comprise more than one species (Morton, 1988).

Most seasonality studies have focused on representative plant species rather than the whole plant community. Sanders and Fitter (1992) suggested that it is essential to follow the seasonal dynamics of AM fungi in whole plant communities and emphasize that true seasonal patterns can be observed when colonization levels are followed for more than a year. Understanding the fungal colonization at the community level provides valuable information especially in a Long-Term Ecological Research (LTER) site where system-level perspective is essential to understanding grassland dynamics. Environmental change impacts many ecosystem processes including mycorrhizal symbiosis. An important global change phenomenon is the accelerating anthropogenic N enrichment. Anthropogenic N additions exceed natural N_2 fixation (Vitousek, 1994).

Mycorrhizal roots have a greater capacity to take up water and withstand the stress of low levels of available moisture better than non-mycorrhizal roots (Gerdemann, 1975). The occurrence of mycorrhizal fungi in soils, its association with forest trees and agricultural crops, influence on plant growth, nutrition uptake and diseases resistance are well documented by various workers (Krishna *et al.*, 1982; Kumar *et al.*, 2002; Rani *et al.*,1999; Gill and Singh, 2002; Singh *et al.*, 2006; Ambika *et al.*, 1994; Parkash *et al.*, 2005).

The presence of arbuscular mycorrhizal fungi (AMF) may be essential for ecosystem sustainability, establishment of plants and maintenance of biological diversity. The participation of AMF in the biodiversity and ecosystem functioning is now being recognized, particularly due to their effect on plant diversity and productivity (Van der Heijden *et al.*, 1998a, 1998b). Several authors have reported positive relationships between plant diversity and AMF colonization (Grime *et al.*, 1987; Van der Heijden *et al.*, 1998a, 1998b; Moreira-Souza *et al.*, 2007). Mycorrhizal fungi are one of the main pathways by which most plants obtain nutrients (Smith and Barker, 2002; Chen *et al.*, 2005) and as such are critical for terrestrial ecosystem functioning (Kernaghan, 2005). The success of reforestation programs may greatly depend on mycorrhizal root colonization of seedlings, which increases their competitiveness due to increase in the initial growth rate (Moreira-Souza *et al.*, 2003). In that sense, the rehabilitation of tropical forests would not be possible only with chemical fertilizers but would also need AMF inoculation (Cuenca *et al.*, 1998).

Arbuscular mycorrhizal fungi constitute an important component of the soil microbial community and are extremely successful fungi that form mutualistic symbioses with about two thirds of all plant species (Trappe, 1987). They improve plant nutrition and promote plant diversity (Van der Heijden *et al.*, 1998a, 1998b), help to control pests and fungal pathogens (Azcon-Aguilar and Barea, 1996) and affect the fitness of plants in polluted environments (Hildebrandt *et al.*, 1999). They can even alter the folia chemistry and influence the life history traits of lepidopteran herbivores (Goverde *et al.*, 2000). In nutrient-poor soil of the humid tropics, many plants are obligately or ecologically dependent on arbuscular mycorrhizal fungi (Gemma *et al.*, 2002). Therefore, it is important to study the biodiversity of arbuscular mycorrhizal fungi as this can improve understanding of tropical forest functioning, plant succession and reforestation in disturbed areas (Zhao *et al.*, 2003).

4. Conclusion

AM fungi are ubiquitous in soils around the world and known to be associated with improved plant growth and vigour for over a century. These symbiotic fungi account for 25 per cent of the combined biomass of soil micro flora and micro fauna in natural ecosystems. So, AM technology will be utilized in establishing the seedlings of some useful forest plant/s species in field conditions. AM fungi will help in absorbing the macronutrients which are essential for the seedlings/plantlets because these macronutrients are not absorbed by plantlets itself. The study of endomycorrhizae occurrence, status and diversity on some useful forest plant species is important from conservation and efficient utilization point of view.

Secondly, there is also more stress on the exploitation of forest vegetation. Indiscriminate exploitation and habitat destruction has consequently resulted in loss of a substantial part of the world's fungal biodiversity, on a regular basis. Due to this overexploitation by uprooting, deforestation, battering, mauling and other external factors like conflagration, the rhizospheric microbial diversity is also affected and results in disappearance of many useful strains of rhizospheric microflora. Therefore, an attempt was also made to conserve rhizospheric microflora (AM fungi *i.e. Glomus* sp. and *Acaulospora* sp.) and some useful forest plant/s species by employing AM technology so that the useful forms of microbes could not be lost irrevocably. With the help of this technology, some of useful forest plant/s species can be cultivated by incorporating with AM inoculation so that exploitation stress should be lessened for meeting the future needs. Accordingly, the native AM fungi and their germplasm preservation to establish a VA Mycorrhizal germplasm Bank at Rain Forest Research Institute, Jorhat is also initiated.

References

Abbott, L.K. and Robson, A.D. 1991. Factors influencing the occurrence of vesicular arbuscular mycorrhizas. *Agric. Ecosyst. Environ.*, 35: 121–150.

Adholeya, A. and Gaur, A. 1994. Estimation of VAM fungal spores in soil. *Mycorrhiza News*, 6(1): 10-11.

Agerer, R., Ammirati, J., Blanz, P., Courtecuisse, R., Desjardin, D.E., Gams, W., Hallenberg, N., Halling, R.E., Hawksworth, D.L., Horak, E., Korf, R.P., Mueller, G.M., Oberwinkler, F., Rambold, G., Summerbell, R.C., Triebel, D. and Watling, R. 2000. Always deposit vouchers. *Mycol. Res.*, 104: 642–644.

Aggarwal, A., Parkash, V., Sharma, D. and Sharma, S. 2005a. Mycorrhizal occurrence, status and diversity of some ethnobotanical plants of Himachal Pradesh. In: Dargan, J.S., Atri, N.S. and Dhingra, G.S. (Eds.). The Fungi - Diversity and Conservation in India, Bishen Singh Mahendra Pal Singh, Dehradun, (U.A.) India, pp. 241-251.

Aggarwal, A., Parkash, V., Sharma, D. and Sharma, S. 2005b. Vesicular Arbuscular Mycorrhizal (VAM) aids in growth of *Prosopis juliflora* and *Tecomella undulata* in arid/semi arid climate. *J. Mycol Pl. Pathol.*, 35(1): 184-187.

Allen, E.B., Allen, M.F., Helm, D.J., Trappe, J.M., Moliva, R. and Rincon, E. 1995. Patterns and regulation of mycorrhizal and fungal diversity. *Plant Soil*, 170: 47-62.

Allen, M.F. 1991. The Ecology of Mycorrhizae. Cambridge University Press, New York.

Ambika, P.K., Katiyar, R.S., Chaudhury, S. and Das, P.K. 1994. The influence of VAM association on growth yield and nutrient uptake in some Mulberry genotypes. *Indian J. Seric.*, 33(2): 166-169.

Azcon-Aguilar, C. and Barea, J.M. 1996. Arbuscular mycorrhizas and biological control of soil-borne plant pathogens - an overview of the mechanisms involved. *Mycorrhiza*, 6: 457-464.

Bagyaraj, D.J. 1992. Vesicular Arbuscular Mycorrhiza: Application in agriculture. *Methods Microbiol.*, 24: 360-373.

Bagyaraj, D.J. and Menge, J.A. 1978. Interactions between VA mycorrhizae and *Azotobacter* and their effects on the rhizosphere microflora and plant growth. *New Phytol.*, 80: 567-573.

Bagyaraj, D.J. and Manjunath, A. 1980. Selection of suitable host for mass production of VA mycorrhizal inoculation. *Plant Soil*, 55: 495-498.

Basumatary, N., Parkash, V., Tamuli, A.K., Saikia, A.J. and Teron, R., 2015. Distribution and diversity of arbuscular mycorrhizal fungi along with soil nutrient availability decline with plantation age of *Hevea brasiliensis* (Willd. ex A. Juss.) Müll. Arg. *The Journal of Biodiversity*, 115: 401-412.

Bayils, G.J.S. 1962. Rhizophagus, the catholic symbiont. *Aust. J. Sci.* 25: 195-209.

Bever, J.D., Morton, J.B., Antonovics, J. and Schultz, P.A. 1996. Host-dependent sporulation and species diversity of arbuscular mycorrhizal fungi in a mown grassland. *J. Ecol.*, 84: 71-82.

Bhat, N.M., Jeyarajan, R. and Ramaraj, B. 1994. Occurrence of vesicular arbuscular mycorrhizae in forest trees at Nilgiris. *Ind. J. For.*, 17: 175-177.

Bhatia, N.P., Sundari, K. and Adholeya, A. 1996. Diversity and selective dominance of vesicular arbuscular mycorrhizal fungi. In: Mukerji, K.G. (Ed.). Concepts in Mycorrhizal Research, Kluwer Academic Publishers, Netherlands, pp. 133-178.

BirdLife International. 2016. Important Bird and Biodiversity Area factsheet: Nongkhyllem Wildlife Sanctuary. http://www.birdlife.org (Accessed on April – August, 2014).

Blaszkowski, J. 2003. Arbuscular mycorrhizal fungi (Glomeromycota), Endogone and Complexipes species deposited in the Department of Plant Pathology, University of Agriculture in Szczecin, Poland. http://zor.zut.edu.pl/Glomeromycota/index.html (Accessed on April – August, 2014).

Brundrett, M. 2004. Diversity and classification of mycorrhizal associations. *Biol. Rev.*, 79: 473- 495.

Brundrett, M.C. 1991. Mycorrhiza's in natural ecosystems. *Adv. Ecol. Res.*, 21: 171–313.

Brundrett, M.C. and Kendrick, B. 1988. The mycorrhizal status, root anatomy, and phenology of plants in a sugar maple forest. *Can. J. Bot.*, 66: 1153-1173.

Brundrett, M., Ashwath, N. and Jasper, D.A. 1996. Mycorrhiza in the Kakad region of tropical Australia I. Propagules of mycorrhizal fungi and soil properties in natural habitats. *Plant Soil*, 184: 159-171.

Champion, H.G. and Seth, S.K. 1968. A Revised Survey of the Forest types of India. Government of India, New Delhi.

Chen, X., Tang, J., Zhi, G. and Hu, S. 2005. Arbuscular mycorrhizal colonization and phosphorus acquisition of plants: effects of coexisting plant species. *Appl. Soil Biol.*, 28: 259- 269.

Cuenca, G., Andrade, Z. and Escalante, G. 1998. Diversity of Glomalean spores from natural, disturbed and revegetated communities growing on nutrient-poor tropical soils. *Soil Biol. Biochem.*, 30: 711-719.

Curtis, J.T. and McIntosh, R.P. 1950. The interrelations of certain analytic and synthetic phyto-sociological characters. *Ecology*, 31: 434-455.

Daniels, B.A. and Trappe, J.M. 1980. Factors affecting spore germination of the vesicular–arbuscular mycorrhizal fungus, *Glomus epigaeus*. *Mycologia*, 72: 457–471.

Fitter, A.H., and Merryweather, J.W. 1992. Why are some plants more mycorrhizal than others? An ecological enquiry. In: Read, D.J., Lewis, D.H., Fitter, A.H. and Alexander, I.J. (Eds.). Mycorrhizas in ecosystems, CAB International, Wallingford, UK, pp. 26–36.

Fitter, A.H. 1977. Influence of mycorrhizal infection on competition for phosphorus and potassium by two grasses. *New Phytol.*, 79: 119-125.

Friese, C.F. and Koske, R.E. 1991. The spatial dispersion of spores of vesicular-arbuscular mycorrhizal fungi in a sand dune: Microscale patterns associated with the root architecture of American beach grass. *Mycol. Res.*, 95: 952–957.

Ganesan, V., Balajee, B., Gopalkrishnan, C. and Mahadevan, A. 1990. Distribution of VAM in Tamil Nadu, In: Jalali, B. S. and Chand, H. (Eds.). Current Trends in Mycorrhizal Research, CCSHAU, Hisar, Haryana, India, pp. 210.

Gemma, J.M., and Koske, R.E. 1988. Seasonal variation in spore abundance and dormancy of *Gigaspora gigantea* and in mycorrhizal inoculum potential of a dune soil. *Mycologia*, 80: 211-216.

Gemma, J.N., Koske, R.E. and Habte, M. 2002. Mycorrhizal dependency of some endemic and endangered Hawaiian plant species. *Am. J. Bot.*, 89: 337-345.

Gerdemann, J.W. 1975. Vesicular arbuscular mycorrhizae. In: Torrey, J.G. and Clarkson, D.T. (Eds.) The development and function of roots, Academic Press, New York, pp. 575-595.

Gerdemann, J.W. and Nicolson, Y.H. 1963. Spores of mycorrhizae *Endogone* species extracted from soil by wet sieving and decanting. *Trans. Brit. Mycol. Soc.*, 46: 235-244.

Gill, T.S. and Singh, R.S. 2002. Effect of *Glomus fasciculatum* and *Rhizobium* inoculation on VA mycorrhizal colonization and plant growth of chickpea. *J. Mycol. Pl. Pathol.*, 32: 162 - 167.

Giovannetti, M. and Gianinazzi- Pearson, V. 1994. Biodiversity in arbuscular fungi. *Mycol. Res.*, 98: 705-715.

Goverde, M., Van der Heijden, M.G.A., Wiemken, A., Sanders, I.R. and Erhardt, A. 2000. Arbuscular mycorrhizal fungi influence life history traits of a lepidopteran herbivore. *Oecologia*, 125: 362-369.

Grime, J.P., Mackey, J.M.L., Hillier, S.H., Read, D.J. 1987. Floristic diversity in a model system using experimental microcosms. *Nature,* 328: 420-422.

Guadarrama, P. and Alvarez-Sanchez, F.J. 1999. Abundance of arbuscular mycorrhizal fungi spores in different environments in a tropical rain forest, Veracruz, Mexico. *Mycorrhiza*, 8: 267-270.

Hawksworth, D.L. 1991. The fungal dimension of biodiversity: magnitude significance and conservation. *Mycol. Res.*, 95: 641-655.

Hildebrandt, U., Kaldorf, M. and Bothe, H. 1999. The zinc violet and its colonization by arbuscular mycorrhizal fungi. *J. Plant. Physiol.*, 154: 709-717.

Hirrel, M.C., Mehravaran, H. and Gerdemann, J.W. 1978. Vesicular-arbuscular mycorrhizae in the Chenopodiaceae and Cruciferae: do they occur? *Can. J. Bot.*, 56: 2813-2817.

Jagpal, R. and Mukerji, K.G. 1987. Large scale cropping on Indian arid lands. In: Proceedings of International Symposium on Dry Land Farming. 17-22 September, 1987, Yang Ling, Shaanxi, China, pp. 56-77.

Janos, D.P. 1980. Mycorrhizae influence tropical sucession. *Biotropica*, 12: 56-64.

Jeffries, P. and Dodd, J.C. 1991. The use of mycorrhizal inoculants in agriculture. In: Arora, D.K., Rai, B., Mukerji, K.G. and Knudson, G.R. (Eds.). Handbook of applied Mycology vol. – 1, Mercel Dekker, Inc. Madison, Ave., New York, pp. 156-186.

Johnson, N.C., Graham, J.H. and Smith, F.A. 1997. Functioning of mycorrhizal associations along the mutualism-parasitism continuum. *New Phytol.*, 135: 575-585.

Johnson, N.C. and Wedin, D.A. 1997. Soil carbon, nutrients and mycorrhizae during conversion of A. dry tropical forest to grassland. *Ecol. Appl.* 7: 171-182.

Kernaghan, G. 2005. Mycorrhizal diversity: Cause and effect? *Pedobiol.*, 49: 511-520.

Koide, R.T. 1991. Nutrient supply, nutrient demand and plant response to mycorrhizal infection. *New Phytol.* 117: 356-386.

Kormanik, P.P., Bryan, W.C. and Schultz, R.C. 1980. Increasing endomycorrhizal fungus inoculum in forest nursery soil with cover crops. *South. J. Appl. For.*, 4: 151-153.

Koske, R.E. 1987. Distribution of VA mycorrhizal fungi along a latitudinal temperature gradient. *Mycologia*, 79: 55-68.

Krishna, K.R., Balakrishna, A.N. and Bagyaraj, D.J. 1982. Interaction between a vesicular arbuscular mycorrhizal fungus and *Streptomyces cinnamomeus* and their effect on finger millet. *New Phytol.*, 92: 401-405.

Krucklemann, H.W. 1975. Effccts of fertilizers, soils, soil tillage and plant species on frequency of *Endogone* Chlamadospores and mycorrhizal infection in arable soils. In: Sanders, F.E., Mosse B. and Tinker, P.B. (Eds.). Endomycorrhizas, Academic press, London, pp. 511-525.

Kumar, K.V.C., Chandrashekar, K.R., Lakshmipathy, R. 2008. Variation in Arbuscular Mycorrhizal Fungi and Phosphatase Activity Associated with *Sida cordifolia* in Karnataka. *WJAS*, 4(6): 770-774.

Kumar, P.P., Reddy, S.R. and Reddy, S.M. 2000. Mycorrhizal dependency of some agro forestry tree species. *Ind. For.*, 126 (4): 397-402.

Kumar, R., Jalali, B.L. and Chand, H. 2002. Influence of vesicular arbuscular mycorrhizal fungi on growth and nutrient uptake in chickpea. *J. Mycol. Plant Pathol.*, 32(1): 11-15.

Lakshmipathy, R., Balakrishna, A.N., Bagyaraj, D.J., Sumana, D.A., and Kumar, D.P. 2004. Evaluation, grafting success and field establishment of cashew rootstock as influenced by VAM fungi. *Ind. J. Expt. Biol.*, 42: 1132-1135.

Lakshmipathy, R., Gowda, B. and Bagyaraj, D.J. 2003. VA Mycorrhizal colonization pattern in RET medicinal plants (*Mammea suriga, Saraca asoca, Garcina* spp., *Embelia ribes* and *Calamus spp.*) in different parts of Karnataka. *Asian J. Microbiol. Biotect. Env. Sci.*, 5: 505-508.

Malloch, D.W., Pirozynski, K.A. and Raven, P.H. 1980. Ecological and evolutionary significance of mycorrhizal symbiosis in vesicular plants. *Proc. of Nat. Acad. of Sci. USA.*, 77: 2113-2118.

Mangan, S.A., Eom, A.H., Adler, G.H., Yavitt, J.B. and Herre, E.A. 2004. Diversity of arbuscular mycorrhizal fungi across a fragmented forest in Panama: insular spore communities differ from mainland communities. Community Ecology. *Oecologia*, 141: 687–700.

Martins, M.A. 1993. The role of the external mycelium of arbuscular mycorrhizal fungi in the carbon transfers process between plants. *Mycol. Res.*, 97: 807-810.

McGonigle, T.P. and Fitter, A.H. 1990. Ecological specificity of vesicular arbuscular mycorrhizal associations. *Mycol. Res.*, 94: 120-122.

McGraw, A. C. and Hendrix, J. W. 1984. Host and fumigation effect on spore population, densities of species of endogonaceous mycorrhizal fungi. *Mycologia*, 76: 122-131.

Mehrotra, A. and Mehrotra, M. D. 2000. Studies on vesicular arbuscular mycorrhizal fungi of forest trees-1. Association of Gigasporaceous fungi, an important feature of some *Acacia* species. *Ind. J. For.*, 23(1): 20-27.

Miller, R.I., Bratton, S.P. and White, P.S. 1987. A regional strategy for reserve design and placement based on an analysis of rare and endangered species' distribution patterns. *Biol. Conserv.*, 39: 255-268.

Mohan, V. and Babber, N. 1997. Status of endomycorrhizal association in economically important tree species of Indian arid zone. *Kavaka*, 25: 11-17.

Mohan, V. and Singh, Y. P. 1997. Endomycorrhizal associations of *Acacia* in nurseries and plantations of Indian arid zone. *Ind. For.*, 4: 323-330.

Moreira-Souza, M., Baretta, D., Tsai, S.M., Gomes-da-Costa, S.M., Cardoso, E.J.B.N. 2007. Biodiversity and distribution of Arbuscular Mycorrhizal fungi in *Araucaria angustifolia Forest. Sci. Agric.* (Piracicaba, Braz.), 64(4): 393-399.

Moreira-Souza, M., Trufem, S.F.B., Gomes-Da-Costa, S.M., Cardoso, E.J.B.N. 2003. Arbuscular mycorrhizal fungi associated with *Araucaria angustifolia* (Bert.) O. Ktze. *Mycorrhiza*, 13: 211-215.

Morton J. 2014 – International Culture Collection of (Vesicular) Arbuscular Mycorrhizal Fungi. http://invam.caf.wvu.edu (Accessed on April – August, 2014).

Morton, J.B., Redecker, D. 2001. Two new families of Glomales, Archaeosporaceae and Paraglomaceae, with two new genera Archaeospora and Paraglomus, based on concordant molecular and morphological characters. *Mycologia*, 93,181–195.

Morton, J.B. 1988. Taxonomy of VA mycorrhizal fungi: classification, nomenclature and identification. *Mycotaxon*, 32: 267-324.

Morton, J.B. and Benny, G.L. 1990. Revised classification of arbuscular mycorrhizal fungi (Zygomycetes): New order, Glomales two new sub-orders Glomineae and Gigasporineae and two new families, Aculosporaceae and Gigasporaceae with emendation of Glomaceae. *Mycotaxon*, 37: 471-491.

Mukerji, K.G. 1996. Taxonomy of endomycorrhizal fungi. In: Mukerji, K.G., Mathur, B., Chamola, B.P. and Chitralekha, P. (Eds.). Advances in Botany, APH Pub. Crop. New Delhi, pp. 211-221.

Muthukumar, T., Udaiyan, K. and Manian, S. 1994. Vesicular arbuscular mycorrhizae in certain tropical wild legumes. *Ann. For.*, 2(1): 33-43.

Nagabhushanam, P., Reddy, S.M. and Reddy, S.R. 1999. VAM fungi associated with some common legume trees of Godavari belt. *Ind. J. For.*, 22 (1/2): 129-131.

Oehl, F., Sieverding, E., Ineichen, L., Mader, P., Boller, T. and Wienmken, A. 2003. Impact of land use intensity on the species diversity of *Arbuscular mycorrhizal* fungi in agro-ecosystems of central Europe. *Appl. Env. Microbiol.*, 69: 2816-2824.

Panwar, J. and Vyas, A. 2002. Arbuscular mycorrhizal association in *Tamarix aphylla* in Indian Thar desert. *Mycorrhiza News*, 14 (2): 14-16.

Parkash, V. 2012. Utilization of Vesicular Arbuscular mycorrhizal diversity for the quality stock production of some useful forest plant/s species of Nongkhyllem Reserve forest, Nongpoh, Meghalaya, India (RFRI-12/2008-09/SFM). Project Completion Report. Rain Forest Research Institute, Jorhat, Assam, India.

Parkash, V. and Saikia, A.J. 2015. Habitational abiotic environmental factors alter Arbuscular mycorrhizal composition, species richness and diversity index in *Abroma augusta* L. rhizosphere. *Plant Pathology and Quarantine*, 5(2), 98–120, Doi 10.5943/ppq/5/2/8.

Parkash, V. Aggarwal, A., Sharma, S. and Sharma, D. 2005. Effect of endophytic mycorrhizae and fungal bioagent on the development and growth of *Eucalyptus saligna* Sm. seedlings. *Bull. Nat. Inst. Ecol.*, 15: 127-131.

Parkash, V. Sharma, S., Kaushik, S. and Aggarwal, A. 2009. Endomycorrhizal association of Khair (*Acacia catechu* Willd.): A traditionally important medicinal plant of Himachal Pradesh. *Mycorrhiza News*, 21(1):15-20.

Phillips, J.M. and Hayman, D. S. 1970. Improved produces for clearing roots and staining parasitic and VAM fungi for rapid assessment of infection. *Trans. Brit. Mycol. Soc.*, 55: 158-161.

Picone, C.M., 2000. Diversity and abundance of Arbuscular mycorrhizal fungus spores in tropical forest and pasture. *Biotropica*, 32: 734-750.

Powell, C.L. and Bagyaraj, D.J. 1984. Effect of mycorrhizal inoculation on the production of blueberry cuttings—A note. *N. Z. J. Agr. Res.*, 27: 467–471.

Rahangdale, R. and Gupta, N. 1999. Vesicular arbuscular mycorrhizal association of biomass tree species in the tropical forest of Madhya Pradesh. *Ind. J. For.*, 22 (1/2): 62-65.

Raj, B. K.V. and Chandrashekar, K.R. 2003. Association of mycorrhizal fungi with the endemic dipterocarps of Sharavathi valley forest (Karnataka) region. *Mycorrhiza News*, 14(4): 18-21.

Rani, A., Dhungana, H.N. and Sharma, G.S. 1995. Occurrence of vesicular arbuscular mycorrhizal fungi in forest nursery seedlings in Assam. In: Adholeya, A. and Singh, S. (Eds.) Mycorrhizae: Biofertilizers For The Future, TERI Lodhi Road, New Delhi, pp. 62-64.

Rani, P., Aggarwal, A. and Mehrotra, R.S. 1999. Growth responses in *Acacia nilotica* inoculated with VAM fungus (*Glomus mosseae*), *Rhizobium* sp. and *Trichoderma harzianum*. *Indian Phytopath.*, 52(2): 151-153.

Ravi, K.B., Prabakaran, J. and Mariappan, S. 1995. Survey of vesicular-arbuscular mycorrhiza in agroforestry trees in alfisol. In: Adholeya, A. and Singh, S. (Eds.) Mycorrhizae: Biofertilizers For The Future, TERI Lodhi Road, New Delhi, pp. 95-99.

Ravikumar, R., Ananthakrishnan, G., Girija, S. and Ganapathi, A. 2001. AM associations of bamboo rhizosphere in different study sites of Tamil Nadu. *Ind. For.*, 7: 804-807.

Read, D.I. 1991. Mycorrhizas in ecosystems. *Experientia*. 47: 376-391.

Read, D.J., Koucheki, H.K. and Hodgson, J. 1976. Vesicular-Arbuscular Mycorrhiza in natural vegetation systems. *New Phytol.*, 77: 641-653.

Rosendahl, S., Rosendahl, C. N. and Sochting, U. 1989. Distribution of VA mycorrhizal endophytes amongst plants from a Danish grassland community. *Agric. Ecos. Environ.*, 29: 329-335.

Sanders, I.R. and Fitter, A.H. 1992. The ecology and functioning of vesicular-arbuscular mycorrhizas in co-existing grassland species Seasonal patterns of mycorrhizal occurrence and morphology. *New Phytol.*, 120: 517–524.

Santhaguru, K. and Sadhana, B. 2000. Vesicular arbuscular mycorrhizal status of *Acacia* species from Madhurai district. *Ann. For.,* 8(2): 266-269.

Santhaguru, K., Gladis Ponmalar, S.B. and Karunakaran, R. 1995. Vesicular arbuscular mycorrhizae in tree legumes and its rhizospheric soils in Alagar hills. *Ind. For.*, 9: 817-823.

Schenck, N.C. and Kinloch, R.A. 1980. Incidence of mycorrhizal fungi on six field crops in monoculture on a newly cleared woodland site. *Mycologia*, 72(3): 445-455.

Schenck, N.C. and Perez, Y. 1990. Manual for the identification of VA mycorrhizal (VAM) fungi. Univ. of Florida, Synergistic Pub., Florida, USA, 241 p.

Senapati, M., Das, A.B. and Das, P. 2000. Association of vesicular arbuscular mycorrhizal fungi with 21 forest tree species. *Ind. J. For.*, 23 (3): 326-331.

Shanker, A., Mathew, J., Neeraj, Kaur, R., Mehrohtra, R.S. and Verma, A. 1990. Mycorrhizal status of some desert plants and their physiological significance. In: Jalali, B.S. and Chand, H. (Eds.). Current trends in mycorrhizal research. CCSHAU, Hisar, Haryana, India, pp. 160-161.

Sharma, S., Parkash, V. and Aggarwal, A. 2008. Glomales I: A monograph of *Glomus* spp. in rhizosphere of sunflower (Glomaceae) of Haryana, India. *Helia*, 31(49), 13–18.

Sharma, S., Parkash, V. and Aggarwal, A. 2009. A monograph of *Acaulospora* spp. (VAM fungi) in rhizosphere of sunflower in Haryana, India. *Helia*, 32(50), 69–76.

Sieverding, E. and Leihner, D.E. 1984. Influence of crop rotation and intercropping of cassava with legumes on VA mycorrhizal symbiosis of cassava. *Plant Soil*, 80: 143-146.

Singh, P.K., Chakrabarti, S. and Khan, M.A. 2006. Arbuscular mycorrhizal association with some common weeds of Mulberry garden. *Ind. J. For.*, 29(1): 91-94.

Singh, R. and Adholeya, A. 2002. AMF biodiversity in wheat agrosystems of India. *Mycorrhiza News*, 14(3): 21-23.

Smith, S.E. and Barker, S.J. 2002. Plant phosphate transporter genes help harness the nutritional benefits of arbuscular mycorrhizal symbiosis. *Trends Plant Sci.*, 75:189-190.

Sylvia, D.M. (1986). Spatial and temporal distribution of vesicular-arbuscular mycorrhizal fungi associated with *Uniola paniculata* in Florida foredunes. *Mycologia*, 78: 728-734.

Tarafdar, J.C. and Rao, A. V. 1990. Survey of Indian arid zone tree species for the occurrence of VAM colonization. In: Jalali, B.S. and Chand, H. (Eds.). Current trends in mycorrhizal research. CCSHAU, Hisar, Haryana, India, pp. 44-46.

Thapar, H.S., Uniyal, K. and Verma, R.K. 1991. Survey of native VAM fungi of sodic soils of Haryana state. *Ind. For.*, 117: 1059-1069.

Thapar, H.S., Vijyan, A.K. and Uniyal, K. 1992. Vesicular arbuscular mycorrhizal associations and root colonization in some important tree species. *Ind. For.*, 3: 207-212.

Tommerup, I.E. 1988. The vesicular-arbuscular mycorrhizas. *Adv. Plant Path.* 6: 81-91.

Trappe, J.M. 1982. Synoptic key to the genera and species of zygomycetous mycorrhizal fungi. *Phytophathol.*, 72: 1107-1108.

Trappe, J.M. 1987. Phylogenetic and ecologic aspects of mycotrophy in the angiosperms from an evolutionary standpoint. In: Safir, G.R. (Ed.). Ecophysiology of VA Mycorrhizal Plants, CRC Press, Boca Raton, Florida, USA, pp. 5-26.

Uniyal, K. 2003. Arbuscular mycorrhizal association of *Populus deltoides*. *Ind. For.*, 4: 527-530.

Van der Heijden, M.G.A., Boller, T., Wiemken, A., Sanders, I.R. 1998a. Different Arbuscular mycorrhizal fungal species are potential determinants of plant community structure. *Ecology*, 79:2082-2091.

Van der Heijden, M.G.A., Klironomos, J., Ursic, M., Moutoglis, P., Streitwolf-Engel, R., Boller, T., Wiemken, A. and Sanders, I. 1998b. Mycorrhizal fungal diversity determines plant biodiversity, ecosystem variability and productivity. *Nature*, 396: 69-72.

Vijaya, T., Kumar, R. V., Reddy, B. V. P., Sastry, P. S. S. and Srivastav, A. K. 1995. Studies on occurrence of endomycorrhiza in some forest soils of Andhra Pradesh. In: Adholeya, A. and Singh, S. (Eds.). Mycorrhizae: Biofertilizers for the future, TERI, Lodhi Road, New Delhi, pp. 45-47.

Vijayalakshmi, M. and Rao, A.S. 1988. Vesicular-Arbuscular mycorrhizal associations of some Asteraceae and Amaranthaceae. *Acta Bot. Indica*, 16(2): 168-174.

Vitousek, P.M. 1994. Beyond global warming: ecology and global change. *Ecology*, 75:1861–1876.

Walker, C. 1992. Systematics and taxonomy of arbuscular endomycorrhizal fungi (Glomales) –A possible way forward. *Agronomie*, 12: 887-897.

Walker, C., Mize, C. W. and McNabb Jr., H. S. 1982. Populations of endogonaceous fungi at two locations in central Iowa. *Can. J. Bot.* 60: 2518-2529.

Zhao, Z.W., Wang, G.H. and Yang, L. 2003. Biodiversity of arbuscular mycorrhizal fungi in tropical rainforests of Xishuangbanna, Southwest China. *Fungal Divers.*, 13: 233-242.

Zhao, Z.W., Xia, Y.M., Qin, X.Z., Li, X.W., Cheng, L.Z., Sha, T. and Wang, G.H. 2001. Arbuscular mycorrhizal status of plants and the spore density of arbuscular mycorrhizal fungi in the tropical rain forest of Xishuangbanna, Southwest China. *Mycorrhiza*, 11: 159-162.

2017, Mycorrhizal Fungi   *Pages 117–133*
Editors: Ashok Aggarwal and Kuldeep Yadav
Published by: ASTRAL INTERNATIONAL PVT. LTD., NEW DELHI

5

Ecological Relevance of Arbuscular Mycorrhizal Technology in Agroecosystem Services: Potentials and Challenges

Shanti Chaya Dutta*

Department of Life Sciences, Dibrugarh University,
Dibrugarh – 786 004, Assam
*Corresponding Author: shanti.chaya@gmail.com

ABSTRACT

Understanding of functionality and ecology of the rhizosphere is a key to crop improvement which could resolve how the plant-microbe framework influences biogeochemical cycling, plant growth and tolerance to abiotic and biotic stresses. For sustainability and productivity of agroecosystems it is crucial to understand the rhizospheric players i.e. the myriad of microorganisms and invertebrates as well as above and below ground processes leading to innumerable interactions between plants, antagonists and mutualistic symbionts. Among the organic input agricultural strategies, rhizosphere engineering with plant growth promoting rhizobacteria and arbuscular mycorrhizal fungi are considered as most powerful tools which can wield direct and indirect effects to ameliorate plant growth. In this review, introduction of AM fungi in farming system has been discussed in relation to assessing its adaptability to a particular soil and climatic condition along with the challenges related to the application of these beneficial fungi in sustainable agriculture. This article outlines the sensitivity of arbuscular mycorrhizal fungi to certain agricultural management systems along with functional variability and plant responsiveness. Here, the key factors shaping the rhizosphere microbiome that could redesign the agroecosystem to increase mycorrhizal efficacy pleading for a sound ecosystem development has also been highlighted.

Keywords: *Plant-microbe interactions, Agricultural sustainability, Mycorrhizal inoculum, Crop improvement.*

1. Introduction

From the mid 20th century, intensive agricultural practices have been gaining acceleration by introduction of new cultivars and increased use of fertilizers and biocides. To maintain productivity while reducing chemical input, bio-inoculants are now-a-days gaining popularity. But the survival of such crop inoculants in rhizosphere is a major point of concern as their mode of interaction with native plants and resident rhizospheric microbial community and their mechanisms involving various nutrient and/or biogeochemical cycles in soil are considered to be cumbersome to understand.

It is assumed that by 2050 the world's population will exceed nine billion (Rodriguez and Sanders, 2015) and as a result global agricultural production will demand almost doubling-up with an implementation or revitalization of eco-friendly technologies in order to safeguard human as well as environmental health. For sustainability and productivity of agro-ecosystems, optimization of turnover and recycling of plant nutrients is considered as a fundamental issue where the framework of plant-microbe plays the role of functional agent as essential drivers of nutrient cycling. Sustainability has emerged as one of the most potent concerns of the present era resulting in an overt emphasis on aversion of chemical fertilizers and use of various organic inputs to agricultural soil. Organic input strategies to support rhizosphere engineering with plant growth promoting rhizobacteria (PGPR) and arbuscular mycorrhizal fungi (AMF) are considered as some powerful tools which can exert direct and indirect effects to ameliorate plant growth and health.

AMF form mutualistic associations with most land plants and are able to control carbon, nitrogen and phosphorus cycling between above and below ground components of soil-plant ecosystem. AMF can colonize and ameliorate nutrient uptake in most of the staple food crops and on a wider perspective the need of the hour is exploitation and promotion of novel AMF taxa with multiple attributes. AMF are regarded as the most overlooked bio-stimulants which could pave the way for the largest input in agriculture particularly in tropics where most soils are acidic and nutrient poor especially in soluble orthophosphate (Pi) (Rodriguez and Sanders, 2015). To amplify the knowledge of the symbiotic determinants of plant-AMF compatibility and more effective use of AMF, ecologists could focus on the understanding of:

☆ Survival and colonizing ability of introduced AMF

☆ Adaptability of AMF to rhizospheric environment and interaction with pre-existing microbial community of that specific environment

☆ Functional variability of AMF and other mycorrhiza helper rhizospheric microbes and variability in plant responsiveness to the beneficial microbes

To overcome the challenge of sustainable intensification of agriculture, the success of bio-inoculants is being underpinned by biotechnological innovations and results revealed by lab based trial are often not replicated in case of field trials. Moreover, due to the absence of native soil microbial pool in sterile soil condition of green house study, the impact on AM colonization may vary as the ubiquitous soil microbes are known to influence AM colonization rate (Dhillion, 1992). Therefore, to prevail over the problem of inconsistency of performance of bio-inoculants, it is crucial to understand the interactions among the soil inhabitants and plant roots which will lead to develop more efficient formulations of bio-inoculants in compatible combinations (*e.g.* bacteria and AM fungi).

2. Use of Microbial Inoculants in Rhizosphere Engineering: A Promising Avenue for Crop Improvement

Microbial inoculants as biopesticides and biofertilizers have been gaining much attention in agriculture from the last two decades (Calvo *et al.*, 2014). Considering the fact that without proper understanding of the features of beneficial bio-inoculants and their interactions with plant-microbe framework, growth and development of plants cannot be properly investigated and planned. In this context, application of multifunctional microbial features manifested by a single microbial strain is used in singe inoculation. On the other hand, development of consortia or mixed inoculants include two or more different microorganisms. Selection and standardization of inoculum of novel multifunctional microbial strains, such as phosphate solubilizing, nitrogen fixing and biocontrol active strains are some adventitious strategies to be used as single inoculants. Some examples of such microbes are *Rhizobium* strains having P solubilizing and N_2-fixing capacity, *Trichoderma* strains having biocontrol activity along with the capacity to solubilize phosphate and other sparingly soluble minerals (Altomere *et al.*, 1999; Peix *et al.*, 2001) and other N_2-fixing bacterial species having the potentiality of minimizing the growth of phytopathogens through siderophore-mediated iron competition (Lemanceau *et al.*, 2009; Nagata *et al.*, 2013; Pii *et al.*, 2015). AMF are another example of beneficial microorganisms for their ability to uptake both inorganic and organic N together with phosphorus and other micro and macro nutrients (Hawkins *et al.*, 2000; Smith and Read, 2008).

Development of consortia or mixed inocula is based on the fact that plants co-exist with large fraction of plant-microbe encounters where microorganisms normally exist in communities and facilitate the cycling and mobilization of plant nutrients by releasing different metabolites. The multiple interactions among AMF and other beneficial rhizospheric bacteria and/or mycorrhiza helper bacteria have opened new avenues in the design of mixed microbial inocula by understanding the complexity of the symbiotic associations (Bianciotto and Bonfante, 2002; Hildebrandt *et al.*, 2002; Finley, 2004; Vassilev *et al.*, 2015). Certain such bacterial groups which could influence the efficacy of the inocula are reported to be present frequently in soil associated with AMF whereas some other bacteria are reported to be inhibited by them (Filion *et al.*, 1999; Johansson *et al.*, 2004; Mansfeld-Giese *et al.*, 2002; Wamberg *et al.*, 2003; Toljander *et al.*, 2007).

It has also been reported that inocula having different types of plant beneficial symbiotic, non-symbiotic and/or free-living bacteria and mycorrhizal fungi may affect the existing autochthonous microbial community of soil (Naiman *et al.*, 2009; de Salamone *et al.*, 2010; Trabelsi and Mhamdi, 2013). Moreover, the phenomenon like quorum sensing and motility also bestow enormous competitive advantages on bacterial survivability (Manefield and Turner, 2002; Gera and Srivastava, 2006). In this complex context, inconsistent and varied responses obtained from field inoculations are largely influenced by stimulatory and inhibitory effects exerted by rhizoexudates of the host plants and the successful establishment of the beneficial microbe-plant interactions (Hartmann *et al.*, 2009). Furthermore, compatibility with the growing conditions and plants in which they are inoculated also play a significant role in determining the fate of the inocula on its way to efficacious functioning (Martinez-Viveros *et al.*, 2010).

3. Delineation of AM Fungi in Agricultural Soil

The occurrence of AMF is ubiquitous in almost all ecosystems (Smith and Read, 2008). However, variability of AMF diversity is affected by a number of interrelated and site specific agronomic factors as well as agricultural practices and tillage system. There have been a considerable number of reports on distribution of AMF in both natural conditions and farming system managements. In agricultural lands, AMF development is intervened by conditions of agricultural soils, use of synthetic fertilizers and pesticides, types of crop rotation *etc.* (Gosling *et al.*, 2006; Jansa *et al.*, 2006; Schneider *et al.*, 2015). Plant assemblage composition is also advocated by some other pioneer workers in influencing AMF community structure (Schneider *et al.*, 2015).

Several authors depict an increased root colonization and spore abundance of AMF in organically managed agro-ecosystems in comparison to conventional tillage system. However, AMF spore abundance and rate of colonization rarely can provide specific information on species diversity and richness and their beneficial traits influencing plant growth and health (Schneider *et al.*, 2015). Helgason *et al.* (2002) proposed that AMF diversity may be responsible for a wide array of benefits at diverse physiological levels and phenological stages. Soil fertility is considered as a key factor which contributes much to biodiversity (Henkin *et al.*, 2006). Agricultural practices may impose strong selection pressures on AMF community (Verbruggen *et al.*, 2013). Previous investigations reported varying results of the effects of organic and mineral fertilizers on AMF activity and community structure in soil. Heavy doses of phosphatic fertilizers are reported to put forth negative impact on AMF colonization (Kahiluoto *et al.*, 2001; Bagyaraj *et al.*, 2015). It is well established that soil P has a direct effect on AMF infectivity (Bolán and Abbott, 1983; Miranda and Harris, 1994; Kurle and Pfleger, 1996). Use of other fertilizers especially N fertilizers also have been reported to exert negative effects on AMF abundance and diversity in some agricultural soils (Treseder and Allen, 2002; Linderman and Devis, 2004;

Karunasinghe *et al.*, 2009; Gosling *et al.*, 2010) but not in some other agroecosystems (Jumpponen *et al.*, 2005; Nyaga *et al.*, 2014). Low input organic farming systems are reported more favorable for AMF colonization (Ryan *et al.*, 1994, Gosling *et al.*, 2006; Kahiluoto *et al.*, 2009). Due to adverse effects of mineral fertilizers and pesticides in conventional farming management systems, greater AMF species richness and abundance occur under organic farming systems as compared to conventional systems (van der Gast *et al.*, 2010; Verbruggen *et al.*, 2010; Dai *et al.*, 2014). Moreover, management intensity and site history is also considered as another two important factors for distribution and diversity of AMF community structure (Hijri *et al.*, 2006; Oehl *et al.*, 2003).

Diversity of AMF has been studied till date by different researchers in a variety of natural and agricultural ecosystems with the characteristic measure of the species richness of a site (Douds Jr. and Milner, 1999). Kivlin *et al.* (2011) carried out a detailed study on global diversity and distribution of AMF by collecting 14,961 DNA sequences from 111 published reports and correlated the community composition to geography, environment and plant biomass. They reported a six fold higher species richness of AMF than previously enumerated records owing to high beta diversity existed among sampling sites.

An understanding of factors affecting AMF diversity and composition of a particular agroecosystem is crucial to manage AMF communities and their benefits to agriculture. A number of previous workers have investigated the effects exerted by two mostly used agricultural practices of conventional and intensive agriculture, namely physical disturbance and fertilizer regime in long and short term experiments (Jansa *et al.*, 2002, 2003; Ohel *et al.*, 2003; Lin *et al.*, 2012; Thian *et al.*, 2013; Stockinger *et al.*, 2014; Peyret-Guzzon *et al.*, 2016). Physical soil disturbance in the form of tillage can affect the existing AMF community as regrowth of extraradical mycelia favour mycorrhizal species to survive under disturbed soils under different tillage (Verbruggen and Kiers, 2010). Verbruggen *et al.* (2010) hypothesized that increased incidence of mycorrhizal species having r-type life history strategy *i.e.* low investment in formation of external mycelium and rapid production of spores result in reduction of mycorrhizal colonization rate in high N and P input agricultural soil (Mäder *et al.*, 2000) and shift in community compositions (Kim *et al.*, 2014). However, Borriello *et al.* (2012) concluded that N fertilization more strongly influences AMF community structure than by physical soil disturbance. Some other authors Alguacil *et al.*(2010), Gianinazzi *et al.* (2002) advocated for shifts in AMF community composition intensified by higher phosphorus fertilization. Disturbance like cultivation practices can be recognized for such alteration in species composition of AMF in cultivated lands including loss of large-spored AMF species under the family Gigasporaceae and increase in relatively smaller sized *Glomus* sp. (Talukdar and Germida, 1993; Johnson, 1993). High P and N content in soil due to regular fertilizer use may also result in decrease of large-spored species (Egerton-Warburton *et al.*, 2001). Peyret-Guzzon *et al.* (2016) advocated that short term effects of physical disturbance to cultivated soil by ploughing and fertilizer usage can exert a strong effect on intraspecific populations of AMF.

4. AMF: Ecological Aspects in Agricultural Relevance

A considerable number of publications cited for different microbial species used as bio-inoculants on different crops and plant species, demonstrate the wide application of microbial inoculants as agricultural inputs and expose the responses of plants to above and below ground plant-microbe interactions mostly addressing the two strategies given by Godfray *et al.* (2010): "closing the yield gap" and "increasing production limits". Beneficial interactions of mutualistic microbial partners of plants could trigger the reduction of yield losses from biotic as well as abiotic above and below ground stresses (Orrel and Bennet, 2013; Vassilev *et al.*, 2015). In order to meet an environment friendly global food demand, biostimulatory effect of AMF is considered as an important aspect. A number of reviews on AMF colonization on many important crop and medicinal plants were evaluated and discussed till date (Smith and Smith, 2011; Zeng *et al.*, 2013; Sbrana *et al.*, 2014; Baum *et al.*, 2015; Rouphael *et al.*, 2015). AMF contribute greatly to current agriculture and plant improvement aiming sustainability, facilitating the cycling and mobilization of essential nutrients especially phosphorus (Treseder, 2013; Bagyaraj *et al.*, 2015; Wiel *et al.*, 2016), nitrogen (Hawkins *et al.*, 2000; Corrêa *et al.*, 2015; Hodge and Storer, 2015) and sulfur (Gahan and Schmalenberger, 2014), enhancing tolerance to stressful environments like salinity, drought, high temperature, water logging condition *etc.* (Nadeem *et al.*, 2014), heavy metal contamination of soil (Cabral *et al.*, 2015) and radionuclotides (Davies *et al.*, 2015), enhancing production of plant secondary metabolites (Pedone-Bonfim *et al.*, 2015) and increasing disease resistance to plants (Baum *et al.*, 2015).

Co-inoculation of AMF with PGPR result in ameliorating nutrient acquisition process by a wide variety of mechanisms (reviewed by Orrel and Bennet, 2013; Calvo *et al.*, 2014; Pii *et al.*, 2015; Vassilev *et al.*, 2015). Lekberg and Koide (2005) observed a linear relationship between rate of mycorrhizal colonization and increase in plant biomass in agricultural systems during their meta-analysis study. Feldmann *et al.* (2009) investigated effects of AMF inocula isolated from corn rhizosphere on seven different host plants and advocated that mycorrhizal colonization influenced growth of the treated plants in a non-linear fashion with a threshold rate at 20 6/ 30 per cent colonization below which no plant benefit were recorded. Hoeksema *et al.* (2010) carried out another meta-analysis study on plant responses to mycorrhizal inoculations and reported 3.1 times higher growth of mycorrhizal plants compared to uninoculated controls.

Rodriguez and Sanders (2015) pointed out two major challenges *i.e.* effective and safe use of AMF in tropical agriculture and proposed that adoption of community and population ecology approaches could resolve these challenges. A considerable number of studies have reported positive effects of AMF on enhanced crop production (Douds *et al.*, 2007; Nyaga *et al.*, 2014), however, majority of previous field inoculations included exotic strains of AMF regardless the effects of existing natural AMF community associated with the particular crop (Izaguirre-Mayoral *et al.*, 2000; Njeru *et al.*, 2014). It may lead to failure in field inoculations as native species have been regarded as more adapted to a specific environment than introduced one (Izaguirre-Mayoral *et al.*, 2000; Klironomos, 2003). Functionality of mycorrhizal species could be highly location specific and therefore, selection of resident strains would be a productive approach in low input farming systems. Thus, introduction of autochthonous soil microbes as biofertilizers in the rhizosphere and augmenting native microbial activities in soil through manipulating agricultural practices could be a sustainable approach to crop improvement.

AMF contribute greatly to current agricultural regimes. In order to achieve the ideal conditions under which AMF can ameliorate plant productivity in a specific soil and climatic condition inherent to a certain region, it is necessary to assess the impact of AMF species of that particular ecosystem. Loss or impoverishment of AMF species in a particular ecosystem results in less ecosystem functioning and this plead for a re-establishment of the AMF community aiming to sustainability targeting the global recession to deal with the demand of environmentally aware clientele (Berruti *et al.*, 2016). One of the prominent strategies to enhance efficiency of the ecosystem is introduction of AMF propagules as inoculants in soil. However, adaptation of AMF to the target soil and establishment of functional symbiosis to the plants are two main key concerns for crop improvement.

In natural environments, a non-mycorrhizal condition is considered as infrequent for majority of plant species (Smith and Smith, 2012), although there exists a marked diversity among the AMF community depending on plant responsiveness, soil status and climatic conditions *etc.* Host specificity (Öpik *et al.*, 2006) and competition for plant roots than competition for soil resources (Maherali and Klironomos, 2012) may play a dominant role in compositional change in below ground AMF community (Dickie *et al.*, 2013). Besides, fine scale environmental variability and seasonal changes may exert site-scale responses to ecosystem development and can overwhelm AMF community compositional shifts during early and mature stages of ecosystem development (Dickie *et al.*, 2013).

Till date, considerable progress has been made in understanding of the microbial ecology of belowground environment and a number of challenges corresponding to some paradigm shift have been proposed in agroecology.

5. Challenges to Inoculum Production of AMF and its Application

The role of AMF in agricultural intensification to cope with the increased demand for bio-energy feedstocks and bio-based products is increasingly recognized. Since the appropriate management of these fungi results into a sustainable and quality production of crops, development and implementation of novel agricultural strategies to inoculate AMF propagules (inoculum) into a target soil is required to ensure desirable ecosystem functioning. Unfortunately, AMF cannot be cultivated in pure cultures away from their host plants as they are the obligate symbionts and thus large scale production of AMF becomes challenging for this contradicting features. Berruti *et al.* (2016) described three main types of AMF inocula comprising soil from root zone of the host plant containing colonized root fragments, AMF spores and hyphae. The main constrain of such type of inocula is lack of precise information about the variability of viable spores, spore abundance and diversity and possibility of transferring weed seeds and pathogens. The second and most commonly used type of inocula contain AMF spores extracted from soil which are used as the starters for crude inoculum production. For the production of crude inoculum, AMF isolates are grown together with a host trap plant in an inert medium optimized for AMF propagation. Such type of inocula contains abundant AMF propagules of same or different species. The third type of inoculum contains infected root fragments alone of a known host plant that have been separated from a trap plant culture.

Quality assurance is one of the main issues in commercial AMF inocula production. The only way to produce non contaminated inocula production is to use transformed roots in axenic conditions (Vosátka *et al.,* 2012). However, the main tailback of this procedure is that not all AMF starter cultures can be cultivated successfully in axenic conditions. Another key checkpoint of *in-situ* host-plant mediated inocula production is testing of presence of potential pathogenic fungi in roots of the trap plants. Here, the use of microscopic studies to check the presence of pathogens, use of plant varieties susceptible to soil-borne pathogens and molecular tests for detecting fungal pathogens are some necessary measures to ensure the quality production of AMF inocula (Vosátka *et al.,* 2012). Moreover, inoculum tuning seems to be necessary for storage and distribution of inocula under a wide range of climatic conditions having a varied temperature without losing viability of fungal spores (Vosátka and Dodd, 2002).

Even though, new methods for massive production of AMF inocula and seed coating technology have been developed in the context of sustainable bio-inoculation (IJdo *et al.,* 2011; Vosátka *et al.,* 2013; vander Heijden *et al.,* 2015), the major challenges for AMF inoculum production restricts to its obligate symbiotic behavior as a certain host plant for growth and competition for their survival. The need of cultivation of the host plant is time and space demanding which is a potent constrain of AMF inocula production. Besides, the setting up of AMF in terms of plant responsiveness, fungal effectiveness is another questionable point whether the used AMF species can really be able to provide the desired ecosystem services to a certain agroclimatic environment. Fester and Sawers (2011) have advocated for two different approaches in establishing a biological process in manmade agroecosystems: adaptation of the processes as far as practicable in the existing ecosystem (the "reductionist" approach) or adaptation of ecosystems to fit with the desirable process (the "holistic" approach). Under reductionist approach AMF are introduced to ameliorate plant health and growth in agroecosystems having low biological diversity. Such approaches results in adaptation of the fungal inocula to that specific environmental condition and target plants. On the other hand, in holistic approach introduced AMF and its ecosystem functioning are dependent on the diversity of rhizospheric microbial communities aiming at conservation and restoration of native AMF diversity of that specific ecosystem. Thus redesigning of the whole production process of an ecosystem relies on either introduction of AMF to a poor rhizospheric microbial community or restoration and conservation of native AMF community structure.

Compatibility of introduced AMF strains with local agricultural environment affects establishment of new taxa. Intense selection pressures of agricultural management practices put forth strong effects on well adapted microbial taxa in the soil (Oehl *et al.*, 2010; Schnoor *et al.*, 2011; Verbruggen *et al.*, 2013) and thus introduced one has to compete with the well adapted local community. Hence, one promising approach is selection of AMF taxa for introduction to a specific environmental condition of agroecosystems, such as tillage environment (Schnoor *et al.*, 2011), soil type and pH (Oehl *et al.*, 2010) and selection of potential hosts (Öpik and Moora, 2012). Selection of potential hosts is now-a-days gaining attention as it seems that some AMF are host or habitat "specialists" and others are "generalists" (Öpik and Moora, 2012). Generally in agricultural fields crops are frequently rotated and weed can, therefore, serve as additional inoculum sources and hence plant-host generalist AMF taxa could be most likely to be introduced.

Inoculum quantity and carrying capacity of agricultural soil to support AMF populations are another two main point of concern for AMF introduction (Verbruggen *et al.*, 2013). Some agricultural soils exhibit reduced carrying capacity for low plant allocation towards AMF and this is thought to crop up for availability of non-host plants, high nutrient contents in soil especially phosphorus and nitrogen and other soil edaphic and climatic factors. For a successful establishment and persistence of introduced AMF inocula, the threshold amount of the inocula applied in a specific soil is crucial as AMF propagules in reduced fitness at low population sizes could easily be outcompeted by native AMF community (Verbruggen *et al.*, 2013).

Another phenomenon determining the inoculum potentiality is the priority effects, the process by which initial fungal taxa ascertain the community composition in soil. Evidences on priority effects on AMF community describe establishment of host plants colonized by different AMF taxa as a decisive factor in determining the resulting community structure (Mummey *et al.*, 2009; Hausmann and Hawkes, 2010). However, it is becoming more apparent that plants colonized by AMF have either prohibitive or stimulatory effect on population dynamics of a community structure by directly or indirectly affecting competition for root or soil space and therefore, those factors affecting the shape of a certain community are important in the context of persistence and success of specific fungal taxa applied as fungal inocula.

6. Concluding Remarks and Perspectives

Exploitation of AMF with other beneficial rhizospheric microbes has been considered as a key concern to address in recent times. Mycorrhizal technology is a promising avenue with multipurpose activities pleading to sustainability of soil as well as improved plant health. A number of reviews on biogenesis and multifunctional role of AMF have been documented till date with special emphasis on crop enhancement. There have been a large number of previous evaluations and summaries about the prospects, utilization and feasibility of introduced and native AMF management through cropping systems. Emphasizing on the sensitivity of AMF to certain agricultural management regimes, functional variability of the fungi and other associated rhizospheric bacteria and plant responsiveness are some recent advances to be focused by researchers on the context of agricultural sustainability.

The main way forward in sustainable utilization of mycorrhizal technology in the field conditions of plant production depends on the scientific knowledge derived from the joint efforts of fundamental and applied research and collaboration of both mycorrhizal science and mycorrhizal industry. Numerous bottlenecks exist in commercial exploitation of mycorrhizal fungi and to overcome these drawbacks further research is necessary to exploit positive effects of mycorrhizal symbiosis. To understand Carbon and nutrient cycles in soil, inclusion of global models on plant-microbe framework have become a major frontier in current research. Heijden *et al.* (2015) advocated for development of such biogeochemical models for prediction of suitable conditions and profitable application of mycorrhizal technology.

Till date major advances have been made in the field of mycorrhizal application in shaping terrestrial ecosystems and plant communities that dominate the major terrestrial biomes of the world. Comparative analysis of different agroclimatic ecosystem functioning will improve our understanding to this plant-microbe association and these new findings will be some potent prerequisites for future sustainable management of agroecosystems.

References

Alguacil, M.M., Lozano, Z., Campoy, J.M., Roldán, A. 2010. Phosphorus fertilisation management modifies the biodiversity of AM fungi in a tropical savanna forage system. *Soil Biol. Biochem.*, 42: 1114–1122.

Altomare, C., Norvell,W.A., Björkman, T., Harman, G.E. 1999. Solubilization of phosphates and micronutrients by the plant-growth-promoting and biocontrol fungus *Trichoderma harzianum* Rifai 1295–22. *Appl. Environ. Microbiol.*, 65: 2926–2933.

Bagyaraj, D.J., Sharma, M.P., Maiti, D. 2015. Phosphorus nutrition of crops through arbuscular mycorrhizal fungi. *Curr. Sci.*, 108: 1288–1293.

Baum, C., El-Tohamy, W., Grudac, N. 2015. Increasing the productivity and product quality of vegetable crops using arbuscular mycorrhizal fungi: A review. *Sci. Horti.* 187:131–141.

Berruti, A., Lumini, E., Balestrini, R., Bianciotto, V. 2016. Arbuscular mycorrhizal fungi as natural biofertilizers: let's benefit from past successes. *Front. Microbiol.*, 6: 15-59.

Bianciotto, V., Bonfante, P. 2002. Arbuscular mycorrhizal fungi: a specialised niche for rhizospheric and endocellular bacteria. *Ant. Van Leeu.*, 81: 365–371.

Bolán, N.S., Abbott, L.K. 1983. Seasonal variation in infectivity of vesicular-arbuscular mycorrhizal fungi in relation to plant response to applied phosphorus. *Aus. J. Soil Res.*, 21: 207–210.

Borriello, R., Lumini, E., Girlanda, M., Bonfante, P., Bianciotto, V. 2012. Effects of different management practices on arbuscular mycorrhizal fungal diversity in maize fields by a molecular approach. *Biol. Fertil. Soils,* 48: 911–922.

Cabral, L., Soares, C.R.F.S., Giachini, A.J., Siqueira, J.O. 2015. Arbuscular mycorrhizal fungi in phytoremediation of contaminated areas by trace elements: mechanisms and major benefits of their applications. *World J. Microbiol. Biotechnol.*, 31: 1655–1664.

Calvo, P., Nelson, L., Kloepper, J.W., 2014. Agricultural uses of plant biostimulants. *Plant Soil*, 383: 3–41.

Corrêa, A., Cruz, C., Ferrol, N. 2015. Nitrogen and carbon/nitrogen dynamics in arbuscular mycorrhiza: the great unknown. *Mycorrhiza*, 25: 499–515.

Dai, M., Hamel, C., Bainard, L.D., St. Arnaud, M., Grant, C.A., Lupwayi, N.Z., Malhi, S.S., Lemke, R. 2014. Negative and positive contributions of arbuscular mycorrhizal fungal taxa to wheat production and nutrient uptake efficiency in organic and conventional systems in the Canadian prairie. *Soil. Biol. Biochem.*, 74: 156–166.

Davies, H.S., Cox, F., Robinson, C.H., Pitmann, J.K. 2015. Radioactivity and the environment: technical approaches to understand the role of arbuscular mycorrhizal plants in radionuclide bioaccumulation. *Front. Plant Sci.*, 6: 580.

de Miranda, J.C.C., Harris, P.J. 1994. Effects of soil phosphorus on spore germination and hyphal growth of arbuscular mycorrhizal fungi. *New Phytol.*, 128: 103–108.

de Salamone, I.E.G., di Salvo, LP, Ortega, J.S.E., Sorte, P.M.F.B., Urquiaga, S., Teixeira, K.R.S. 2010. Field response of rice paddy crop to *Azospirillum* inoculation: physiology of rhizosphere bacterial communities and the genetic diversity of endophytic bacteria in different parts of the plants. *Plant Soil*, 336: 351–362.

Dhillion, S. 1992. Dual inoculation of pretransplant stage *Oryza sativa* L. plants with indigenous vesicular-arbuscular mycorrhizal fungi and *Fluorescent pseudomonas* spp. *Biol. Fertil. Soils*, 13: 147–151.

Dickie, I.A., Martínez-García, L.B., Koele, N., Grelet, G-A., Tylianakis, J.M., Peltzer, D.A., Richardson, S.J. 2013. Mycorrhizas and mycorrhizal fungal communities throughout ecosystem development. *Plant Soil*, 367: 11–39.

Douds, D.D. Jr, Nagahashi, G., Reider, C., Hepperly, P.R. 2007. Inoculation with arbuscular mycorrhizal fungi increases the yield of potatoes in a high P soil. *Biol. Agric. Hortic.*, 25: 67–78.

Douds, D.D. Jr., Millner, P.D. 1999. Biodiversity of arbuscular mycorrhizal fungi in agroecosystems. *Agric. Ecosyst. Environ.*, 74: 77–93.

Egerton-Warburton, L.M., Graham, R.C., Allen, E.B., Allen, M.F. 2001. Reconstruction of the historical changes in mycorrhizal fungal communities under anthropogenic nitrogen deposition. *Proc. R. Soc. Lond. B Biol. Sci.*, 268: 2479–2484.

Feldmann, F., Gillessen, M., Hutter, I., Schneider, C. 2009. Should we breed for effective mycorrhiza symbioses? In: Feldmann, F., Alford, D.V., Furk, C. (Eds.). *Crop plant resistance to biotic and abiotic factors*. Deutsche Phytomedizinische Gesellschaft, Braunschweig, pp. 507–522.

Fester, T., Sawers, R. 2011. Progress and challenges in agricultural applications of arbuscular mycorrhizal fungi. *Critic. Rev. Plant Sci.*, 30: 459–470.

Filion, M., St-Arnaud, M., Fortin, J.A. 1999. Direct interaction between the arbuscular mycorrhizal fungus *Glomus intraradices* and different microorganisms. *New Phytol.*, 141: 525–533.

Finley, R. 2004. Mycorrhizal fungi and their multifunctional roles. *Mycologist* 18:91–96.

Gahan, J., Schmalenberger, A. 2014. The role of bacteria and mycorrhiza in plant sulfur supply. *Front. Plant Sci.*, 5: 723.

Gera, C., Srivastava, S. 2006. Quorum-sensing: the phenomenon of microbial communication. *Curr. Sci.*, 90: 566–677.

Gianinazzi, S., Schüepp, H., Barea, J.M., Haselwandter, K. 2002. Mycorrhizal technology in agriculture: from genes to bioproducts. Birkhäuser Verlag, Switzerland.

Godfray, H.C.J., Beddington, J.R., Crute, I.R., Haddad, L., Lawrence, D., *et al.*, 2010. Food security: the challenge of feeding 9 billion people. *Sci.*, 327: 812–818.

Gosling, P., Hodge, A., Goodlass, G., Bending, G.D. 2006. Arbuscular mycorrhizal fungi and organic farming. *Agric. Ecosyst. Environ.*, 113: 17–35.

Gosling, P., Ozaki, A., Jones, J., Turner, M., Rayns, F., Bending, G.D. 2010. Organic management of tilled agricultural soils results in a rapid increase in colonisation potential and spore populations of arbuscular mycorrhizal fungi. *Agric. Ecosyst. Environ.*, 139: 273–279.

Hartmann, A., Schmid, M., van Tuinen, D., Berg, G. 2009. Plant-driven selection of microbes. *Plant Soil*, 321: 235–257.

Hausmann, N.T., Hawkes, C.V. 2010. Order of plant host establishment alters the composition of arbuscular mycorrhizal communities. *Ecol.*, 91: 2333–2343.

Hawkins, H.J., Johansen, A., George, E. 2000. Uptake and transport of organic and inorganic nitrogen by arbuscular mycorrhizal fungi. *Plant Soil*, 226: 275–285.

Helgason, T., Merryweather, J.W., Denison, J., Wilson, P., Young, J.P.W., Fitter, A.H., 2002. Selectivity and functional diversity in arbuscular mycorrhiza of co-occuring fungi and plant from a temperate deciduous woodland. *J. Ecol.*, 90: 371–384.

Henkin, Z., Sternberg, M., Seligman, N.G., Noy-Meir, I. 2006. Species richness in relation to phosphorus and competition in a Mediterranean dwarf-shrub community. *Agric. Ecosyst. Environ.*, 113:277–283.

Hijri, I., S korová, Z., Oehl, F., Ineichen, K., Mäder, P., Wiemken, A., Redecker, D. 2006. Communities of arbuscular mycorrhizal fungi in arable soils are not necessarily low in diversity. *Mol. Ecol.*, 15: 2277–2289.

Hildebrandt, U., Janetta, K., Bothe, H. 2002. Towards growth of arbuscular mycorrhizal fungi independent of a plant host. *Appl. Environ. Microbiol.*, 68: 1919–1924.

Hodge A., Storer, K. 2015. Arbuscular mycorrhiza and nitrogen: implications for individual plants through to ecosystems. *Plant Soil*, 386:1–19.

Hoeksema, J.D., Chaudhary, V.B., Gehring, C.A., Johnson, N.C., Karst, J., Koide, R.T., Pringle, A., Zabinski, C., Bever, J.D., Moore, J.C., Wilson, G.W.T., Klironomos, J.N., Umbanhowar, J. 2010. A meta-analysis of context-dependency in plant response to inoculation with mycorrhizal fungi. *Ecol. Lett.*, 13: 394–407.

IJdo, M., Cranenbrouck, S., Declerck, S. 2011. Methods for large-scale production of AM fungi:past, present, and future. *Mycorrhiza*, 21: 1–16.

Izaguirre-Mayoral, M.L., Carballo, O., Carreno, L., de Mejia, M.G. 2000. Effects of arbuscular mycorrhizal inoculation on growth, yield, nitrogen, and phosphorus nutrition of nodulating bean varieties in two soil substrates of contrasting fertility. *J. Plant Nutr.*, 23: 1117–1133.

Jansa, J., Mozafar, A., Anken, T., Ruh, R., Sanders, I.R., Frossard, E. 2002. Diversity and structure of AMF communities as affected by tillage in a temperate soil. *Mycorrhiza*, 12: 225–234.

Jansa, J., Mozafar, A., Kuhn, G., Anken, T., Ruh, R., Sanders, I.R., Frossard, E. 2003. Soil tillage affects the community structure of mycorrhizal fungi in maize roots. *Ecol. Appl.*, 13: 1164–1176.

Jansa, J., Wiemken, A., Frossard, E. 2006. The effects of agricultural practices on arbuscular mycorrhizal fungi. *Geol. Soc. Spec. Pub.*, 266: 89–115.

Johansson, J.F., Paul, L.R., Finlay, R.D. 2004. Microbial interactions in the mycorrhizosphere and their significance for sustainable agriculture. *FEMS Microbiol. Ecol.*, 48: 1–13.

Johnson N.C. 1993. Can fertilization of soil select less mutualistic mycorrhizae? *Ecol. Appl.*, 3: 749–757.

Jumpponen, A., Trowbridge, J., Mandyam, K., Johnson, L. 2005. Nitrogen enrichment causes minimal changes in arbuscular mycorrhizal colonisation but shifts community composition-evidence from rDNA data. *Biol. Fertil. Soil*, 41: 217–224.

Kahiluoto, H., Ketoja, E., Vestberg, M., Saarela, I. 2001. Promotion of AM utilization through reduced P fertilization 2. Field studies. *Plant Soil*, 231: 65–79.

Kahiluoto, H., Ketoja, E. Vestberg, M. 2009. Contribution of arbuscular mycorrhiza to soil quality in contrasting cropping systems. *Agric. Ecosyst. Environ.*, 134: 36–45.

Karunasinghe, T.G., Fernando, W.C., Jayasekera, L.R. 2009. The effect of poultry manure and inorganic fertilizer on the arbuscular mycorrhiza in coconut. *J. Natl. Sci. Found Sri Lanka*, 37: 277–279.

Kim, Y.C, Gao, C , Zhong, Y., IIc, X.II., Chen, L., Wan, S.Q., Guo, L.D. 2014. Arbuscular mycorrhizal fungal community response to warming and nitrogen addition in a semi-arid steppe ecosystem. *Mycorrhiza*, 25: 267–276.

Kivlin S.N., Hawkes C.V., Treseder K.K. 2011. Global diversity and distribution of arbuscular mycorrhizal fungi. *Soil Biol. Biochem.*, 43: 2294–2303.

Klironomos, J.N. 2003. Variation in plant response to native and exotic arbuscular mycorrhizal fungi. *Ecol.*, 84: 2292–2301.

Kurle, J. E., Pfleger, F. L. 1996. Management influences on arbuscular mycorrhizal fungal species composition in a corn–soybean rotation. *Agron. J.*, 88: 155–161.

Lekberg, Y., Koide, R.T. 2005. Is plant performance limited by abundance of arbuscular mycorrhizal fungi? A meta analysis of studies published between 1988 and 2003. *New Phytol.*, 168: 189–204.

Lemanceau, P., Bauer, P., Kraemer, S., Briat, J-F. 2009. Iron dynamics in the rhizosphere as a case study for analyzing interactions between soils, plants and microbes. *Plant Soil*, 321: 513–535.

Lin, X., Feng, Y., Zhang, H., Chen, R.,Wang, J, Zhang, J., Chu, H. 2012. Long term balanced fertilization decreases arbuscular mycorrhizal fungal diversity in an arable soil in north China revealed by 454 pyrosequencing. *Environ. Sci. Technol.*, 46: 5764–5771.

Linderman, R.G., Davis, E.A. 2004. Evaluation of commercial inorganic and organic fertilizer effects on arbuscular mycorrhizae formed by *Glomus intraradices*. *Hortic. Technol.*, 14: 196–202.

Mäder, P., Edenhofer, S., Boller, T.,Wiemken, A., Niggli, U. 2000. Arbuscular mycorrhizae in a long-term field trial comparing low-input (organic, biological) and high-input (conventional) farming systems in a crop rotation. *Biol. Fertil. Soils*, 31: 150–156.

Maherali, H., Klironomos, J.N. 2012. Phylogenetic and trait-based assembly of arbuscular mycorrhizal fungal communities. *PLoS One*, 7: e36695.

Manefield, M., Turner, S.L. 2002. Quorum sensing in context: out of molecular biology and into microbial ecology. *Microbiol.*, 148: 3762–3764.

Mansfeld-Giese, K., Larsen, J., Bodker, L. 2002. Bacterial populations associated with mycelium of the arbuscular mycorrhizal fungus *Glomus intraradices*. *FEMS Microbiol. Ecol.*, 41: 133–140.

Martinez-Viveros, O., Jorquera, M.A., Crowley, D.E., Gajardo, G., Mora, M.L. 2010. Mechanisms and practical considerations involved in plant growth promotion by rhizobacteria. *J. Soil. Sci. Plant Nutr.*, 10: 293–319.

Mummey, D.L., Antunes, P.M., Rillig, M.C. 2009. Arbuscular mycorrhizal fungi preinoculant identity determines community composition in roots. *Soil Biol. Biochem.*, 41: 1173–1179.

Nadeem, S.M., Ahmad M., Zahir Z.A., Javaid A., Ashraf M. 2014. The role of mycorrhizae and plant growth promoting rhizobacteria (PGPR) in improving crop productivity under stressful environments. *Biotech. Adv.*, 32: 429–448.

Nagata, T., Oobo, T., Aozasa, O. 2013. Efficacy of a bacterial siderophore, pyoverdine, to supply iron to *Solanum lycopersicum* plants. *J. Biosci. Bioeng.*, 115: 686–690.

Naiman, A.D., Latronico, D.A., de Salamone, I.E.G. 2009. Inoculation of wheat with Azospirillum brasilense and Pseudomonas fluorescens: impact on the production and culturable rhizosphere microflora. *Eur. J. Soil Biol.*, 45: 44–51.

Njeru, E.M., Avio, L., Sbrana, C., Turrini, A., Bocci, G., Ba'rberi, P., Giovannetti, M. 2014. First evidence for a major cover crop effect on arbuscular mycorrhizal fungi and organic maize growth. *Agron. Sustain. Dev.*, 34: 841–848.

Nyaga, J., Jefwa, J.M, Muthuri, C.W., Okoth, S.A., Matiru, V.N.,Wachira, P. 2014. Influence of soil fertility amendment practices on ex-situ utilization of indigenous arbuscular mycorrhizal fungi and performance of maize and common bean in Kenyan highlands. *Trop. Subtrop. Agroecosyst.*, 17: 129–141.

Oehl, F., Laczko, E., Bogenrieder, A., Stahr, K., Bosch, R., van der Heijden, M.G.A., Sieverding, E. 2010. Soil type and land use intensity determine the composition of arbuscular mycorrhizal fungal communities. *Soil Biol. Biochem.*, 42: 724–738.

Oehl, F., Sieverding, E., Ineichen, K., Mäder, P., Boller, T., Wiemken, A. 2003. Impact of land use intensity on the species diversity of arbuscular mycorrhizal fungi in agroecosystems of central Europe. *Appl. Environ. Microbiol.*, 69: 2816–2824.

Öpik, M., Moora, M., Liira, J., Zobel, M. 2006. Composition of root-colonizing arbuscular mycorrhizal fungal communities in different ecosystems around the globe. *J. Ecol.*, 94: 778–790.

Orrell, P., Bennett, A.E. 2013. How can we exploit above–below ground interactions to assist in addressing the challenges of food security? *Front. Plant Sci.*, 4: 432.

Pedone-Bonfim, M.V.L., da Silva F.S.B., Maia L.C. 2015. Production of secondary metabolites by mycorrhizal plants with medicinal or nutritional potential. *Acta Physiol. Plant* 37: 27.

Peix, A., Rivas-Boyero, A.A., Mateos, P.F., Rodriguez-Barrueco, C., Martínez-Molina, E., Velazquez, E. 2001. Growth promotion of chickpea and barley by a phosphate solubilizing strain of *Mesorhizobium mediterraneum* under growth chamber conditions. *Soil Biol. Biochem.*, 33: 103–110.

Peyret-Guzzon, M., Stockinger, H., Bouffaud, M-L., Farcy, P.,Wip D.,Redecker, D. 2016. Arbuscular mycorrhizal fungal communities and *Rhizophagus irregularis* populations shift in response to short-term ploughing and fertilisation in a buffer strip. *Mycorrhiza*, 26: 33–46.

Pii, Y., Mimmo, T., Tomasi, N., Terzano, R.,Cesco, S., Crecchio, C., 2015. Microbial interactions in the rhizosphere: beneficial influences of plant growth-promoting rhizobacteria on nutrient acquisition process- A review. *Biol. Fertil. Soils*, 51: 403–415.

Rodriguez, A. Sandoro, I.R. 2015. The role of community and population ecology in applying mycorrhizal fungi for improved food security. *ISMEJ.*, 9: 1053–1061.

Rouphaela, Y., Frankenb, P., Schneiderc, C., Schwarzd, D., Giovannettie, M., *et al.*, 2015. Arbuscular mycorrhizal fungi act as biostimulants in horticultural crops. *Sci. Horti.*, 196: 91–108.

Ryan, M.H., Chilvers, G.A., Dumaresq, D.C. 1994. Colonization of wheat by VA-mycorrhizal fungi was found to be higher on a farm managed in an organic manner than on a conventional neighbor. *Plant Soil*, 160: 33–40.

Sbrana, C., Avio, L., Giovanetti, M. 2014. Beneficial mycorrhizal symbionts affecting the production of health-promoting phytochemicals. *Electrophoresis*, 35: 1535–1546.

Schneider, K.D., Lynchb, D.H., Khoslaa, K.D.K., Jansac, J., Voroneya, R.P. 2015. Farm system management affects community structure of arbuscular mycorrhizal fungi. *Appl. Soil Ecol.*, 96: 192–200.

Schnoor, T.K., Lekberg, Y., Rosendahl, S., Olsson, PA. 2011. Mechanical soil disturbance as a determinant of arbuscular mycorrhizal fungal communities in semi-natural grassland. *Mycorrhiza*, 21: 211–220.

Smith, S.E., Read, D.J. 2008. Mycorrhizal Symbiosis. Academic Press, London.

Smith, S.E., Smith, F.A. 2011. Roles of arbuscular mycorrhizas in plant nutrition and growth: new paradigms from cellular to ecosystem scales. *Annu. Rev. Plant Biol.*, 62: 227–250.

Smith, S.E., Smith, F.A. 2012. Fresh perspectives on the roles of arbuscular mycorrhizal fungi in plant nutrition and growth. *Mycologia*, 104: 1–13.

Stockinger, H., Peyret-Guzzon, M., Bouffaud, M.L., Koegel, S., Redecker, D. 2014. The largest subunit of RNA polymerase II as a new marker gene to study assemblages of arbuscular mycorrhizal fungi in the field. *PLoS One*, 9: e107783.

Talukdar, N.C., Germida, J.J. 1993. Occurrence and isolation of vesicular-arbuscular mycorrhizae in cropped field soils of Saskatchewan, Canada. *Can. J. Microbiol.*, 39: 567–575.

Thian, H., Drijber, R.A., Zhang, J.L., Li, X.L. 2013. Impact of long term nitrogen fertilization and rotation with soybean on the diversity and phosphorus metabolism of indigenous arbuscular mycorrhizal fungi within the roots of maize (*Zea mays* L.). *Agric. Ecosyst. Environ.*, 164: 53–61.

Toljander, J.F., Lindahl, B.D., Paul, L.R., Elfstrand, M., Finlay, R.D. 2007. Influence of arbuscular mycorrhizal mycelial exudates on soil bacterial growth and community structure. *FEMS Microbiol. Ecol.*, 61: 295–304.

Trabelsi, D., Mhamdi, R. 2013. Microbial inoculants and their impact on soil microbial communities: a review. *Biomed. Res. Int.*, 86: 32–40.

Treseder, K.K. 2013. The extent of mycorrhizal colonization of roots and its influence on plant growth and phosphorus content. *Plant Soil*, 371: 1–13.

Treseder, K.K., Allen, M.F. 2002. Direct nitrogen and phosphorus limitation of arbuscular mycorrhizal fungi, a model and field test. *New Phytol.*, 155: 507–515.

van de Wiel, C.C.M., van der Linden, C.G., Scholten, O.E. 2016. Improving phosphorus use efficiency in agriculture: opportunities for breeding. *Euphytica*, 207: 1–22.

van der Gast, C.J., Gosling, P., Tiwari, B., Bending, G.D. 2010. Spatial scaling of arbuscular mycorrhizal fungal diversity is affected by farming practice. *Environ. Microbiol.*, 13: 241–249.

van der Heijden, M.G.A., Martin, F.M., Selosse, M-A., Sanders, I.R. 2015. Mycorrhizal ecology and evolution:the past,the present,and thefuture. *New Phytol.*, 205: 1406–1423.

Vassilev, N., Vassileva, M., Lopez, A., Martos, V., Reyes, A., Maksimovic, I., Eichler-Löbermann, B., Malusà, E. 2015. Unexploited potential of some biotechnological techniques for biofertilizer production and formulation. *Appl. Microbiol. Biotechnol.*, 99: 4983–4996.

Verbruggen, E., Kiers, E.T. 2010. Evolutionary ecology of mycorrhizal functional diversity in agricultural systems. *Evol. Appl.*, 3: 547–560.

Verbruggen, E., Roling, W.F.M., Gamper, H.A., Kowalchuk, G.A., Verhoef, H.A., van der Heijden, M.G.A. 2010. Positive effects of organic farming on below-ground mutualists: large-scale comparison of mycorrhizal fungal communities in agricultural soils. *New Phytol.*, 186: 968–979.

Verbruggen, E., van der Heijden, M.G.A., Rillig, M.C., Kiers, E.T. 2013. Mycorrhizal fungal establishment in agricultural soils:factors determining inoculation success. *New Phytol.*, 197: 1104–1109.

Vosátka, M., Dodd, J.C. 2002. Ecological considerations for successful application of arbuscular mycorrhizal fungi inoculum. In: Gianinazzi, S., Schuepp, H., Barea, J.M., Haselwandter, K. (Eds.). Mycorrhizal technology in agriculture. Birkhauser Verlag, Basel, pp. 235–248.

Vosátka, M., Látr, A., Gianinazzi, S., Albrechtová, J. 2012. Development of arbuscular mycorrhizal biotechnology and industry: current achievements and bottlenecks. *Symbiosis*, 58: 29–37.

Wamberg, C., Christensen, S., Jakobsen, I., Müller, A.K., Sørensen, S.J., 2003. The mycorrhizal fungus (*Glomus intraradices*) affects microbial activity in the rhizosphere of pea plants (*Pisum sativum*). *Soil Biol. Biochem.*, 35: 1349–1357.

Zeng, Y., Guo, L-P., Chen, B-D., Hao, Z-P.,Wang, J-Y., *et al.*, 2013. Arbuscular mycorrhizal symbiosis and active ingredients of medicinal plants: current research status and prospective. *Mycorrhiza*, 23: 253–265.

— *Part II* —
Mass Multiplication of
Arbuscular Mycorrhizal Fungi

2017, Mycorrhizal Fungi
Editors: Ashok Aggarwal and Kuldeep Yadav
Published by: ASTRAL INTERNATIONAL PVT. LTD., NEW DELHI

Pages 137–155

6

AMF Spore Propagation: Conventional and Recent Advancements

*Gopal Selvakumar[1,#], Ramasamy Krishnamoorthy[2,#],
Kiyoon Kim[3], Charlotte C. Shagol[4], Manoharan Melvin Joe[3],
Denver Walitang[3], Mak Chanratana[3] and Tongmin Sa[3,*]*

[1]*Horticultural and Herbal Crop Environment Division,
National Institute of Horticultural and Herbal Science, Wanju, South Korea*
[2]*Department of Agricultural Microbiology,
Agricultural College and Research Institute,
Tamil Nadu Agricultural University, Madurai, India*
[3]*Department of Agricultural Chemistry, Chungbuk National University,
Cheongju, Chungbuk 361-763, Republic of Korea*
[4]*Department of Agronomy, Benguet State University,
La Trinidad, Benguet 2601, Philippines*
[#]*These authors contributed equally to this work*
[*]*Corresponding Author: tomsa@chungbuk.ac.kr*

ABSTRACT

Arbuscular mycorrhizal fungi (AMF) are obligate biotroph and known for their ability to improve plant growth by enhancing nutrient uptake. Due to their environmental friendly application for plant growth, interest on cultivation of AMF have increased. Large scale production of AMF have been attempted for many decades. AMF mass production in the fumigated soil substrate was attained large interest. However, the quality of inoculum produced is very low and the possibility of soil pathogen spread is very high. So to improve AMF inoculum quality, alternative in vitro methods for AMF propagation have evolved. In vitro methods such as hairy root, autotrophic, aeroponics, hydroponics and slide method were used. In hairy root method, a naturally genetic transformed roots are used as host to mass produce AMF in Petri plates, whereas in autotrophic method, live plants were used as a host in Petri dishes. In hydroponic method, liquid medium is used to growth AMF and mass produce. Slide method can be used as an effective method to mass produce single culture of AMF species. Even though the in vitro methods produce high quality inoculum, adaptable by the farmers are very low. Soil based substrate method received a great interest as it is less artificial and readily applicable. On-farm production technique of AMF propagation using the indigenous microorganisms has shown to produces millions of AMF propagules.

Keywords: *AMF propagation, Slide method, On-farm production method, Hairy root, Autotrophic culture system.*

1. Introduction

Arbuscular mycorrhizal fungi (AMF) are widespread in soil and their importance in natural and semi-natural ecosystem are widely recognized and materialized by plant growth establishment, nutrient uptake and resistance against biotic and abiotic stress (Smith and Read, 2008). Mass production or propagation of arbuscular mycorrhizal fungi requires an association with plant roots for their growth. AMF produce propagules such as spores, colonize roots and hyphae and these propagules can act as initial inoculum. AMF cultures are necessary for taxonomic identification, research and practical application (Brundrett and Jasper, 1999).

Trap culture method of AMF spore development is widely followed to obtain a mixed inoculum. Although the culture development in this method are higher compared to other methods, the culture developed in this method does not belong to single species and are not pure. Depending on the dominant species and host plant used, the propagation of AMF varies. The freshly collected soil samples can be maintained in this method to minimize the loss or viability of the AMF spores (Brundrett and Jasper, 1999). This chapter deals with the problem in AMF mass culturing and different methods used for large scale production of AMF (Figure 6.1).

2. Problems in AMF Mass Production

Obligate biotrophic nature of AMF hindered the inoculum production without host plants. Many factors affect the spore production including host plant involved, substrate used, environmental factors, fertilizers used and the origin of the AMF culture (Selvakumar *et al.*, 2016). Among the many substrates, soil is most widely used. Soil from the rhizosphere of indigenous AMF can be used as an effective medium as long as the soil is sterilized completely to remove the other unwanted microorganisms or pathogens associated with it. Selection of host is another important parameter in AMF mass production. Non-specific host plant tend to decrease the mycorrhizal dependency and eventually reduction in spore production (Estrada *et al.*, 2013). International culture collections such as The International

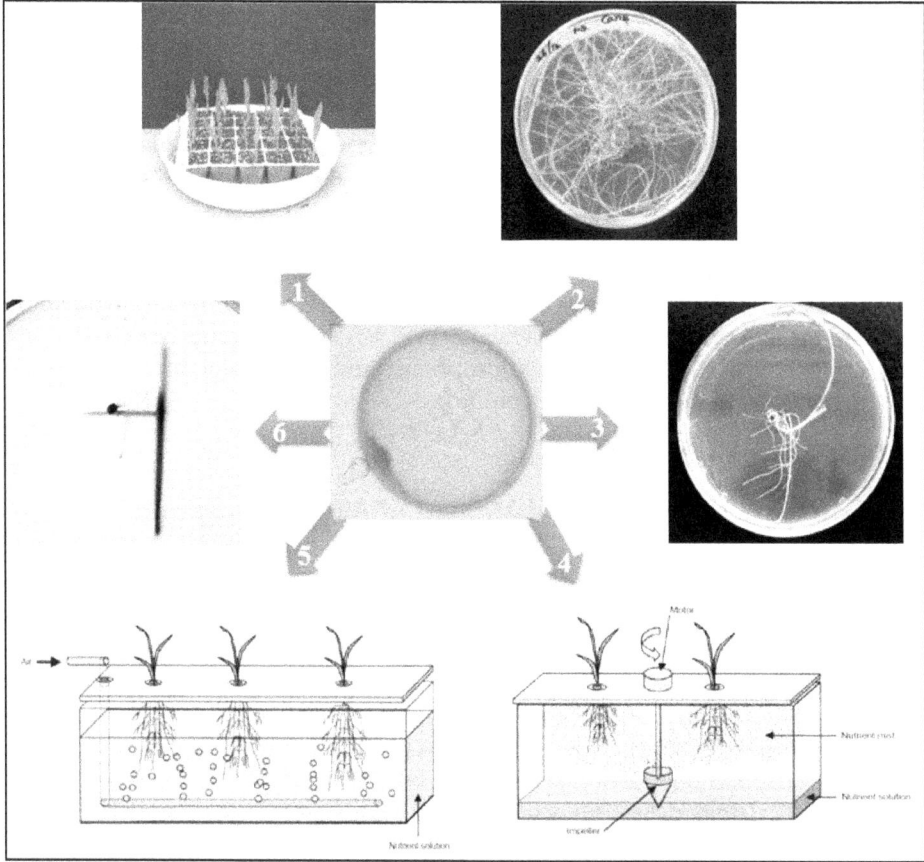

Figure 6.1: Different Methods of AMF Mass Production.
1: Soil based conventional method; 2: Hairy root; 3: Autotrophic culture system; 4: Aeroponics; 5: Hydroponics and 6: Slide method.

Bank of Glomeromycota (BEG), An International Culture Collection of Arbuscular Mycorrhizal Fungi (INVAM) and Glomeromycota In vitro Collection (GINCO) use different host plants such as sudan grass and red clover. Other commonly used host plants were Bahia grass (*Paspalum notatum Flugge*), maize (*Zea mays* L.), Leek (*Allium* spp.), Alfalfa (*Medicago sativa*) and onion. The INVAM culture collection center reported that the number of spores got reduced after successive propagation cycles in some pots. They suggested that change in host plant (for instance, shift from C3 to C4 host) may overcome this problem.

Environmental factors such as cold, water, drought and salt significantly influence spore production. Many authors reported that presence of AMF spores in high saline soils get reduced dramatically (Krishnamoorthy *et al.*, 2014, Estrada *et al.*, 2013, Ruiz-Lozano *et al.*, 2012). Although AMF known to improve plant growth under salt stress, their performance tend to reduce. Salt tolerant or indigenous AMF spore collected from salt affected soil may help better understanding and propagation under saline conditions. Manipulation of nutrient regimes has shown to impact AMF propagule production regardless of AMF and host plant used (Douds *et al.*, 2006; Sylvia and Schenck, 1983). Nutrient present in the substrate as well as the addition of macro and micro nutrients may alter the mycorrhizal dependency of the host plant and eventually affects AMF propagule production. Although it is unclear that to till which extent the addition of nutrients may influence AMF performance, adequate nutrients may support initial symbiosis of AMF and host plants. Douds and Schenck (1990) reported that manipulation of nutrient regimes in the soil influenced the AMF sporulation. They also found that increase in the nutrients other than phosphorous enhanced the AMF root colonization and spore count. Other important factor influencing AMF production is source of inoculum. Artificial medium like hairy root, hydrophonic and aerophonic methods do not have problem with the source of inoculum.

3. AMF Propagation Methods

3.1. Soil Based Conventional Method of AMF Propagation

Substrate based method of AMF propagation is more widely used and is most reliable technique as it is less artificial compared to other types of methods (Figure 6.1). Depending on the type of technique employed, the starter inoculum varied from isolated spores of single AMF (Panwer *et al.*, 2007) to mixtures of multiple AMF cultures (Gaur and Adholeya, 2002). The commonly used substrates are soil and sand (Sylvia and Schenck, 1983), pure sand (Millner and Kitt, 1992), other substitutes such as peat (Ma *et al.*, 2007), Glass beads (Neumann and George, 2005), vermiculite (Douds *et al.*, 2006), perlite (Lee and George, 2005), compost (Douds *et al.*, 2005) and calcinated clay (Plenchette *et al.*, 1982). These substrates can be used as medium for AMF spore production solely or in combination with other substrates. Different methods of cultivation techniques were followed for the last three decades to mass produce AMF in substrate based method. Some of the successful methods are discussed below.

The importance of the spore size of the substrate used in the AMF cultivation is described by Gaur and Adholeya (2000). They conducted two different experiments to check the impact of pore size and substrate medium for AMF mass production. In the first experiment, different sizes of pore size substrate was used. The result showed that substrate with the pore size of 0.78-0.50 mm and 0.50-0.25 mm produced high number of infective propagules compared to much larger pore size substrates and more smaller pore size substrates. This result suggests that substrates with better soil aeration, drainage, oxygen supply may favour AMF production. In the second experiment, the authors found that among the substrates tested, sand and clay-brick substrates were able to show high number of mycorrhizal parameters. In their experiment, they were able to produce upto 700 spores per 100 ml of substrate.

To obtain a monosporic culture of AMF, Fracchia *et al.* (2001) developed a new technique. They used mixture of vermiculite and perlite in Petri dish and one surface sterilized spore was inoculated with 10 ml of 10 mM 2-(N-morpholin) ethane sulphonic acid (MES) buffer (pH7). *Trifolium repens, Sorghum halepensis* and *Wedelia glauca* were used as a host plants. After successful germination the contents were transferred to Gel-Gro medium. The setup was kept for three weeks to allow initial mycorrhizal establishment. After three weeks, the content were transferred to pots containing steam sterilized soil and maintained for 8 weeks to obtain monosporic cultures. The authors suggested that this can permit to obtain a monosporic culture of AMF with high percentage of success.

The large scale application of AMF is far off vision as the significant development of this obligate biotroph is achieved only in the presence of host plant (Sahay *et al.*, 1998). In order to improve the recognition between host plant and AMF, Selvaraj and Kim (2004) proposed a new method. Root exudates were collected from maize seedlings and sterilized with bleach. Sucrose (25 per cent) and agar (25 per cent) were thoroughly mixed and shaped into globular structures. In the center of the globular structure, 10-20 spores and 1 ml of root exudate were injected and the hole was sealed with sucrose and agar mixture and maintained in the plate for 10 weeks. AMF spore inoculated in soil used as a control. After 10 weeks, it was found that sucrose agar with root exudate produced a high number of spores (270 spores per 100 g of globules) compared to soil inoculum (200 spores per 100 g of soil).

3.2. Substrate Based on-Farm Production Method

An alternative to *in vitro* propagation methods, on-farm production method for AMF spores has been established in late 1980s. Sieverding (1987, 1991) developed this technique and produced effective inoculum strain of *Glomus manihotis* in Columbia. In this method, initially 25 m² field plot was used after fumigating with methyl bromide and other fumigant. Use of this methyl bromide was banned in 2005 as it may cause harmful effect on ozone layer. Fumigation of the soil can be used to kill the indigenous AMF which may interfere with introduced AMF spores. Fumigation also kills weed seeds as it might contaminate the soil-based inoculum and also fumigation kills pathogens present in the soil. After fumigation of the soil, *G. manihotis* was inoculated in the holes drilled and then the seeds of host *Brachiaria decumbens* were sown. Alternatively, pre-colonized host plant seedlings can be sowed to minimize the need of starter inoculum. After four months, soil and roots were harvested to a depth of 20 cm to analyse AMF spore production.

The post-harvest analysis showed that higher spore production were in the fumigated soil as compared to unfumigated soil (Sieverding, 1987). After the increased success rate of this methods, it's been widely used with slight modifications according to the purposes. Dodd *et al.* (1990a,b) followed this technique and successfully propagated three AMF spores *i.e.* G. manihotis (250 spores g⁻¹), *Glomus occultum* (250 spores g⁻¹) and *Entrophospora Columbiana* (10 spores g⁻¹).

In order to improve the quantity of the inoculum produced in on-farm production method, Adholeya and co-workers (Gaur, 1997; Gaur *et al.*, 2000) developed an advance on-farm production technique using raised beds of soil 60 cm × 60 cm × 16 cm. After soil fumigation, AMF spores are inoculated into furrow in the beds. The starter inoculum is either indigenous spores or introduced spores. The difference with Sieverding's method is that the hosts are grown for three years with rotation of crops such as *Sorghum Sudanese*, *Zea mays* and *Daucus carota* for every four months. After three years of propagation, AMF production yielded upto 2.5 10⁶ propagules per bed. Later Gaur *et al.* (2000) and Gaur and Adholeya (2002) modified this technique to reduce the host crop cycle and produce AMF in shorter time. Same raised bed method is used with the exception that substrate is 2:1 soil and leaf compost mixture. Forage crops and vegetable crops were sown as host plants for four months. The AMF spores yield was 40-fold fewer than the previous 3 year method. However, the inoculum was effective giving 51-119 per cent increase in shoot weight of vegetables (Gaur *et al.*, 2000).

Douds and co-workers (Douds *et al.*, 2006) modified this on-farm production method to provide more effective AMF inoculum using raised bed enclosures containing compost or soil diluted with vermiculite. The enclosures are made using silt fence or black plastic sheeting. Before transplanting the *Paspalum notatum* (bahia grass) host plants into enclosures, host plants are inoculated with AMF and grown for 15 weeks in greenhouse. Plants are grown in a pots containing field soil, sand, vermiculite and calcined clay. The enclosures are filled with yard clipping compost and vermiculite mixture. The vermiculite is added to dilute the nutrient rich compost and soil. The colonized plants are transferred into enclosures and maintained for 6 weeks. Douds *et al.* (2006) reported that AMF yielded upto 530 propagules cm⁻³ and the spores are produced upto 80 cm⁻³. The compost and vermiculite mixture produced the high number of AMF propagules, Douds *et al.* (2010) modified the technique by diluting the field soil with yard clipping compost, vermiculite, perlite or peat based horticultural potting media. The inoculum used are indigenous spores. The spores produced in this method are several hundred fold higher than the original soil.

3.3. Aeroponics

Aeroponics is a form of hydroponics in that the roots that are colonized by AMF are bathed in the mist of nutrient solution. Spraying of water droplets increases the aeration and the liquid film around the roots allows gas exchange. Hung and Sylvia, (1988) mass produced AMF using aeroponics culture and tested the colonization and storability of the produced inoculum. Inoculum produced by aeroponics culture showed more than 40 per cent colonization of Bahia grass roots even after 9 months of storage. A study conducted by Mohammad *et al.* (2000), to determine the effect of microdroplet size on the cultivation of AMF in aeroponics. Results of this study showed that the microdroplet size of 1 μm in diameter produced by ultrasonic nebulizer technology was more effective in culturing *G. intraradices* than that of the microdroplet produced by atomizing disk. Recently, de Santana *et al.* (2014) used sweet potato plants as a host in aeroponics system. *Claroideoglomus etunicatum* and *Glomus clarum* colonized sweet potato roots by 49 per cent and 43 per cent respectively. And the inoculum produced by aeroponics system was viable even after 10 months of storage period (de Santana *et al.*, 2014).

In both hydroponics and aeroponics precolonized plants are used. For pre-inoculation plant seedlings and the surface disinfected AMF spores/colonized roots are precultured in a seedling tray/pots. After the plants roots were colonized by AMF, the plants were transferred to the substrate-free production systems such as hydroponics and aeroponics. In substrate-free production systems, the nutrient solutions are regularly changed. Besides the prevention of mineral depletion, the periodic changes in the nutrient solution reduced the problems that occurs by build up of undesirable toxin and contamination in the medium.

The main advantages of substrate free cultivation system is the production of inoculum which are free from substrate particles. In addition, the produced inoculum can be directly used, and the spores can be easily separated from the roots. Nutrient content and pH of the nutrient solution can be monitored and/or manipulated in substrate-free cultivation system for higher inoculum production. Lack of substrate ensures the extensive root growth, which facilitate the higher amount of AMF colonized root inoculum and ensures clean AMF propagules (Abdul-Khaliq *et al.,* 2001). The disadvantages are the development of algae growth in nutrient solution and microbial contamination. Carrier free condition could affect the spore production rate and the percentage of AMF spore germination also get affected.

3.4. Hydroponics and Nutrient Film Technique (NFT)

Hydroponic AMF mass culture can be done in two ways, by static culture or by the Nutrient Film Technique (NFT), which continuously circulates nutrient solution through a trough in which plants grow. A rectangular container filled with an aerated nutrient solution is used. AMF colonized growing plants are floated on supports, and their roots are immersed in the nutrient solution. Hawkins and George (1997) used *Triticum aestivum, Sorghum bicolor* and *Linum usitatissimum* as a host plants and the colonization percentage was 73, 36 and 65 respectively. Static culture gives rise to root aeration problems, pH shifts and accumulation of plant and microbial metabolites. All of these perceived problems were thought to result in poor fungal development and thus prevent commercial AMF production.

NFT are widely used in commercial horticulture production. In this method the plant roots lie in a shallow layer of rapidly flowing nutrient solution and as root mat develops, the upper layer above the liquid retains a film of moisture around the roots. The main advantage of this method is that, this system provides enough air and nutrient solution to the AMF without disturbing the hyphal growth. *Gigaspora margarita* inoculum produced by NFT using Mung bean as a host plant was found to colonize upto 84 per cent of maize roots in pot culture condition (Mathew and Johri, 1987). AMF inoculum production using *Glomus mosseae* was found to colonize the lettuce plants up to 85 per cent under the NFT system (Lee and George, 2005).

3.5. Slide Method–Pure Culture Development of AMF from Single Spore

Single spore inoculation is important to obtain a pure cultures of AMF. When single spore is used, mycorrhizal development and propagation would be very slow. However, this single spore inoculation is very useful technique to produce the pure cultures of AMF in large scale. This technique will also enable us to study specific or individual effect of AMF on plant growth and development (Panwar *et al.,* 2007). In this method, spore germination and hyphae elongation can be visualized under microscope.

Figure 6.2: Steps Involved in the Development of AMF Pure Culture using Single Spore.

1: Collection of rhizosphere soil; 2: Development of trap culture; 3: AMF spore isolation from trap culture; 4: Separation and surface disinfection of spores; 5: Inoculation of spores in hairy root/single spore mass culturing methods; 6: AMF root colonization in the spore mass culturing methods.

Mass culturing of AMF from single spores could be done as shown in Figure 6.2. First the rhizosphere sample from the field has to be taken and trap culture will be made. Trap culture is to increase the AMF population which are less in field collected samples. After successful establishment of the trap cultures AMF spores will be isolated by wet sieving and decanting method. The isolated spores will be surface disinfected with chloramine T and antibiotic solutions. Then the surface disinfected spores will be placed near the root which is developed in slide method. Plant and AMF symbiosis will be developed and the pure culture of AMF from single spore can be developed.

Recently, Selvakumar *et al.* (2016) modified the method developed by Panwar *et al.* (2007) to produce large amount of single AMF culture in slide method. Single spore collected from the soil is inoculated near the root zone of the sorghum seedlings in the glass slide containing moisture filter paper. The setup is kept in a falcon tube containing root exudates collected from growing sorghum seedlings. Spore germination and early root infection is monitored under microscope. After successful colonization, the setup is transferred to pots contain autoclaved soil and maintained for 240 days. After 120 days the crop cycle is changed by sowing new sorghum seeds. In this method, upto 391 spores per 100 g of soil were produced after 240 days.

3.6. AMF Mass Culture on *in vitro* Agar Medium

Obligate symbiotic nature of arbuscular mycorrhizal fungi hampered *in vitro* artificial medium cultivation. Since 1970s three major breakthroughs were occurred in this field, first is *in vitro* culturing of AMF in hairy root was initiated by Mosse and Hepper 1975, second was during 1986, when Strullu and Romand achieved the first sub-culture of AMF. In 1988, Becard and Fortin (1988) adopted the *Agrobacterium rhizogenus* transformed hairy roots for AMF cultivation. Establishment of *in vitro* cultures has greatly influenced our understanding on arbuscular mycorrhizal symbiosis, molecular responses under different stress conditions (Gonzalez-Chavez *et al.*, 2011). In addition to this, it results in production of pure and high quality inoculum. Inoculum produced by these techniques are highly pure and the quality is excellent. However, the problem in this technique are high investment cost, not all the AMF are successfully cultivable and not easily adoptable by farmers (Gianinazzi and Vosatka, 2004). Voets *et al.* (2009) proved that *in vitro* grown seedlings (autotrophic culture system) may be efficient in AMF colonization and spore production in short time compared to hairy root methods. Recently, *in vitro* developed tissue culture plants were inoculated with AMF and colonized, which enhance the plant adaptation in the main field (Koffi and Declerck, 2015).

3.7. Hairy Root/Root Organ Culturing

Several attempts have been made in the past to overcome AMF mass culturing. *Agrobacterium rhizogenes* is a gram negative soil inhabiting bacteria, which produces 'hairy roots' as a result of the modified hormonal balance of the tissue that makes them vigorous and allows it to grow rapidly on artificial media. After the development of hairy roots AMF spores/propagules are introduced to establish plant AMF symbiosis. *In vitro* developed AMF cultures are the powerful tool to study the symbiosis between AMF and plants. In addition, this system was useful for the production of pure, concentrated and aseptic AMF inoculum.

Root-organ cultures were first developed by White and coworkers (White, 1943). In this method authors used excised roots on synthetic mineral media supplemented with vitamins and a carbohydrate source. Use of AMF in root culture was first developed by Mosse and Hepper (1975). The authors used *Lycopersicum esculentum* Mill. (tomato) and *Trifolium pratense* L. (red clover) root cultures to establish *in vitro* mycorrhiza with *Glomus mosseae*. A natural genetic transformation of plants by *Agrobacterium rhizogenes* produces a condition known as hairy roots (Riker *et al.*, 1930). These hairy roots are widely used for various fundamental studies and in addition, hairy roots were used for AMF mass culturing (Mugnier and Mosse, 1987). In this method, mycorrhizal fungi produced spores when the concentration of nutrient in the medium are reduced. To overcome this, bi-compartmental system was developed (St-Arnaud *et al.*, 1996). In addition, many experiments were conducted to increase the efficiency AMF spore production. In that, gel replacement method gained more attention and successful for higher spores production in hairy root cultures (Douds Jr, 2002).

In bi-compartmental system a divider will be placed in the center so that the petri plate will be separated into two compartments. Bi-compartment was first developed by St-Arnaud *et al.* (1996), which consisted of filling the root compartment with sucrose minimal medium and adding one third of this volume to the sucrose free medium in another compartment and then a gel slop was made along the middle petri dish wall. The gel slop act as a connector between the two compartments of the petri dish. However, this kind of connector may not be effective for slow growing AMF strains, because longer time required to colonize the root compartment and reach the next compartment. The gel slope connector tend to dry and the hyphae will not cross from the root compartment. To simplify this Dalpe and Seguin (2010) developed a new method called paper bridge method. This is made by placing the sterile folded filter paper on the top of the divider, which physically link the root and fungal compartments.

AMF sporocarps are used as inoculum for the *in vitro* mass culture using *Daucus carota* hairy roots as host (Bi *et al.*, 2004). A single sporocarp was able to produce forty seven sporocarp after 6 months of incubation. Potato and carrot hairy roots were used for the mass culturing of *Glomus intraradices* under *in vitro* condition. Colonization percentage under in vitro condition was 52.88 per cent in potato and 82.89 per cent in carrot hairy root (Puri and Adholeya, 2013).

3.8. Autotrophic Culture System

Autotrophic culture system is another method of *in vitro* mycorrhization of plantlets. Roots of the plantlets are inside the petri dish and the shoots develop in open air condition. This method has advantage over hairy root method. In hairy root method non-photosynthetic plant root materials will be used and in autotrophic method, photosynthetic plant material will be used. This system can be used to continuously culture AMF and also serve as a powerful tool to study various aspects of AMF and plant interactions. Two weeks old clover seedlings were transferred to gel gro medium which contained the single germinated AMF spores and maintained for 8 weeks. More than 45 per cent of the clover roots are colonized by the single spore (Fracchia *et al.*, 2001). This method help to develop monosporic AMF autotrophic culture system. Voets *et al.* (2005) developed autotrophic culture system for *in vitro* mycorrhization using potato plantlets.

3.9. Mycorrhization of Tissue Culture Plants

Photoautotrophic micropropagation method was developed in last 1980s (Kozai and Iwanami 1988). Micropropagation technique is characterizes as the controllable microenvironment in which plantlets are grown photoautotrophically with enriched CO_2 concentration, the optimal air exchange and low photosynthetic photon flux density. It is estimated that about 40 billion plants are required per year for the re-afforestation. Micropropagation techniques may useful to produce large quantities of uniform high-value forestry and horticultural crops. Micropropagation presents several advantages such as high multiplication rate, small space requirement, season independence and free from pest and diseases.

In vitro mycorrhization refers to the inoculation of AMF to the roots of plantlets developed by micropropagation. Generally, micropropagated plantlets are first grown in *in vitro* condition which is devoid of microorganisms. Therefore, no change for AMF colonization prior to transplantation without artificial introduction. Hence, it has been suggested that the mycorrhization of plantlets could be better protected against various biotic and abiotic stress that occurs in field conditions (Vestberg *et al.*, 2004). Furthermore, by introducing AMF in plantlets may reduce the fertilizers and pesticide uses. The steps involved in the introduction of AMF during micropropagation was illustrated in Figure 6.3.

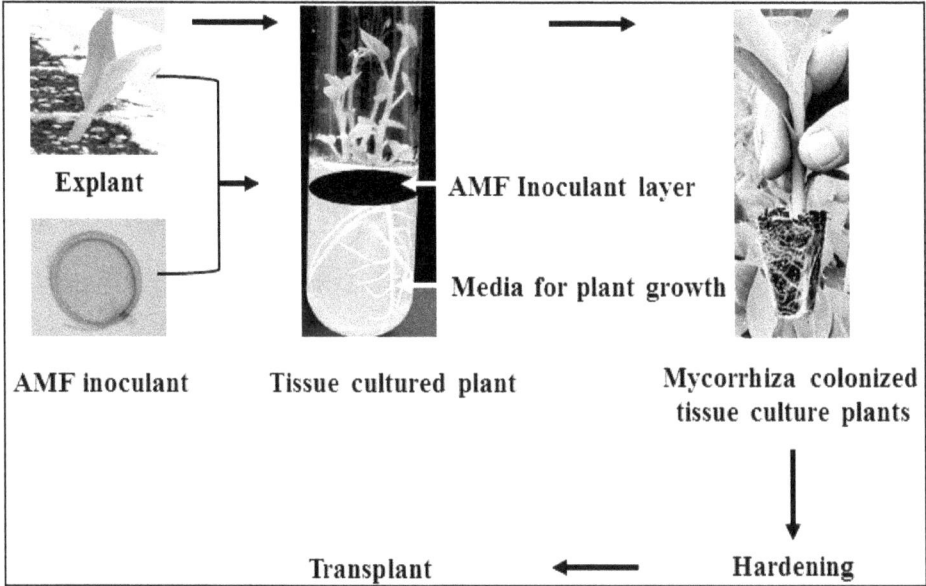

Figure 6.3: Mycorrhization of Micropropagated Tissue Culture Platelets.

Many researchers have suggested that the mycorrhization is beneficial for the establishment and health of propagated plantlets post-transplantation. Mycorrhization of tissue culture banana significantly improved the acclimatization compared to the non mycorrhization plantlets (Koffi and Declerck, 2015). In addition, mycorrhizal colonized banana plantlets has significantly increased the Pseudostem height, Pseudostem diameter, shoot and root dry weight than that of non-mycorrhizal plantlets after 7 weeks of acclimatization (Koffi and Declerck, 2015).

Another study focused on the mycorrhization latex-producing *Hevea brasiliensis* plantlets with dense extraradical mycelium network of the arbuscular mycorrhizal fungus *Rhizophagus irregularis* MUCT 41833 developed from a mycelium donor plant (*Medicago truncatula* A17). Intraradical structures such as hyphae, arbuscules and spores/vesicles were observed in the root of *Hevea brasiliensis* plantlets (Sosa-Rodriguez *et al.*, 2013). These studies shows that the *in vitro* mycorrhization of *in vitro* produced plantlets acclimatization and development during the ex vitro cultivation could be improved. AMF spore produced using different methods are given in Table 6.1.

Table 6.1: AMF Propagules Produced in different Culturing Methods and the Efficiency AMF Colonization in Host Plants

Mass Culture Method	AMF Species	Plant Material Used	Spores	Colonization (per cent)	References
Slide method	Gigaspora margarita S-9	Sorghum sudangrass hybrid	28.67/100g soil	82.83	Selvakumar et al., 2016
	Claroideoglomus lamellosum S-11	Sorghum sudangrass hybrid	391.33/100g soil	33.87	Selvakumar et al., 2016
	Gigaspora margarita S-23	Sorghum sudangrass hybrid	235.33/100g soil	73.17	Selvakumar et al., 2016
Aeroponics	G. intraradices	Hegari	~ 175,000 propagules/gram	NA	Mohammad et al., 2000
	Claroideoglomus etunicatum	Sweet potato	5.380 g^{-1} root	49.2	de Santana et al., 2014
	Glomus clarum	Sweet potato	1.661 g^{-1} root	43.6	de Santana et al., 2014
Hydroponics	Glomus intraradices	Phaseolus vulgaris	NA	40	Tajini et al., 2009
	Glomus sp.	Onion	3,738 spores per plant	NA	Jarstfer et al., 1988
Hairy root/Root	Glomus intraradices	Potato root	21936/100ml medium	52.88	Puri and Adholeya, 2013
Organ Culture	Glomus intraradices	Carrot root	24853/100ml medium	82.89	Puri and Adholeya, 2013
	Glomus intraradices	carrot	10000	74.7	Tiwari and Adholeya, 2003
	Glomus intraradices	clover	2000	30.0	Tiwari and Adholeya, 2003
	Glomus intraradices	carrot	21991	36.0	Douds Jr, 2002
Sucrose-agar globule	Gigaspora gigantea	Zingiber officinale	410	90	Selvaraj and Kim, 2004
with root exudates	Glomus fasciculatum	Zingiber officinale	468	93	Selvaraj and Kim, 2004
Autotrophic system	Gigaspora rosea	Clover	NA	50	Fracchia et al., 2001
	Gigaspora sp.	Clover	NA	52	Fracchia et al., 2001
	Glomus mosseae	Clover	NA	48	Fracchia et al., 2001
	Glomus sp.	Clover	NA	45	Fracchia et al., 2001
Soil based methods	Glomus spp.	Plantago lanceolata	163 spores/g soil	NA	Gryndler et al., 2003
	Glomus intraradices	Zea mays	600 propagules/100 g substrate	NA	Gaur and Adholeya 2000

Contd...

Table 6.1–*Contd...*

Mass Culture Method	AMF Species	Plant Material Used	Spores	Colonization (per cent)	References
	Rhizophagus clarus	Sorghum bicolor	162/100cm^{-3} soil	59.4	Schlemper and Sturmer, 2014
	Claroideoglomus etunicatus	Sorghum bicolor	240/100cm^{-3} soil	10.6	Schlemper and Sturmer, 2014
Nutrient film technique	Glomus mosseae	Lettuce	NA	86.3	Lee and George, 2005
	Gigaspora margarita	Mung been	NA	84	Mathew and Johri, 1987

NA: Not available.

4. Limitation in AMF Mass Production

Although AMF is well known for their application in agriculture, horticulture and eco-environmental protection, *in vitro* propagation may not be sufficient for large scale application. In the past, AMF culture techniques such as pot cultures, hydroponic culture, aeroponic and *in vitro* culture methods were developed with little consideration of large scale production and the quality of inoculum. Although pot culture method was shown to have considerable success, presence of other unwanted microorganisms limits their application. The limitation in the hydroponic and aeroponic culture methods are difficult to master the common producers and hard to control the quality of AMF propagules. Other methods such as *in vitro* aseptic culture and organ culture need complex technology with no commercial inoculant output (Liu and Yang, 2008). Most of the AMF mass production techniques were focused on the practical use.

5. Uses of Mass Produced Mycorrhizal Inoculum in Field Application

Chen *et al.* (2001) proposed this glass bead compartment cultivation system. In which they checked the AMF spore efficiency in nutrient and trace metal uptake using a container made of Plexigals (Acrylic). The container was divided into five different compartments using different pore sizes of screens. The first and last compartment was received inoculum containing spores, root and hyphae along with host plant and was separated using 1 mm pore size screen. The second and fourth compartment was separated from the middle compartment using 30 μm sieve and received glass beads or coarse river sand. The middle compartment received only glass beads. This method is very useful to monitor the mycorrhizal inoculum efficiency on nutrient uptake of other heavy metal uptakes.

6. Conclusion and Future Perspective

The propagation of AMF is complex process and time consuming. The *in vitro* propagation techniques are very useful to study the morphological and physiological characters of this obligate endosymbiont. Hairy root method is useful to study the molecular mechanism of symbiosis between AMF and host plants. Spores produced in this method are useful to study their effect in greenhouse experiments. Spore produced in the autotrophs are similar to those of produced in hairy root with the exception that the host plants are not in the autotrophs rather than genetically modified hairy roots. Contamination by other microorganisms can be easily controlled in hydroponic method. Although hairy root, autotrophic, hydrophonic methods are very useful to propagate single AMF species without other microbial contaminations, the quantity of the outcome is not enough for large scale application. Although slide method produce AMF spores with high quality in soil, their application limits to experimental purposes alone. All these methods can yield high quality of AMF spores which can be used for experimental purposes. On-farm production method is appropriate technique for production of AMF in large scale. On-farm production technique is also very advantageous as it is least artificial and uses indigenous AMF spore for propagation.

References

Abdul-Khaliq, Gupta, M.L. and Alam, A., 2001. Biotechnological approaches for mass production of arbuscular mycorrhizal fungi: current scenario and future strategies. In: Mukerji, K.G., Manoharachary, C., Chamola, B.P. (Eds.), Techniques in Mycorrhizal Studies. Kluwer Academic Publishers, The Netherlands, pp. 299–312.

Becard G. and Fortin J.A. 1988. Early events of vesicular-arbuscular mycorrhiza formation on Ri T-DNA transformed roots. *New Phytol.*, 108: 211–218.

Bi, Y., Li, X., Wang. H. and Christie, P. 2004. Establishment of monoxenic culture between the arbuscular mycorrhizal fungus Glomus sinuosum and Ri T-DNA-transformed carrot roots. *Plant Soil*, 261: 239–244.

Brundrett, M.C. and Jasper, D.A. 1999. Glomalean mycorrhizal fungi from tropical Australia. *Mycorrhiza*, 8: 305-314.

Chen, B., Christie, P. and Li, X. 2001. A modified glass bead compartment cultivation system for studies on nutrient and trace metal uptake by arbuscular mycorrhiza. *Chemosphere*, 42: 185-192.

Dalpe, Y. and Seguin, S. 2010. A "paper bridge" system to improve *in-vitro* propagation of arbuscular mycorrhizal fungi. *Botany*, 88: 617–620.

de Santana, A.S., Cavalcante, U.M.T., de Sa Barreto Sampaio, E. and Costa Maia, L. 2014. Production, storage and costs of inoculum of arbuscular mycorrhizal fungi (AMF). *Braz. J. Bot.*, 37: 159–165.

Dodd, J.C., Arias, I., Koomen, I. and Hayman, D.S. 1990a. The management of populations of vesicular-arbuscular mycorrhizal fungi in acid-infertile soils of a savannah ecosystem. II. The effect of pre-cropping and inoculation with VAM-fungi on plant growth and nutrition in the field. *Plant Soil*, 122: 229-240.

Dodd, J.C., Arias, I., Koomen, I. and Hayman, D.S. 1990b. The management of populations of vesicular-arbuscular mycorrhizal fungi in acid-infertile soils of a savannah ecosystem. II. The effects of pre-crops on the spore populations of native and introduced VAM fungi. *Plant Soil*, 122: 241-247.

Douds, D.D., Nagahashi, G. and Hepperly, P.R. 2010. On-farm production of inoculum of indigenous arbuscular mycorrhizal fungi and assessment of diluents of compost for inoculum production. *Biores. Tech.*, 101: 2326-2330.

Douds, Jr D.D. 2002. Increased spore production by Glomus intraradices in the split-plate monoxenic culture system by repeated harvest, gel replacement, and resupply of glucose to the mycorrhiza. *Mycorrhiza*, 12: 163–167.

Douds, D.D. and Schenck, N.C. 1990. Increased sporulation of vesicular-arbuscular mycorrhizal fungi by manipulation of nutrient regimens. 1990. *Appl. Environ. Microbiol.*, 56: 413-418.

Douds, D.D. Jr, Nagahashi, G., Pfeffer, P.E., Kayser, W.M. and Reider, C. 2005. On-farm production and utilization of arbuscular mycorrhizal fungus inoculum. *Can. J. Plant Sci.*, 85: 15–21.

Douds, D.D. Jr, Nagahashi, G., Pfeffer, P.E., Reider, C. and Kayser, W.M. 2006. On-farm production of AM fungus inoculum in mixtures of compost and vermiculite. *Biores. Tech.*, 97: 809–818.

Estrada, B., Aroca, R., Maathuis, F.J.M., Barea, J.M. and Ruiz-Lozano, J.M. 2013. Arbuscular mycorrhizal fungi native from a Mediterranean saline area enhance maize tolerance to salinity through improved ion homeostasis. *Plant Cell Environ.*, 36: 1771-1782.

Fracchia, S. Menendez, A. Godeas, A. and Ocampo, J.A. 2001. A method to obtain monosporic cultures of arbuscular mycorrhizal fungi. *Soil Biol. Biochem.*, 33: 1283-1285.

Gaur, A. 1997. Inoculum production technology development of vesicular-arbuscular mycorrhizae. Ph. D. thesis. University of Delhi, Delhi, India.

Gaur, A. and Adholeya A. 2002. Arbuscular mycorrhizal inoculation of five tropical fodder crops and inoculum production in marginal soil amended with organic matter. *Biol. Fertil. Soils*, 35: 214-218.

Gaur, A. and Adholeya, A. 2000. Effects of the particle size of soil-less substrates upon AM fungus inoculum production. *Mycorrhiza*, 10: 43-48.

Gaur, A., Adholeya, A. and Mukerji, K.G. 2000. On-farm production of VAM inoculum and vegetable crops in marginal soil amended with organic matter. *Tropical Agric.*, 77: 21-26.

Gianinazzi, S. and Vosátka, M. 2004. Inoculum of arbuscular mycorrhizal fungi for production systems: science meets business. *Can. J. Bot.*, 82: 1264–1271.

González-Chávez, M.C., Ortega-Larrocea, M.P., Carrillo-González, R., López-Meyer, M., Xoconostle-Cázares, B., Gomez, S.K., Harrison, M.J., Figueroa-López, A.M. and Maldonado-Mendoza, I.E. 2011. Arsenate induces the expression of fungal genes involved in As transport in arbuscular mycorrhiza. *Fungal Biol.*, 115: 1197-1209.

Gryndler, M., Jansa, J., Hršelová, H., Chvátalové, I., Vosátka, M. 2003. Chitin stimulates development and sporulation of arbuscular mycorrhizal fungi. *Appl. Soil Ecol.*, 22: 283–287.

Hawkins, H.J. and George, E. 1997. Hydroponic culture of the mycorrhizal fungus *Glomus mosseae* with *Linum usitatissimum* L., *Sorghum bicolor* L. and *Triticum aestivum* L. *Plant Soil*, 196: 143-149.

Hung, L. and Sylvia, D.M. 1998. Production of Vesicular-Arbuscular Mycorrhizal Fungus Inoculum in Aeroponic Culture. *Appl. Environ. Microbiol.*, 54: 353-357.

Jansa, J., Mozafar, A., Anken, T., Ruh, R., Sanders, I.R. and Frossard, E. 2002. Diversity and structure of AMF communities as affected by tillage in a temperate soil. *Mycorrhiza*, 12: 225-234.

Jarstfer, A.G., Farmer-Koppenol, P. and Sylvia, D.M. 1988. Tissue magnesium and calcium affect arbuscular mycorrhiza development and fungal reproduction. *Mycorrhiza*, 7: 237-342.

Koffi, M.C. and Declerck, S. 2015. *In vitro* mycorrhization of banana (*Musa acuminata*) plantlets improves their growth during acclimatization. *In Vitro Cell. Dev. Biol.-Plant*, 51: 265-273.

Kozai, T. and Iwanami, Y. 1988. Effects of CO_2 enrichment and sucrose concentration under high photon flux on plantlet growth of carnation (*Dianthus caryophyllus* L.) in tissue culture during the preparation stage. *J. Jpn. Soc. Hortic. Sci.*, 57: 279–288.

Krishnamoorthy, R., Kim, K., Kim, C. and Sa, T. 2014. Changes of arbuscular mycorrhizal traits and community structure with respect to soil salinity in a coastal reclamation land. *Soil Biol. Biochem.*, 72: 1-10.

Lee, Y.J. and George, E. 2005. Development of a nutrient film technique culture system for arbuscular mycorrhizal plants. *hortsci.*, 40: 378–380.

Liu, W.K. and Yang, Q. 2008. Integration of mycorrhization and photoautotrophic micropropagation in vito: feasibility analysis for mass production of mycorrhizal transplants and inoculants of arbuscular mycorrhizal fungi. *Plant Cell Tiss. Org. Cult.*, 95: 131-139.

Ma, N., Yokoyama, K. and Marumoto, T. 2007 Effect of peat on mycorrhizal colonization and effectiveness of the arbuscular mycorrhizal fungus Gigaspora margarita. *Soil Sci. Plant Nutr.*, 53: 744–752.

Mathew, J. and Johri, B.N. 1987. Vesicular-arbuscular mycorrhizal inoculum through nutrient film technique (NFT) grown moong (*Phaseolus mungo*). In: Sylvia, D.M., Hung, L.L. and Graham, J.H. (Eds.) Proceedings of the Seventh North American Conference on Mycorrhiza, edited by Florida: University of Florida. 364 pp.

Millner, P.D. and Kitt, D.G. 1992. The Beltsville method for soilless production of vesicular–arbuscular mycorrhizal fungi. *Mycorrhiza*, 2: 9-15.

Mohammad, A., Khan, A.G. and Kuek, C. 2000. Improved aeroponic culture of inocula of arbuscular mycorrhizal fungi. *Mycorrhiza*, 9: 337–339.

Mosse, B. and Hepper, C.M. 1975. Vesicular–arbuscular infections in root-organ cultures. *Physiol. Plant Pathol.*, 5: 215–223.

Mugnier, J. and Mosse, B. 1987. Spore germination and viability of a vesicular-arbuscular mycorrhizal fungus, Glomus mosseae. *Trans. Br. Mycol. Soc.*, 88: 411–413.

Neumann, E. and George, E. 2005. Extraction of extraradical arbuscular mycorrhizal mycelium from compartments filled with soil and glass beads. *Mycorrhiza*, 15: 533–537.

Panwer, J., Tarafdar, J.C., Yadav, R.S., Saini, V.K., Aseri, G.K. and Vyas, A. 2007. Technique for visual demonstration of germinating arbuscular mycorrhizal spores and their multiplication in pots. *J. Plant Nutr. Soil Sci.*, 170: 659-663.

Plenchette, C., Furlan, V. and Fortin, J.A. 1982. Effects of different endomycorrhizal fungi on 5 host plants grown on calcined montmorillonite clay. *J. Am. Soc. Hortic. Sci.*, 107: 535–538.

Puri, A. and Adholeya, A. 2013. A new system using *Solanum tuberosum* for the co-cultivation of Glomus intraradices and its potential for mass producing spores of arbuscular mycorrhizal fungi. *Symbiosis,* 59: 87–97.

Riker, A.J., Banfield, W.M., Wright, W.H., Keitt, G.W. and Sagen, H.E. 1930. Studies on infectious hairy root of nursery apple trees. *J. Agric. Res.,* 41: 507–540.

Ruiz-Lozano, J.M., Porcel, R., Azcon, C. and Aroca, R. 2012. Regulation by arbuscular mycorrhizae of the integrated physiological response to salinity in plants: new challenges in physiological and molecular studies. *J. Exp. Bot.,* 63: 4033-4044.

Sahay, N.S., Sudha, Archana, S. and Varma, A. 1998. Trends in endomycorrhizal research. *Ind. J. Exp. Boil.,* 36: 1069-1086.

Schlemper, T.R. and Sturmer, S.L. 2014. On farm production of arbuscular mycorrhizal fungi inoculum using lignocellulosic agrowastes. *Mycorrhiza,* 24: 571-580.

Selvakumar, G., Krishnamoorthy, R., Kim, K. and Sa, T. 2016. Propagation technique of arbuscular mycorrhizal fungi isolated from coastal reclamation land. *Eur. J. Soil Biol.,* 74: 39-44.

Selvaraj, T. and Kim, H. 2004. Use of sucrose-agar globule with root exudates for mass production of vesicular arbuscular mycorrhizal fungi. *J. Microbiol.,* 42: 60-63.

Sieverding, E. 1991. Vesicular-arbuscular mycorrhiza management in tropical agrosystems. Deutsche Gesellschaft fur Technische Zusammansabeit (GT2) GmbH. Eschbon, Germany.

Sieverdling, E. 1987. On-farm production of VAM inoculum. In: Sylvia, D.M., Hung, L.L. and Graham, J.H. (Eds.) Proceedings of the 7th North Amer. Conf, on Mycorrhiza, Gainesville, FL, p. 284.

Smith, S.E. and Read, D.J. 2008. Mycorrhizal symbiosis, 3rd edn. Academic, London.

Sosa-Rodriguez, T., Dupré de Boulois, H., Granet, F., Gaurel, S., Melgarejo, L.M., Carron, M.P. and Declerck, S. 2013. *In Vitro Cell. Dev. Biol.- Plant,* 49: 207-215.

St-Arnaud, M., Hamel, C., Vimard, B., Caron, M. and Fortin, J.A. 1996. Enhanced hyphal growth and spore production of the arbuscular mycorrhizal fungus Glomus intraradices in an in vitro system in absence of host roots. *Mycol. Res.,* 100: 328–332.

Strullu, D.G. and Romand, C. 1986. Methode dobtention dendomycorhizes avesicules et arbuscules en conditions axeniques. C.R. Acad. Sci., Ser. III Sci. Vie 303: 245–250.

Sylvia, D.M. and Schenck, N.C. 1983. Application of superphosphate to mycorrhizal plants stimulates sporulation of phosphorus tolerant vesicular–arbuscular mycorrhizal fungi. *New Phytol.,* 95: 655–661.

Tajini, F., Suriyakup, P., Vailhe, H., Jansa, J. and Drevon, J.J. 2009. Assess suitability of hydroaeroponic culture to establish tripartite symbiosis between different AMF species, beans, and rhizobia. *BMC Plant Biol.,* 9: 73.

Tiwari, P. and Adholeya, A. 2003. Host dependent differential spread of Glomus intraradices on various Ri T- DNA transformed roots in vitro. *Mycol. Prog.*, 2: 171–177.

Vestberg, M., Kukkonen, S., Saari, K., Parikka, P., Huttunen, J., Tainio, L. Devos, N. Weekers, F., Kevers, C., Thonart, P., Lemoine, M.C., Cordier, C., Alabouvette, C. and Gianinazzi, S. 2004. Microbial inoculation for improving the growth and health of micropropagated strawberry. *Appl. Soil Ecol.*, 27: 243–258.

Voets, L., Enrique de la Providencia, I., Fernandez, K., IJdo, M., Cranenbrouck, S. and Declerck, S. 2009. Extraradical mycelium network of arbuscular mycorrhizal fungi allows fast colonization of seedlings under *in vitro* conditions. *Mycorrhiza*, 19: 347–356.

Voets, L., de Boulois, H.D., Renard, L., Strullu, D.G., Declerck, S., 2005. Development of an autotrophic culture system for the in vitro mycorrhization of potato plantlets. *FEMS Microbiol. Lett.*, 248: 111–118.

White, P.R. (Ed). 1943. A handbook of plant tissue culture. J. Cattel, Lancaster, Pa.

2017, Mycorrhizal Fungi
Editors: Ashok Aggarwal and Kuldeep Yadav
Published by: ASTRAL INTERNATIONAL PVT. LTD., NEW DELHI

Pages 157–174

7

Mass Multiplication of Arbuscular Mycorrhizal Fungi

Sapana Sharma[1]*, Sandeep Sharma[2], Ashok Aggarwal[3], Vivek Sharma[1], M.J. Singh[1] and Sunita Kaushik[4]

[1]*Regional Research Station, Punjab Agricultural University, Ballowal Saunkhri, Punjab*
[2]*Department of Soil Science, Punjab Agricultural University, Ludhiana, Punjab*
[3]*Department of Botany, Kurukshetra University, Kurukshetra, Haryana*
[4]*Department of Botany, D.A.V. College for Girls, Yamunanagar, Haryana*
*Corresponding Author: sapnasharma13@gmail.com

ABSTRACT

Numerous techniques have been developed in the past few decades for the mass production of arbuscular mycorrhizal (AM) fungi. The main obstacle behind the mass production techniques is the obligatory nature of these biotropic fungi and species level identification is not possible at early stage of development. Currently, in vitro cultivation methods such as hydroponic system and root organ culture have been widely used for mass production of AM fungi. But traditional method i.e. mass production in soil based media and living host is very popular and economical for the rapid production of these AM fungi. The aim of this review article is to highlight the recent and advanced methods used for mass production of AM fungi.

Keywords: Soil, Biofertilizers, Host, Colonization, Symbiosis.

1. Introduction

Arbuscular mycorrhiza (AM) is the most abundant kind of mycorrhiza described as "universal plant symbiosis". They are found in every taxonomic group of plants and the list of species not colonized is much shorter than the colonized ones. Lack of host specificity is even more critical characteristic feature of this symbiotic association. This association occurs in wide variety of hosts and habitats. The mutualistic nature of interaction differentiates it from other plant fungus associations. This symbiosis is primarily characterized by inorganic components from fungus to plant and organic components from plant to fungus. AM fungal symbionts are widespread and consist of aseptate hyphae belonging to the order Glomales of class Zygomycetes (Morton and Benny, 1990). Mycorrhizal associations involve 3-way interactions between host plants, mutualistic fungi and soil factors. They are beneficial microorganisms that are very essential as a booster for plant growth. However, the role of AM fungi in the soil is less well understood. Many soils of India are nutrient deficient or have substantial constraints to plant growth. The mycorrhizal symbiosis provides soil fertility and soil stability to the degraded (nutrient deficient) soils. From sociological point of view, many research and developmental companies are producing and commercializing AM fungi as a biofertilizer. As the AM fungi are eco-friendly, beneficial to soil, plants and other microfauna the plant growers now turn to use the AM fungi rather than chemical fertilizers. Hence, undoubtedly, mycorrhizal use and application in future have a potentially important role in agriculture, forestry and in socio-economics. The interest of scientists in manipulation of soil microorganisms and their metabolic processes arose from the numerous studies showing that these microorganisms have a great potential for improvement of soil quality, degradation or immobilization of xenobiotic compounds, plant growth and plant disease resistance.

They are beneficial to plants and provide nutrients like phosphorus, zinc, copper, potassium and calcium and helps the plants in the uptake of nutrients from low concentration zone around roots.

AM fungi are characterized by the presence of their unique extra radical mycelium branched haustoria like structure within the cortical cells, known as arbuscules (Smith and Read, 2008). The main role of arbuscules is to increase the surface of roots during nutrient transfer (Akhtar *et al.,* 2011). AM fungi colonize the plant roots and penetrate into surrounding soil, extends the root depletion zone and the root system (Akhtar and Panwar, 2011).

Different structures are formed in this symbiotic association like mycelium, arbuscules, vesicles, auxillary cells and spores.

Hyphae

Hyphae are filamentous network which ramify to form different shapes like H-shape, Y-shape and perform different functions. These hyphae grow along the roots, colonize in the soil and form new spores which are defined as fertile hyphae. *Arbuscules*: Arbuscules are formed by all glomalean fungi inside a cell within a plant root. These are the places where the plant and fungus exchange food and nutrients with each other (Strack *et al.,* 2003). Carbon and phosphorus and other nutrients may also be exchanged through hyphae that ramify inside the root, but it is likely that the arbuscules are the major site for nutrient exchange. Arbuscules look like minute sea anemones because they have many small projections that extend inside the plant cells. They are formed by repeated branching of hypha when it enters a cell of plant root. Arbuscules last for few days before they are dissolved and digested by the host plants.

Vesicles

Vesicles are swollen end cells either between root cells or within cell wall. Vesicles are also structures formed inside the plant root. They look like an oval bag and act as storage locations (Mosse, 1973) for fungal food reserves. Only three of the six genera of AM fungi (*Glomus*, *Acaulospora* and *Entrophospora*) form vesicles. They can act as infective propagules and remain viable for long periods (Diop *et al.*, 1994) and contain large amount of lipids (triacylglycerides).

Spores

Spores are between 10 and 1000 μm in size and form as swellings on subtending hypha in the soil or roots. They are of different colours. They function as storage structures, propagules, resting stages and reproductive structures.

Auxiliary Cells

These cells are found only in sub-order Gigasporinae. Auxiliary cells are cluster of thin walled cells. They function as temporary storage structure of carbon compounds.

Gallaud (1905) divided AM fungi into two structural classes *i.e.* Paris type and Arum Type named after the plants in which they are first described.

(a) *Paris Type*: This type of mycorrhizal association is reported in 41 angiospermic families and characterized by the presence of extensive intracellular hyphal coils and absence of intercellular phase. Hyphae spread by intracellular growth through cortical cells. Vesicles are intracellular and arbuscules are relatively few in number and may be restricted to a single layer of cells in the cortex. Resultant colonies have cumulated appearance.

(b) *Arum Type*: Characteristic feature of Arum type mycorrhizal is the presence of extensive intercellular hyphal growth, vesicles are intercellular or intracellular and development of arbuscules is terminal on intracellular hyphal branches and the resultant colonies are linear in appearance.

Host Specificity

AM associations occur in wide spectrum of tropical and temperate tree species. They are virtually ubiquitous and have wide ecological range. The fungi are considered to have low specificity of association with host plant species. The specificity of fungal response could contribute to the maintenance of diversity with in the AM fungal community. It is estimated that 90 per cent of plant species live in symbiosis with mycorrhizal fungi (Smith and Read, 1997). This association is not found in few plants namely members of the families Amaranthaceae, Pinaceae, Betulaceae, Cruciferae, Chenopodiaceae, Cyperaceae, Juncaceae, Polygonaceae and Orchidaceae.

When an AM fungus comes in contact with root of host plant then plant secretes some exudates which have been thought to stimulate the germination of AM spores. During its limited independent growth, the main carbon storage compounds of the fungus, are mobilized (Gasper *et al.*, 1997). This mobilization fuels the development of coenocytic germ tubes and provides carbon skeletons for anabolism. Asymbiotic growth is maintained for 1 or 2 weeks, during which germ tube development may reach several centimeters. However, if symbiosis is not successfully established within the limited period, AM fungi arrest their growth. Arrest of growth is accompanied by germ tube septation and nuclear autolysis, after which fungal propagules re-enter a state of dormancy and have the ability to regerminate several times. Growth arrest before complete depletion of carbon stores may be a strategy to increase the chances of finding an appropriate root to colonize.

Signaling

If and when the asymbiotically growing AM fungus does contact a host root, a series of signaling events occurs between the partners, which lead to the "acceptance". Over the years, evidence has accumulated that roots emit a volatile signal that stimulates the directional growth of the AM fungus towards them. One of the prime factors for this volatile signal is CO_2, which can stimulate extensive hyphal growth of some AM fungi *in vitro*.

Penetration

It is characterized by localized production of wall-degrading enzymes by the fungus and by the exertion of hydrostatic pressure by the hyphal tip. Hydrolytic enzymes seem to be involved in the penetration and development of AM fungi in plant roots. Cellulase, pectinase and xyloglucanase activities have been found in colonized roots and in the external mycelium of AM fungi (Garcia-Garrido *et al.*, 1992). Differences in cellulase and pectinase activities between some *Glomus* isotypes have been observed (Garcia-Rommera *et al.*, 1991). It is, therefore, possible that variations in colonization capacities of host tissues may be related to the ability of the fungi to produce hydrolytic enzymes (Gianinazzi-Pearson, 1994). Close observations have revealed that as the main hypha of diameter 20-30 µm approaches a root, it puts out a characteristic fan-shaped complex of lateral branches. Recent studies have indicated that topographical and biochemical signals on the root surface may be necessary for appressorium formation (Gadkar *et al.*, 2001) which aid in fungal penetration.

Growth of AM Fungi inside the Root

The fungus then develops extensively between and within root exodermal and cortical cells, and forms intraradical structures including arbuscules and lipid rich vesicles. Colonization of root cells induces dramatic changes in the cytoplasmic organization in host; vacuole fragmentation, transformation of the plasma membrane to a periarbuscular membrane covering the arbuscule, increase of the cytoplasm volume and numbers of cell organelles, as well as movement of the nucleus into a central position. The plastids form a dense network covering the symbiotic interface. A number of phytohormones (cytokinins, abscisic acid) as well as various secondary metabolites have been examined. Vesicles are initiated soon after the initiation of arbuscular colonization, but continue to develop when the arbuscules senesce.

Sporulation

Root colonization is accompanied by the development of an extraradical mycelium that includes characteristic branched structures (Mosse and Hepper, 1975). The timing of the onset of sporulation varies with species. It often occurs within 3 to 4 weeks after the onset of mycorrhizal colonization. These may be involved in the uptake of mineral nutrients by extraradical hyphae. The external spores develop on some of these branched structures completing the fungal life cycle.

2. Ecology of Mycorrhizal Fungi

AM fungi are ubiquitous and abundant in cultivated soil as compared to virgin soil (Chaudhary and Panja, 2007). Their presence goes on decreasing with the increasing depth of the soil. Distribution of AM fungi is affected by climatic and edaphic factors *e.g. G. mosseae* is found in acidic soil and *Gigaspora* in sandy soils. Soil often contains spores of many fungi. Soil management practices like chemical fertilizers applications may also change the mycorrhizal population (Mosse and Bowen, 1968). Reynolds *et al.* (2003) discussed many aspects of this plant microorganism interaction, and included the fact that it ishighly dependent on the soil environment, pH, moisture availability, nutrient availability and presence or absence of other microbes.

3. Taxonomy of Mycorrhizal Fungi

The presence of AM fungi is generally not restricted by taxa but the type and features of the ecosystem may well determine their distribution. Around 170 morphologically distinct species of Glomales belonging to 6 genera (*Glomus, Acaulospora, Gigaspora, Scutellospora, Sclerocystis* and *Entrophospora*) are credited to form AM in one or other plant species (Morton, 1997). Out of these, *Glomus, Acaulospora* and *Gigaspora* seem to predominate across different plant ecologies. *Sclerocystis* was earlier included in a separate family but now it is included under Glomaceae and two new genera *i.e. Paraglomus* and *Archaeospora* have been created to accommodate a few species of *Glomus*, based on phylogenetic divergence in the evolutionary conserved rDNA sequence between the species (Morton and Redecker, 2001). Several other researchers have made efforts to clarify AM fungi on the basis of morphological, biochemical and molecular characteristics (Reddy *et al.*, 2005; Manoharachary, 2004; Walker and Schussler, 2004, Schussler *et al.*, 2001; Redecker *et al.*, 2000; Morton and Redecker, 2001). Walker and Schussler (2004) have described nine genera *i.e. Acaulospora, Archaeospora, Diversispora, Entrophospora, Gigaspora, Glomus, Pacispora, Paraglomus* and *Scutellospora*.

4. Multiplication of AM Fungi

Since, isolation and selection of AMF species are effective for growth promotion and rising of pure culture of these species is difficult, a suitable host is required to maintain pure culture of AM inoculum. The beneficial uses of AM inoculum in agriculture and raising nurseries have been reported (Muthukumar *et al.*, 2001; Smith and Read, 1997). One most important consideration in inoculum production is the choice of the fungal isolates which are capable of growth promotion of target plant. Another one is the selection of host plant species upon which the fungus will grow (Ryan and Graham, 2002; Dalpe and Montreal, 2004). The plants that naturally facilitate the higher colonization of AM can be generally considered to be used as stock plants. Genney *et al.* (2001) observed that the degree of AM colonization was related to host density in the field. Host can influence diversity of mycorrhizal fungi under controlled conditions (Franke-Synder *et al.*, 2001). Although many AM fungi are thought to have a broad host range, the appropriate test plants for trap cultures should be evaluated to ensure maximum detection of fungal species in specific soils. Differential responses of AM fungi to host plant species may also play an important role in regulation of species composition and diversity of AM fungal communities (Eom *et al.*, 2000). Inspite of this, the degree of importance of these two above considerations also depends upon the growth system. Mass multiplication of AM fungi varies greatly with root structure and habitat of host plants.

With the advancement in technology, the preparation of AM inoculum should proceed further towards commercial application. The techniques like hydroponics, aeroponics or soil-less culture produce high quality inoculum with the number of propagules many times greater than the indigenous ones. The lesser cost of inoculum and ease of production are making these technologies applicable in less-developed agricultural areas as well as in highly industrialized agricultural systems. Irrespective of any production technique which may vary from most advanced to the simplest one, certain requirements of AM inoculum are known to be indisputably universal, though some of them are purely practical and majority of these principles are based on general host–parasite considerations. The inoculum ought to be infective in the field conditions. Besides that, it should also show its utility in the desired growth response for the targeted crop under production conditions. This is not all, the inoculum has to be pathogen-free, concentrated and should possess a shelf–life that allows series of changes, distribution and use without losing the latent capacity of propagules and their quantity. The parameter of infectivity is the power to enter and extend the surface and enter the roots of the host. Therefore prior examination or testing or certification is required.

Relative efficacy of mycorrhizal fungi is a measure of their power to promote plant growth or provide momentum for stress tolerance. Effectivity must be examined under field conditions which include pathogens, fertility levels, indigenous AM fungi, suitable growth media or soil. Concentration of the inoculum, position of inoculum (Abbott and Robson, 1984), colonization of inoculum (Wilson and Trinick, 1983), all affect the ability of the inoculated AM fungus to infect and increase growth of the crop (Hepper *et al.*, 1988).

The need for a concentrated inoculum is unquestionable. The suitable inoculum is such which is small in dose and large in reaction. Concentrated inoculum is easy to store, convenient to transport and suitable to apply. The concentrated inoculum can be produced by pot culture and aeroponic technique.

Undesirable organisms cause harmful effects. AM inoculum should not spread unwanted organisms such as plant pathogens or harmful bacteria. Unadulterated inoculum is therefore required to maintain purity.

A variety of inoculum formulations for AM fungi are known to occur. A mixture of spores, colonized roots, hyphae and soil from pot cultures grown in sterilized soil constitute a form of widely reported inoculum composition (Gemma *et al.*, 2002).

The most common application method places the inoculum below the seed or seedling before planting (Jackson *et al.*, 1972). The basic principle is that the inoculum must be in the root zone. The roots of seedlings must grow through the inoculum.

AM fungi vary much in their power to promote plant growth and must be screened for booster strains before they are used for inocula production (Muthukumar and Udaiyan, 2007). Booster strains are those which quickly enter and colonize the root system and help in nutrient translocation. There should be competition with the indigenous AM fungi (Sylvia *et al.*, 1993); should be able to colonize newly formed roots and form extensive hyphal network in the soil.

To reduce the chances of incompatibility between fungus and host or fungus and soil, the AM fungi should be isolated from rhizosphere of the plant for which the inoculum is to be used. Rhizosphere is the most accessible and most abundant source of inoculum for starter culture (Muthukumar and Udaiyan, 2007).

AM are maintained and mass produced in pot cultures on suitable host plants. The host plant selected should be suitable to agro-climate conditions of the area, having thick root system for sizeable sporulation and infection, annual in growth habit and adaptable to polyhouse conditions. The host plants also may stimulate selectively or limit sporulation of certain AM fungal species suggesting varied affinities between hosts and symbiont (Al-Raddad, 1995).

The composition of soil mixture is crucial for ensuring good mycorrhization and sporulation. Addition of substrate should also enhance growth of the host plant. The substrate should be such as to avoid detrimental effect on the inoculum; it should be light in weight to ensure economy in transportation cost; possess good water holding capacity, less leaching of essential nutrients and easy to be removed from root surface.

Container size has the power to affect AM inoculum. Pot size should match the potential volume of the root system (Ferguson and Menge, 1982a). Light quality and irradiance are very influential for colonization and spore production (Ferguson and Menge, 1982b). Proper irradiance provides good results. Lower light intensities caused reduced sporulation (Daft and El-Giahmi, 1978). Water has also been shown to affect VAM fungal sporulation. Unsaturated and non stressed water conditions are best for spore production (Nelson and Safir, 1982). A correlation between water content and spore number existed across a natural soil moisture gradient in the field.

Temperature apparently can affect sporulation (Ferguson and Menge, 1982a). The increased temperature does not keep pace with colonization and growth response (Furlan and Fortin, 1973). Pesticides and biocides greatly reduce VAM colonization and sporulation but some increase it (Sreenivasa and Bagyaraj, 1989). Several workers have tried and suggested different hosts under different soil conditions and substrates in addition to soil-sand potting mixture to promote the AM spore production.

Table 7.1

Sl.No.	Substrate	Host	Reference
1.	Soil	Rice (*Oryza sativa*)	Yeasmin *et al.* (2007)
2.	Marginal soil amended with organic matter	Peanut, Sorghum and Maize	Carrenho *et al.* (2002)
3.	Compost and vermicompost	Sorghum	Hameeda *et al.* (2007)
4.	Saline alkali soil + farmyard manure	Sorghum	Raghuwanshi and Upadhyaya (2004)
5.	Farmyard manure	Maize	Srivastava *et al.* (2001)
6.	Natural soil	*Hordeum vulgare, Triticum aestivum, Phaseolus vulgaris* and *Phaseolus mungo*	Chaurasia and Khare (2005)
7.	Fly ash	Bajra, Maize, Sorghum, Sudan grass	Neeraj and Yadav (2007); Sharma, 2004
8.	Compost + soilrite	Maize, Rice, Onion	Gupta *et al.* (2006)
9.	Sterile sand: soil (1:1)	Onion (*Allium Cepa*)	Murugan and Selvaraj (2003)
10.	Fly ash + Soil	*Allium cepa*	Sheela and Sundaram (2003)
11.	Sand soil	Onion, *Cenchurus*	Nehra *et al.* (2003)
12.	Sand soil, Loam, solirite, farm yard manure	*Sesbania aculeate*, Maize, Chickpea, Sorghum	Gill and Singh (2001)
13.	Fly ash	*Allium cepa, Arachis hypogea, Vigna mungo*	Selvam and Mahadevan (2002)
14.	Solarization of soil	Ragi (*Eleusine coracana*)	Mohandas *et al.* (2004)
15.	Mixture of compost and vermiculite	Bahia grass	Douds *et al.* (2008)
16.	Green leaf manure	Maize	Javaid and Riaz (2008)
17.	Sand peat medium	Maize	Tarbell and Koske (2007)
18.	Apple pomace	Wheat, lemon grass, lily grass	Chauahan *et al.* (2013)
19.	Sugarcane ash and Baggase	Lemon Grass	Tanwar *et al.* (2013)
20.	Mustrad seed waste	Wheat, Barley	Mangla *et al.* (2012)
21.	Sugarcane ash and Baggase	Barley, Maize, Onion	Sharma *et al.* (2015)
22.	Hoagland solution	Rhodes grass	Bhowmik *et al.* (2015)
23.	Groundnut and soybean solid waste	Maize, Barley, Wheat	Alpa (2015)

Inoculum production technology is drawing the attention of scientists and researchers nowadays. Different practices like use of waste substrates are being tried for mass culture of AM fungi. Although different workers have done this work by using different substrates and hosts as shown in the Table 7.1.

5. Techniques Used in the Multiplication of AM Fungi

5.1. Selection of Host Plant

The most important factor in choosing a host plant is selecting a plant that supports mycorrhizal growth. Crops such as spinach, sugarbeet, lupine and the members of mustard family do not form a symbiosis with AM fungi. Mainly monocots and plants with extensive root system are very good hosts for propagation of AM fungi. Additionally, to prevent the spread of pathogens, the host plant should be from a different family than the inoculated crop.

5.2. Preparation of Starter Inoculum

Pure culture or starter culture of AM fungi can be raised by funnel technique (Menge and Timmer, 1982) using different host plants. The funnel is first filled with small amount of sterilized sand and soil mixture (1:1) up to neck portion. Seven to ten spores of AM fungi are added to this mixture. Then seeds of host plant are sown in the funnel and watered regularly. After 45 days, seedling roots will be analyzed for AM colonization (Phillips and Hayman, 1970) and soil sample are studied for spore quantification. After this the seedlings and soil can be transferred to bigger earthen pots containing sterilized sand: soil for the multiplication of individual spores again by using trap plant.

5.3. Mass Production of AM Fungi

There are three major well known systems adopted widely in the mass production of AM fungi. These are substrate based production system, substrate free production system and *in-vitro* production system.

5.3.1. Substrate Based Production System

This method is also known as the classical method for the production of AM fungi. In this method first the plants and their associated symbionts are cultivated in soil or sand based substrate. After the initial production of AM fungal inoculums, these fungi are propagated for the mass multiplication by using a single species or a consortium of identified AM fungal species in clay or plastic pots or scaled up to medium-size bags and containers and large raised or grounded beds (Gaur and Adholeya, 2002). The whole system setup is cultivated under controlled or semi-controlled condition in greenhouses or plant growth chambers to easily control the humidity and temperature. The starter inoculum usually consist of a single or a consortium of spores and infected root segments. In order to prepare the after inoculum, the root segments are dried and chopped into fine pieces to obtain the mixed inoculum, while, wet sieving and decanting techniques were used to obtain the single spores. Mixed inoculums were commonly used for the production of those AM fungal species which may produce intra-radical spores and vesicles (Klironomos and Hart, 2002).

Advantages and Disadvantages of Substrate Based Production System

Substrate based production systems preserve the mass production of single or consortia of AM fungal species. In this type of system, the nutrient supplies to the AM fungus and plant can be monitored and regulated properly. This system may provide controlled culture conditions but there might also be a chance for superfluous contaminants.

5.3.2. Substrate Free Production System

At the present time, variety of substrate free cultivation system or nutrient flow techniques is known. All these available techniques may differ from each other in the mode of aeration and application of the nutrient solution. In the static type of system, the nutrient solution is aerated through an aeration pump to avoid the roots suffering from oxygen deprivation. The pumps must be switched on periodically to minimize the flow of nutrient solutions and stuffed of air bubbles, which might be damage the expansion of the delicate extraradical hyphae (Ijodo *et al.*, 2011). The nutrient flow technique has been initially introduced by Mosse and Thompson (1981) and recently by Lee and George (2005). The nutrient flow technique is an alternative system in which a thin nutrient solution covers the roots and increases the relative area for gas exchange and conquers problems due to insufficient aeration into the inclined channels where the plant roots and AM fungus develop.

Aeroponics is a kind of hydroponics system that involves the dipping of roots of host plant and AM fungal propagules in nutrient solution fog. Spraying of micro-droplets increases the aeration of the medium, and the liquid film surrounding the roots imparts gas exchange.

Advantages and Disadvantages of Substrate Free Production System

The main advantage of this system is the production of substrate free inoculum. The root pieces with a high density of infective propagule could be directly used as inoculum. The liquid nutrient solutions are highly prone to the growth and development of algal contaminants. Moreover, the spore production rates could also be affected by lack of a carrier substrate.

5.3.3. *In vitro* Production System

In vitro production system of AM fungi was first established by Mosse and Hepper (1975). Afterward, the root organ culture system was introduced by Becard and Fortin (1988) using TDNA transformed root of *Daucus carota.* Mass scale production of AM fungi was achieved by root organ culture in small containers (Tiwari and Adholeya, 2003) in an airlift bioreactor.

Advantages and Disadvantages of In vitro Production System

The lack of unwanted microorganisms makes this system more appropriate for the mass production of high quality of AM fungal inoculums. In this system there is always requirement for monitoring and regulating the cultures. To make this system cost effective skilled technicians and laboratory equipments were also required.

However, in-vitro plant cultures need regular additions of culture medium which might increase the risks of cross contamination.

6. Methods of Application of AM Inoculums

Mycorrhizal spores, pieces of colonized crop roots and viable mycorrhizal hyphae function as active propagules of AM fungi that can be used as inoculum to colonize other plants.

The three methods, which can be adopted to apply mycorrhiza in the fields are:

(i) Adding AM inoculum at the time of sowing: AMF culture can be added at the time of sowing of seeds with ploughing

(ii) Adding AM inoculum at the time of transplanting: One of the most important factor in the application of AM inoculum is that the inoculum must be placed in the soil, where new roots will grow through it. Colonization will succeed only if the fungi are properly placed and root must be healthy and growing. An ideal AM endophyte should have the ability to infect host plant early in the growth period, efficiently exploit the soil, transfer nutrients readily to the host, multiply rapidly, competes effectively with indigenous AM fungi and infect a wide range of host plant (Mehrotra, 2005).

(iii) Seed treatment: Seeds can also be treated with mycorrhizal culture before sowing which will enhance their germination percentage.

Basic concepts of effective strain selection involve efficient production of biomass, universal formulation, long term storage ability and convenient method of application.

The number of AM propagules also includes the external hypha growing out from the surface of mycorrhizal roots and colonized root pieces (Gaur and Adholeya, 2000). The production of recently formed extraradical AM mycelia is an important parameter since it can be directly related to the capacity of plant to take up nutrients and to improve the soil structure and stability.Particularly addition of organic matter can have a beneficial effect on the growth of indigenous AM fungi in nutrient limited soil (Carvaca *et al.*, 2002; Gaur and Adholeya, 2002). It has also been reported that organic amendments enhance spore production (Doud *et al.*, 1997), extraradical proliferation of hyphae (Joner and Jakobsen, 1995) and improve colonization of roots (Muthukumar and Udaiyan, 2000). Organic matter addition to the soil in eroded sites could be an approach to enhance the beneficial effect of AM fungi on soil stabilization and plant establishment.

Spore germination is generally stimulated by root exudates from mycotrophic plants however some non-mycorrhizal plants, such as *Brassica* species have exudates that reduce the germination of AM spores (Smith and Read, 1997).

7. Application of AM Fungi

Application of mass production of inoculum, its proper handling and distribution of viable inoculum at the farmer's level through networking of strong field force will support and give impetus to programme of agro-forestry, wasteland development and energy plantation in rural areas. Further refinement of several rate limiting factors will lead to development of the technology, which involves the extraction of potentially viable propagules from soil in less time and space and optimization of growth conditions. However, one of the main tasks for both producers and researchers is to raise awareness in the public about potentials of mycorrhizal technology for sustainable plant production and soil conservation. Most of the areas of AM inoculum production and application technology need further careful study. In response to more concern for environmental quality, innovative approaches need to be introduced into the agricultural systems. High production cost and obligate symbiotic nature of AM fungi have dampened enthusiasm and impede research. Supplementing the nutrient requirements of crops through AM inocula for sustaining soil fertility and crop production would spur commercial production and the development of new formulations. Mycorrhiza in combination with other beneficial microorganisms has a promising option and should allow more holistic approaches to rhizosphere health in future and help in a drive towards minimizing excessive chemical usage. This should draw attention of agricultural scientists, plant physiologists, forest managers, policy-makers and different government and non-government officials towards nature farming systems and utilizing the potential of AM in production system, though it is assumed that the inability to grow AM fungi in pure sterile culture is a major disadvantage in study of AM fungi. This may be true for certain areas of AM fungal research like fungal nutrition and genetics; this is not an important problem in relation to their application. Presumably, soil amended with substrates and AM fungal spores can be a good ameliorative solution for reclaiming acidic soils. AM fungi are and will be a major component in any sustainable plant productive systems, which can be explicitly used for the maintenance of species diversity and productivity.

8. Conclusion

The isolation and mass multiplication studies raise the concerns about the quality control practices, inoculum viability, potential loss of desired AMF species in inoculum through subsequent culturing and the possibility for contamination of stock cultures. A standard method of screening and certification of effectiveness should be considered before distribution at commercial level which will prevent the distribution of low quality inoculum. However one of the main tasks for producers and researchers is to raise awareness in the public about potentials of mycorrhizal technology for sustainable plant production and soil conservation.

References

Abbott, L.K., and Robson, A.D. 1984. The effect of root density, inoculum placement and infectivity of inoculum on the development of vesicular arbuscular mycorrhizae. *New Phytol.*, 97: 285-299.

Akhtar, M.S. and Panwar, J. 2011. Arbuscular mycorrhizal fungi and opportunistic fungi: Efficient root symbionts for the management of plant parasitic nematodes. *Adv. Sci. Eng. Med.* 3: 165-175.

Akhtar, M.S., Siddiqui, Z.A. and Weimken, A. 2011. Arbuscular mycorrhizal fungi and Rhizobium to control plant fungal diseases. In: Lichtfouse, E. (Ed.) Alternative Farming Systems, Biotechnology, Drought Stress and Ecological Fertilization. Sustainable Agriculture Reviews 6, Springer, Dordrecht, The Netherlands, pp. 263-292.

Al-Raddad, A. 1995. Mass Production of *Glomus mosseae* spores. *Mycorrhiza*, 5: 229-231.

Becard, G.and Fortin J.A. 1988. Early events of vesicular arbuscular mycorrhiza formation on Ri T-DNA transformed roots. *New Phytol.*, 108: 211-218.

Bhowmik, S.N., Yadav, G.S. and Datta, M. 2015. Rapid mass multiplication of Glomus mosseae inoculums as influenced by some biotic and abiotic factors. *Bangl. J. Bot.* 44: 209-214.

Caravaca, F. J., Barea, M. and Roldan, A. 2002. Synergistic influence of an arbuscular mycorrhizal fungus and organic amendment on *Pistacia lentiseus* L. seedling afforested in a degraded semi-arid soil. *Soil Biol. Biochem.*, 34: 1139-1145.

Carrenho, R., Sandra, F. B. and Bononi, V. L. R. 2002. Effects of using different host plants on the detected biodiversity of arbuscular mycorrhizal fungi from an agroecosystem. *Revista Brasil. Bot.*, 25: 93-101.

Chaudhary, B. and Panja, B. 2007. Diversity and integration in mycorrhizas: Meaning to plant ecology. In: Tiwari, M. and Sati, S.C. (Eds.) The Mycorrhizae: Diversity, Ecology and Applications, Daya Pub. House, Delhi, pp. 36-56.

Chauhan, S., Kaushik S., Bajaj N., and Aggarwal A. 2013. Inoculum Production of *Acaulospora laevis* using Fresh and decomposed Apple Pomace as Substrate. *Int. Res. J. Bio. Sci.* 2: 32-36.

Chaurasia, B. and Khare, P. K.2005. *Hordeum vulgare*: A suitable host for mass production of arbuscular mycorrhizal fungi from natural soil. *Appl. Ecol. Env. Res.*, 4: 45-53.

Daft, M.J., and El-Giahmi, A.A. 1978. Effect of arbuscular mycorrhiza on plant growth. *New Phytol.*, 80: 365-372.

Dalpe, Y. and Montreal, M. 2004. Arbuscular mycorrhizal inoculum support sustainable cropping systems. *Crop Manag.*, 10, 1094–1104. 10.1094/CM-2004-0301-09-RV.

Diop, T. A. Plenchette, C. and Strullu, D. G. 1994. Dual axenic culture of sheared root inocula of vesicular arbuscular mycorrhizal fungi associated with tomato roots. *Mycorrhiza*, 5: 17-22.

Douds, D. D., Galvez, L., Franke-Synder, M., Reider, G. and Drinkwater, L. E. 1997. Effect of compost addition and crop rotation upon VAM fungi. *Agric. Eco. Env.*, 65: 257-266.

Douds, D. D., Nagahashi, G., Reider, C., Hepperly, P. R. 2008. Choosing a mixture ratio for the on-farm production of AM fungus inoculum in mixtures of compost and vermiculite. *Compost Sci. Util.*, 16: 52-60.

Eom, A. H., Hartnett, D. C. and Wilson, G. W. T. 2000. Host plant species effects on arbuscular mycorrhizal fungal communities in tallgrass piairie. *Oecologia*, 122: 435- 444.

Ferguson, J.J. and Menge, J. A. 1982a. Factors that affect production of endomycorrhizal inoculum. *Proc. State Hort. Soc.*, 95: 37-39.

Ferguson, J.J. and Menge, J.A. 1982b. The influence of light intensity and artificially extended photoperiod upon infection and sporulation of *Glomus fasciculatum* on sudan grass and on root exudation of sudan grass. *New Phytol.*, 92: 183-191.

Ferguson, J.J. and Woodhead, S.H. 1982. Production of endomycorrhizal inoculum. An increase and maintenance of vesicular arbuscular fungi. In: Schenck, N.C. (Ed.) Methods and Principles of Mycorrhizal Research, APS Press, St. Paul, pp. 47-54.

Franke- Synder, M., Douds, D. D., Galvez, L., Phillips, J. G., Wagoner, P., Drinkwater, L. and Morton.B. 2001. Diversity of communities of arbuscular mycorrhizal (AM) fungi present in conventional versus low input agricultural sites in eastern Pennsylvania, USA. *Appl. Soil Ecol.*, 16: 35-48.

Furlan, V., and Fortin, J.A. 1973. Formation of endomycorrohizae by *Endogone calospora* on *Allium cepa* under three temperature regimes. *Nat. Can. J.*, 100: 467-477.

Gadkar, V., Davaid-Schwartz, R., Kunik, T. and Kapulnik, Y. 2001. Arbuscular mycorrhizal fungal colonisation. Factors involved in host recognition. *Plant Physiol.*, 127: 1493-1499.

Gallaud, I. 1905. Etudes surles mycorrhizes endophytes. *Rev. Gen. Bot.*, 17: 5.

Garcia-Garrido, J.M., Garcia-Romera, I. and Ocampo, J.A. 1992. Cellulase activity in lettuce and onion plants colonized by the vesicular-arbuscular mycorrizal fungus *Glomus mosseae*. *Soil Biol. Biochem.*, 25: 503-504.

Garcia-Rommera, I.M., Garcia-Garrido, J.M. and Ocampo., J.A. 1991. Pectolytic enzymes in the vesicular-arbuscular mycorrhizal fungus *Glomus mosseae*. *FEMS Microbiol.*, 78: 343-346.

Gasper, L., Pollero, R. and Cabello, M. 1997. Partial purification and characterization of a lipolytic enzyme from spores of the arbuscular mycorrhizal fungus *Glomus versiforme*. *Mycologia*, 89: 610-614

Gaur, A. and Adholeya, A. 2000. Effects of particle size of soil less substrates upon AM fungus inoculum production. *Mycorrhiza*, 10: 43-48.

Gaur, A. and Adholeya, A. 2002. Arbuscular mycorrhizal inoculation of five tropical fodder crops and inoculum production in marginal soil amended with organic matter. *Biol. Fertil. Soils*, 35: 214-218.

Gemma, J.N., Koske, R.E. and Habte, M. 2002. Mycorrhizal dependency of some endemic and endangered Hawaiian plant species. *Plant Bot. Ecol.*, 89: 337-345.

Genney, D. R., Hartley, S. H. and Alexander, I. J. 2001. Arbuscular mycorrhizal colonization increases with host density in a healthland community. *New Phytol.*, 152: 355-363.

Gianinazzi-Pearson, V. 1994. Morphofunctional compatibility in interactions between roots and arbuscular endomycorrhizal fungi: molecular mechanisms, genes expression. *In*: Kihmito, K., Singh, R.P. and Singh, U.S. (Eds.) Pathogenesis and host-parasite specificity in plant disease: histological, genetic, biochemical and molecular basis, Oxford Pergamon Press, UK, pp: 251-263.

Gill, T. S. and Singh, R. S. 2001. Effect of host and substrates on the development of VA mycorrhizal fungi for root colonization, spore production and population and plant growth responses in chickpea. *Indian Phytopath.*, 55: 210-212.

Gupta, N., Ruataray, S. and Basak, V. C. 2006. The growth and development of AMF and its effects on the growth of maize under different soil composition. *Myco. News*, 18: 15-23.

Hameeda, B., Harini, G., Rupela, O. P. and Reddy, G. 2007. Effect of composts or vermicomposts on sorghum growth and mycorrhizal colonization. *Afr. J. Biotechnol.*, 6: 9-12.

Hepper, C.M., Azcon-Aguilar, C., Rosendahl, S., and Sen, R., 1988. Competition between three species of *Glomus* used as spatially separated introduced and indigenous mycorrhizal inocula for *Allium porrum* L. *New Phytol.*, 110: 207-215.

IJdo, M., Cranenbrouck S., Declerck S. 2011.Methods for large-scale production of AM fungi: past, present, and future. *Mycorrhiza*, 21: 116.

Jackson, N.E., Francklin, R.E., and Miller, R.H. 1972. Effects of VAM on growth and phosphorus content of three agronomic crops. *Soil Sci. Soc. Am. Proc.*, 36: 64-67.

Javaid, A. and Riaz, T. 2008. Effects of application of green leaf manure on growth and mycorrhizal colonization of maize. *Allelopathy Journal*, 21: 339-347.

Joner, E. J. and Jacobsen, I. 1995. Growth and extracellular phosphatase activity of arbuscular mycorrhizal hyphae as influenced by soil organic matter. *Soil Biol. Biochem.*, 27: 1153-1159.

Klironomos, J.M. and Hart, M.M. 2002. Colonization of roots by arbuscular mycorrhizal fungi using different sources of inoculum. Mycorrhiza; 12:181-184.

Lee, Y.J. and George, E. 2005. Development of a nutrient film technique culture system for arbuscular mycorrhizal plants. *Hort Sci.*, 40: 378-380.

Manoharachary, C. 2004. Biodiversity, taxonomy, ecology, conservation and biotechnology of arbuscular mycorrhizal fungi. *Indian Phytopath.*, 57: 1-6.

Mehrotra, V. S.2005. A premier biological tool for managing soil fertility. In: Mehrotra, V.S. (Ed.) Mycorrhiza: Role and Applications, Allied Publ. Pvt. Ltd., New Delhi, 1-65.

Mohandas, S., Chandre Gowda, M. J. and Manamohan, M. 2004. Popularization of arbuscular mycorrhizal (AM) inoculum production application on farm. *Acta Hort.*, 638: 279- 83.

Morton, J. B. 1997. Names and authorities of fungi in Glomeronycota. http//invam. caf.wvu.edu.

Morton, J. B. and Redecker, D. 2001. Two new families of Glomales, Archaeosporaceae and Paraglomaceae with two new genera, *Archaeospora* and *Paraglomus* based on concordant molecular and morphological characters. *Mycorrhiza*, 93: 181-195.

Morton, J.B. and Benny, G.L. 1990. Revised classification of arbuscular mycorrhizal fungi (Zygomycetes), A new order Glomales, two new suborders, Glomineae and Gigasporineae and two new families, Acaulosporaceae and Gigasporaceae. *Mycotaxon*, 37: 471-491.

Mosse B. and Hepper C. 1975. Vesicular-arbuscular infections in root organ cultures. *Physiol. Plant Pathol.*, 5: 215-218.

Mosse, B. 1973. Advances in the study of vesicular arbuscular mycorrhiza. *Ann. Rev. Phytopathol.*, 11: 171-196.

Mosse, B. and Bowen, G. D. 1968. The distribution of endogone spores in some Australian and New Zealand soils and in an experimental field soil at Rothmasted. *Trans. British Mycol. Soc.*, 51: 485-492.

Mosse, M. and Thompson, J.P. 1981.Production of mycorrhizal fungi. US Patent No. 4294037.

Murugan, R. and Selvaraj, T. 2003. Reaction of kashini (*Cichorium itybus* L.) to different native arbuscular fungi. *Myco. News*, 15: 10-13.

Muthukumar, T. and Udaiyan, K. 2000. Influence of organic manures on arbuscular mycorrhizal fungi associated with *Vigna unguiculata* (L.) Walp. in relation to tissue nutrients and soluble carbohydrate in root under field conditions. *Biol. Fertil. Soils*, 31: 114-120.

Muthukumar, T. and Udaiyan, K. 2007. Arbuscular mycorrhizal fungi in forest tree seedling production. In: Tiwari, M. and Sati, S.C. (Eds.) The Mycorrhizae: Diversity, Ecology and Applications, Daya Pub. House, Delhi, pp. 200-223.

Muthukumar, T., Udaiyan, K. and Rajeshkannan, V. 2001.Response of neem (*Azadirachta indica* A. Juss) to indigenous mycorrhizal fungi, phosphate solublizing and asymbiotic nitrogen- fixing bacteria under tropical nursery conditions. *Biol. Fertil. Soils*, 34: 417-426.

Neeraj and Yadav, K. 2007. Flyash amended soil improves colonization and spore count of arbuscular mycorrhizal fungal cultures. *Myco. News*, 19: 15-18.

Nehra, S., Pandey, S. and Trivedi, P. C. 2003. Interaction of arbuscular mycorrhizal fungi and different levels of root-knot nematode on ginger. *Indian Phytopath.*, 56: 297-299.

Nelsen, C.E., and Safir, G.R. 1982. Increased drought tolerance of mycorrhizal onion plants caused by improved phosphorus nutrition. *Planta*, 154: 407-413.

Phillips, J. H. and Hayman, D. S. 1970. Improved procedures for clearing roots and staining parasitic and vesicular-arbuscular mycorrhizal fungi for rapid assessment of infection. *Trans. British Mycol. Soc.,* 55: 158-161.

Raghuwanshi, R. and Upadhyay, R. S. 2004. Performance of vesicular arbuscular mycorrhizae in saline-alkali soil in relation to various amendments. *World J. Microbiol. Biotechnol.,* 20: 1-5.

Reddy, S. R., Pindi, P. K. and Reddy, S. M. 2005. Molecular methods for research on arbuscular mycorrhizal fungi in India: Problems and prospects. *Curr. Sci.,* 89: 1699-1709

Redecker, D., Morton, J. B. and Bruns, T. D. 2000. Molecular phylogeny of the arbuscular mycorrhizal fungi *Glomus sinuosum* and *Sclerocystis coremiodes*. *Mycologia*, 92:282-285.

Reynolds, H. L., Packer, A., Bever, J. D. and Clay, K. 2003. Grassroots ecology: plant microbe- soil interactions as drivers of plant community structure and dynamics. *Ecol*. 84(9): 2281-2291.

Ryan, M. H. and Graham, J. H. 2002. Is there a role for arbuscular mycorrhizal fungi in production agriculture? *Plant Soil*, 244: 263-271.

Selvam, A. and Mahadevan, A. 2002. Effect of ash, pond soil and amendments on the growth and arbuscular mycorrhizal colonization of *Allium cepa* and germination of *Arachis hypogea*, *Lycopersicon esculentum* and *Vigna mungo* seeds. *Soil Sediment Contam.,* 11: 673-686.

Sharma, S., Sharma, S. and Aggarwal, A. 2015. Screening of different hosts and substrates for inocula production of arbuscular mycorrhizal fungi. *Mycorrhiza News,* 27: 6-9.

Sheela, M. A. and Sundaram, M. D. 2003. Role of VA mycorrhizal biofertilizers in establishing black gram (*Vigna mungo* L.) var. T-9 in abandoned ash ponds of Neyeli Thermal Power Plant. *Myco. News*, 15: 13-16.

Smith, S. E. and Read, D. J. 1997. Vesicular- arbuscular mycorrhizas. In: Smith, S.E. and Read, D.J. (Eds.) Mycorrhizal Symbiosis, Academic Press, London, pp. 9-160.

Smith, S.E., Read D.J. Mycorrhizal Symbiosis. 3rd ed. Academic Press, London, 2008.

Sreenivasa, M.N. and Bagyaraj, D.J. 1989. Use of pesticides for mass production of vesicular arbuscular mycorrhizal inoculum. *Plant Soil*, 119: 127-132.

Strack, D., Fester, T., Hause, B., Schliemann, W. and Walter, M.H., 2003. Arbuscular mycorrhiza: Biological, chemical and molecular aspects. *J. Chem. Ecol.,* 29: 1955-1979

Sylvia, D.M., Wilson, D.O., Graham, J.N., Maddai, J.J., Millner, P., Morton, J.B., Skipper, H.D., Wright, S.F. and Jarafster, A.G. 1993. Evaluation of vesicular arbuscular mycorrhizal fungi in diverse plants and soils. *Soil Biol. Biochem.,* 25: 705-713.

Tanwar, A., Aggarwal A., Yadav, A. and Prakash, V. 2013. Screening and selection of efficient host and sugarcane bagasse as substrate for mass multiplication of *Funneliformis mosseae. Biol. Agril. Hort.*, 29: 107-117.

Tarbell, T. J. and Koske, R. 2007. Evaluation of commercial arbuscular mycorrhizal inocula in sand/peat medium. *Mycorrhiza*, 18: 51-56.

Tiwari, P. and Adholeya, A. 2003. Host dependent differential spread of *Glomus intraradices* on various Ri T-DNA transformed root *in vitro. Mycol. Prog.*, 2: 171-177.

Walker, C. and Schussler, A., 2004. Nomenclature, classification and new taxa in the Glomerymycota. *Mycol. Res.,* 108: 979-982.

Yeasmin, T., Zaman, P., Rahman, A., Absar, N. and Khanum, N.S. 2007. Arbuscular mycorrhizal fungus inoculum production in rice plants. *Afr. J. Agric. Res.*, 2: 463-467.

— *Part III* —
Arbuscular Mycorrhizal Fungi as Biofertilizers

2017, Mycorrhizal Fungi
Editors: **Ashok Aggarwal and Kuldeep Yadav**
Published by: ASTRAL INTERNATIONAL PVT. LTD., NEW DELHI

Pages 177–193

8

Arbuscular Mycorrhizal Fungi as Biofertilizer and its Application to Different Horticultural Crops

Chaya P. Patil[1],* and K.A. Chandalinga[2]

[1]*Department of Agricultural Microbiology,*
[2]*Department of Horticulture,*
KRC College of Horticulture, Arabhavi – 591 218
*Corresponding Author: patilpradeepc@yahoo.co.in

ABSTRACT

Green revolution started on a soil rich in organic carbon and the response to applied fertilizers were spectacular. With passage of time, green revolution is showing symptoms of fatigue and response to applied fertilizers have slurted dwindling. Decline in food production, degeneration in native fertility and deterioration in environmental quality are the three gigantic problems in present scenario of agriculture. Excessive use of plant protection chemicals and imbalanced use of fertilizers have further resulted in escalation of the above problems including exorbitant cost on cultivation. Therefore, farming technologies such as organic, biodynamics, homu, use of bioformulations such as panchagavya, amrit pani, jeevamrutha, agnihotra, rishi krishi etc. for agricultural sustainability and marketability are gaining a new momentum not only in India, but also all over the world. To revive the soil health and microorganisms which are the core supporters to achieve sustainable production system, the soil environment needs to be made congenial for persistence of useful microbial population responsible for continuous availability of nutrients from natural sources. The search for an effective microorganism which has a role in germination, nutrient uptake, growth of plants, productivity, and tolerance to abiotic and biotic stresses led us for the use of Arbuscular mycorrhizal fungi (AMF). AMF are obligate symbionts having mutualistic association with plant roots. There is tremendous scope for the use of AMF in all the ecosystems particularly semi-arid and arid ecosystems for the horticultural crops. AMF inoculation offers noble additive effects to crops owning to: increased productivity, increased crop uniformity, reduced time to market, reduced transplant losses, increased disease resistance, improved crop marketability, accelerated growth rates, accelerated budding and flowering, increased revenue reduced fertilizer and fungicide applications. Auxins, cytokinins, gibberellins, and vitamins have been shown to be produced by mycorrhizal fungi in pure culture and the effects of these compounds on plant growth and development are well documented. The tremendous scope for use of AMF in horticultural crops makes it a novel biofertilizer. It has effect on root geometry, enhanced uptake of nutrients, controls root rotting nematodes and pathogens viz., Pythium, Phytopthora, Fusarium, Sclerotium.

Keywords: *AM fungi, Biofertilizer, Horticultural crops, Nutrient acquisition.*

1. Introduction

India has a wide variety of climate and soil on which a large range of crops such as fruits, vegetables, ornamental plants, spices, medicinal and aromatic plants *etc.* are grown. Development of high yielding varieties and high production technologies and their adoption in areas of assured irrigation paved the way towards food security through green revolution in the sixties. It has however, gradually become clear that horticultural crops for which the Indian topography and agro climates are well suited is an ideal method of achieving sustainability of small holdings, increasing employment, improving environment, providing an enormous export potential and above all these achieving nutritional security. As a result, due emphasis on diversification to horticultural crops is being given since the last two decades.

The horticulture scenario of our country is rapidly changing. The production and productivity of horticultural crops have increased manifold. Production of fruits has tripled in the last 50 years. The productivity has gone up by three times in banana and papaya. Today, horticultural crops cover about 25 per cent of total agricultural exports of the country. The corporate sector is also showing greater interest in horticulture. A major shift in consumption pattern of fresh and processed fruits and vegetables is expected in the coming century. There will be greater technology adoption both in traditional horticultural enterprise as well as in commercial horticulture sectors.

Agriculture has changed dramatically, especially since the end of World War II. Food productivity soared due to new technologies, mechanization, increased chemical use and government policies. These changes allowed fewer farmers with reduced labour demands to produce the majority of the food. Green revolution started on a soil rich in organic carbon and the response to applied fertilizers were spectacular. With passage of time, green revolution is showing symptoms of fatigue and response to applied fertilizers have started dwindling. Decline in food production, degeneration in native fertility and deterioration in environmental quality are the three gigantic problems in present scenario of agriculture (Pathak and Ram, 2003). Excessive use of plant protection chemicals and imbalanced use of fertilizers have further resulted in top soil depletion, groundwater contamination including exorbitant cost on cultivation. Integrated plant nutrient management (IPNM) was considered as a remedy to the above problems and to ameliorate the Indian soils from multi-nutrient deficiencies. Thus, the combined and cogent use of organics has become essential part of agriculture. Unfortunately, limited availability of biomass, cattle dung and farmer's apathy for preparation and use of organic manure, IPNM practices are not being adopted as per expectations (Pathak and Ram, 2003).

The increasing concern about the environment and socio- economic impact of chemical agriculture has led to seek alternative practices for agricultural sustainability and marketability by progressive farmers. Therefore, chemical free traditional farming technologies such as organic, biodynamics, homa, panchagavya, amrit pani, agnihotra, rishi krishi, jeevamrutha *etc.*, are gaining a new momentum not only in India, but all over in world (Singh *et al.*, 2007). These systems offer a means to address self- reliance, rural upliftment and conservation of natural resources.

To revive the soil health and microorganisms which are the core supporters to achieve sustainable production system, the soil environment needs to be made congenial for persistence of useful microbial population responsible for continuous availability of nutrients from natural sources. Experiencing the adverse effects of inorganic input dependent agriculture, the concept of organic/sustainable/natural farming is gaining momentum (Duragannavar, 2005).

In spite of significant achievements in horticultural Research and Development, a number of challenges still need to be met. They are:

☆ Inadequate supply of quality planting material

☆ Heavy losses caused by several biotic and abiotic stresses

☆ Several unresolved chronic disorders.

As a result, the productivity per unit area is low, resulting in high cost of production. Further, the quality of produce in many cases is far away from satisfactory. The post-harvest losses continue to be high. Full advantage has yet to be taken of several frontier areas *e.g.*, biotechnology, protected cultivation, computer aided management of inputs, integrated nutrient management, leaf nutrient standards, biofertilizers, bioformulations, integrated pest management and mycorrhiza. There is also need for change both in the content and approach of research which can be taken up in partnership with private sector on aspects like production of hybrids, green house production of flowers, biotechnology, value addition and export. The future growth of horticulture industry will largely depend on new and globally competitive technologies. As such, ambitious research programme is warranted for horticultural crops in the following thrust areas:

a) Use of organics in sustainable production

b) Quality assurance and production to suit export standards

c) Reduction in post-harvest losses

The agro climatic zones of India have a wide range of annual rainfall distribution which ranges from very high rainfall areas to very low (scanty) rainfall areas. The establishment of fruit orchards particularly in low rainfall needs intensive care during establishment of the fruit crops since these areas experience high evaporative demands with erratic rainfall and frequent occurrence of drought. As established orchard is a long time investment, needs careful planning since initially made mistakes may considerably reduce the returns from the enterprise.

The sustainability in the income of the farmers can only be achieved through growing rain fed fruit, vegetable and flower crops. As a consequence, the area under high value fruit crops is rapidly expanding in semi-arid and arid ecosystems besides transitional and high rainfall areas. The increase in area and productivity is the noteworthy accomplishment of techniques like use of chemical fertilizers, pesticides, insecticides, *etc.*, which were the main components of green revolution.

The search for alternatives or supplements to fossil fuel based on inorganic fertilizers brought not only by the likelihood of future price increase for chemical fertilisers but also by the need to maintain long term soil productivity and ecological sustainability. Biofertilisers - a cost effective renewable energy source, play a crucial role in plant nutrition, increase growth productivity of plant besides giving ability to tolerate abiotic and biotic stresses. Although these microorganisms are available in the rhizosphere of the plant, their population, root colonization efficiency, host specificity decides the necessity of artificial inoculation. The work carried out by scanty research at Kittur Rani Channamma College of Horticulture, Arabhavi for the past few years with different horticultural crops has resulted in identification of efficient strains of Arbuscular Mycorrhizae (AM) and microbial consortia (Containing nitrogen fixers, 'P' solubilisers, cellulose decomposers, biocontrol agents and PGPR organisms) useful for horticultural crops. The search for an effective microorganism which has a role in germination, nutrient uptake, growth of plants, productivity, and tolerance to abiotic and biotic stresses led us for the use of AM fungi (AMF).

Arbuscular mycorrhizal fungi (AMF) are obligate symbionts having mutualistic association between plant roots and fungal mycelia. There is tremendous scope for the use of AMF in all the ecosystems particularly semi-arid and arid ecosystems for the fruit crops mainly because AMF inoculated transplanted plants respond better in the field under drought and water-logged conditions over non-mycorrhizal plants (Shivaputra *et al.*, 2004a and 2004b). AMF inoculation offers noble additive effects to crops owing to:

- ☆ Increased productivity
- ☆ Increased crop uniformity
- ☆ Reduced time to market
- ☆ Reduced transplant losses
- ☆ Reduced fertilizer and fungicide applications
- ☆ Increased disease resistance
- ☆ Improved crop marketability

☆ Accelerated growth rates

☆ Accelerated budding and flowering

☆ Increased revenue

2. Role of VAM Fungi in Plant Propagation

2.1. Germination

To explain the mechanisms and role of AMF in initiation of early germination of seeds with increased germination percentage and germination index, a number of sequential events occur beginning with the presence of chlamydospores in the rhizosphere (germinating media) of seeds. Needless to say, initiation of AM association begins with the germination of the spore and contact of fungal hyphae with root epidermis followed by a molecular and chemical dialogue that occurs between AM fungi and host (Giorannetti *et al.,* 1993 and Harborne and Williams, 2000). Plant exudates from the appropriate host are necessary to initiate the symbiosis. The leachates from seeds might influence chlamydospores to initiate or promote quick germination and hyphal growth. Further, spore germination and some independent hyphal growth of AM fungi can occur in the absence of roots and this saprophytic phase is crucial to AMF life cycle since survival depends on the ability to rapidly and effectively infect suitable hosts. Thus, AM fungi germinate and some independent hyphal growth is achieved till the emergence of radical, which is much earlier than the plumule emergence, certainly the radical gets infected as soon as it emerges. Further, certain introduced AM fungi being efficient root colonisers establish well in the rhizosphere (Manjunath *et al.,* 1983). Highest root infections by AM fungi have effects on further root and shoot growth (Shivashankar and Iyer, 1988).

Auxins, cytokinins, gibberellins (GA$_3$) and vitamins have been shown to be produced by mycorrhizal fungi in pure culture (Miller, 1971; Crafts and Millar, 1974; Slankis, 1975) and the effects of these compounds on plant growth and development are well documented (Torrey, 1976). Gibberellins have been found to be effective in increasing germination of seeds of many fruit crops (Ram, 1997). Thus the increase in germination percentage with higher germination index and lower time span could be due to the production of GA$_3$ by AM fungi (Santosh, 2004; Patil *et al.,* 2004; Venkat, 2004). Gange *et al.* (1993) experimentally demonstrated that AM fungal presence affects seed germination in secondary succession site. Santosh (2004) while working on mango noted an increased germination per cent with the inoculation of particular AM fungi (*Gigaspora margarita, Acaulospora laevis, Glomus monosporum* and *Glomus fasciculatum*). Despite the lack of absolute specificity accorded to the AM symbiosis, the symbiotic efficiency is under genetic control and is affected by host and fungal species. The superiority and inferiority of AM fungal strains after inoculation indicate the extent of AM fungal association with the host and is a morphogenetically determined phenomenon. In spite of the fact that AM fungi are not host specific, they vary widely in their effectiveness. Variation in the nature of mycorrhizal effects on the hosts has been reported by Janos (1980) and Johnson *et al.* (1997).

Inoculation of twelve different VAM fungal species resulted in the early germination, increased germination percentage and vegetative growth of two marigold varieties plants *i.e.* shoot length, number of leaves, length of longest root, number of roots, root volume, root to shoot ratio, fresh shoot weight, fresh root weight over uninoculated ones. Among the different VAM species studied higher root colonization was recorded with *Gigaspora margarita* in Local variety and *Glomus mosseae* in Double Orange variety of Marigold (Vijetha, 2013).

2.2. Rooting of Cuttings

Induction of rooting of cuttings is one of the important horticulture techniques for mass multiplication of clonally propagated plants. The benefits from AM Fungi due to root colonisation are thought to be highest when colonisation occurs as early as possible during plant growth. In mass multiplication of plants particularly at the time of root formation AM inoculum should be added during the formation of root primordial growth *i.e.* prior to the adventitious root formation in propagation of plants through cuttings. Increase in rooting of cuttings by adding AM fungal inoculum to rooting media has also been reported by Linderman and Call (1977). According to Scagel (2000), addition of AM fungal inoculum into rooting media was equal or better than the rooting response obtained by using rooting hormone. Further, he also noticed that combination of rooting hormone and AM fungi generally produced a better percentage of rooted cuttings with more roots than cuttings treated with rooting hormone alone. Since AM fungal colonisation has been shown to increase the survival and growth of several plant species, the root colonisation resulting from inoculation could result in higher quality cuttings since higher quality withstands the stress of transplanting and increases growth during later stage of plant development.

A research carried out at the Pukekohe Horticultural Research Station, MAF, Pukekohe, New Zealand, kiwi fruit (*Actinidia deliciosa* var. *deliciosa*) cuttings when grown in peat-pumice mix (amended with three rates of phosphorus fertilizer and five rates of mycorrhizal inoculum) for 130 days under glasshouse condition increased shoot length and weight by 294 per cent and 265 per cent, respectively, with no significant interaction between inoculation and fertilizer in mycorrhizal treated plants (Powell, 1986).

A study conducted at the Department of Agronomy and Soil Science, University of Hawaii, Honolulu, USA, cassava (*Manihot esculenta*) plants when raised in the greenhouse either from small cuttings or large cuttings in a subsurface oxisol and were inoculated with *Glomus aggregatum* at target soil solution 'P' concentrations of 0.003-0.2 mg per litre, it was observed that VAM colonization exceeded 60 per cent level irrespective of cutting size or soil 'P' level. Plants from large cuttings grew faster than those from small cuttings. The latter were very highly dependent on mycorrhiza while large cutting plants were only marginally dependent. This appears to be related to the high P reserve in large cuttings and hence low requirement of soil 'P' by the plant until the 'P' reserve in the cuttings is significantly depleted (Habte and Byappanahalli, 1994).

Production of pomegranate rooted cuttings with inoculation of *Acaulospora laevis* and *Trichoderma harzianum* resulted in more than 90 per cent rooting and less than five per cent transplantation shock (Patil and Patil, 2002). More than 90 per cent of rooting can be achieved in Fig cuttings by inoculating the cuttings with five grams of either *Glomus leptotichum* or with *A. laevis* with minimum (<5 per cent) transplantation shock (Patil, 2016).

2.3. Effect of AM Fungi on Grafting

In production of horticultural plant materials through vegetative means of propagation, grafting forms a major technique of propagation which involves rootstock that forms the root system of scion. The rootstock growth has pronounced effect on the formation and success of graft union. The vigour and growth of rootstock depends mainly on the formation of efficient root system since the rapid developmental functioning of root system is crucial to the successful growth and development of rootstock and seedlings. Though the root system morphology is genetically determined and varies among the species, it can be modified by many environmental factors, including nutrient availability and temperature. Root plasticity is also influenced by AM fungi, although variations in the effects have been reported (Berta *et al.*, 1994). Further, root system morphology influences root function (Harper *et al.*, 1991), it is of significance (interest) to understand the effect of AM fungi on the first phase of root development in root stocks. It has been observed that early AM fungal infection is beneficial for the performance of rootstocks. As the root system structure depends on meristematic activity of root apices, all the factors that affect root development must have directly or indirectly effect root apical meristem activity and structure. Furthermore, because of modification in rooting patterns, physiology implies to change in gene expression and protein metabolism (Schellenbum *et al.*, 1992).

Since, AM fungi induce enhancement in phosphate assimilation leading to mycorrhizal plant development, it is presumed that morphological and physiological changes in roots could be due to changes in P content. Internal P content influences root geometry, assimilate partitioning to lateral roots, and an elevated P content enhances lateral root formation (Amijee *et al.*, 1989; Adalsteinsson and Jensen, 1989). In a study with *Andropogon gerardii*, both AM fungi and P enhanced growth and increased the number and diameter of primary, secondary and tertiary roots, but decreased specific root length.

Investigations carried out on mango (Santosh, 2004); Citrus (Venkat, 2004) and Jamun (Patil, 2004) indicated that the AM rootstocks had higher values for stem girth (diameter), diameter of seminal and lateral roots and high vigour of AM plants in comparison to non inoculated ones. Further, it was also noted that the desired graftable size of AM fungi inoculated stock would be attained much earlier to non-inoculated rootstocks. When these rootstocks are subjected for grafting, they showed higher success per cent of graft take with higher survival percentage after six months of grafting over non-inoculated rootstocks. These results concur with findings of Ramesh (1997) in cashew. The higher graft-take results due to increased stem diameter, branched root system, higher values for seminal and lateral roots, which resulted in increased vigour of the plants helps in early sprouting of scion thereby increasing photosynthetic activity of plant. However, there exists differential responses to AM fungal species as occurs for germination.

Early attainment of graft-able size of rootstocks facilitates early grafting in the season which is a boon for nurserymen particularly in the areas of low humid tropics where evapro-transpiration demand is high during the later parts of the season. The use of AM fungi can be effectively exploited in this area.

3. Effect of AM Fungi on Vegetative Parameters

Rosemary plants raised in presence of AM fungi in polythene bags showed an increase in plant growth (shoot length, number of leaves and biomass) and root parameters over those grown in non- AMF inoculated soil. The extent of growth and root biomass varied with the AM fungi used. Based on the plant growth and root biomass per plant, *Glomus monosporum* was found to be the best AM symbiont for inoculating *Rosmarinus officinalis* followed by *Sclerocystis dusii* (Praveenkumar *et al.*, 2014).

The effectiveness of 12 AM fungi on Tulsi (*Ocimum sanctum*) showed an increase in growth parameters over those grown in the absence of soil inoculation with AM fungi. The extent of growth and root biomass varied with the different AM fungi used. Based on the plant growth and root biomass per plant, *Glomus monosporum* is the best AM symbiont for inoculating *Ocimum sanctum* followed by *Glomus leptotichum* (Praveenkumar *et al.*, 2014).

4. Effect of AM Fungi on Nutrient Content

AM fungi mediated increased root geometry, nutrient assess and supply, resulting in the development of extramatricular hyphae might contribute to improved growth resulting in increased nutrient uptake. Hooker *et al.* (1992) were first to demonstrate both direct nutrient uptake and indirect growth effects of fungal inoculation on plants. AM fungi had beneficial effects of mango, citrus, papaya and jamun seedlings. These effects were evident from the increased vegetative parameters, *viz.*, seedling vigour, rootstock height, stem diameter, number of leaves, and other parameters when compared to uninoculated control. Increase in the growth of the these plants may be attributed to the beneficial synthesis of the hormones and growth factors by AM fungi leading to increased cell multiplication and cell division with overall increase in the vegetative parameters.

The improved vegetative parameters and rootstock growth in the field and pot plants can be attributed to increased AM fungal association in terms of per cent root colonisation and spore count leading to increased surface area for absorption and uptake of nutrients in the rootstocks. Besides this, AM fungi are also known to release growth substances, growth regulators, hormones and enzymes (acid phosphatase) in the rhizosphere, which help in the conversion of insoluble nutrients to soluble form and increase their availability to plants resulting in increased contents of major nutrients like N, P and K and micronutrients like Fe, Mg, Mn, Mo and Co (Adivappar *et al.*, 2004; Hatch, 1937). Later findings of many workers with a variety of woody and herbaceous horticultural species have further confirmed the findings of Hatch

(Holevas, 1966; Marx *et al.*, 1971; Deal *et al.*, 1972; Larue *et al.*, 1975; Plenchette *et al.*, 1981; Granger *et al.*, 1983; Sukhada, 1988; Sukhada *et al.*, 1995; Adivappar *et al.*, 2004, Shivaputra *et al.*, 2004b and Patil *et al.*, 2004).

5. Effect of AM Fungi on Yield Parameters

Yield is a manifestation of yield contributing characters. Increase in the yield by inoculation of mycorrhizae has been reported by Pande and Mishra (1975), Shrinivas (1998) and Durgannavar (2005) in papaya. The differences in these yield components could be traced back to the physiological characters both in vegetative and reproductive phase of crop growth. Further, it is well-documented that infection of plant roots by VAM fungi has beneficial effects on vegetative parameters of host plant (Waterer and Coltman, 1998).

Higher yields can be achieved by the plant only when the plants have sufficient carbohydrates and other required metabolites. The increase in yield can also be attributed to higher translocation of carbohydrates from other parts to reproductive parts during development. The source to sink translocation can be more effective only when the rates of photosynthesis are high (Johnson, 1984). Further, the photosynthetic capacity of a plant depends upon chlorophyll content in leaves. Higher photosynthetic rates might be an indirect effect of 'P' on ATP and ADP ratio or a direct action on RuBP carboxylase activity (Johnson, 1984). Accordingly, further improvement of photosynthesis by AM fungal inoculated plants could be related to higher tissue levels of 'P'. Similar higher content of 'P' in papaya and banana plants inoculated with AM fungi was reported by Shrinivas (1998) and Manjunath (2000). In experiments, using ^{32}P, it was found that AM fungal inoculated plants accumulate and store 'P'. Further, the rate of inflow of 'P' into mycorrhizal roots will be much higher than that of non-mycorrhizal plants (Sanders and Tinker, 1973). This increase in absorption of 'P' by mycorrhizal plants could be brought about by increased physical exploration of the soil, thereby making positionally unavailable nutrients available which can be achieved by decreasing the distance for diffusion of phosphate ions and through increasing surface area and absorption. Sanders and Tinker (1973) have attributed the increased absorption of 'P' by mycorrhizal plants to increase in surface area of absorption, while Bolan (1991) had studied mycorrhizal roots on a unit weight basis and observed much higher amounts of 'P' in mycorrhizal plants, which suggested that fungal hyphae have higher affinity for phosphate ions.

Inoculation of AM fungi has significant influence on the production of leaf, which is the photosynthetic site. Plants inoculated with AM fungi recorded higher number of leaves both in field as well as potted plants. Undoubtedly, the vegetative growth depends on the availability of nutrients and water, which is also stated by Nemec and Vu (1990).

Further, improvement in the rate of photosynthesis by AM fungal colonisation (Johnson and Pflefer, 1992) may be evoked by a number of physiological changes such as increase in plant hormones (Miller, 1971; Allen *et al.*, 1980 and Allen *et al.*, 1982), which may directly elevate photosynthetic rates through stomatal opening

(Incoll and Whitelam, 1977), enhancing ions transfer and regulating leaf chlorophyll levels (Richmond and Lang, 1957).

The inoculation of *Glomus fasciculatum* (VAM) and use of bioformulations enhances yield, colonization and chlamydospore count in turmeric (*Curcuma longa* L.) cv. Salem. Significantly higher number of rhizomes per clump (33.31), fresh rhizome yield (30.10 t/ha), cured yield (9.25 t/ha) and processing percentage (30.45) result in the crop inoculated with G. *fasciculatum* (VAM) compared to the uninoculated RDF (23.48 t/ha and 6.00 t/ha, respectively). Maximum root colonization (84.63 per cent) and number of chlamydospores per 50 g of soil (901.15) occur in the crop inoculated with G. *fasciculatum* (VAM) compared to the uninoculated control. Among nine bio-formulation treatments, the highest number rhizomes per clump (44.00), fresh rhizome yield (35.67 t/ha), cured yield (12.47 t/ha), processing percentage (34.96), root colonization (100 per cent), number chlamydospores per 50 g of soil (1348.33), lowest per cent disease index (PDI) for anthracnose disease (24.33, control: 61.33) at 180 DAP was recorded by the application of G. *fasciculatum* (VAM)+ RDF+ Panchagavya (3 per cent) + Amritpani (3 per cent) + mulch (Sugarcane trash)+ *Trichoderma harzianum* (2.5 per cent) spray on mulch+ Agnihotra ash (5 g)+ Triambakam homa ash (5 g) followed by above treatment without ashes (34.07 t/ha, 100 per cent and 12 respectively) compared to the lowest values recorded in RDF (21.03 t/ha 3.33 per cent 248 respectively) receiving 180:90:90 kg NPK/ha and 25 t FYM/ha. Thus higher yields could be obtained by the application of bio-inoculant (G. *fasciculatum*) and bio-formulations in turmeric (Chandalinga *et al.*, 2016).

Organic and inorganic fertilizers differentially influence on sprouting, growth, flowering and nutrient status in Heliconia (*Heliconia* sp.) cv. Golden Torch under protected cultivation. Inoculation with *Glomus fasciculatum, Trichoderma harzianum* and use of bioformulations *viz.*, Panchagavya, Amrit pani, Agnihothra ash, dry mulch and 2 kg FYM/m^2 record significantly less number of days for initiation of sprouting (60.67), 50 per cent sprouting (76.30), flower bud formation (208 days) and first flowering (217 days). The complete sprouting (92 days) and high sprouting percentage (94.45 per cent) can be achieved by inoculating with *Glomus fasciculatum + Trichoderma harzianum*. The treatment helps to enhance other parameters like plant height, number of tillers and number of flowers per plant. Whereas maximum chlorophyll 'a' (0.722 mg/g), 'b' (0.373 mg/g) and total chlorophyll (1.095 mg/g), flower yield per plant (3.07), flower yield per meter square (64.73 flowers), higher content of nitrogen (133.73 kg/ha), phosphorus (35.84 kg/ha) and potassium (284.78 kg/ha) in soil and nitrogen (1.72 per cent), phosphorus (0.37 per cent) and potassium (3.09 per cent) in leaves (Sushma, *et al.*, 2012).

The growth and yield of gerbera (*Gerbera jamasonii* L.) var. Rosalin can be achieved by using 75 per cent RDF + *Glomus fasciculatum + Trichoderma harzianum +* Panchagavya + Amrut pani + Dry mulch+ Agnihothra ash. The treatment favourably influenced flower parameters like minimum number of days taken for 50 per cent flowering, maximum stalk length, flower diameter, number of flowers per plant and flower yield per m^2 (Sushma, 2016).

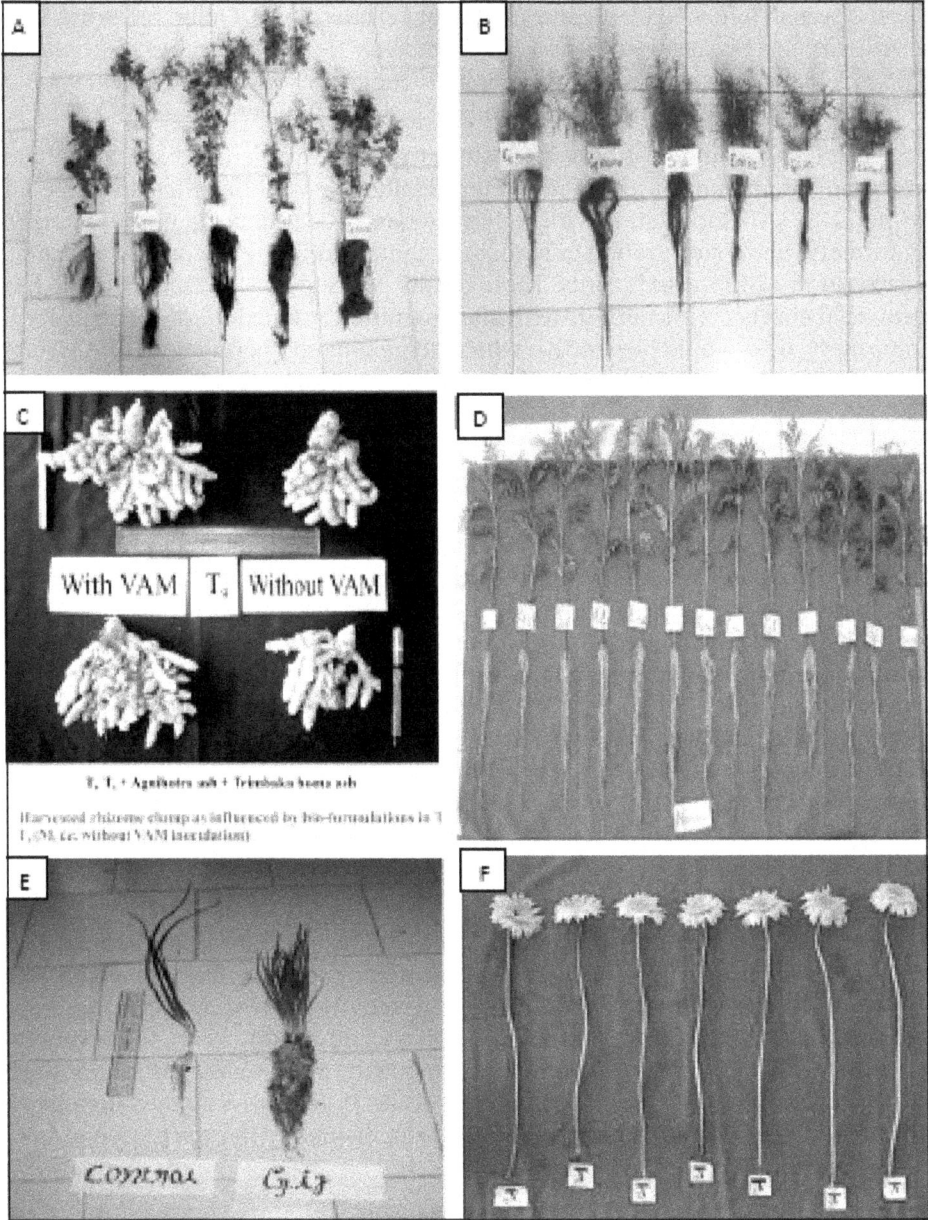

Figure 8.1: Effect of VAM Fungi on different Horticultural Crops.
A: Tusli, B: Rosemary; C: Turmeric; D: Marigold; E: Onion; F: Gerbera.

Nutrition affects growth, flowering and yield in African marigold (*Tagetes erecta* L.) for two seasons which indicate that individual flower weight, flower diameter, number of petals/flower, flower yield, chlorophyll, xanthophyll content and shelf life will be higher with RDF in the first season and higher with INM during second season followed by organic treatment (Vijetha, 2013). Onion receiving RDF + *Azospirillum brasilense* + *Azotobacter chroococcum* + VAM (*Gigaspora gigantia*) + PSB + *T. harzianum* resulted in significantly higher bulb yield in pooled values and significantly lower per cent disease index at harvest (Praveenkumar, 2015).

The effect of inoculation with AMF on growth, nutrient uptake, yield and curcumin production of turmeric under glasshouse conditions results in better response to AMF inoculation with higher biomass production and nutrient uptake in turmeric, the concentration of curcumin, contained in the rhizome of turmeric, increased in AMF treated plants. These results indicate that AMF inoculation has beneficial effects on turmeric growth and curcumin production. AMF inoculation to turmeric field would be effective when indigenous soil populations of AMF are low or native AMF are no longer effective (Yamawaki *et al.,* 2013).

6. AM Fungi and Biotic Stress

The role of AM fungi in the control of plant pathogen has been the subject of several reviews. In these reviews, arbuscular mycorrhizal associations have been suggested as bio-control agents and the general conclusion is that they can reduce or even suppress damage caused by soil borne plant pathogens. In tropics the AM fungi are found to be the major symbionts. This prophylactic ability displayed by some AM-fungi could be exploited in cooperation with other rhizospheric microorganisms which have the antagonistic abilities against plant pathogens and that are being used as bio-control agents (Linderman, 1994). Microbial antagonists of fungal pathogens can improve VA-mycorrhizal formation. Thus, accurate management of these interactions by tailoring appropriate rhizosphere systems relevant to plant growth and health could be an integrated approach for sustainable fruit production (Barea and Jeffries, 1995). The beneficial influence of colonisation by mycorrhiza in reducing the incidence of fungal pathogens have been studied in avocado (Davis *et al.,* 1978), citrus (Schenck and Nicolson, 1977) and guava (Srivastava *et al.,* 2001).

Land degradation due to salinity, waterlogging, erosion *etc.,* are serious and growing problems in India and other countries. Excessive NaCl levels in soil inhibit mycorrhizal formation and restrict the activity of most mycorrhizal fungi, but some can tolerant these conditions (Juniper and Abbott 1993). Observations in natural ecosystems have shown that plants with mycorrhizal associations are often less common than non-mycorrhizal species in soils which are waterlogged or saline, but some mycorrhizal plants are normally present in even the worst soils (Brundrett, 1991).

The tremendous scope for use of VA Mycorrhizal fungus in horticultural crops makes it a novel biofertilizer/bioinoculant. It has effect on root geometry, enhance uptake of nutrients, controls root rotting nematodes and pathogens *viz., Pythium, Phytophthora, Fusarium, Sclerotium etc.,* increase biomass, quality, yield and shelf life by way of production of growth hormones *viz.,* IAA, Cytokinins *etc.*

References

Adalsteinsson, S. and Jensen, P. 1989. Modification of root geometry in Winter Wheat by phosphorus deprivation. *J. Plant Physiol.,* 135: 513-517.

Adivappar, N., Patil, P. B., Patil, C.P., Swamy, G. S. K. and Athani, S. I. 2004. Effect of AM fungi on growth and nutrient content of container grown papaya plants. In: Pathak, R.K., Ram, K., Khan, R.M. and Ram, R.A. (Eds.). Organic Farming in Horticulture. CISH, Lucknow, pp. 166-169.

Allen, M.F., Moore, T.S. and Christensen, M. 1980. Phytoharmone changes in *Bouteloua gracillis* infected by vesicular arbuscular mycorrhiza cytokinin increase in the host plants. *Can. J. Bot.*, 58: 371-374.

Amijee, F., Tinker, P.B. and Stribely, D.P. 1989. Effect of phosphorus on the morphology of VA mycorrhizal root system of leek (*Allium porrum*). *Plant and Soil*, 119:334-336.

Barea, J.M. and Jeffries, P. 1995. Arbuscular mycorrhiza in sustainable soil plant system. In: Verma, A. and Hous, B. (Eds.). Mycorrhiza Structure, Function, Molecular Biology and Biotechnology, Springer-Verlag, Heidelberg, pp. 521-560.

Berta, G., Trotta, A., Fusconi, A., Hooker, J.E., Munpo, M., Atkinson, D., Govannetti, M., Morini, S., Fortuna, P., Tisserant, B., Gianinazzi-Pearson, V. and Gianinazzi, S. 1994. Arbuscular mycorrhizal induced changes to plant growth and root system morphology in *Prunus cerasifera. Tree Physiol.*, 15: 281-293.

Bolan, N.S. 1991. A critical review on the role of mycorrhizal fungi in the uptake of phosphorus by plants. *Plant and Soil*, 134: 189-207.

Brundrett, M. C. and Abbott, L. K. 1991. Roots of jarrah forest plants. I. Mycorrhizal associations of shrubs and herbaceous plants. *Afr. J. Bot.*, 39: 445-457.

Chandalinga, Patil, C.P., Hegde, N.K., Kallappa Narode, Arif, A.A. and Suresh, H.L. 2016. Effect of *Glomus fasciculatum* (VAM) and bioformulations on yield, disease incidence, root colonization and chlamydospore count in turmeric (*Curcuma longa* L.) cv. Salem (Abstract). In: National Seminar on Chilli and Turmeric: Challenges and Opportunities. January 8-9, University of Horticultural Sciences, Bagalkot, Karnataka, pp. 96.

Crafts, C.B. and Millar, C.O. 1974. Detection and identification of cytokinins produced by mycorrhizal fungi. *Plant Physiol.*, 54: 586-588.

Davis, R.M., Menge, J.A. and Zentmeyer, G.A. 1978. Influence of vesicular arbuscular mycorrhizae on *Phytophthora* root rot of three crop plants. *Phytopathol.*, 68: 1614-1617.

Deal, D.R., Boothroyd, C.W. and Mai, W.F. 1972. Replanting of vine yards and its relationship to vesicular-arbuscular mycorrhiza. *Phytopathol.*, 62: 172-175.

Duragannavar, M. P. 2005. Effect of bioformulations on growth and yield of papaya cv. Red lady. M. Sc. (Hort.) Dissertation, Dept. of Fruit Science University of Agricultural Sciences, Dharwad, India.

Gange, A.C., Brown, V.K. and Sinclair, G.S. 1993. Vesicular arbuscular mycorrhizal fungi: A determinant of plant community structure in early succession. *Funct. Ecol.*, 7: 616-622.

Giorannetti, M., Qavio, L., Sbrana, C. and Citernesi, A. S. 1993. Factors affecting apperessorium development in vesicular arbuscular mycorrhizal fungus *Glomus mosseae* (Nicol. and Gerd.) Gerd. and Trappe. *New Phytol.*, 123: 114-122.

Granger, R.L., Plenchette, C. and Fortin, J.A. 1983. Effect of vesicular arbuscular (VA) endomycorrhizal fungus (*Glomus epigaeum*) on the growth and leaf mineral content of two apple clones propagated *in-vitro. Can. J. Plant Sci.*, 63: 551-555.

Habte, M. and Byappanahalli, M.N. 1994. Dependency of cassava (*Manihot esculenta* Crantz.) on vesicular-arbuscular mycorrhizal fungi. *Mycorrhiza*, 4: 241–245.

Harborne, J. B. and Williams, C.A. 2000. Advances in flavonoid research since 1992. *Phytochemistry*, 55: 481-504.

Harper, J.L., Jones, M. and Hamilton, N.R. 1991. Evolution of roots and the problems of analyzing their behavior. In: Atkinson, D. (Ed.). Plant Root Growth: An Ecological Prospective, Blackwell Scientific Publications, Oxford, pp. 3-24.

Hatch, A.B. 1937. The physical basis of mycotrophy in the genus *Pinus. Black Rock Forestry Bulletin*, 6: 1-168.

Holevas, C. D. 1966. The effect of vesicular-arbuscular mycorrhizae on the uptake of soil phosphorus by strawberry (*Fragaria* sp. var. Cambridge Favourite). *J. Hortic. Sci.*, 41: 57-64.

Hooker, J.F., Munro, M. and Atkinson, D. 1992. Vesicular arbuscular mycorrhizal fungi induced alteration in polar root system morphology. *Plant and Soil*, 145: 207-214.

Incoll, L.D. and Whitelam, G.C. 1977. The effect of kinetin on stomata of the grass *Anthephera pubescens* Neer. *Planta*, 137: 243-245.

Janos, D.P. 1980. Mycorrhiza influence tropical succession. *Biotropica*, 12: 56-64.

Johnson, C.R. 1984. Phosphorus nutrition on mycorrhizal colonization, photosynthesis, growth and nutrient composition of *Citrus aurantium. Plant Soil*, 80: 35-42.

Johnson, N.C. and Pflefer, F.L. 1992. Vesicular-arbuscular mycorrhizae and cultural stresses. In: Bethlenfalvay, G.J. and Linderman, R.G. (Eds.). Mycorrhizae in Sustainable Agriculture, ASA Publication, Madison, WI, pp. 71-99.

Johnson, N.C., Graham, J.H. and Smith, F.A. 1997. Functioning of mycorrhizal associations along the mutualism-parasitism continuum. *New Phytol.*, 135: 575-585.

Juniper, S. and Abbott, L. 1993. Vesicular arbuscular mycorrhizas and soil salinity. *Mycorrhiza*, 4: 45-57.

Larue, J.H., Mcclellen, W.D. and Peacock, W.L. 1975. Mycorrhizae and soil microbial interactions. In: Bethlenfalvay, G.J. and Linderman, R.G. (Eds.). Mycorrhizae in Sustainable Agriculture, ASA Publication, Madison, WI, pp. 45-70.

Linderman, R.G. 1994. Vesicular arbuscular mycorrhizae and soil microbial interactions. In: Bethlenfalvay, G.J. and Linderman, R.G. (Eds.). Mycorrhizae in Sustainable Agriculture. ASA Publication, Madison, WI, pp. 45-70.

Linderman, R.G. and Call, G.A. 1977. Enhanced rooting of woody plant by mycorrhizal fungi. *J. Am. Soc. Hortic. Sci.*, 102: 529-532.

Manjunath, A. 2000. Effect of vesicular arbuscular mycorrhizal species and phosphorus levels on growth and yield of papaya cv. Sunset Solo. M.Sc. (Hort.) Dissertation, Dept. of Fruit Science, University of Agricultural Sciences, Dharwad, India.

Manjunath, A., Mottan, R. and Bagyaraj, D.J., 1983. Response of citrus to vesicular arbuscular mycorrhizal inoculation in unsterile soil. *Can. J. Bot.*, 61: 2729-2732.

Marx, D.H., Bryan, W.C. and Campbell, W.A. 1971. Effect of endomycorrhiza by *Endogone mosseae* on growth of citrus. *Mycologia*, 63: 1222-1226.

Miller, C.O. 1971. Cytokinin production by mycorrhizal fungi In: Hacskaylo, E. (Ed.). Mycorrhizae, GPO, Washington, pp. 168-174.

Nemec, S. and Vu, J.C.V. 1990. Effect of soil, phosphorus and *Glomus intraradices* on growth, non-structural carbohydrates and photosynthetic activity of *Citrus aurantium*. *Plant Soil*, 128: 257-263.

Pande, S.P. and Mishra, A.P. 1975. Mycorrhizae in relation to growth and fruiting of *Litchi chinensis*. *J. Indian Bot. Soc.*, 54: 280-293.

Pathak, R.K. and Ram, R.A. 2003. Biodynamic Agriculture. CISH, Lucknow, p. 42.

Patil, C.P. 2016. Package of practices, University of Horticultural Sciences, Bagalkot, India.

Patil, P.B. and Patil, C.P. 2002. Package of practices, University of Agricultural Sciences, Dharwad, India.

Patil, P.B., Patil, C.P., Swamy, G.S.K., Athani, S.I., Adiveppar, N., Manjunath, V.G. and Mankani, S. 2004. Influence of different AM fungi on growth of papaya under field condition. In: Pathak, R. K., Ram, K., Khan, R. M. and Ram, R. A. (Eds.). Organic Farming in Horticulture, CISH, Lucknow, pp.185-187.

Plenchette, C., Furlan, V. and Fortin, J.A. 1981. Growth stimulation of apple trees in unsterilized soil under field conditions with VA mycorrhizal inoculation. *Can. J. Bot.*, 59: 2003-2008.

Powell, C.V. 1986. Effect of mycorrhizal inoculation and phosphorus fertilizer on the growth of hardwood cuttings of kiwifruit (*Actinidia deliciosa* cv. Hayward) in containers. *N. Z. J. Agric. Res.*, 29: 263–268.

Praveenkumar, D.A. 2015. Studies on organic production and storage of onion (*Allium cepa* L.) var. Arka Kalyan. Ph. D. (Hort.) Dissertation, Department of Horticulture, University of Horticultural Sciences, Bagalkot, India.

Praveenkumar, D.A., Chandalinga, A. and Patil, C.P. 2015. Tulsi (*Ocimum sanctum*) shoot and root parameters as influenced by Vesicular Arbuscular Mycorrhiza (VAM). *An. Plant Soil Res.*, 17: 5-7.

Praveenkumar, D.A., Patil, C.P., Chandalinga, A. and Vijetha, 2014, Effect of Vesicular Arbuscular Mycorrhiza (VAM) on shoot and root Parameters of Rosemary (*Rosmarinus officinalis*). *Trends Biosci.*, 7: 2928-2931.

Ram, S. 1997. Propagation. In: Litz, R.E. (Ed.). The Mango Botany, Production and Uses. Cab International, p. 367.

Ramesh, N. 1997. Biofertilisers for raising quality rootstocks for softwood grafting in cashew. M.Sc. Dissertation, Dept. of Spices and Plantation, University of Agricultural Sciences, Bangalore, India.

Richmond, A. and Lang, A. 1957. Effect of kinetin on protein content and survival of detached xanthium leaves. *Science*. 125: 650-651.

Sanders, F.E. and Tinker, B.P. 1973. Phosphate flow in mycorrhizal roots. *Pestic. Sci.*, 4: 385-395.

Santosh, 2004. Enhancement of germination, growth, graft-take and stress tolerance of mango rootstocks using bioformulations. M.Sc. (Hort.) Dissertation, Dept. of Fruit Science, University of Agricultural Sciences, Dharwad, India.

Scagel, C. F. 2000. Effect mycorrhizal crops, 3rd International Symposium on Adventitous root formation, Veldhoven, NL, pp. 1-16.

Schellenbum, L., Gianinazzi, S. and Gianinazzi-Pearson, V. 1992. Comparison of acid soluble protein synthesis in the roots of endomycorrhizal wild type *Pisum sativum* and corresponding isogenic mutants. *J. Plant Physiol.*, 141: 2-6.

Schenck, N.C. and Nicolson, T.H. 1977. A zoosporic fungus occurring on species of *Gigaspora margarita* and other vesicular arbuscular mycorrhizal fungi. *Mycologia*, 69: 1049.

Shivaputra, S.S., Patil, P.B., Swamy, G.S.K., Patil, C.P. and Athani, S. I. 2004a. Interaction effect of AM and vermiculture on drought tolerance of papaya. In: Pathak, R. K., Ram Kishun, Khan, R. M. and Ram, R. A. (Eds.). National Organic Farming in Horticulture, CISH, Lucknow, pp. 170-172.

Shivaputra, S.S., Swamy, G.S.K., Patil, C.P., Patil, P.B. and Athani, S.I., 2004b, Combined effect of AM fungi and vermicompost on growth of papaya. In: Pathak, R.K., Ram Kishun, Khan, R.M. and Ram, R.A. (Eds.). National Organic Farming in Horticulture, CISH, Lucknow, p. 253-255.

Shivashankar, S. and Iyer, R. 1988. Influence of vesicular-arbuscular mycorrhiza on growth and nitrate reductase activity of black pepper. *Indian Phytopathol.*, 41: 428-433.

Shrinivas, M. 1998. Response of papaya to vesicular arbuscular mycorrhizal fungi at graded levels of phosphorus. M. Sc. (Hort.) Dissertation, Dept. of Fruit Science, University of Agricultural Sciences, Dharwad, India.

Singh, B., Ranjan, S. and Ramchandra, 2007. Response of Panchagavya and Manchurian Mushroom Tea on floral characters in tuberose (*Polyanthus tuberosa* Linn.) cv. Pearl Double. *J. Ornamental Hort.*, 10: 250-254.

Slankis, V. 1975. Hormonal relationship in mycorrhizal development. In: Marx, G. C. and Kozlowski, T. T. (Eds.). Ectomycorrhizae. Academic Press, New York, pp. 231-298.

Srivastava, A. K., Ahmed, R., Kumar, S. and Sukhada, M. 2001. Role of VA-mycorrhiza in the management of wilt disease of guava in the alfisols of Chotanagpur. *Ind. Phytopathol.*, 59: 78-87.

Sukhada, M. 1988. Response of papaya (*Carica papaya*) to VAM fungal inoculation. In: Mahadevan, N., Raman and Natarajan, H. (Eds.), Mycorrhizae for Green Asia. First Asian Conference on Mycorrhizae. Madras, p. 260.

Sukhada, M., Shivananda, T. N. and Iyengar, B. R. V. 1995. Uptake of ^{32}P labelled superphosphate by endomycorrhizal papaya (*Carica papaya* cv. Coorg Honey Dew). *J. Nuc. Agric. Biol.*, 24: 220-224.

Sushma, B., 2016, Effect of organic and inorganic treatments on yield and quality of Gerbera var. Rosalin. M. Sc. (Hort) Dissertation, Dept. of Floriculture and Landscape Architecture, University of Horticultural Sciences, Bagalkot, India.

Sushma, H. E., Reddy, B. S., Patil, C. P. and Kulkarni, B. S. 2012. Effect of organic and inorganic nutrients on sprouting, growth, flowering and nutrient status in Heliconia (*Heliconia* sp.) cv. Golden Torch. *Karnataka J. Agric. Sci.*, 25: 370-372.

Torrey, J. A. 1976. Root hormones and plant growth. *Annu. Rev. Plant Physiol.*, 27: 435-459.

Venkat, 2004. Exploitation of Rangpur lime as a rootstock for different citrus species. M. Sc. (Hort.) Dissertation, Dept. of Fruit Science, University of Agricultural Sciences, Dharwad, India.

Vijetha, 2013. Effect of organic and inorganic nutrition on growth and flowering of marigold (*Tagetes erecta* L.). M. Sc. (Hort.) Dissertation, Dept. of Floriculture and Landscape Architecture, University of Horticultural Sciences, Bagalkot, India.

Waterer, D. R. and Coltman, R. E. 1998. Response of mycorrhizal bell peppers to inoculation timing, phosphorus and water stress. *Hort. Sci.*, 24: 688-690.

Yamawaki, K., Matsumura, A., Hattori, R., Tarui, A., Hossain, M. A., Ohashi, Y. and Daimon, H. 2013. Effect of inoculation with arbuscular mycorrhizal fungi on growth, nutrient uptake and curcumin production of turmeric (*Curcuma longa* L.). *Agric. Sci.*, 4: 66-71.

2017, Mycorrhizal Fungi
Editors: Ashok Aggarwal and Kuldeep Yadav
Published by: ASTRAL INTERNATIONAL PVT. LTD., NEW DELHI

Pages 195–220

9

Arbuscular Mycorrhizal Fungi: An Efficient Tool as a Biofertilizer in Sustainable Agriculture and Crop Productivity

B.N. Reddy* and A. Hindumathi

Mycology and Plant Pathology Laboratory, Department of Botany,
Osmania University, Hyderabad – 500 007, Telangana
*Corresponding Author: reddybn1@yahoo.com

ABSTRACT

Use of costly fertilizers especially phosphate fertilizers and pesticides used extensively to improve plant growth and productivity contribute to soil pollution that possibly can lead to groundwater contamination besides adversely affecting the non-target species including some beneficial microorganisms and thereby threatening the ecological balance. Farmers in the tropics are generally poor and cannot afford phosphate fertilizers due to their heavy cost. Mycorrhizal benefits are greatest and most obvious under low fertility conditions that exist in the developing countries. AM fungi promote plant growth, primarily through the enhanced uptake of P. Under these circumstances AM fungi can prove to be the most effective alternative to fertilizer for enhancing plant health and soil quality. This microbial system can be used as a potential biological alternative to expensive chemical fertilizers and pesticides for improving plant growth and productivity both in agriculture as well forestry. However, AM fungal inoculum *production in large scale is a major problem. Exploitation of AM fungi in commercial agriculture is possible only while mycorrhizal inoculum is produced in large quantities and appropriate strategies are adopted for successful application of inoculum in field conditions.*

Keywords: *Arbuscular mycorrhizal fungi, Biofertilizer, Mass multiplication, Nutrient uptake, Phosphorus.*

1. Introduction

Mycorrhizae (Frank, 1885) are highly evolved, intimate associations between plant roots and a unique group of soil borne fungi (Gerdemann, 1968). The mycorrhizal condition of plants was shown to be the rule and non mycorrhizal condition is the exception in nature (Gerdemann, 1969). Some plant species are so dependent on mycorrhizal fungi that they cannot establish in nature or maintain normal growth without them.

Mycorrhizal symbiosis is highly interdependent, mutualistic relationship characterized by bidirectional movement of nutrients where the host plant receives inorganic mineral nutrients from the soil through the fungal hyphae and the fungus obtains photosynthetically derived carbon compounds from the host plant (Harley and Smith, 1983), thereby providing a linkage between the plant root and the soil.

Mycorrhizal fungi benefit the host plant primarily by increasing the absorption capacity of the root system and translocate inorganic nutrients mainly phosphorus, other complexed compounds and minor elements through extensive network of hyphae external to the root (Gerdemann, 1968; Van der Heijden *et al.,* 1998) more effectively than the plant roots. Mycorrhizal fungi are associated with more than 92 per cent of the plant families (80 per cent of the plant species) belonging to vascular and non-vascular plants (Wang and Qui, 2006) majority of agricultural crops.

There are several different types of mycorrhizal associations classified into ectomycorrhizal, ectendomycorrhizal and endomycorrhizal based on the presence of extraradical or intraradical hyphal structures (Bonfante and Perotto, 2000). At least seven different types of mycorrhizal associations have been recognised based on the distinct morphological patterns and largely upon their external or endophytic nature of the symbiont. They are ectomycorrhizas, ectendomycorrhizas, arbutoid mycorrhizas, monotropoid mycorrhizas, vesicular arbuscular mycorrhizas, orchid mycorrhizas and ericoid mycorrhizas (Figure 9.1). Ectomycorrhizas or ectotrophic mycorrhizas are characterized by the formation of a mantle or sheath around the root surface and a Hartig net between the root cells. Ectomycorrhizae are associated with the roots of around 10 per cent plant families, majority of which are forest trees (Wang and Qui, 2006) and fungal partner belonging to Basidiomycota, Ascomycota and Zygomycota.

Ectendomycorrhizas, arbutoid and monotropoid mycorrhizas are similar to ectomycorrhizal associations and are grouped under ectomycorrhizal associations, but have specialized anatomical features. Endomycorrhizas are characterized by the

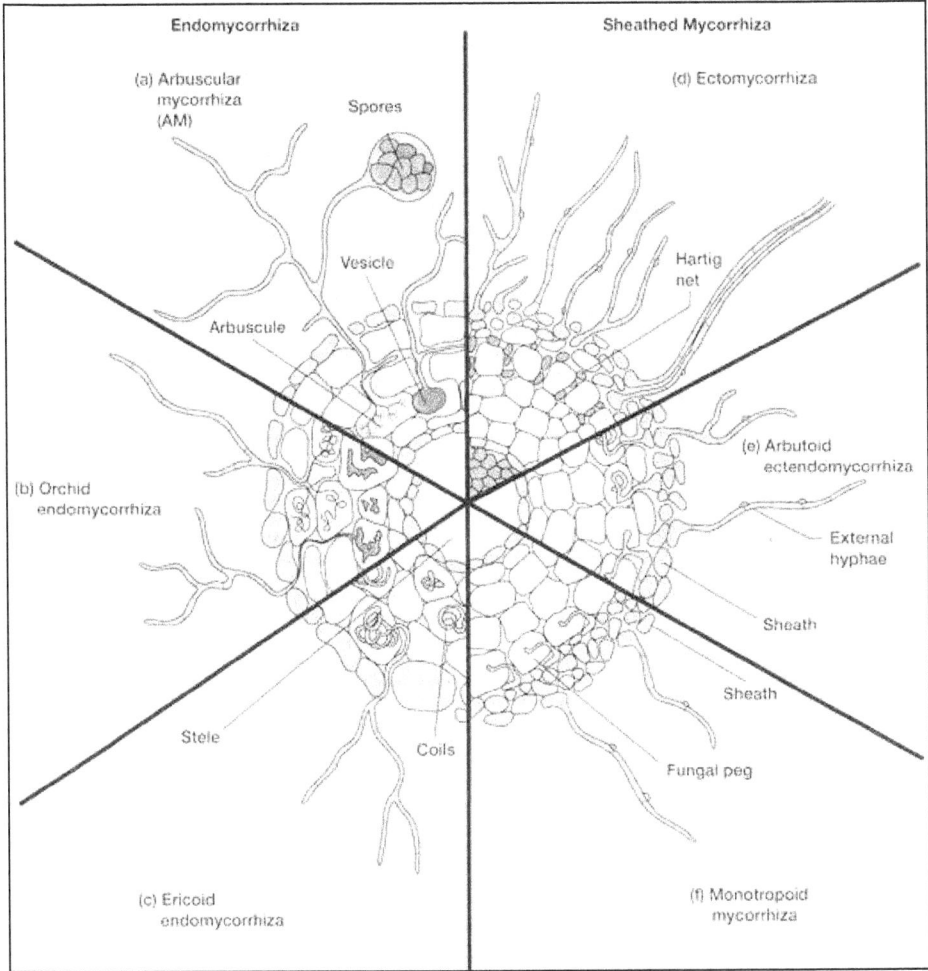

Figure 9.1: Root Cross Section Illustrating different Types of Mycorrhizal Relationships that Exist within Plants (Prescot *et al.,* 2005).

formation of loose network of hyphae in the soil surrounding the root and extensive hyphal growth within the root cortex (intracellular).

Endotrophic mycorrhizae or endomycorrhizae have been further classified as arbuscular, ericoid, and orchid mycorrhizas. Orchid and ericoid mycorrhizas of endomycorrhizae are formed by septate fungi (Basidiomycota). The endomycorrhiza are separated into two groups: those produced by aseptate fungi (Zygomycota), the most common type vesicular arbuscular mycorrhizas (VAM) or arbuscular mycorrhizas (AM). These are characterized by the formation of inter- and intracellular hyphae, intracellular arbuscules and inter- and intracellular vesicles within the root cortex cells. Their presence led to the former common name vesicular arbuscular mycorrhizal (VAM) fungi but the term arbuscular mycorrhiza (AM) is now preferred (Friberg, 2001) because not all fungi produce vesicles. But there is some disagreement about the two terms as arbuscules are not always present in the mycorrhizal roots. Vesicles are not formed by the genera belonging to the order Gigasporales but are found in the other genera of the Glomerales (Isaac, 1992).

2. Arbuscular Mycorrhiza

The arbuscular mycorrhizal fungi (AMF) are the most common obligate symbiotic fungi, belonging to phylum Glomeromycota (Schüßler *et al.,* 2001). This association is geographically ubiquitous, occurring in arctic, temperate and tropical regions over a broad ecological range from aquatic to desert environment (Gerdemann, 1975). The evolution of symbiotic fungi was thought to have existed at least 470 million years ago from fossil records of ordovician age.

2.1. Taxonomy

The taxonomy of AM fungi was largely based on the morphological and anatomical characteristics of their spores and sporocarps, spore germination, *etc.* Some of the AM fungal species isolated from the rhizosphere soils of our studies and identified based on their morphological characters are represented in Figures 9.2 and 9.3. From among these AM fungal species *Acaulospora alpina, A. myriocarpa, G. australe, G. diaphanum, G. heterosporum, G, manihotis, G. microaggegatum, G. multicaule* and *Gigaspora rami sporopora* are the first reports from Telangana state (Hindumathi and Reddy, 2016b). Now, several modern methods like serology, isozyme variation by electrophoresis (Hepper *et al.,* 1988), fatty acid variation (Bentivenza and Morton, 1994), molecular techniques such as DNA based methods (Helgason *et al.,* 1999; Schüßler *et al.,* 2001) have aided in a clearer phylogenetic analysis than was possible using morphological and microscopic identification.

The most recent phylogenetic synthesis period (2000 to present) started with the proposal of new classification based on genetic characters using sequences of the multicopy rRNA genes. However, with the recent advent of molecular and phylogenetic studies, AMF have been elevated and assigned their own phylum Glomeromycota. The classification of arbuscular fungi used in the website presented in Table 9.1 is given by Schüßler *et al.* (2001) with emendations of Oehl and Sieverding (2004), Walker and Schüßler (2004), Sieverding and Oehl (2006), Spain *et al.* (2006), Walker *et al.* (2007a, b) and Palenzuela *et al.* (2008).

2.2. Occurrence

Because of wide spread occurrence of AM fungi in nature and their numerous benefits to plants, researchers in several disciplines have shown interest in studying mycorrhizae. Many tropical plantation crops like coffee, tea, rubber, citrus, cocoa, oil palm, many tropical timber trees and all temperate fruit trees are reported to have AM association.

The occurrence of AM association was reported in several tropical trees and other plants (Thaper and Khan, 1973). The occurrence of AM fungal association has been reported in the roots of several plants including grasses (Nicolson, 1959), vegetable crops (Hayman, 1975; Manjunath and Bagyaraj, 1984; Reddy *et al.,* 2006b; Satya Vani, 2012; Satya Vani *et al.,* 2014a), field crops (Hayman, 1975; Bagyaraj

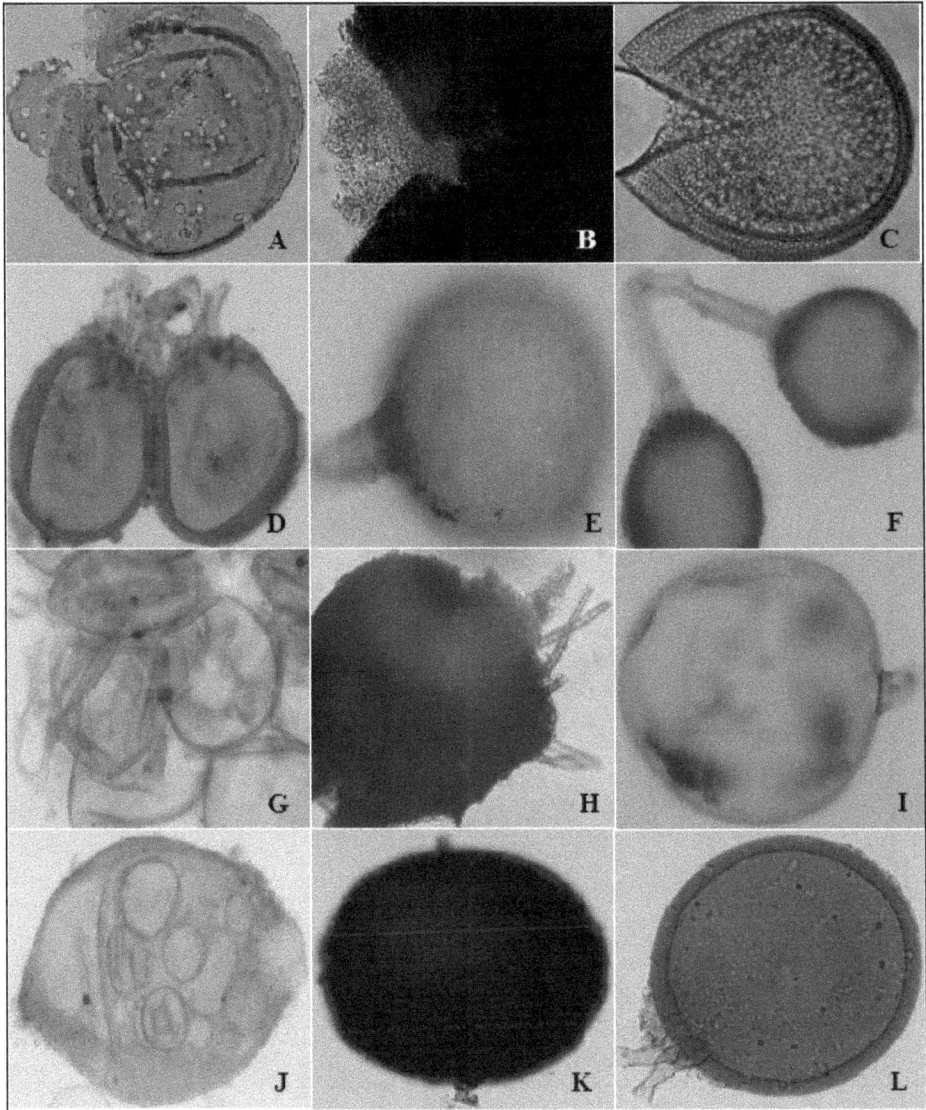

Figure 10.9

A. *Acaulospora alpina;* B: *A. myriocarpa;* C: *A. scrobiculata;* D: *Glomus ambisporum;* E: *G. australe;* F: *G. diaphanum;* G: *G. fasciculatum;* H: *G. heterosporum;* I: *G. manihotis;* J: *G. Microaggregatum;* K: *G. multicaule;* L: *G. multisubstensum.*

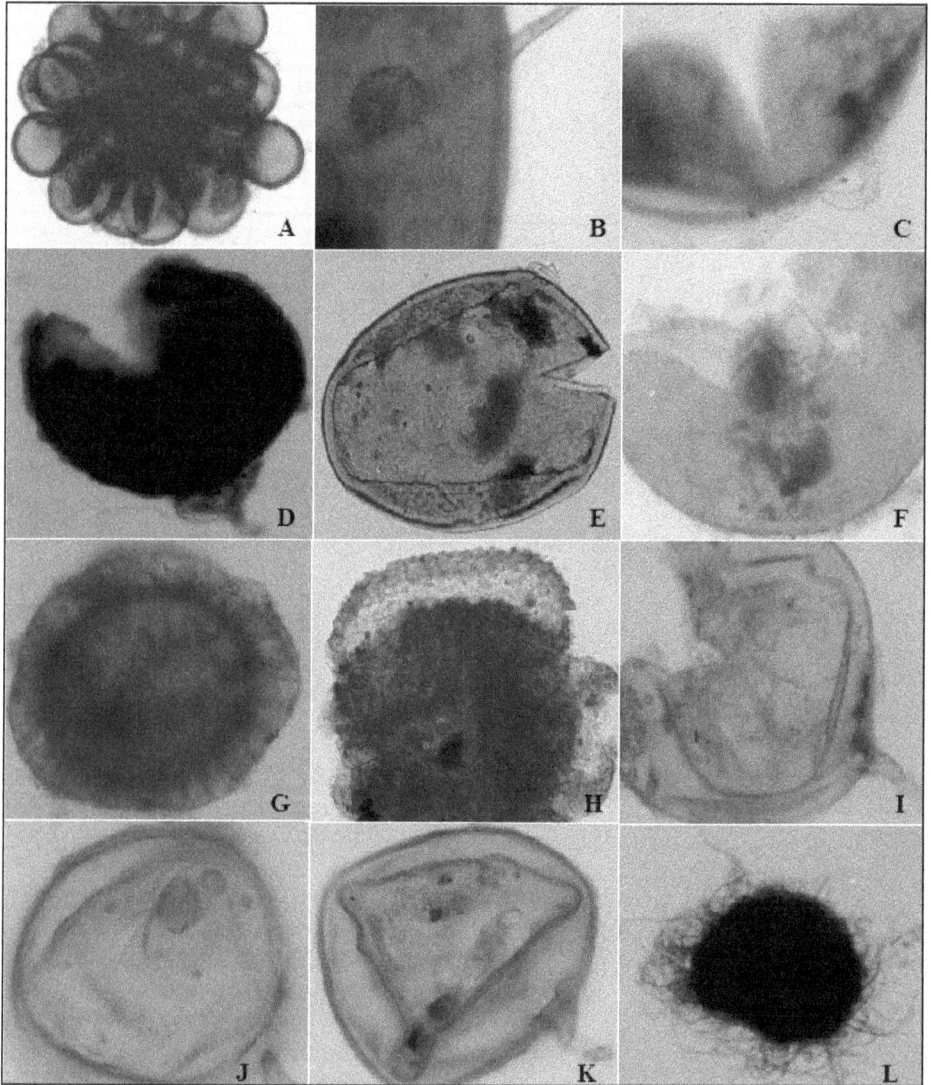

Figure 9.3

A: *Glomus rubiforme*; B: *Gigaspora gigantea*; C: *Gig. rami sporopora*; D: *Entrophospora nigra*; E: *S. pellucida*; F: *Scutellospora sp.*; G: *Entrophospora schenkii*; H: *Entrophospora sp.*; I: *Funneliformis caledonius*; J: *F. geosporum*; K: *F. mosseae*; L: Unidentified genus/sp1.

Table 9.1: Classification of Glomeromycota from Class to Genus Level

Phylum: Glomeromycota
Class: Glomeromycetes
    Order: Glomerales
        Family: Glomeraceae
            Genus: *Glomus*
            Genus: *Funneliformis*
            Genus: *Septoglomus*
            Genus: *Simiglomus*
        Family: Entrophosporaceae
            Genus: *Claroideoglomus*
            Genus: *Albahyphae*
            Genus: *Viscospora*
            Genus: *Entrophospora*
    Order: Diversisporales
        Family: Diversisporaceae
            Genus: *Diversispora*
            Genus: *Redeckera*
            Genus: *Otospora*
            Genus: *Tricospora*
        Family: Sacculosporaceae
            Genus: *Sacculospora*
        Family: Pacisporaceae
            Genus: *Pacispora*
        Family: Acaulosporaceae
            Genus: *Kuklospora*
            Genus: *Acaulospora*
    Order: Gigasporales
        Family: Scutellosporaceae
            Genus: *Orbispora*
            Genus: *Scutellospora*
        Family: Dentiscutataceae
            Genus: *Fuscutata*
            Genus: *Dentiscutata*
            Genus: *Quatunica*
        Family: Racocetraceae
            Genus: *Cetraspora*
            Genus: *Racocetra*

Contd...

Table 10.1–*Contd...*

Class	Classification of Glomeromycota from class to Genus level
	Family: Gigasporaceae
	Genus: *Gigaspora*
Class: Archaeosporomycetes	
	Order: Archaeosporales
	Family: Ambisporaceae
	Genus: *Ambispora*
	Family: Archaeosporaceae
	Genus: *Archaeospora*
	Genus: *Intraspora*
	Family: Geosiphonaceae
	Genus: *Geosiphon*
Class: Paraglomeromycetes	
	Order: Paraglomerales
	Family: Paraglomeraceae
	Genus: *Paraglomus*

et al., 1979b), cereal crops (Hindumathi, 1999; Hindumathi and Reddy, 2011b, 2012b; Reddy *et al.*, 2006a, 2007), tropical legumes (Krishna and Bagyaraj, 1982; Rao and Parvathi, 1982; Hindumathi and Reddy, 2011a) and in some oil seed plants (Rao and Parvathi, 1982; Hindumathi and Reddy, 2015, 2016b).

With few exceptions, arbuscular mycorrhizal fungi are cosmopolitan in distribution, frequency is markedly influenced by a variety of environmental and edaphic factors. Soil physico-chemical factors and microbiological components of the soil are reported to play significant role in the distribution, density, composition and activity of AM fungi (Hindumathi, 1999; Sreevani and Reddy, 2005; Reddy *et al.*, 2006a, b, 2007; Hindumathi and Reddy, 2011a, 2012b; Satya Vani *et al.*, 2014a).

Among the numerous techniques (sedimentation, floatation bubbling and sucrose density layers or gradient centrifugation) employed to recover AM fungal spores from soil debris the most basic and standard method is wet sieving and decanting (Gerdemann and Nicolson, 1963). *Glomus* species were observed to be consistently isolated in high number while, species of other genera were observed relatively few in numbers.

3. Arbuscular Mycorrhizal Root Colonization

The spores of most species do not require host factors for germination and initiation of the hyphal growth, but continuous hyphal growth, differentiation into infection structures and host penetration are reported to be affected by plant signals (Bécard and Piche, 1989). Three major parameters such as specificity, infectivity and effectivity determine root colonization.

The process and rate of colonization determines the effectiveness of an AM fungus or a mycorrhizal association. As the infection spreads with in the root cortical cells of the host, extra-radical hyphae grow out into the soil, play an important role in nutrient acquisition and furthermore, form a source of secondary colonization (Harley and Smith, 1983). The development of the fungus inside the root is described as the intramatrical phase and the fungus that proliferates in the soil outside the root as extramatrical phase. The other important structures involved in the colonization of roots are spores and extraradical auxiliary bodies produced in the soil and unique structures such as hyphae, arbuscules, vesicles produced inside the roots (Figure 9.4).

Arbuscules are considered as the major sites of carbon needed for energy and nutrient exchange between the fungus and host plant. Vesicles are storage organs, store phosphorus as phospholipids and sometimes help in vegetative reproduction. The establishment, development, survival and performance of AM fungi is affected by soil fertility, cropping patterns, environmental factors, host plant genotype.

Maximum root colonization and sporulation is most prevalent in soils of low fertility. External input of mineral N or P to the soil decreased mycorrhizal development in several legumes (Abbot and Robson, 1977) and non-legumes (Krishna and Bagyaraj, 1982).

Plant growth hormones like auxins and other compounds such as flavonoids, phenolics, carbon dioxide are known to play an important role in spore germination, development, proliferation and stimulation of hyphal growth and mycorrhizal colonization.

The association/colonization of AM fungi is usually not detected by naked eye as there are no external or morphological root changes. Root colonization structures (Figure 9.4) are visible only when they are cleared, stained and examined under microscope following the most commonly and frequently adopted method (Phillips and Hayman, 1970). Roots heavily pigmented with phenolics and other secondary metabolites require a post-clearing bleaching step with freshly prepared alkaline solution of hydrogen peroxide. Detection and quantification of AM fungal colonization in roots is very essential for mycorrhizal research. A range of light microscopy-based techniques, biochemical and molecular techniques were used for identification and/or quantification of AM fungi in roots. Non-vital staining with various stains such as trypan blue, cotton blue, aniline blue, ink, vinegar, Chlorazol Black E (CBE) are some of the methods used to visualize AM fungi in roots (Vierheilig *et al.*, 2005).

4. Mycorrhizal Dependency

Plants in natural ecosystems have varying degrees of dependence on mycorrhizal associations, based on the availability of nutrients in the soil in which they naturally occur. Mycorrhizal dependency is a measure of the benefit provided by mycorrhizae and depends on relative contribution of root and mycorrhizal mediated nutrient uptake to plants. AM fungi are not host specific, any plant species can be infected by an AM fungal species. The magnitude of response to colonization varied with different cultivars of sorghum (Hindumathi and Reddy, 2011c).

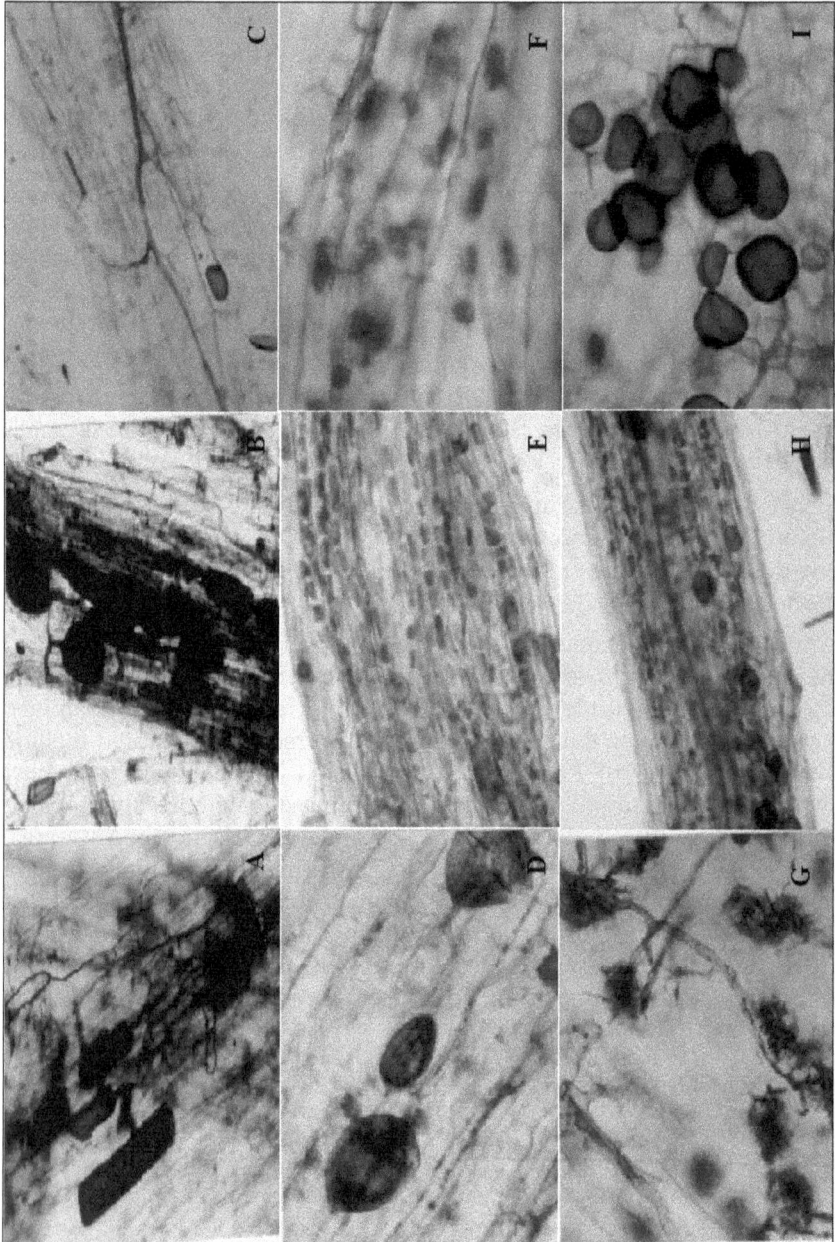

Figure 9.4

a-d: Colonization of root cortical cells of sorghum by arbuscular mycorrhizal fungi showing mycelium and different shaped vesicles in agricultural field conditions; e-h: Mycelium and arbuscules; i: Sporocarp in root tissue (***Source***: Hindumathi and Reddy, 2011c).

5. Potential Benefits of AM Fungi

It is well established fact that AM fungi play an important role in plant health by improving nutrient (especially P) and water uptake and provide protection against soil borne plant pathogens. The fungal hyphae in ERH phase extending into the soil serve as extensions of the root systems, that are both physiologically and geometrically more effective in the absorption of nutrients especially, phosphorus (P) than the roots themselves. Minerals more than 4 cm distant from the nearest host root can be absorbed by the hyphae and translocated to the root. AM fungal hyphae are not only structurally efficient for extraction of nutrients from exchange sites in soil, they also produce exogenous enzymes such as phosphatases, phytases and nitrate reductase, which are important, in uptake and metabolism of nutrients (Ho and Trappe, 1980).

It was demonstrated that AM fungal inoculation increased mineral nutrient uptake with consequent increase in plant growth and seed yield over control plants in soybean (Bagyaraj *et al.*, 1979a; Hindumathi and Reddy, 2012a); brinjal (Satya Vani, 2012, Satya Vani *et al.*, 2015), tomato, chilli (Satya Vani, 2012). Other major benefits include improved tolerance to drought and salinity (Augé, 2004; Augé *et al.*, 2015), high soil temperatures, to adverse soil pH, alleviate heavy metal toxicity (Lingua *et al.*, 2008; Meier *et al.*, 2015), increased uptake of macro nutrients (N, K and Mg) other than P, as well as uptake of micro nutrients and overcome transplantation shock compared to non-mycorrhizal plants.

Furthermore, AM fungi improve soil structure through the secretion of proteinaceous substance called glomalin (Steinberg and Rillig, 2003). It can have a direct effect on the ecosystem, as they improve the soil aggregation by forming structure of macroaggregates through physical binding of soil particles and organic material (Rillig and Mummey, 2006; Leifheit *et al.*, 2014, 2015; Rillig *et al.*, 2015). Such aggregates enhance carbon and nutrient storage and create conducive environment for survival and growth of soil microorganisms. They also influence soil porosity, which promotes aeration and water movement, essential for better root growth, root development and microbial activity thereby, drive the structure of plant communities and productivity.

Janos (1983) suggested that combined interaction of mycorrhizal and saprophytic hyphae enable mycorrhizae to remove mineralized N and P from soil nutrient pool. Because of all these attributes, mycorrhizae are now considered important in the establishment of plants in inhospitable sites like coal and copper mine wastes, borrow pits, and badly eroded locations. They are also non-polluting to the environment.

6. Nutrient Uptake

The beneficial effects of AM fungal inoculation on plant growth and yield promotion have generally been attributed to improved nutrition uptake, mobilization of nutrients (Abbot and Robson, 1977), production of ectoenzymes (Tarafdhar and Claassen, 1988), plant protection against pathogen infection (Dehne, 1982).

The most important role of AM fungi is to absorb nutrients from the soil and transfer them to the host. It is now well established that AM fungi can improve 'P nutrition of the host particularly in low fertility due to exploration of the soil by external hyphae to greater volume beyond root hair and phosphorus depletion zone (Gerdemann, 1975) resulting in more efficient absorption of 'P'. These depletion zones limits the rate of 'P' uptake by non-mycorrhizal plants but gives greater advantage to the non-mycorrhizal plants because of the ability of the AM fungal extraradical hyphae to extend past this nutrient depletion zone to enhance absorption capacity (Liu *et al.*, 2000; Sylvia *et al.*, 2001). Radio tracer experiments on phosphate labeled with ^{32}P showed that hyphae of AM fungi derived their extra phosphate from the labile pool rather than dissolving the insoluble phosphate. The absorbed 'P is probably converted into polyphosphate granules in the external hyphae (Callow *et al.*, 1978) and passed to the arbuscules for transfer to the host. This flow of phosphates is known to occur in the presence of acid phosphatases (Gianinazzi *et al.*, 1979) during arbuscule life span or senescence.

Exogenous enzymes, like phosphatases produced by AM fungal extraradical hyphae hydrolyse unavailable sources of 'P' and release 'P' from organic 'P' complexes and facilitate absorption of 'P' especially under humid tropical conditions (Koide and Kabir, 2000; Carlile *et al.*, 2001). In the recent times, it has been shown that mycorrhizal fungi could play a significant role in the establishment of plantlets by providing ample nutrients through improved root proliferation.

Many parts of India have red soil lands for agriculture. The red soils are usually phosphorus deficient which seems to be a limiting factor in crop production. The most effective mechanism could therefore be the one in which the plant is made capable of greater phosphate uptake, by using AM fungi, so as to reduce the quantum of addition of phosphate fertilizers.

Increased uptake of 'P' is not the only effect of mycorrhizal fungi on plant growth but they also stimulate uptake of Zn, Cu, S, K and Ca although not as markedly as 'P'. In the many experiments, AM fungi have shown to improve uptake of Ca and S (Sharma, 1990; Satya Vani, 2012), Cu (Sharma, 1990), Zn (Jamal *et al.*, 2002; Satya Vani, 2012), Ni (Jamal *et al.*, 2002), S (Pacovsky, 1986), Pb (Pawlowska and Charvat, 2004), and Fe (Caris *et al.*, 1998), *etc.* However, Kothari *et al.* (1991) reported decrease in the uptake of Mn in AM fungal infected plants.

AM fungal extraradical hyphae obtained nitrogen in different forms such as amino acids, peptides, ions (No_3^- or NH_4^+) and recalcitrant organic nitrogen forms (Ames *et al.*, 1983; Hawkins *et al.*, 2000; Giri and Mukerji, 2004). The uptake of ^{15}N labeled N by mycorrhizal plants was much higher than non-mycorrhizal plants (Barea *et al.*, 1987; Johansen *et al.*, 1992, 1994). Hawkins *et al.* (2006) and Ames *et al.* (1983) reported that the extraradical hyphae of different *Glomus* spp. can assimilate and metabolise both organic and inorganic sources of N perhaps by glutamate synthatase activity.

Azcon *et al.* (2003) stated that concentration of P and N in the soil can determine the rate of other micro (Fe, Cu, Mn, Zn) and macronutrient (K, Ca) uptake by mycorrhizal plants. It was confirmed by Liu *et al.* (2000) in their study with mycorrhizal maize plants uptake micro nutrients (Cu, Zn, Mn and Fe) was significantly influenced by soil 'P' nutrition.

AM fungi also play an important role in the water economy of plants. Their association improves hydraulic conductivity of roots and this improvement is one of the contributing factors towards better uptake of water by the plants. It has been suggested that mycorrhizal fungi help the plants in better absorption of water by the roots by exploiting in wider zones of soil and resulting in a better performance (Safir *et al.*, 1971, Kehri and Chandra, 1990). It has been noted that mycorrhizal plants show a better survival than non-mycorrhizal ones in extremely dry condition. The most established benefit from AM fungus to the host plant is through the widespread network of mycelium that penetrate deeper and wider in the soil in search of water and nutrients thereby widening the zone of activity.

Mycorrhizal symbiosis plays an important role in the tropical agriculture because the soils are phosphorus deficient as well as phosphorus fixing. The soil phosphate (P) availability is the most limiting factor in legume growth and biological N-fixation. Low-P soils will limit legume growth to a greater extent than low-N soils due to the fact that legumes can utilize both atmospheric N as well as soil N. The formation of AM fungal symbiosis with legume can overcome this limitation.

Nodules require relatively large amounts of 'P' than the surrounding root tissue indicating the high demand for 'P' by the nodules. Thus, a 'P' deficiency can lead to a reduction in both nodulation and symbiotic N-fixation. It has been well documented that AM fungal inoculation in legumes enhance nodulation (Carling *et al.*, 1978). It has been suggested that 'P' level influences not only mycorrhizal infection frequency but also process of nodulation in legume spp. since legumes are poor competitors for soil phosphates. It was demonstrated that mycorrhizal nodulated plants exhibited higher levels of nitrogenase and nitrate reductase compared to non-mycorrhizal plants (Carling *et al.*, 1978).

AM fungal plants have been shown two fold increase in the efficiency of 'P' fertilizers and also help utilization of 'P' fertilizer like bone meal and rock phosphate more efficiently than non-mycorrhizal plants. Inoculation of AM fungi in legume-Rhizobium symbiosis (with/without pathogen) resulted in better nodulation, biological N_2-fixation, enhanced plant growth and nutrition, and biological control of root rot pathogens and increased soil nitrogen content (Bagyaraj *et al.*, 1979a; Hindumathi and Reddy, 2012a; Hindumathi *et al.*, 2016a). Experimental evidence reported that dual inoculation with AM fungi and Rhizobium strain in tripartite symbiosis enhanced nutrient uptake in soybean (Hindumathi and Reddy, 2012a), chickpea (Champavat, 1990), cowpea (Islam *et al.*, 1980), mungbean (Hindumathi *et al.*, 2016a), groundnut (Krishna and Bagyaraj, 1982), clovers (Morton *et al.*, 1990). Legumes cultivated in soil with low P were most responsive to dual inoculation of AM fungi and Rhizobium as increased 'P' availability stimulates biological N_2-fixation and growth of the host legume.

7. Disease Resistance

It is well established that AM fungal colonization in plant roots increases plant tolerance to pathogen attack thereby acting as biocontrol agents (Azcon-Aguillar and Barea, 1996). Evidence exists that mycorrhizal plants are very much capable of resisting the parasite invasion and providing protection against infection by soil-borne plant pathogens.

Primarily, the ability of AM fungi to enhance plant vigour due to increased nutrient uptake enables it to resist pathogen infection. Biocontrol of Fusarium wilt of cotton, jute, tomato Macrophomina root rot of cowpea was achieved by application of AM fungi (Bagyaraj 1984; Caron *et al.*, 1986). AM fungi mediated biocontrol potential of *Pythium aphanidermatum* on tomato (Reddy *et al.*, 2006c), Verticillium wilt in different cultivars of brinjal, tomato, chilli (Satya Vani, 2012), tomato (Satya Vani *et al.*, 2014b), charcoal rot in different cultivars of sorghum (Hindumathi, 1999) proved to increase plant tolerance to the pathogen.

Several mechanisms or combination of mechanisms are reported to involve in bioprotection of plants by AM fungi against soil borne plant pathogens. These include (1) direct competition, (2) mechanism mediated by alteration in plant growth, nutrition and morphology, (3) biochemical and molecular changes in mycorrhizal plants that induce plant resistance, and (4) alteration in the soil microbiota and development of pathogen antagonism. Although some of the mechanisms suggested might play no role, it is generally agreed that bioprotection through mycorrhization is the result of combination of several of the mechanisms and not of a single one. It is evidenced that the nutritional effect of the AM symbiosis is only one among several aspects of the mycorrhizal effect on pathogens. Moreover it has been suggested that the nutrient uptake by the fine extraradical mycelium of the AM fungi could compensate for a pathogen-reduced root system (Singh *et al.*, 2000). Colonization of roots by AM fungi at an early stage of plant growth is known to reduce the severity of disease and protect the roots from certain root pathogens.

AM fungal root colonization has been reported to confer both local and systemic resistance against pathogens. AM fungi have been known to elicit compounds only in low amounts typically involved in plant defense reaction, that could act locally or transient by making the root more prone to react against the pathogen. Mycorrhizal infection in plants have been reported to stimulate biosynthesis and accumulation of compounds phytoalexins, peroxidases, pathogenesis related (PR) proteins, phenol compounds (Pozo *et al.*, 2002; Dumas-Gaudot *et al.*, 2000) that will enhance mechanisms of defense to subsequent pathogen infection. The increased lignification of root endodermal cells induced by AM colonization has been suggested to play an important role in the plant defense mechanism (Dehne, 1982). The *Fusarium* hyphae within mycorrhizal roots exhibited a high level of structural disorganization, characterized by the massive accumulation of phenolic-like compounds and the production of hydrolytic chitinases. This reaction was not induced by non-mycorrhizal roots, suggesting that the activation of plant defense responses by mycorrhiza formation provides certain protection against the pathogen (Azcón-Aguilar and Barea, 1996).

8. Mycorrhiza in Agriculture

Mycorrhizal symbiosis may play an important role in the tropical agriculture because the soils in tropics are phosphorus deficient as well as phosphorus fixing.

The use of costly phosphate fertilizers has increased to a very large extent. Farmers are generally poor in tropics and cannot afford their heavy cost and also there is a pollution problem. Under these circumstances, there is a need for system which can improve fertilizer efficiency to the agriculture and will be a great asset in developing countries. According to Mosse (1973), around 75 per cent of 'P' applied to the crops is not utilized by them but gets converted to forms unavailable to plants. Under such conditions, AM fungi can be very effective, because of their ability to utilize extremely small quantities of fertilizer. These marginal agricultural lands could be more productive if AM fungi having the ability to utilize extremely small quantities of fertilizers are added to the soil. It was indicated that the indigenous AM fungi are not efficient and inoculation of efficient strains may be rewarding (Powell *et al.*, 1980). A number of factors have been shown to govern the response of the host to the bioinoculant *viz.*, inoculum type and density, infection potential, site of placement, spread and competitive abilities, *etc.* Due to this reason AM fungi are being regarded as biofertilizers and their use in agriculture would be encouraging.

Four major factors can determine the success of the inoculation: the crop species involved, the size and effectiveness of the indigenous AM fungal population, the fertility of the soil and cultural practices. It is also important to know which plant derives more benefits from mycorrhizal symbiosis.

A combination of legumes, rhizobia and AM fungus brings a significant improvement in plant growth through increased availability of 'P' together with higher N-fixation in soil. Thus, this dual combination may prove the cheapest way to enrich tropical soils with nitrogen. This tripartite association, legume-AM fungi and nitrogen fixing bacteria has shown to be effective in increasing growth, nodulation and yield in many legumes.

The synergistic effects of free living nitrogen fixing bacteria, AM or dinitrogen fixers, AM and phosphate solubilizers have been investigated by many workers and their studies revealed improvement in plant growth (Bagyaraj and Menge, 1978).

In view of increased awareness of the numerous benefits of biofertilizers, the potential of mycorrhiza as a tool for improving growth of agricultural and horticultural plants can play a key role in the future agroeconomic scene as a partial substitute to costly phosphate fertilizers.

9. Mycorrhiza in Forestry

The ability of AM fungi to increase plant growth and establishment make it a suitable symbiont to be used in forestry. The beneficial effects of AM fungal inoculation in forest nurseries are being recognized. It was reported that inoculation of *G. fasciculatum* on a fast growing legume *Sesbania sesbano* resulted in a tremendous increase in plant biomass over control. *Leucaena leucocephala*, a tropical tree used by foresters is strongly mycorrhiza-dependent, as it lacks root hairs. *Leucaena* inoculated with mycorrhizal fungi improved plant growth, 'P' uptake and N_2-fixation. Positive response was shown by AM fungi in re-establishment of plant cover over sandy area. It has been suggested that AM fungal infection could be of importance in plants colonizing industrial waste. AM fungi also play important role in sand dune management as they improve physical structure of soil. The importance of AM fungi is being investigated by many workers.

10. Inoculum Production

One of the major problems of AM application as 'bio-fertilizer' in deriving the benefits on agricultural crops is the absence of technology to produce inoculum quickly and in large quantities preferably *in vitro*. AM fungi being an obligate symbiont cannot be cultured in artificial media under laboratory conditions (Mosse, 1973). Therefore, the techniques used for inoculum production are different from those generally employed for other biotechnologically interesting fungi. Although AM fungi cannot be cultured on agar media, methods of multiplication on host plant using root-bit inoculum and AM fungi infested soil are in practice. The only method of AM fungal inoculum production is in association with host plant, which is considered a major disadvantage. Producing AM fungal inoculum on living plant is challenging and complex feature. At present, the only technique successful in mass production of AM inoculum is by traditional soil 'pot culture' method using living host plant. AM fungi can also be propagated by growing plants in aeroponic or hydrophonic culture technique (Jarstfer and Sylvia, 1992), plants grown in solid substrate, in surface calcined montmorillonite clay or root organ culture (Becard and Piche, 1992) either as organ (tissue) or solution cultures. Attempts are also being made at various laboratories to culture the AM fungi and achieve spore production *in vitro* (without host or root culture in laboratory condition). There are different types of AM fungal inoculum required for different purposes, which include: spore inoculum, infected root inoculum, peat based inoculum, soil based inoculum, carrier material (expanded clay).

10.1. Soil Based Inoculum

Soil based inoculum produced in 'pot cultures' using host plant contain all AM fungal structures is highly infective. The selection of the host plant used to multiply AM fungi in inoculum production can have a significant influence on fungal sporulation, rapid and abundant root formation and consequently on inoculum levels. The host plant should be well adapted to local ambient conditions, should allow good AM colonization and should not be easily susceptible to diseases. A wide variety of plants like sorghum, sudan grass, *Cenchrus cilliaris*, maize, groundnut, *Allium cepa* and other fibrous monocots were frequently used as alternate or collateral host for mass multiplication of AM fungal propagules. But the response to AM association may vary in different plants.

Earlier reports suggested high light intensity and long day lengths improved mycorrhizal colonization or spore production in maize, alfalfa, onion. Temperatures around 30°C are required for greater spore formation and root colonization. Therefore, to maximize, AM inoculum production these conditions have to be maintained.

Production of 'clean' and viable inoculum in large quantities poses a great difficulty. Some trials have been made in this line by Menge (1983), Mosse (1962), Sreenivasa and Bagyaraj (1988), Bhownik and Singh (2004). Some modifications in these methods have proved beneficial in increasing AM fungal inoculum production (Sreenivasa and Bagyaraj, 1988).

Briefly, AM fungal soil based root inoculum was produced using sorghum as collateral host because of its extensive root system which will support more mycorrhizal formation (Reddy *et al.*, 2006d). Surface sterilized spores of desired AM fungi were inoculated near the root tips of sorghum planted in pots containing sterilized alfisol sand and soil ('P' deficient) mixture in the ratio of 1:1. Hoagland nutrient solution (Hoagland and Arnon, 1950) containing all nutrients except 'P' was given to the plants at regular intervals of 15 days and water holding capacity was maintained by adding sterile distilled water whenever required. At different intervals root bits were observed for AM fungal root colonization. As soon as 70 – 80 per cent root infection is established, the root bits and soil mixture can be used as inoculum.

The root system together with adhering soil are finely chopped and mixed with soil from pots, air dried to remove free water before storing. The cultures are packed in polythene bags and stored at 4-5°C. By this the cultures could be effectively stored for about 4yrs. One gram of soil containing 20 spores or 1gm root with 70 – 80 per cent infection are the standard units for inoculation purpose.

Due to potential of mycorrhizal fungi to enhance nutrient uptake, this benefit has however brought about the suggested use of AM inoculation instead of chemical fertilizer for plant productivity, plant growth, disease resistance, restoration of polluted soils or in revegetation (Cardoso and Kuyper, 2006; Khan, 2006).

10.2. Spore Inoculum

The spore inoculum of AM fungi is used as inoculum generally for experiments where the plants are grown *In vitro* conditions. Nopamornbodi *et al.* (1987) applied 200 AM spores in water suspension under each soybean seed and obtained increased yields in field conditions. However, large-scale production of spores is difficult and tedious task. It may be feasible for nursery raised crop.

10.3. Infected Root Inoculum

The production of infected root inoculum in large-scale is possible in aerophonic culture (Sylvia and Hubbel, 1986). Infected roots contain internal and external mycelium (may have spores) can colonize the host in 1-2 days after inoculation. *In vitro* production of some AM fungi on tissue culture roots has been demonstrated (Nopamornbodi *et al.*, 1988). This production process is difficult and also experiences problems like: infected roots introduced as inoculum acts as nutrient source for several saprophytic and parasitic microorganisms, short survival time and requirement of inoculum in large quantities.

10.4. Peat Based Inoculum (Nutrient Film Technique)

The sphagnum moss peat was mixed with three times its weight of water and lime to obtain pH suitable for the mycorrhizal fungus. AM inoculum is added on to the peat and compressed into blocks of 4 x 4 x 4cm and lettuce is grown on peat block for 2 – 5 wks. The blocks are then transferred to nutrient film technique (NFT) channels (Cooper, 1985). The plants are allowed to grow in NFT system for 8 – 10 wks during which mass production of spore takes place. Peat blocks are then allowed to dry, chopped and used as inoculum. The shelf life is around 6 months. During production process contamination with root pathogens is possible.

10.5. Carrier Material (Expanded Clay)

For mass multiplication, light expanded clay aggregates have been used as a carrier material in Germany (Dehne and Backhaus, 1986). Clay granules of 4-10mm are optimum for reproduction of AM fungi. They sporulate within the porous inorganic particles. The inoculum can be produced in a relatively short time. This being inorganic material, there is less contamination with root pathogens. The inoculum is light, easy to transport and shelf life is 5yrs. The product is granular and hence can be used with agricultural machinery.

References

Abbott, L.K., Robson, A.D. 1977. Growth stimulation of subterranean clover with vesicular–arbuscular mycorrhizas. *Aust. J. Agric. Res.*, 29: 639-649.

Ames, R.N., Reid, P.P., Porter, L.K. and Cambardella, C. 1983. Hyphal uptake and transport of nitrogen from two 15N-labelled sources by *Glomus mosseae*, a vesicular-arbuscular mycorrhizal fungus. *New Phytol.*, 95: 381-396.

Augé, R.M. 2004. Arbuscular mycorrhizae and soil/plant water relations. *Can. J. Soil Sci.*, 84: 373-381.

Augé, R.M., Toler, H.D., Saxton, A.M. 2015. Arbuscular mycorrhizal symbiosis alters stomatal conductance of host plants more under drought than under amply watered conditions: a meta-analysis. *Mycorrhiza*, 25: 13-24.

Azcün, R., Ambrosano, E. and Charest. C. 2003. Nutrient acquisition in mycorrhizal lettuce plants under different phosphorus and nitrogen concentration. *Plant Sci.*, 165: 1137-1145.

Azcon-Aguilar, C. and Barea, J.M. 1996. Arbuscular mycorrhizae and biological control of soil-borne plant pathogens: An overview of the mechanisms involved. *Mycorrhiza*, 6: 457-464.

Bagyaraj, D.J. and Menge, J.A. 1978. Interaction between VA mycorrhiza and *Azotobacter* and their effects on rhizosphere microflora and plant growth. *New Phytol.*, 80: 563-573.

Bagyaraj, D.J. 1984. Biological interactions with VA mycorrhizal fungi. In: Powell, C.L, Bagyaraj, D.J. (Eds.). VA mycorrhiza. CRC, Boca Raton, Fla., pp. 131-153.

Bagyaraj, D.J., Manjunath, A. and Patil, R.B. 1979a. Interaction between a vesicular-arbuscular mycorrhiza and *Rhizobium* and their effects on soybean in the field. *New Phytol.*, 82: 141-145.

Bagyaraj, D.J., Manjunath, A. and Patil, R.B. 1979b. Interaction of vesicular-arbuscular mycorrhiza with root knot nematode in tomato. *Plant Soil*, 51: 397.

Barea, J.M., Azcon-Aguilar, C. and Azcün, R. 1987. Vesicular-arbuscular mycorrhiza improve both symbiotic N 2 fixation and N uptake from soil as assessed with a ^{15}N technique under field conditions. *New Phytol.*, 106: 717-725.

Bécard, G. and Piché, Y. 1989. Fungal growth stimulation by CO_2 and root exudates in vesicular-arbuscular mycorrhizal symbiosis. *Appl. Environ. Microbiol.*, 55: 2320–2325.

Bécard, G. and Piché, Y.1992. Establishment of VA mycorrhizae in root organ culture: Review and proposed methodology. In: Methods in Microbiology. Vol. 24. Norris, J.R, Read, D.J. and Varma, A.K. (Eds.). Academic Press, UK, pp. 89-108.

Bentivenga, S.P. and Morton, J.B. 1994. Systemics of glomalean endomycorrhizal fungi: Current views and future directions. In: Mycorrhizae and Plant Health. (Eds.). Pfleger, F.L. and Linderman, R.G. APS press, Minnesota. pp. 283-308.

Bhowmik, S.N. and Singh, C.S. 2004. Mass multiplication of AM inoculum: Effect of plant growth-promoting rhizobacteria and yeast in rapid culturing of *Glomus mosseae. Curr. Sci.*, 86: 705-709.

Bonfante, P. and Perotto, S. 2000. Outside and inside the roots: cell-to-cell interactions among arbuscular mycorrhizal fungi, bacteria and host plants. In: Podila G.K., Douds, D.D. Jr. (Eds.). Current advances in mycorrhizae research. APS Press, St. Paul, MN, pp. 141-155.

Callow, J. A., Capaccio, L.C.M., Parish, G. and Tinker, P.B. 1978. Detection and estimation of polyphosphate in vesicular-arbuscular mycorrhiza. *New Phytol.*, 80: 125-134.

Caris, C., Hördt, W., Hawkins, H.J., Römheld, V. and George, E. 1998. Studies of iron transport by arbuscular mycorrhizal hyphae from soil to peanut and sorghum plants. *Mycorrhiza*, 8: 35–39.

Cardoso, I.M. and Kuyper, T.W. 2006. Mycorrhizas and tropical soil fertility. *Agric. Ecosyst. Environ.*, 116: 72-84.

Carling, D.E., Richie, W.G., Brown, M.F. and Tinker, P.B. 1978. Effects of vesicular-arbuscular mycorrhizal fungus on nitrate reductase and nitrogenase activities in nodulating and non-nodulating soybeans. *Phytopathol.*, 68: 1590-1596.

Caron, M., Fortin, J.A. and Richard, J. 1986. Effect of *Glomus intraradices* on infection by *Fusarium oxysporum*. f. sp. *radicis-lycopersici* tomatoes over a 12 week period. *Can. J. Bot.*, 64: 552-556.

Champawat, R.S. 1990. Response of chickpea (*Cicer arietinum*) to *Rhizobium* and vesicular-arbuscular mycorrhizal dual inoculation. *Acta Microbiol. Pol.*, 39: 163-169.

Cooper, A.J. 1985. 22 new ABC's of NFT. Hydrophonics Worldwide: State of the Art in Soilless Crop Production. (Eds.). Adam, J. and Savage. Int. Center for Special Studies, Honolulu, HI, pp. 180-185.

Dehne, H.W. 1982. Interaction between vesicular-arbuscular mycorrhizal fungi and plant pathogens. *Phytopathol.*, 72: 1115-1119.

Dehne, H.W. and Backhaus, G.F. 1986. The use of vesicular-arbuscular mycorrhizal fungi plant protection. I. Inoculum production. Z. Pflkrankh Pflschutz, 93: 415-424.

Dumas-Gaudot, E., Armelle, G., Cordier, C., Gianinazzi, S. and Gianinazzi-pearson, V. 2000. Modulation of host defence systems. In: Arbuscular Mycorrhizas: Physiology and Function. (Eda.) Kapulnik, Y., Douds, J.D.D. Kluwer Academic Publishers, Netherlands, pp. 173-200.

Frank, A. B. 1885. Under die auf wurz elymboise beruhende Ernahrug grewisser Baume durch unterirdische pilze. *Aer. Dt. Bot. Ges.* 3: 128-145.

Friberg, S. 2001. Distribution and diversity of arbuscular mycorrhizal fungi in traditional agriculture on the Niger inland delta, Mali, West Africa. *CBM's Skriftserie,* 3: 53-80.

Gerdemann, J.W. 1968. Vesicular-arbuscular mycorrhizae and plant growth. Ann. Rev. *Phytopathol.,* 6: 397-418.

Gerdemann, J.W. 1969. Fungi that form the arbuscular-vesicular type of endomycorrhizae. In: Hacskaylo, E. (Ed.) Proc. 1st N. Amer. Conf. Mycorrhizae. pp. 9-17.

Gerdemann, J.W. 1975. Vesicular arbuscular mycorrhizae. In: Torrey, J.G. and Clarkson, D.T. (Eds.) The Development and Function of Roots, Academic Press, New York. pp. 575-595.

Gerdemann, J.W. and Nicolson, T.H. 1963. Spores of mycorrhizal *Endogone* species extracted from soil by wet sieving and decanting. *Trans. Br. Mycol. Soc.,* 46: 235-244.

Giri, B. and Mukerji, K.G. 2004. Mycorrhizal inoculant alleviates salt stress in *Sesbania aegyptiaca* and *Sesbania grandiflora* under field conditions: Evidence for reduced sodium and improved magnesium uptake. *Mycorrhiza,* 14: 307-312.

Gianinazzi, S., Gianinazzi-Pearson, V. and Dexheimer J. 1979. Enzymatic studies on the metabolism if VA-mycorrhiza. III. Ultrastructural localization of acid and alkaline phosphatase in onion roots infected by *Glomus mosseae* (Nicol. and Gerd.). *New Phytol.,* 82: 127-132.

Harley, J.L. and Smith, S.E. 1983. Mycorrhizal symbiosis. Academic press, London.

Hawkins, H., Johansen, A. and George, E. 2000. Uptake and transport of organic and inorganic nitrogen by arbuscular mycorrhizal fungi. Plant and Soil. 226: 275-285.

Hayman, D.S. 1975. The occurrence of mycorrhiza in crops as affected by soil fertility. In: Sanders, F.E., Mosse, B. and Tinker, P.B. (Eds.). Endomycorrhizas. Academic Press, London. pp. 495–509.

Helgason, T., Fitter, A.H. and Young, J.P.W. 1999. Molecular diversity of arbuscular mycorrhizal fungi colonising Hyacinthoides non-scripta (bluebell) in a semi natural woodland. *Mol. Ecol.,* 8: 659-666.

Hepper, C.M., Sen, R., Azcon-Aguilar, C. and Grace, C. 1988. Variation in certain isozymes amongst different geographical isolates of the vesicular-arbuscular mycorrhizal fungi *Glomus clarum, Glomus monosporum* and *Glomus mosseae. Soil Biol. Biochem.*, 20: 51-59.

Hindumathi, A. 1999. Role of arbuscular mycorrhizae in plant growth and biocontrol of charcoal rot in sorghum. Ph.D. Thesis Dept of Botany, Osmania University, Hyderabad, India.

Hindumathi, A. and Reddy, B. N. 2011a. Occurrence and distribution of arbuscular mycorrhizal fungi and microbial flora in the rhizosphere soils of mungbean [*Vigna radiata* (L.)] and soybean (*Glycine max* (L.) Merr.] from Adilabad, Nizamabad and Karimnagar districts of Andhra Pradesh state, India. *Adv. Biosci. Biotechnol.*, 2: 275-286.

Hindumathi, A. and Reddy, B.N. 2011b. Influence of Arbuscular Mycorrhizal Fungi on plant growth and nutrition of Sorghum. In: Proc. of II Asian PGPR Congress at Beijing, China, August 21-24.

Hindumathi, A. and Reddy, B.N. 2011c. Dependency of Sorghum on Arbuscular Mycorrhizal colonization for growth and development. *J. Mycol. Plant Pathol.*, 41: 537-542.

Hindumathi, A. and Reddy, B.N. 2012a. Synergistic effect of arbuscular mycorrhizal fungi and *Rhizobium* on the growth and charcoal rot of soybean [*Glycine max* (L.) Merr.]. *World J. Sci. Technol.*, 2: 63-70.

Hindumathi, A. and Reddy, B.N. 2012b. Systematics and Occurrence of Arbuscular Mycorrhizal Fungi. Lap Lambert Academic Publishing. GmbH and Co. K.G. Dudweiler Landstr., Germany. pp. 168.

Hindumathi, A. and Reddy, B.N. 2015. Species diversity and population density of arbuscular mycorrhizal fungi associated with *Carthamus tinctorius* L. rhizosphere soils of Telangana, India. *Mycorrhiza News,* 7: 5-17.

Hindumathi, A. and Reddy, B.N., Sabitha Rani, A., Reddy, A.N. 2016a. Associative effect of arbuscular mycorrhizal fungi and *Rhizobium* on plant growth and biological control of charcoal rot in green gram [*Vigna radiata* L. (Wilczek)]. In: Bhima, B. and Anjana Devi, T. (Eds). Microbial Biotechnology: Technological Challenges and Developmental Trends. Apple Academic Press. USA. pp. 155-70.

Hindumathi, A. and Reddy, B.N. 2016b. Dynamics of arbuscular mycorrhizal fungi in the rhizosphere soils of safflower from certain areas of Telangana State, India. *Ind. Phytopath.* 69: 67-73.

Hoagland, D.R. and Arnon, D.I. 1950. The water culture method for growing plants without soil. California Agric. Exp. Station, Circular- 347.

Isaac, S. 1992. Fungal–plant interactions. Chapman and Hall, London.

Islam, R., Ayanaba, A. and Sanders F.E. 1980. Response of cowpea (*Vigna unguiculata*) to inoculation with VA mycorrhizal fungi and to rock phosphate fertilization on some unsterilized Nigerian soil. *Plant Soil*, 54: 107-109.

Jamal, A., Ayub, N., Usman, M. and Khan, A.G. 2002. Arbuscular mycorrhizal fungi enhance zinc and nickel uptake from contaminated soils by soyabean and lentil. *Internat. J Phytoremed.*, 4: 205-221.

Janos, D.P. 1983. Tropica mycorrhizas, nutrient cycles and plant growth. In: Sutton, S.L., Whitmore, T.C. and Chadwick, A.C. (Eds.). Tropical rain forest: ecology and management. Blackwell, Oxford, pp. 327-345.

Johansen, A., Jakobsen, I. and Jensen, E.S. 1992. Hyphal transport of ^{15}N-labelled nitrogen by vesicular-arbuscular mycorrhizal fungus and its effect on depletion of inorganic soil N. *New Phytol.* 122: 281-288.

Kehri, H. K. and Chandra, S. 1990. Mycorrhizal association in crops under sewage farming. *J. Ind. Bot. Soc.*, 69: 267-270.

Khan, A.G. 2006. Mycorrhizoremediation-an enhanced form of phytoremediation. *J. Zhejiang University Sci. B.*, 7: 503-514.

Koide, R.T. and Kabir, Z. 2000. Extraradical hyphae of the mycorrhizal fungus *Glomus intraradices* can hydrolyse organic phosphate. *New Phytol.*, 148: 511—517.

Kothari, S.K., Marschner, H. and Römheld, V. 1991. Contribution of the VA mycorrhizal hyphae in acquisition of phosphorus and zinc by maize grown in a calcareous soil. *Plant Soil*, 131: 177-185.

Krishna, K.R. and Bagyaraj, D.J. 1982. Influence of VA mycorrhiza on growth and nutrition of *Arachis hypogea*. *Legume Res.*, 5: 18-22.

Leifheit, E.F., Verbruggen, E. and Rillig, M.C. 2015. Arbuscular mycorrhizal fungi reduce decomposition of woody plant litter while increasing soil aggregation. *Soil Biol. Biochem.*, 81: 323–328.

Leifheit, E.F., Veresoglou, S.D., Lehmann, A., Morris, E.K. and Rillig, M.C. 2014. Multiple factors influence the role of arbuscular mycorrhizal fungi in soil aggregation—a meta-analysis. *Plant Soil*, 374: 523-537.

Lingua, G., Franchin, C., Todeschini, V., Castiglione, S., Biondi, S. and Burlando, B. 2008. Arbuscular mycorrhizal fungi differentially affect the response to high zinc concentrations of two registered poplar clones. *Environ. Pollut.*, 153: 137-147.

Liu, A., Hamel, C., Hamilton, R.I. and Ma, B.L. 2000. Acquisition of Cu, Zn, Mn and Fe by mycorrhizal maize (*Zea mays* L.) grown in soil at different P and micronutrient levels. *Mycorrhiza*, 9: 331-336.

Manjunath, A. and Bagyaraj, D.J. 1984. Response of pigeonpea and cowpea to phosphate and dual inoculation in vesicular-arbuscular mycorrhiza and *Rhizobium*. *Trop. Agric.*, 61: 48-52.

Meier, S., Cornejo, P., Cartes, P., Borie, F., Medina, J. and Azcón, R. 2015. Interactive effect between Cu-adapted arbuscular mycorrhizal fungi and biotreated agrowaste residue to improve the nutritional status of Oenothera picensis growing in Cu-polluted soils. *J. Plant Nutr. Soil Sci.*, 178: 126–135.

Menge, J.A. 1983. Utilization of vesicular arbuscular mycorrhizal fungi in agriculture. *Can. J. Bot.*, 61: 1015-1024.

Morton, J.B., Yarger, J.E. and Wright, S.F. 1990. Soil solution phosphorus requirement for nodulation and nitrogen fixation in mycorrhizal and nonmycorrhizal red clover (*Trifolium pratense* L.). *Soil Biol. Biochem.*, 22:128–129.

Mosse, B. 1973. Advances in the study of Vesicular-arbuscular mycorrhiza. *Ann. Rev. Phytopathol.*, 11: 171-196.

Mosse, B. 1962. The establishment of vesicular mycorrhizal under aseptic conditions. *J. Gen. Microbiol.*, 27: 509 – 520.

Nicolson, T.H. 1959. Mycorrhiza in the Gramineae. I. Vesicular-arbuscular endophytes, with special reference to the external phase. *Trans. Br. Mycol. Soc.*, 42: 421-438.

Nopamornbodi, O., Thumsurakul, S. and Vasuvat, Y. 1987. Survival of VA mycorrhizal fungi after paddy rice. In: Sylvia, D.M., Hung, L.L. and Graham, J.H. (Eds.). Mycorrhizae in the Next Decade, Practical Applications and Research Priorities, pp. 53.

Nopamornbodi, O., Rajansriwong, Q. and Thumsurakul, S. 1988. In: Mahadevan, A., Raman, N. and Natarajan, K. (Eds.). Mycorrhiza for Green Asia. Alamau Printers, Madras. pp. 315-316.

Oehl, F. and Sieverding, E. 2004. *Pacispora,* a new vesicular arbuscular mycorrhizal fungal genus in the glomeromycetes. *J. Appl. Bot.*, 78: 72-82.

Pacovsky, 1986. Micronutrient uptake and distribution in mycorrhizal or phosphorus-fertilized soybeans. *Plant Soil*, 95: 379-388.

Palenzuela, J., Ferrol, N., Boller, T., Azcón-Aguilar, C. and Oehl, F. 2008. *Otospora bareai*, a new fungal species in the *Glomeromycetes* from a dolomitic shrub-land in the National Park of Sierra de Baza (Granada, Spain). *Mycologia*, 100: 282–291.

Pawlowska, T.E. and Charvat, I. 2004. Heavy metal stress and developmental patterns in arbuscular mycorrhizal fungi. *Appl. Environ. Microbiol.*, 70: 6643-6649.

Philips, J.M. and Hayman, D.S. 1970. Improved procedures for cleaning roots and staining parasitic and vesicular-arbuscular mycorrhizal fungi for rapid assessment of infection. *Trans. Br. Mycol. Soc.*, 55: 158-161.

Powell, C.L. 1980. Mycorrhizal infectivity of eroded soils. *Soil Biol Biochem.*, 12: 247-250.

Pozo, M.J., Slezack-Deschaumes, S. and Dumas-Gaudot, E. 2002. Plant defense responses induced by arbuscular mycorrhizal fungi. In: Gianinazzi, S., Scüepp, H., Barea, J.M., Haselwandter, K. (Eds.) Mycorrhizal Technology in Agriculture: From Genes to Bioproducts, Birkhäuser Verlag, Basel, pp. 103-111.

Prescot, L.M., Harley, J.P. and Klein, D.A. 2005. Microbiology. McGraw-Hill, New York.

Quilambo, O.A. 2003. The vesicular-arbuscular mycorrhizal symbiosis. *Afr. J. Biotech.*, 2: 539-546.

Rao, A.S. and Parvathi, K. 1982. Development of VA mycorrhiza in groundnut and other hosts. *Plant Soil*, 66: 133-137.

Reddy, B.N., Hindumathi, A. and Raghavender, C.R. 2006a. Influence of Physico-chemical factors on arbuscular mycorrhizal population associated with sorghum. *Ind. J. Bot. Res.*, 2: 75-82.

Reddy, B.N., Sreevani, A. and Raghavender, C.R. 2006b. Association of AM fungi in three solanaceous vegetable crops. *Ind. J. Mycol. Pl. Pathol.*, 36: 52-56.

Reddy, B.N., Raghavender, C.R. and Sreevani, A. 2006c. Approach for enhancing mycorrhiza-mediated disease resistance of tomato damping-off. *Ind. Phytopath.*, 59: 299-304.

Reddy, B.N., Hindumathi, A. and Raghavender, C.R. 2006d. Mass multiplication of mycorrhizal inoculum: use of sorghum roots for rapid culturing of *Glomus fasciculatum*. *Natl. Acad. Sci. Lett.*, 29: 355-359.

Reddy, B.N., Hindumathi, A. and Raghavender, C.R. 2007. Occurrence and systematics of arbuscular mycorrhizal fungi associated with sorghum. *J. Phytol. Res.*, 20: 11-22.

Rillig, M.C. and Mummey, D.L. 2006. Mycorrhizas and soil structure. *New Phytol.*, 171: 41-53.

Rillig, M.C., Aguilar-Trigueros, C.A., Bergmann, J., Verbruggen, E., Veresoglou, S.D. and Lehmann, A. 2015. Plant root and mycorrhizal fungal traits for understanding soil aggregation. *New Phytol.*, 205: 1385-1388.

Safir, G.R., Boyer, J.S. and Gerdemann, J.W. 1971. Mycorrhizal enhancement of water transport in soybean. *Sci.*, 172: 581-583.

Satya Vani, M. 2012. AM fungi as bio-fertilizer and bio-control agent of Verticillium wilt of some solanaceous crops. Ph.D. Thesis. Dept of Botany, Osmania University, Hyderabad, India.

Satya Vani, M., Hindumathi, A., Reddy, B.N. 2014a. Arbuscular myorrhizal fungi associated with rhizosphere soil of Brinjal cultivated in Andhra Pradesh, India. *Int. J. Curr. Microbiol. App. Sci.*, 3: 519-529.

Satya Vani, M., Hindumathi, A. and Reddy, B.N. 2014b. Influence of arbuscular mycorrhizal fungi on plant growth promotion and biological control of Verticillium wilt of Tomato (*Lycopersicum esculentum*). *Int. J. Pharm. Bio. Sci.*, 5: 1000-1009.

Satya Vani, M., Hindumathi, A. and Reddy, B.N. 2015. Application of arbuscular mycorrhizal fungi to improve plant growth in *Solanum melongena* L. *Ann. Biol. Res.*, 6: 21-28.

Schüßler, A., Schwarzott, D. and Walker, C. 2001. A new phylum, the Glomeromycota: Phylogeny and evolution. *Mycol. Res.*, 105: 1413-1421.

Sharma, A.K. 1990. Nature of interaction between *Rhizobium* and *Glomus caledonium* in chick pea (*Cicer arietinum* L.). *Nat. Acad. Sci.*, 60: 81-85.

Sieverding, E. and Oehl, F. 2006. Revision of *Entrophospora* and description of *Kuklospora* and *Intraspora*, two new genera in the arbuscular mycorrhizal *Glomeromycetes*. *J. Appl. Bot. Food Quality*, 80: 69-81.

Singh, A., Sharma, J., Rexer, K.H. and Varma, A.K. 2000. A plant productivity determinants beyond minerals, water and light. *Piriformospora indica* a revolutionary plant growth promoting fungus. *Curr. Sci.*, 79: 101-106.

Spain, J.L., Sieverding, E. and Oehl, F. 2006. *Appendicispora*, a new genus in the arbuscular mycorrhizal-forming *Glomeromycetes*, with a discussion of the genus *Archaeospora*. *Mycotaxon*, 97: 163–182.

Sreenivasa, M.N. and Bagyaraj, D.J. 1988. *Chloris gayana* (Rhodes grass), a better host for the mass production of *Glomus fasciculatum*, *Plant Soil*, 106: 289-290.

Sreevani, A. and Reddy, B.N. 2005. Arbuscular mycorrhizal fungi with tomato (*Lycopersicum esculentum* Mill.) as influenced by soil physico chemical properties. *Philippine J. Sci.*, 133: 115-129.

Steinberg, P.D. and Rillig, M.C. 2003. Differential decomposition of arbuscular mycorrhiza fungal hyphae and glomalin. *Soil Biol. Biochem.*, 35: 191-194.

Sylvia, D., Alagely, A., Chellemi, D. and Demchenko, L. 2001. Arbuscular mycorrhizal fungi influence tomato competition with Bahiagrass. *Biol. Fert. Soil.*, 34: 448-452.

Sylvia, D.M. and Hubbel, D.H. 1986. Growth and sporulation of vesicular – arbuscular mycorrhizal fungi in aeroponic and membrane systems. *Symbiosis*, 1: 259-267.

Tarafdar, J.C. and Claassen, N. 1988. Organic phosphorus compounds as a phosphorus source for higher plants through the activity of phosphatases produced by plant roots and microorganisms. *Biol. Fert. Soil.*, 5: 308-312.

Tarafdar, J.C. and Marschner, H. 1994. Phosphatase activity in the rhizosphere and hyphosphere of VA mycorrhizal wheat supplied with inorganic and organic phosphorus. *Soil Biol. Biochem.*, 26: 387-395.

Thaper, H.S. and Khan, S.N. 1973. Studies on endomycorrhiza of some forest species. In: Ind. Natn. Sci. Acad. B. Forest Res. Inst., Dehradun, India, pp. 687-694.

Van der Heijden, M.G.A., Boller, T., Wiemken, A. and Sanders, I.R. 1998. Different arbuscular mycorrhizal fungal species are potential determinants of plant community structure. *Ecol.* 79: 2082-2091.

Vierheilig, H., Scheweiger, P. and Brundrett, M. 2005. An overview of methods for the detection and observation of arbuscular mycorrhizal fungi in roots. *Physiol. Plant.*, 125: 393-404.

Walker, C. and Schüßler, A. 2004. Nomenclatural clarifications and new taxa in the Glomeromycota. *Mycol. Res.*, 108: 981-982.

Walker, C., Vestberg, M., Demircik, F., Stockinger, H., Saito, M., Sawaki, H., Nishmura I. and Schüßler, A. 2007a. Molecular phylogeny and new taxa in the Archaeosporales (Glomeromycota): *Ambispora fennica* gen. sp. nov., Ambisporaceae fam. nov., and emendation of *Archaeospora* and Archaeosporaceae. *Mycol. Res.*, 111: 137-13.

Walker, C., Vestberg, M. and Schüßler, A. 2007b. Nomenclatural clarifications in Glomeromycota. *Mycol. Res.*, 111: 253-255.

Wang, B. and Qui, Y.L. 2006. Phylogenetic distribution and evolution of mycorrhizas in land plants. *Mycorrhiza*, 16: 299-363.

2017, Mycorrhizal Fungi
Editors: Ashok Aggarwal and Kuldeep Yadav
Published by: ASTRAL INTERNATIONAL PVT. LTD., NEW DELHI

Pages 221–244

10

Phosphorus Management in Agricultural Soils by Mycorrhizal Fungi

Richa Raghuwanshi[1]* and Amrita Saxena[2]

[1]*Department of Botany, Mahila Mahavidyalaya,*
Banaras Hindu University, Varanasi – 221 005
[2]*Department of Botany,*
Banaras Hindu University, Varanasi – 221 005
*Corresponding Author: richabhu@yahoo.co.in

ABSTRACT

Plant nutrition is affected by certain macro-nutrients and micro-nutrients. Phosphorous is one of the major nutrient required by almost all life forms for proper cell structure and metabolism. Plants also depend hugely on P-supply for proper growth and differentiation. Next to Nitrogen, in terms of availability, it plays an important role in almost all the major metabolic processes in plant. For optimum productivity, P-requirement for plants is ~30 $\mu mol\ l^{-1}$ of which approximately 1 $\mu mol\ l^{-1}$ is generally found in the soils. To ensure optimum P-requirement of plants, P-fertilizers are added to soil. In soil, two forms of P are found, organic and inorganic P. Plants can utilize the inorganic P only, which is also not sufficient to meet the required amount. Organic matter in form of humus constitutes an important reservoir of P in soils accounting for 20-80 per cent of the total available P in the soils. But, only 0.1 per cent of the total P exists in soluble form that can be utilized by the plants through roots. Microbes and mycorrhizae play an eminent role in solubilizing the fixed, unavailable P-forms *and releasing them into the rhizospheric soil solution thereby making it available for plants uptake. Mycorrhizae mainly form symbiotic associations with roots of many trees species, responsible for their growth and development by absorbing and accumulating N, P, K and Ca in the plant roots. Their association with the plant roots is mutual and it employ two main mechanisms for solubilizing P, one is by increasing the surface area of the roots thereby exposing them more to the soil solution and the other by modifying the rhizosphere region biochemically and hence making P available for plants uptake. An array of transport proteins is required to move Pi across the membrane of AM fungus and plant cells. In recent years, number of different plant-fungal Pi transport proteins have been identified which has made tremendous progress for better understanding of the symbiotic Pi transfer. However, there are some aspects that regulate directly or indirectly the association and the efficiency of AM symbiosis with the plant roots like host specificity and phytohormones. More studies and applications regarding promotion of the use of mycorrhiza in agricultural practices may infer better understanding of the symbiotic relationship thereby leading to increased crop productivity.*

1. Introduction

Plants are the only life-forms which have the capacity of utilizing the solar energy for food production. Apart from the solar energy, various factors are required for the efficient plant nutrition *i.e.* water, minerals *etc*. Some factors are directly utilized by the plant in their naturally available form while others are converted to a usable form, before utilized by the plant. There exist macronutrients and micronutrients which play essential role in different metabolic processes affecting plant nutrition and development. Phosphorous is one of the major key element required by almost all the life forms for proper cell structure and metabolism. It is an important macronutrient required for efficient plant nutrition. Next to Nitrogen, in terms of availability, it plays an important role in almost all the major metabolic processes in plant. Starting from the cellular metabolism including nutrient uptake to transfer of energy at elemental level, it plays an inevitable role in preservation of genetic information as well. Different metabolic processes in plants, like photosynthesis, signal transduction, biosynthesis of macromolecules and respiration are directly or indirectly influenced by the availability of phosphorous (Khan *et al.*, 2010). Even the nitrogen fixation in legumes is under the influence of its availability (Saber *et al.*, 2005). Due to its eminent role in all the vital processes, it is a major limiting factor for the overall growth of plant.

Phosphorus availability in several soils is ~1 µmol l^{-1}, but for optimum productivity, P-requirement for plants is ~30 µmol l^{-1}(Adhya *et al.*, 2015). Phosphorus needs are currently met from geological sedimentary rock. India has an estimated 250mt of low grade rock phosphate (25-30 per cent P_2O_5) which is unsuitable for manufacturing phosphate fertilizers. At the same time the projected P- requirement to ensure food security to the exploding population is estimated to increase by 2.3 per cent p.a. While the estimated global P-reserves of 71 Gt (Van Kauwenbergh, 2010) expects to last in 300 years, the issues lies in the cost and quality of P-extracted, besides the eutrophication. 75-80 per cent of P added as fertilizers to meet the increasing needs of crop production gets fixed in the soil and only the remaining is utilized by the plant.

Rhizospheric microbes play an inevitable role in mobilizing the inorganic and organic P present in soil, by ways like secreting organic acids which help in solubilising the fixed P. Mycorrhizal fungi are well documented for enhancing plant growth through uptake of immobile nutrients such as P, Zn, Cu, *etc.* from soil. The other beneficial attributes include their role in hormone production, greater ability to withstand abiotic stress like drought and biotic stress like pathogens and synergistic interaction with beneficial microorganisms. The activities of Arbuscular Mycorrhizal Fungi (AMF) and plant roots are closely integrated as a result of co-evolution over at least 450 million years (Smith and Read, 2008). AMF though well reported for their growth enhancing activity globally, they are much effective in phosphorus deficient soils prevailing in tropics. The commonly reported and much worked out genera of AMF include *Glomus, Scutellospora, Gigaspora, Acaulospora, Entrophospora*. Research shows that inoculation with efficient AMF not only increases growth and yield of crop plants but also reduces the application of phosphatic fertilizer by nearly 50 per cent, especially in marginal soils deficient in nutrients (Bagyaraj *et al.*, 2015). Goals of increasing crop production through sustainable agriculture can be partly met through application of suitable AM fungal inoculums (preferably indigenous) and augmenting AM activities through other Phosphate Solubilizing Microbes (PSM) keeping the agricultural practices in favour of these fungi.

2. P availability to Plants

The soluble inorganic P occurring as primary orthophosphate ($H_2PO_4^-$) and secondary orthophosphate (HPO_4^{-2}) are taken by plants. Of these two forms, plants prefer to take P in the monovalent H_2PO_4 form (Furihata, 1992). Although being present in organic or inorganic forms in soil, plants are unable to uptake phosphorous to their full potential as in inorganic form; it is in its fully oxidised state after reacting with calcium, iron or aluminium forming insoluble mineral complexes. Most of these are formed after frequent application of chemical fertilizers. Such insoluble, precipitated forms cannot be absorbed by plants (Rengel and Marschner, 2005). Organic matter in form of humus constitutes an important reservoir of P in soils accounting for 20-80 per cent of the total available P in the soils (Richardson, 1994). But, only 0.1 per cent of the total P exists in soluble form that can be utilized by the plants through roots (Zhou *et al.*, 1992). It is now undoubtedly established that P is the limiting factor restraining plant productivity in both terrestrial and freshwater ecosystem (Simpson *et al.*, 2011).

Most of the P gets converted to the unavailable form due to its fixation by either of the two processes. Firstly, its association with the surface soil minerals and secondly, its precipitation by free Al^{3+} and Fe^{3+} in the acidic soils and with Ca^{2+}in the alkaline or neutral soils (Havlin *et al.*, 1999). As most of the available P in soil gets converted to the unusable form, available P has to be supplemented from outside commonly in form of phosphate fertilizers. This addition leads to adverse environmental impacts affecting soil health and degradation of terrestrial, marine and fresh water resources (Tilman *et al.*, 2001) apart from adding financial cost to agricultural production. Other ill-effects of addition of P fertilizers may include eutrophication of surface waters leading to algal bloom (Schindler *et al.*, 2008) and deprivation of beneficial soil microbiota leading to reduced soil fertility (Gyaneshwar *et al.*, 2002) thereby causing reduced crop yield.

Though P fertilizers are added to soil to compensate its availability to plants, according to an interesting characteristic of P geochemistry, only 1 per cent of the total soil P (400-4000 kg P/ha in the top 30 cm) is incorporated to living plant biomass in each growing season (10-30 kg P/ha) reflecting the low availability for plant uptake (Blake *et al.*, 2000; Quiquampoix and Mousain, 2005). This highlights the excess use of P fertilizers in order to meet the requirement of plant nutrition. As, the geological reserves of P are exhaustible and non-renewable, it has been estimated that with the current rate of use, world's known reserves of high quality rock P may get depleted by the end of this century (Cordell *et al.*, 2009) as most of the P is mobilized from geological reserves for fertilizer production. It has been estimated that anthropogenic activities have amplified global P cycling by ~400 per cent relative to pre-industrial times, several folds higher that carbon (~13 per cent) or nitrogen (~100 per cent) (Falkowski *et al.*, 2000).

3. Role of P in Plant Growth

P though critical for plant growth and constituting up to 0.2 per cent of plant dry weight, is one of the most difficult nutrients for plants to acquire. Role of P in plant growth and development is indispensable. The major development and nutrition takes place during the early crop growth, which requires sufficient P availability. The significance for the same has been studied for different crop species (Grant *et al.*, 2005). Enhanced P nutrition at early growing stage in maize has been reported to augment the dry matter of the grain as compared at the later developmental stages (Parewa *et al.*, 2010). A greater difference in dry matter accumulation has been observed in maize under P deficiency during early stages of growth (Plenets *et al.*, 2000). Under P deficiency at early growing stages of maize, up to 60 per cent reduction has been recorded in above ground dry matter accumulation, while only a slight difference in dry matter accumulation was recorded at the time of harvest and grain yield. Similarly, P supply during early growing stages of wheat and barley showed enhanced effect on final grain yield (Smith *et al.*, 2011). Apart from having an additive impact on the grain yield and dry matter accumulation, P availability has also been linked to the shoot growth. Early P deficiency causes decline in shoot growth due to the probable decline in stimulation of root growth (Mollier and Pellerin, 1999). Hence, an initial reduction in growth related to P deficiency leads to an ultimate effect on the final crop yield, which remains unaltered throughout the growing period of a crop.

4. P Dynamics in Rhizosphere and Factors Affecting P Uptake by Plants

Rhizosphere is the most active region and critical zone in terms of interactions between soil, plant and microbes. Plant roots play a major role in modifying rhizosphere composition by releasing root exudates, which may include organic compounds such as mucilage, organic acids, phosphatases and specific signalling substances that are vital for regulating various physiological activities in rhizosphere

(Rakshit *et al.*, 2002; Rai *et al.*, 2013). These different chemical and biological processes in the rhizosphere directly or indirectly influence crop productivity by either determining the mobilization and acquisition of soil nutrients and microbial populations or by controlling the nutrient use efficiency of crop plants (Zhang *et al.*, 2010). The P uptake by roots results in rapid depletion of P in the rhizosphere creating a gradient of P concentration in a radial direction away from the roots surface (Rakshit and Bhadoria, 2009). The low solubility and low mobility of P in soil causes such gradient and even though the total P content in soil exceeding the plant requirement, its availability to plants is restricted. It has been estimated that in order to meet the plant demand, soluble P in the rhizosphere soil solution should be replaced 20 to 50 times per day by delivering P from the bulk soil towards rhizosphere (Marschner, 1995). Hence, it could be said that P dynamics in rhizosphere is basically in control of plant root growth and function along with the soil structure (Neuman and Romheld, 2002). This creates a P cycling in the rhizospheric region which is influenced by different factors.

Nature of solid phases, soil pH and biological activity controls this P cycling (Pierzynski *et al.*, 2005). The oxalate extractable P, Fe and Al estimates P accumulation from fast and slow reactions of phosphate ions with oxi-hydroxides (Lookman *et al.*, 1995). The soil pH is mainly responsible for regulating the distribution of inorganic P in the soil (Kovar and Claassen, 2005). However, the total P content in the soil is also influenced by the biological activities in terms of C and N cycling (Stewart and Tiessen, 1987) and also to the different climatic conditions (Delgado-Baquerizo, *et al.*, 2013). Also, there exist interrelated transformations between the soil inorganic P and soil organic P species, as plants and other life forms utilize mainly inorganic P and organic P undergoes hydrolysis to replenish inorganic P in the soil, thus creating a balance (Stewart and Sharpley, 1987).

Hence, owing to the low mobility and low solubility along with high fixation by the soil matrix, P availability to plants is mainly controlled by two processes:

☆ Root architecture of the plants as well as with the mycorrhizal associations. The presence of different nutrients like C, N, P *etc.* determines the root surface area or the root morphology which in turns absorb P form the soil. The more P held by the plant may cause greater root surface area for better P acquisition (Bloom *et al.*, 1995),

☆ Various rhizopsheric chemical and biological processes also influence the rate of P acquisition and availability to the plants.

4.1. Role of Microbes in P Solubilisation

As discussed in previous sections, biological activity in the rhizosphere play vital role in regulating the distribution and acquisition of soluble P by plants. Numerous different microbes aid in improving the mobilization of lowly available forms of soil P by solubilizing the insoluble P complexes into the rhizosphere soil solution and thus making it available to the plants (Tilak *et al.*, 2005). Such microbes including bacteria, fungi, actinomycetes and mycorrhizae are grouped as phosphate solubilizing micro-organisms (PSMs). These beneficial microbes solubilise the phosphates fixed in soil thereby helping plants in P uptake which leads to enhanced plant nutrition resulting in higher crop yield. Due to their plant growth augmenting property, these microbes are often used as biofertilisers, either as single inoculants or in consortia. Some of the common and major PSMs have been listed in Table 10.1.

A whole diverse group of bacterial genera have been reported to possess P-solubilising ability (Sharma *et al.*, 2013). Not only the common genera like *Bacillus* sp. and *Psuedomonas* show PS activity, many N-fixing bacteria have also been reported as potent P solubilizers having an added benefit of supplying both N and P to the host plants. Some examples include, *Azotobacter* (Kumar *et al.*, 2001), *Enterobacter* and *Klebsiella* (Chung *et al.*, 2005) and *Rhizobium legumiosarum* (Abril *et al.*, 2007). Bacterial isolates from the stressed environments like halophile *Kushneria sinocarni*, has been shown to possess PS activity (Zhu *et al.*, 2011). Such strains may open new avenues for utilizing microbes in salt affected agricultural soils.

Of the total fungal population existing in the soil, about 0.1-0.5 per cent is constituted by PS fungi (Kucey, 1983). Fungal strains have been reported to show greater advantage as PSMs when compared to bacterial counterparts. Firstly, fungal isolates do not lose P-solubilizing ability on repeated culturing under lab conditions thereby being a good source for genetic alterations. Secondly, they are able to extend to longer distances unlike bacterial isolates and hence, serve as a better P source provider to the plants roots (Kucey, 1983). The P-solubilising fungi produce greater amount of acids than bacteria and hence can solubilize P more efficiently than bacterial strains (Venkateshwarlu *et al.*, 1984). The most common PS fungi include that from the genera *Aspergillus* and *Penicillium* (Reyes *et al.*, 2002) apart from the well-studied genus *Trichoderma* (Altomare *et al.*, 1999; Singh *et al.*, 2006). Other characteristic examples of PS fungi include *Rhizoctonia solani, Sclerotinia solani, Phoma sp, Mucor* spp., *Alternatia* spp., *Fusarium* spp. (Sharma *et al.*, 2013). The commonly found PS fungi in agricultural soils like *Penicillium sp, Mucor* spp., and *Aspergillus* spp. have been shown to increase plant growth by 5-20 per cent after inoculation (Gunes *et al.*, 2009).

Apart from bacterial and fungal PSMs, another rising microbial group includes the actinomycetes from the genera *Actinomyces* and *Streptomyces*, that have an added benefit of tolerating extreme environmental conditions and production of antibiotics and phytohormone-like compounds which all together enhance plant growth to many fold (Hamdali *et al.*, 2008).Combined application of PSM and AMF along with rock phosphate could improve crop yield in nutrient deficient soil (Sabannavar and Lakshman, 2009).

Symbiotic association of plant roots with mycorrhizal fungi has gained tremendous attention for its effective P-solubilising ability along with an overall modification of rhizosphere composition in benefit to the plant nutrition uptake and growth. Due to their extra ordinary ability to promote the overall growth and yield of plant, they are promoted to be used as biofertilisers either singly or in combination with other bioinoculants for optimum results (Widada *et al.*, 2007). The common AMF used as biofertilisers include species from the genera *Acaulospora, Archaeospora, Enterophospora, Gerdemannia, Geosiphon, Gigaspora, Glomus, Paraglomus* and *Scutellospora* (Rai *et al.*, 2013; Sharma *et al.*, 2013).

Table 10.1: Different Classes of Phosphate Solubilizing Microbes (PSMs)

Class	Group	Phosphate Solubilizing Microbes (PSMs)	References
Bacteria		*Xanthomonas* spp., *Serratia phosphoticum*, *Erwinia* spp., *Micrococcus* spp., *Escherichia intermedia*	Delvasto *et al.* (2008)
		Bacillus megaterium, *B. circulans*, *B. subtilis*, *B. polymyxa*, *B. sircalmous P. putida*, *P. striata*, *P. fluorescens*, *P. calcis*	
	N-fixing bacteria	*Brevibacterium* spp., *Nitrosomonas* spp., *Enterobacter asburiae*, *Azospirillum brasilense*, *Rhizobium meliloti*, *Bradyrhizobium* spp.	
	S-reducing Bacteria	*Beggiatoa*, *Thiomargarita*, *Thiobacillus ferroxidans*, *T. thioxidans*	Kaviyarasi *et al.* (2011)
	Proteobacteria	*Burkholderia cenocepacia* FeSu 01, *Burkholderia caribensis* FeGI 03, *Burkholderia ferrariae* FeGI 01	
Fungi		*Aspergillus awamori*, *A. niger*, *A. tereus*, *A. flavus*, *A. nidulans*, *A. foetidus*, *A. wentii*	Fenice *et al.* (2000); Khan and Khan (2002); Reyes *et al.* (2002); Sharma *et al.* (2013)
		Penicillium digitatum, *P. lilacinum*, *P. balaji*, *P. funicolosum*	Singh *et al.* (2006); Saxena *et al.* (2015)
		Trichoderma viride, *T. harzianum*, *T. virens*	
	Nemato fungus	*Arthrobotrys oligospora*	Duponnois *et al.* (2006)
		Rhizoctonia solani, *Rhizopus* spp., *Mucor* spp., *Sclerotium rolfsii*, *Alternaria teneius*	Sharma *et al.* (2013)
Yeast		*Yarrowia lipolytica*, *Schizosaccharomyces pombe*, *Pichia fermentans*, *Candida* spp., *Torula thermophila*, *Schwanniomyces occidentalis*	Vassilev *et al.* (2001); Rai *et al.* (2013); Sharma *et al.* (2013)
Mycorrhizae	Ecto-mycorrhizae	*Pisolithus tinctorius*	Schwartz *et al.* (2006)
		Piriformospora indica	Tejesvi *et al.* (2010)
	Arbuscular mycorrhiza	*Acaulospora* spp., *Archaeospora* spp., *Enterophospora* spp., *Gerdemannia* spp., *Geosiphon* spp., *Gigaspora* spp., *Glomus* spp., *Paraglomus* spp. and *Scutellospora* spp.	Rai *et al.* (2013); Adhya *et al.* (2015)
Actinomycetes		*Actinomyces* spp., *Streptomyces* spp., *Nocardia* spp., *Streptoverticillium* spp., *Thermoactinomycetes* spp., *Micromonospora* spp.	Sharma *et al.* (2013)
Cyanobacteria		*Anabena* spp., *Calothrix braunii*, *Nostoc* spp., *Scytonema* spp.	Sharma *et al.* (2013)

4.2. Role of Mycorrhizae in P Uptake

Mycorrhiza constitutes a distinct morphological structure which is found to establish mutualistic associations with the roots of more than 80 per cent plant species including important crops and forest trees (Rai *et al.*, 2013). Plants growing in P, N, Zn, Cu, Fe, S and B deficient soil tend to form symbiotic associations with mycorrhizal fungi. These plants may range from herbs, shrubs, trees, aquatic, xerophytes, epiphytes, hydrophytes or terrestrial ones (Zhu *et al.*, 2011). The significance of utilizing and popularizing mycorrhizal fungi as biofertilisers is the multifarious role they play apart from P-solubilisation, which includes augmented plant yield and resistance against biotic and abiotic stresses.

Two different types of mycorrhizal associations are illustrated- Arbuscular Mycorrhizae (AM Fungi) and ecto-mycorrhizae (EM Fungi). Arbuscular mycorrhizas (Vesicular-Arbuscular Mycorrhizas, VAM or AM) are associations where fungi produce arbuscules, hyphae and vesicles within root cortex cells. These associations are defined by the presence of arbuscules. They belong to the phylum Glomeromycota, which has three classes (Glomeromycetes, Archaeosporomycetes and Paraglomeromycetes) with five orders (Glomerales, Diversisporales, Gigasporales, Paraglomerales and Archaeosporales), 14 families and 26 genera (Sturmer, 2012). Fungi in the roots spread by linear hyphae or coiled hyphae. Ectomycorrhizas (ECM) are associations where fungi form a mantle around roots and a Hartig net between root cells. These associations are defined by Hartig net hyphae which grow around cells in the epidermis or cortex of short swollen lateral roots.

ECM mainly form symbiotic associations with roots of many trees species (Anderson and Cairney, 2007) resulting in their growth and development by absorbing and accumulating N, P, K and Ca more rapidly when compared to the non-mycorhizal roots. ECM aid in breaking the mineral complexes and organic substances thereby freeing the P and transferring nutrients for uptake by trees. Also, enhanced tolerance of trees to drought, high soil temperature, soil toxins and extreme soil pH has also been attributed to the role of ECM association with the trees. The most common ECM used as bioinoculant and popularly used for production of biofertiliser is *Pisolithus tinctorius* which has a wide host range and its inoculum can be produced and applied as vegetative mycelium in a peat or clay carrier (Schwartz *et al.*, 2006). Another ECM used as biofertiliser is *Piriformospora indica* which has the ability of plant growth promotion along with tolerance to biotic and abiotic stresses causing increased biomass (Tejesvi *et al.*, 2010).

Endomycorrhizae or AM fungi form associations inside plant roots where plant roots provide carbohydrates to the fungus and fungus in turn supply nutrients and water to plant roots (Adholeya *et al.*, 2005). AM fungi belong to nine genera: *Acaulospora, Archaeospora, Enterophospora, Gerdemannia, Geosiphon, Gigaspora, Glomus, Paraglomus* and *Scutellospora*. AM fungi are widespread group and are found from the arctic to tropics and are present in most agricultural and natural ecosystems.

Ions are mainly classified as mobile when their transfer in soil occurs through mass flow. However, phosphorus moves as per concentration gradient *i.e.* diffusion, which makes it highly immobile element available to plants. The plants roots absorb phosphorus at a much higher rate than the rate of phosphorus diffusion, which creates a phosphorus depleting zone around the roots. The uptake of phosphorus by plants thus depends upon the soil area explored by plants roots. AM infection in plants roots increase this area by extending their hyphal network in soil, termed as 'effective root zone'. It is to be understood here that AMF does not solubilize the unavailable inorganic phosphorus sources, but only takes up extra phosphate which are not accessible to plant roots. However, the major soil organic phosphorus source *i.e.* phytate is acted upon by acid phosphatase produced by AMF, which releases H_2PO_4 ions readily taken up by plants (Joner *et al.*, 2000). Encouraging results have been obtained in field studies using indigenous AMF (Maiti *et al.*, 2012) where P acquisition was enhanced in upland rice in plots grown with maize–horse gram/ rice rotation (2 years rotation of maize relay cropped by horse gram in first year and rice in second year) than farmers' rotations. Judicial use of AM-supportive rice-based crop rotation can reduced phosphatic fertilizer use by 33.3 per cent in rice without affecting the grain yield (Maiti and Baranwal, 2012).

As mycorrhizal fungi are more efficient in not only P-solubilisation but also in uptake of other important plant nutrients like K, N, Ca, Zn, S, B *etc.*, along with providing resistance to plants towards biotic and abiotic stresses, interest in using these fungi as biofertilisers are gaining momentum, owing to their prominent role in plant growth, health and productivity

5. Mechanism of P-uptake by Mycorrhiza

Plants and fungi take up P as negatively charged Pi ions $H_2PO_4^-$. Besides this the concentration of P in root cells is about 1,000-fold higher compared to the soil solution and the cell membrane has an inside-negative electric potential. Pi uptake, therefore, requires metabolic energy and involves high-affinity transporter proteins in the Pht1 family (Bucher, 2007). P-uptake by plants is thus an active process. Expression of genes encoding high-affinity Pi transporters (PiTs) in root epidermis is maximal in the root apex and root hairs (Gordan-Weeks *et al.*, 2013) and declines in more mature regions. AM colonization, and hence the potential operation of the AM pathway, occurs behind the root apex. Hyphal inflow of P (uptake per unit length of hyphae per unit time) is around 18×10^{-14} mol cm^{-1} s^{-1} *i.e.* about six times more compared to non-mycorrhizal root (Bagyaraj *et al.*, 2015). This not only aids in the higher inflow rate of P but as there is an increased spatial exploitation of soil by mycorrhizal hyphae, AM plants are able to overcome their P requirement more effectively than non-mycorrhizal plants. Fungal hyphae being much smaller in diameter than roots access narrower soil pores and hence increase the soil volume explored (Drew *et al.*, 2003; Smith and Read 2008; Schnepf *et al.*, 2011).

Mechanisms of Pi release from fungus to the interfacial apoplast are obscure, but uptake into the plant is increasingly well understood. AM-inducible plant PiT genes, which are different from those in the direct pathway, are expressed, sometimes exclusively, in the colonized cortical cells (Bucher, 2007; Javot *et al.*, 2007). These PiTs are involved in the uptake of Pi released by the fungi and have been shown to occur in potentially all AM plants investigated, regardless of their responsiveness to AM fungal colonization. Additionally, H⁺-ATPases energize the plant plasma membrane surrounding the intracellular fungal structures, facilitating active Pi uptake (Smith and Read, 2008).

Recognition of the symbiotic partner and the early signals exchanged during AM symbiosis is the same as in legume-*Rhizobium* symbiosis (Hata *et al.*, 2010) referred to as "common symbiosis genes". Similar to the Nod factors reported to play eminent role in rhizobium-legume symbiosis, biochemical compounds called Myc factors have been suggested to be secreted by the mycorrhizal hyphae, which is perceived by host plant roots for activation of common symbiosis pathway (SYM) (Kosuta, 2003; Roberts *et al.*, 2013). The underlying signalling for AM-specific induction of genes is largely unknown, but lysophosphatidylcholine has been identified as a key for the activation of AM-specific PiT genes in potato (*Solanum tuberosum*) and tomato (Drissner *et al.*, 2007).

AM symbiosis is ecologically crucial in terms of nutrient absorption, as its association, either in or on the roots of a host plant makes them a major source for nutrient acquisition besides the roots of the plants (Bagayoko *et al.*, 2000). The mode of action of AM symbiosis can be broadly grouped under two major mechanisms as follows:

Increased Exploitation of Soil Area

Roots associated with AM have the ability to explore more volume of soil than the non-mycorrhizal associated roots (Lambers *et al.*, 2008). The external hyphae of the AM extend beyond the soil P depletion zone gaining access to the greater volume of undepleted soil (Lambers *et al.*, 2011). It has been reported that the hyphae can extend up to 10 cm from the roots surface of the plants (Jakobsen *et al.*, 2005). Also, the smaller diameter of the hyphae (20-50 μm) enables easy access to miniscule soil pores which are otherwise difficult for root hairs to penetrate. Thus, these characteristics of the external hyphae of AM with the plant roots form a meshed network more efficient for P uptake in soil. Also, AM symbiosis has been reported to show increase in lateral root formation and induce root branching in host plant enabling them for better nutrient absorption form the soil (Lynch, 2007). AM associated plants can absorb more P at lower concentration in the soil solution than the non-mycorrhizal associated plants (Lynch and Brown, 2001; Rakshit and Bhadoria, 2009). Due to a very limited concentration around the hyphae having a smaller radius than the roots and the root hairs, the P concentration in the soil solution, around the hyphae, is always higher than that formed around the roots in the soil P depletion zone. This enables the hyphae to absorb P even in low soil P levels without having a higher affinity for P (Barber, 1984). However, it has been reported that the AM hyphae possess higher P affinity (lower K_m) than the roots (Zhu *et al.*, 2010).

Biochemical Modification of the Rhizosphere

Mycorrhizae cause biochemical alterations and physiological changes in the rhizosphere that leads to the improved P accessibility in the rhizosphere (Khan *et al.*, 2010; Richardson and Simpson, 2011). In neutral to calcareous soils, a greater solubilisation of P is observed due to mycorrhizal alterations where a reduced pH (around~6.3) is recorded by increasing proton efflux or pCO_2 activity around the rhizosphere (Rigou and Mignard, 1994; Bowen and Rovira, 1999; Bago and Azcon-Auilar, 1997). In case of acidic soils, Fe or Al bound P is made available to the plants by producing citric acid and siderophores by the mycorrhiza (Richardson and Simpson, 2011). Also, the action of different extracellular enzymes like alkaline phosphatases, have been shown to play eminent role in solubilising the bound P thereby making it accessible for the uptake of the plants (Tarafdar and Marschner, 1994). Organic anion exudation by mycorrhiza in the rhizosphere by releasing anions like citrate, malate and oxalate has been shown to play vital role in bioavailability of soil P (Hinsinger, 2001). These released anions block the sorption sites or replace P in sparingly soluble complexes with Al, Fe or Ca thereby mobilizing the available P, which otherwise would have got bound (Richardson *et al.*, 2011). Another method of affecting P mobilization by mycorrhiza is by producing carboxylates, which influence ligand exchange, ligand promoted distribution of P bearing minerals like Al/Fe oxides (Wang *et al.*, 2010). Though the proper mechanisms have not yet been elucidated in detail.

6. Genetic Framework Responsible for P-solubilization by Mycorrhiza

A symbiotic P-uptake pathway which involves P acquisition from soil by extra radical hyphae of AM and its subsequent transfer to plant root cells, require an array of transport proteins to move Pi across the membrane of AM fungus and plant cells. In recent years, number of different plant-fungal Pi transport proteins have been identified which has made tremendous progress for better understanding of the symbiotic Pi transfer (Karandashov and Bucher, 2005; Bucher, 2007).These transporter proteins belong to three distinct protein families which regulate Pi partioning and recycling throughout a plant's life (Loth-Pereda *et al.*, 2011).

During the initial studies of Pi transport by studying the germ tubes of AM fungus *Gigaspora margarita*, the Pi uptake kinetics revealed the presence of two Pi uptake systems functional; a high affinity active transport system (Km 1.8-3.1 µM) and a low affinity passive transport system (Km 10.2-11.2 mM) (Thomson *et al.*, 1990). The low affinity protein transporters are reported to control the remobilization of acquired P (Smith *et al.*, 2001) while the high affinity active protein transporters regulate the P acquisition from the soil.

Pi/II+ symporters belonging to Pht1 gene family have been reported to play crucial role in regulating Pi uptake into the root symplasm via transport across the plasma membrane (Bucher, 2007; Casieri *et al.*, 2013). Plants employ two major pathways for P-uptake, either through the 'direct pathway' or through the 'mycorrhizal pathway' depending upon their association with the AM fungus. Many Pht1 genes have been reported to express strongly in rhizodermal cells, including root hairs and also in cortical cells which relate to their role in the direct pathway of P acquisition in non-mycorrhizal associated plants (Daram *et al.*, 1998; Chiou *et al.*, 2001; Walder *et al.*, 2015). In the mycorrhizal pathway, the extra-radical hyphae absorb Pi from soil and translocate it to the roots into the arbuscules, from where it is further released into the peri-arbuscular space (Smith *et al.*, 2011). From here, the specifically induced Pht1 transporters come into action and transfer the Pi across the plant's peri-arbuscular membrane (Casieri *et al.*, 2013). The Pi transporters involved in the mycorrhizal pathway are referred to as AM inducible Pi transporters and they have been reported in many plant species of monocots and dicots (Loth-Pereda *et al.*, 2011). However, all the Pht1 transporters can be clustered into two subgroups: group I and group III (Bucher, 2007). The members of subgroup I have been reported to be expressed in the arbuscle-containing cortical cells during AM symbiosis, as evident by immune localization experiments (Harrison *et al.*, 2002; Javot *et al.*, 2007). The members of subgroup III are generally reported to express in plant root cells and specifically induced in cortical cells in case of AM symbiosis (Rausch *et al.*, 2001; Maeda *et al.*, 2006). Interestingly, these transporters are not only specifically induced on AM symbiosis, but have been reported to play important role in mycorrhizal Pi acquisition as manifested by several studies using mutants with reduced or inhibited AM-induced Pht1 gene expression (Maeda *et al.*, 2006; Javot *et al.*, 2007; Yang *et al.*, 2012).

Diverse plant species have been reported to exhibit PHT1 proteins with exclusive or induced expression in arbusculated cells (Yang and Paszkowski, 2011). The rice ORYsa: PHT1:11 (PT11) represents the first AM specific Pi transporter isolated from plants (Paszkowski *et al.*, 2002). Similarly, the first AM-induced Pi transporter gene was identified from potato (*Solanum tuberosum*) as StPT3 which encodes for a high affinity Pi transporter (Rausch *et al.*, 2001). PT11 homolog of *Medicago trunculata* (MEDtr:PHT1;4 [PT4]) localized at the periarbuscular membrane (Harrison *et al.*, 2002) has been demonstrated to be important for AM symbiosis as its mutation caused impairment of both development of AM symbiosis and symbiotic Pi uptake (Javot *et al.*, 2007).

Functional experiments on Pi transporters have indicated that these are required for symbiotic Pi uptake in AM associated plants and also for successful establishment of AM symbiosis. Any loss or alteration in their abilities has prominent impact beyond just nutrient exchange. The mutants of Pi transporters have impaired mycorrhizal growth benefits and also some experiments demonstrated the importance of symbiotic Pi transfer for successful development of AM hyphae within the plant roots (Nagy *et al.*, 2005, 2006; Javot *et al.*, 2007; Wegmueller *et al.*, 2008; Grace *et al.*, 2009; Nagy *et al.*, 2009).

Pht1 genes have also been reported to play crucial role in regulating different physiological functions in the plants apart from its role in symbiotic Pi transport as in leaf senescence, which may be partially important for growth and development of perennial plant species. Also, its expression pattern recorded in different plant tissues further support their role in regulating different growth and developmental processes in plant with major impact on the plant's symbiosis with AM and EM fungal species (Loth-Pereda *et al.*, 2011). Though, recent studies on Pi transporters have given stimulating insight into the mechanism of Pi acquisition through AM symbiosis, further studies are required to give better understanding of the different signalling pathways involved in Pi uptake and symbiosis development.

7. Factors Regulating P Acquisition during AM Symbiosis

AM symbiosis has been reported to play eminent role in upgrading the P-acquisition ability of the host plant. Numerous factors apart from physiological and biochemical characteristics of soil are correlated for successful establishment of the symbiotic association with the plant roots. There are some aspects that regulate directly or indirectly the association and the efficiency of AM symbiosis with the plant roots. These elements thereby influence the overall P-uptake by the plant roots.

Host-Specificity

The AM plant relations though need to be further worked out is found to be specific in the way that different AM fungi deliver different amounts of P. Diversity of responses of AMF exists in the naturally co-occurring plants and AM fungi from the same site (Klironomos, 2003). AM symbioses may have ecological benefits that cannot be predicted from AM growth responses determined for plants grown singly in pots with single AM fungal genotypes. The same AM fungus does not deliver the same proportion of total P to different plant species (Munkvold *et al.*, 2004; Smith *et al.*, 2004). Effective AM colonization depends upon a number of factors like the host plant, AMF genus, soil phosphorus status and physical conditions of the soil. Indigenous AMF are reported to be more effective than the exogenous strains (Raghuwanshi and Upadhyay, 2004). Differences exist between AMF species regarding plant growth promotion, mycelia growth pattern, and spore production.

Phosphorus uptake per unit hyphal length seems to be more conserved on AMF species level (Koch *et al.*, 2004). AM contribution is highly dependent on soil P- content as it decreases with increasing soil P supply, as the direct uptake of P increases (Nagy *et al.*, 2009). This is due to the decreasing percentage of root length colonized. Different responses of AM strains on plant growth have not been conclusively co-related to per cent colonization. This might be because colonization by different AM fungi show different growth responses in a single AM plant species (Klironomos, 2003; Munkvold *et al.*, 2004, Smith *et al.*, 2004) and colonization by the same AM fungus may not cause the same growth responses in different plant species/crop varieties, thereby showing their specificity. It is clear, therefore, that there exists considerable functional diversity among plant-AM fungal symbioses in terms of benefits (P supply to the plant, in this context) and costs (C supply to the fungus). Agronomic practices too are important as they may lower inoculum load in soil and subsequent colonization. For example, frequent use of P fertilizer, long fallow periods, cultivation of non-host crops (especially members of the Brassicaceae), or frequent soil tillage disrupts AM fungal hyphae networks in soil (Jansa *et al.*, 2006).

The finding that wheat, barley and tomato can receive a large proportion of total P as hidden P uptake via the AM pathway, even though they generally do not take up more total P when AM, shows that the contribution of direct uptake must be lower than in NM plants. In an extreme case, the direct pathway was completely inoperative in tomato when colonized by the AM fungus *Glomus intraradices* (Smith *et al.*, 2004).

Phytohormones

Phytohormones play a crucial role in plant response to AMF. They are involved in root development and in sugar signalling (Rouached *et al.*, 2010) during plant Pi-starvation. Auxin and ethylene are key phytohormones in regulating lateral root and root hair development, which is affected by Pi starvation (Rubio *et al.*, 2009) as well as by AM colonization. Exogenous application of cytokinin represses the Pi-starvation genes in plants. Strigolactone secreted by plant roots induce hyphal branching in the germinating AM fungal spores (Akiyama *et al.*, 2005), arbuscles formation in cortical cells (Zang *et al.*, 2010), both of which contribute to strong AM symbiosis with plants. Strigolactone has recently been shown to act in concert with auxin to differentially regulate lateral root development and shoot branching in *Arabidopsis*, depending on the Pi level in the growth medium (Ruyter-Spira *et al.*, 2011). Plants growing in high P-soil show lower Strigolactone content leading to poor AM colonization (Balzergue *et al.*, 2011).There are many shared components between AM symbiosis and Pi-starvation signaling pathways that are interconnected with sugar and phytohormone signaling. This offers a great potential for cross talk between the direct and AM pathways, but specific regulatory elements responsible for such cross talk have not yet been identified (Smith *et al.*, 2011).

8. Conclusion

Agricultural industry faces numerous environmental and biotic stresses which becomes a major hitch in fulfilling the food requirement causing a decline in crop productivity at an unprecedented rate. Also, the dependence on chemical fertilisers and pesticides has further added to the woes of the farmers with an additional economic input required for crop cultivation. Not only that, the continuous magnification of toxic and hazardous chemicals in the food web is life-threatening to almost all the life forms. The alternate way for a safer and eco-friendly strategy is the utilization of biofertilizers on a larger scale and their in-cooperation into the modern agricultural practices that can prove useful in restoring environmental homeostasis. Mycorrhiza has been shown to have significant impact on plant growth and nutrient uptake. Particularly, its central role in P uptake, its allocation and cycling which effect numerous physiological plant processes is unblemished.

However, plant responses to P-limitation can only be understood by considering interactions of different biotic and abiotic factors with plant species richness and production. The use of mycorrhiza in low P concentration soils can add sustainability to the agriculture, by aiding plant adaptations to low P-content in soils. This may also reduce the demand for P-fertilisers thereby averting the high input cost put forth for crop cultivation. To conclude, more studies and applications are required for developing efficient strategies to promote the use of mycorrhiza in agricultural practices and also for breeding cultivars with adapted root systems or exudation strategies that may promote increased crop productivity.

References

Abril, A., Zurdo-Pineiro, J.L., Peix, A., Rivas, R. and Velazquez, E. 2007. Solubilization of phosphate by a strain of *Rhizobium leguminosarum* bv. *Trifolii* isolated from

Phaseolus vulgaris in El Chaco Arido soil (Argentina). In: Velazquez, E. and Rodriguez-Berrueco, C. (Eds.). Developments in Plant and Soil Sciences. Springer, The Netherlands, pp. 135–138.

Adholeya, A., Tiwari, P. and Singh, R. 2005. Large scale inoculum production of arbuscular mycorrhizal fungi on root organs and inoculation strategies. In: Declerck, S., Strullu, D.J. and Fortin, A. (Eds.) Soil Biology-Volume 4, *In vitro* Culture of Mycorrhizae. Springer-Verlag Berlin Heidelberg, pp. 315-338.

Adhya, T.K., Kumar, N., Reddy, G., Podile, A.R., Bee, H. and Samantaray, B. 2015. Microbial mobilization of soil phosphorus and sustainable P management in agricultural soils. *Cur. Sci.* 108: 1280-1287.

Akiyama, K., Matsuzaki, K. and Hayashi, H. 2005. Plant sesquiterpenes induce hyphal branching in arbuscular mycorrhizal fungi. *Nature,* 435: 824–827.

Altomare, C., Norvell, W.A., Borjkman, T. and Harman, G.E. 1999. Solubilization of phosphates and micronutrients by the plant growth promoting and biocontrol fungus *Trichoderma harzianum* Rifai 1295–22. *Appl. Environ. Microbiol.,* 65: 2926–2933.

Anderson, C.I. and Cairney, W.G.J. 2007. Ectomycorrhizal fungi: exploring the mycelial frontier. *FEMS Microbiol. Rev.,* 31: 388-406.

Bagayoko, M., George, E., Romheld, V. and Buerkert, A. 2000. Effect of mycorrhizal fungi and phosphorus on growth and nutrient uptake of millet, cowpea and sorghum on a West African soil. *J. Agri. Sci.,* 135: 399–407.

Bago, B. and Azcon-Aguilar, C. 1997. Changes in the rhizospheric pH induced by arbuscular mycorrhiza formation in onion (*Allium cepa* L). *Z. Pflanz. Bodenkunde,* 160: 333–339.

Bagyaraj, D.J., Sharma, M.P., Maiti, D. 2015. Phosphorus nutrition of crops through arbuscular mycorrhizal fungi. *Cur. Sci.,* 108 (7): 1288-1293.

Balzergue, C., Puech-Pagès, V., Bécard, G. and Rochange, S.F. 2011. The regulation of arbuscular mycorrhizal symbiosis by phosphate in pea involves early and systemic signaling events. *J. Exp. Bot.,* 62: 1049–1060.

Barber, S.A. 1984. Soil Nutrient Bioavailability. A Mechanistic Approach. John Wiley and Sons, New York, USA.

Blake, L., Mercik, S., Koerschens, M., Moskal, S., Poulton, P.R., Goulding, K.W.T., Weigel, A., and Powlson, D.S. 2000. Phosphorus content in soil, uptake by plants and balance in three European long-term field experiments. *Nutr. Cycl. Agroecosys.,* 56: 263–27.

Bloom, A J, Chapin, F.C. and Mooney, H.A. 1995. Resource limitation in plants: An economic analogy. *Ann. Rev. Ecol. Sys.,* 16: 363–392.

Bowen, G.D. and Rovira, A.D. 1999. The rhizosphere and its management to improve plant growth. *Adv. Agron.,* 66: 1–102.

Bucher, M. 2007. Functional biology of plant phosphate uptake at root and mycorrhiza interfaces. *New Phytol.,* 173: 11–26.

Casieri, L., Ait-Lahmidi, N., Doidy, J., Veneault-Fourrey, C., Migeon, A., Bonneau, L., Courty, P.E., Garcia, K., Charbonnier, M. and Delteil, A. 2013. Biotrophic transportome in mutualistic plant–fungal interactions. *Mycorrhiza*, 23: 597–625.

Chiou, T.J., Liu, H. and Harrison, M.J. 2001. The spatial expression patterns of a phosphate transporter (MtPT1) from *Medicago truncatula* indicate a role in phosphate transport at the root/soil interface. *Plant J.*, 25: 281–293.

Chung, H., Park, M., Madhaiyan, M., Seshadri, S., Song, J., Cho, H. and Sa, T. 2005. Isolation and characterization of phosphate solubilizing bacteria from the rhizosphere of crop plants of Korea. *Soil Biol. Biochem.*, 37: 1970–1974.

Cordell, D., Drangert, J.O. and White, S. 2009. The story of phosphorus: global food security and food for thought. *Glob. Environ. Chang.*, 19: 292–305.

Daram, P., Brunner, S., Persson, B.L., Amrhein, N. and Bucher, M. 1998. Functional analysis and cell-specific expression of a phosphate transporter from tomato. *Planta*, 206: 225–233.

Delgado-Baquerizo, M., Maestre, F.T. and Gallardo, A. 2013. Decoupling of soil nutrient cycles as a function of aridity in global drylands. *Nature*, 502: 672–689.

Delvasto, P., Valverde, A., Ballester, A., Muñoz, J.A., González, F., Blázquez, M.L., Igual, J.M. and García-Balboa, C. 2008. Diversity and activity of phosphate bioleaching bacteria from a high-phosphorus iron ore. *Hydrometallurgy*, 92: 124–129.

Drew, E.A., Murray, R.S., Smith, S.E. and Jakobsen, I. 2003. Beyond the rhizosphere: growth and function of arbuscular mycorrhizal external hyphae in sands of varying pore sizes. *Plant Soil*, 251: 105–114.

Drissner, D., Kunze, G., Callewaert, N., Gehrig, P., Tamasloukht, M., Boller, T., Felix, G., Amrhein, N. and Bucher, M. 2007. Lyso-phosphatidylcholine is a signal in the arbuscular mycorrhizal symbiosis. *Science*, 318: 265–268.

Duponnois, R., Kisa, M. and Plenchette, C. 2006. Phosphate solubilizing potential of the nemato fungus *Arthrobotrys oligospora*. *J. Plant Nutr. Soil Sci.*, 169: 280–282.

Falkowski, P., Scholes, R.J., Boyle, E., Canadell, J., Canfield, D., Elser, J., Gruber, N., Hibbard, K., Ho¨gberg, P., Linder, S., Mackenzie, F.T., Moore III, B., Pedersen, T., Rosenthal, Y., Seitzinger, S., Smetacek, V. and Steffen, W. 2000. The global carbon cycle: a test of our knowledge of Earth as a system. *Science.*, 290: 291–296.

Fenice, M., Seblman, L., Federici, F. and Vassilev, N. 2000. Application of encapsulated *Penicillium* variabile P16 in solubilization of rock phosphate. *Biores. Tech.*, 73: 157–162.

Furihata, T., Suzuki, M. and Sakurai, H. 1992. Kinetic characterization of two phosphate uptake systems with different affinities in suspension-cultured *Catharanthus roseus* protoplasts. *Plant Cell Physiol.*, 33: 1151–1157.

Grace, E.J., Cotsaftis, O., Tester, M., Smith, F.A. and Smith, S.E. 2009. Arbuscular mycorrhizal inhibition of growth in barley cannot be attributed to extent of colonization, fungal phosphorus uptake or effects on expression of plant phosphate transporter genes. *New Phytol.*, 181: 938–949.

Grant, C., Bittman, S., Montrea, M., Plenchette, C. and Morel, C. 2005. Soil and fertilizer phosphorus: Effects on plant P supply and mycorrhizal development. *Can. J. Plant Sci.*, 85: 3–14.

Gunes, A., Ataoglu, N., Turan, M., Esitken, A. and Ketterings, Q.M. 2009. Effects of phosphate-solubilizing microorganisms on strawberry yield and nutrient concentrations. *J. Plant Nut. Soil Sci.*, 172: 385–392.

Gyaneshwar, P., Naresh, K.G., Parekh, L.J. and Poole, P.S. 2002. Role of soil microorganisms in improving P nutrition of plants. *Plant Soil*, 245: 83–93.

Hamdali, H., Bouizgarne, B., Hafidi, M., Lebrihi, A., Virolle, M.J. and Ouhdouch, Y. 2008. Screening for rock phosphate solubilizing Actinomycetes from Moroccan phosphate mines. *Appl. Soil Ecol.*, 38:12–19.

Harrison, M.J., Dewbre, G.R. and Liu, J. 2002. A phosphate transporter from *Medicago truncatula* involved in the acquisition of phosphate released by arbuscular mycorrhizal fungi. *Plant Cell*, 14: 2413–2429.

Hata, S., Kobae, Y. and Banba, M. 2010. Interactions between plants and arbuscular mycorrhizal fungi. *Int. Rev. Cell. Mol. Biol.*, 281: 1–48.

Havlin, J., Beaton, J., Tisdale, S.L. and Nelson, W. 1999. Soil fertility and fertilizers. An introduction to nutrient management. Prentice Hall, Upper Saddle River, NJ.

Hinsinger, P. 2001. Bioavailability of soil inorganic P in the rhizosphere as affected by root induced chemical changes: a review. *Plant Soil*, 237: 173-195.

Isherwood, K.F. 2000. Mineral Fertilizer Use and the Environment. International Fertilizer Industry Association/United Nations Environment Program, Paris.

Jakobsen, I., Leggett, M.E. and Richardson, A.E. 2005. Rhizosphere microorganisms and plant Phosphorus Uptake. In: Sims, J.T. and Sharpley, A.N. (Eds). Phosphorus, Agriculture and the Environment. American Society for Agronomy, Madison, pp. 437–494.

Jansa, J., Weimken, A. and Frossard, E. 2006. The effects of agricultural practices on arbuscular mycorrhizal fungi. In: Frossard, E., Blum, W., Warkentin, B. (Eds.). Function of Soils for Human Societies and the Environment. Geological Society, London, pp. 89–115.

Javot, H., Penmetsa, R.V., Terzaghi, N., Cook, D.R. and Harrison, M. J. 2007. A phosphate transporter indispensable for the arbuscular mycorrhizal *Medicago truncatula* symbiosis. *Proc. Nat. Acad. Sci. USA.*, 104: 1720–1725.

Javot, H., Pumplin, N. and Harrison, M. J. 2007. Phosphate in the arbuscular mycorrhizal symbiosis: transport properties and regulatory roles. *Plant Cell Environ.*, 30: 310–322.

Joner, E.J., Briones, R. and Leyval, C. 2000. Metal-binding capacity of arbuscular mycorrhizal mycelium. *Plant Soil*, 226: 227–234.

Karandashov, V. and Bucher, M. 2005. Symbiotic phosphate transport in arbuscular mycorrhizas. *Trends Plant Sci.*, 10: 22–29.

Kaviyarasi, K., Kanimozhi, K., Madhanraj, P., Panneerselvam, A. and Ambikapathy, V. 2011. Isolation, identification and molecularcharacterization of phosphate solubilizing Actinomycetes isolatedfrom the coastal region of Manora, Thanjavur (Dt). *Asian J. Pharm. Tech.*, 1: 119–122.

Khan, M.R. and Khan, S.M. 2002. Effect of root-dip treatment with certain phosphate solubilizing microorganisms. *Biores. Technol.*, 85: 213–215.

Khan, M.S., Zaidi, A., Ahemad, M., Oves, M. and Wani, P. A. 2010. Plant growth promotion by phosphate solubilising fungi – current perspective. *Arch. Agron. Soil Sci.*, 56: 73–98.

Klironomos, J.N. 2003. Variation in plant response to native and exotic arbuscular mycorrhizal fungi. *Ecol.*, 84: 2292–2301.

Koch, A.M., Kuhn, G., Fontanillas, P., Fumagalli, L., Goudet, I. and Sanders, I.R. 2004. High genetic variability and low local diversity in a population of arbuscular mycorrhizal fungi. *Proc. Nat. Acad. Sci. USA*, 101: 2369–2374.

Korb, J.E., Johnson, N.C. and Covington, W.W. 2003. Arbuscular mycorrhizal propagule densities respond rapidly to ponderosa pine restoration treatments. *J. Appl. Ecol.*, 40: 101–110.

Kosuta, S. 2003. Diffusible factor from arbuscular mycorrhizal fungi induces symbiosis-specific expression in roots of *Medicago truncatula*. *Plant Physiol.*, 131: 952–962.

Kovar, J.L. and Claassen, N. 2005. Soil–root interactions and phosphorus nutrition of plants. In: Sims, T.J. and Sharpley, A.N. (Eds.). Phosphorus: agriculture and the environment, Agronomy Series #46. Soil Science Society of America, Madison, WI, pp. 379–414.

Kucey, R.M.N. 1983. Phosphate solubilizing bacteria and fungi in various cultivated and virgin Alberta soils. *Can. J. Soil Sci.*, 63: 671–678.

Kumar, V., Behl, R.K. and Narula, N. 2001. Establishment of phosphate- solubilizing strains of *Azotobacter chroococcum* in the rhizosphere and their effect on wheat cultivars under greenhouse conditions. *Microbiol. Res.*, 156: 87–93.

Lambers, H., Finnegan, P.M., Laliberte, E., Pearse, S.J., Ryan, M.H., Shane, M.W. and Veneklaas, E. J. 2011. Phosphorus nutrition of Proteaceae in severely phosphorus-impoverished soils: Are there lessons to be learned for future crops? *Plant Physiol.*, 156: 1058–1066.

Lambers, H., Raven, J.A., Shaver, G.R. and Smith, S.E. 2008. Plant nutrient-acquisition strategies change with soil age. *Trends Ecol. Evol.*, 23: 95–103.

Lookman, R., Freese, D., Merckx, R., Vlassak, K. and Van Riemsdijk, W. H. 1995. Long-term kinetics of phosphate release from soil. *Environ. Sci. Technol.*, 29: 1569–1575.

Loth-Pereda, V., Orsini, E., Courty, P.E., Lota, F., Kohler, A., Diss, L., Blaudez, D., Chalot, M., Nehls, U., Bucher, M. and Martin, F. 2011. Structure and expression profile of the phosphate Pht1 transporter gene family in mycorrhizal *Populus trichocarpa*. *Plant Physiol.*, 156: 2141–2154.

Lynch, J.P. and Brown, K.M. 2001. Topsoil foraging-an architectural adaptation of plants to low phosphorus availability. *Plant Soil*, 237: 225–237.

Lynch, J.P. 2007. Roots of the second green revolution. *Aust. J. Bot.*, 55: 1–20.

Maeda, D., Ashida, K., Iguchi, K., Chechetka, S.A., Hijikata, A., Okusako, Y., Deguchi, Y., Izui, K. and Hata, S. 2006. Knockdown of an arbuscular mycorrhiza-inducible phosphate transporter gene of *Lotus japonicus* suppresses mutualistic symbiosis. *Plant Cell Physiol.*, 47: 807–817.

Maiti, D. and Barnwal, M. K. 2012. Optimization of phosphorus level for effective arbuscular-mycorrhizal activity in rainfed upland rice based cropping system. *Ind. Phytopath.*, 65: 334–339.

Maiti, D., Variar, M. and Singh, R.K. 2012. Rice based crop rotation for enhancing native arbuscular mycorrhizal (AM) activity to improve phosphorus nutrition of upland rice (*Oryza sativa* L.). *Biol. Fert. Soils*, 48: 67–73.

Marschner, H. 1995. Mineral Nutrition of Higher Plants, Ed 2. Academic Press, London.

Mollier, A. and Pellerin, S. 1999. Maize root system growth and development as influenced by phosphorus deficiency. *J. Exp. Bot.*, 50: 487–497.

Munkvold, L., Kjøller, R., Vestberg, M., Rosendahl, S. and Jakobsen, I. 2004. High functional diversity within species of arbuscular mycorrhizal fungi. *New Phytol.*, 164: 357–364.

Nagy, R., Drissner, D., Amrhein, N., Jakobsen, I. and Bucher, M. 2009. Mycorrhizal phosphate uptake pathway in tomato is phosphorus-repressible and transcriptional regulated. *New Phytol.*, 181: 950–959.

Nagy, R., Karandashov, V., Chague, V., Kalinkevich, K., Tamasloukht, M., Xu, G., Jakobsen, I., Levy, A.A., Amrhein, N. and Bucher, M. 2005. The characterization of novel mycorrhiza-specific phosphate e-transporters from *Lycopersicon esculentum* and *Solanum tuberosum* uncovers functional redundancy in symbiotic phosphate transport in solanaceous species. *Plant J.*, 42: 236–250.

Nagy, R., Vasconcelos, M.J., Zhao, S., McElver, J., Bruce, W., Amrhein, N., Raghothama, K.G. and Bucher, M. 2006. Differential regulation of five Pht1 phosphate transporters from maize (*Zea mays* L.) *Plant Biol. (Stuttg).*, 8: 186–197.

Neumann, G. and Romheld, V. 2002. Root-induced changes in the availability of nutrients in the rhizosphere. In: Waisel, Y., Eshel, A. and Kafkafi, U. (Eds). Plant Roots: The Hidden Half. Marcel Dekker, New York, pp. 617–649.

Parewa, H.P., Rakshit, A., Rao, A.M., Sarkar, N.C. and Raha, P. 2010. Evaluation of maize cultivars for phosphorus use efficiency in an Inceptisol. *Int. J. Agri. Environ. Biot.*, 3: 195–198.

Paszkowski, U., Kroken, S., Roux, C. and Briggs, S.P. 2002. Rice phosphate transporters include an evolutionarily divergent gene specifically activated in arbuscular mycorrhizal symbiosis. *Proc. Nat. Acad. Sci. USA*, 99: 3324–13329.

Pierzynski, G.M., McDowell, R.W. and Sims, J.T. 2005. Chemistry, cycling, and potential movement of inorganic phosphorus in soils. In: Sims, T.J. and Sharpley, A.N. (Eds.), Phosphorus: agriculture and the environment, Agronomy. Series # 46. Soil Science Society of America, Madison, WI, pp. 53–86.

Plenets, D., Mollier, A. and Pellerin, S.I. 2000. Growth analysis of maize field crops under phosphorus deficiency. II. Radiation use efficiency, biomass accumulation and yield components. *Plant Soil*, 224: 259–272.

Quiquampoix, H. and Mousain, D. 2005. Enzymatic hydrolysis of organic phosphorus. In: Turner, B.L., Frossard, E., Baldwin, D.S. (Eds.). Organic phosphorus in the environment. CAB International, Wallingford UK, pp. 89–112.

Raghuwanshi, R. and Upadhyay, R.S. 2004. Performance of vesicular-arbuscular mycorrhizae in saline-alkali soil in relation to various amendments. *World J. Microb. Biot.*, 20: 1-5.

Rai, A., Rai, S. and Rakshit, A. 2013. Mycorrhiza-mediated phosphorus use efficiency in plants. *Environ. Exp. Bot.*, 11: 107–117.

Rakshit, A. and Bhadoria, P.B.S. 2009. Influence of arbuscular mycorrhizal hyphal length on simulation of P influx with the mechanistic model. *Afr. J. Microbiol. Res.* 3: 1–4.

Rakshit, A., Bhadoria, P.B.S. and Das, D.K. 2002. An overview of mycorrhizal symbioses. *J. Interacademecia*, 6: 570–581.

Rakshit, A., Bhadoria, P.B.S. and Mittra, B.N. 2002. Nutrient use efficiency for bumper harvest. *Yojana*, Delhi, May, pp. 12–15.

Rausch, C., Daram, P., Brunner, S., Jansa, J., Laloi, M., Leggewie, G., Amrhein, N. and Bucher, M. 2001. A phosphate transporter expressed in arbuscule-containing cells in potato. *Nature*, 414: 462–466.

Rengel, Z. and Marschner, P. 2005. Nutrient availability and management in the rhizosphere: exploiting genotypic differences. *New Phytol.*, 168: 305–312.

Reyes, I., Bernier, L. and Antoun, H. 2002. Rock phosphate solubilization and colonization of maize rhizosphere by wild and genetically modified strains of *Penicillium rugulosum*. *Microb. Ecol.*, 44: 39–48.

Richardson, A.E. and Simpson, R.J. 2011. Soil microorganisms mediating phosphorus availability. *Plant Physiol.*, 156: 989–996.

Richardson, A.E., Lynch, J.P., Ryan, P.R., Delhaize, E. and Smith, F.A. 2011. Plant and microbial strategies to improve the phosphorus efficiency of agriculture. *Plant Soil*, 349: 121–156.

Richardson, A.E. 1994. Soil microorganisms and phosphorus availability. In: Pankhurst, C.E., Doubeand, B.M. and Gupta, V.V.S.R. (Eds). Soil biota:

management in sustainable farming systems. CSIRO, Victoria, Australia, pp. 50–62.

Rigou, L. and Mignard, E. 1994. Factors of acidification of the rhizosphere of mycorrhizal plants. Measurement of pCO2 inthe rhizosphere. *Acta Bot. Gall.,* 141: 533–539.

Roberts, N.J., Morieri, G., Kalsi, G., Rose, A., Stiller, J., Edwards, A., Xie, F., Gresshoff, P.M., Oldroyd, G.E., Downie, J.A. and Etzler, M.E. 2013. Rhizobial and mycorrhizal symbioses in *Lotus japonicus* require lectin nucleotide phosphohydrolase, which acts upstream of calcium signaling. *Plant Physiol.,* 161: 556–567.

Rouached, H., Arpat, A.B. and Poirier, Y. 2010. Regulation of phosphate starvation responses in plants: Signaling players and cross-talks. *Mol. Plant Biol.,* 3: 288–299.

Rubio, V., Bustos, R., Irigoyen, M.L., Cardona-López, X., Rojas-Triana, M.and Pazres. J. 2009. Plant hormones and nutrient signaling. *Plant Mol. Biol.* 69: 361–373.

Ruyter-Spira, C., Kohlen, W., Charnikhova, T., van-Zeijl, A., van-Bezouwen, L., de- Ruijter, N., Cardoso, C., Lopez-Raez, J.A., Matusova, R. and Bours, R. 2011. Physiological effects of the synthetic strigolactoneanalog GR24 on root system architecture in *Arabidopsis*: another belowground role for strigolactones? *Plant Physiol.,* 155: 721–734.

Sabannavar, S. J. and Lakshman, H.C. 2009. Effect of rock phosphate solubilization using mycorrhizal fungi and phosphor-bacteria on two high yielding varieties of *Sesamum indicum* L. *World J. Agri. Sci.,* 5: 470–479.

Saber, K., Nahla, L.D. and Chedly, A. 2005. Effect of P on nodule formation and N fixation in bean.*Agron. Sustain. Dev.,* 25: 389–393.

Saxena, A., Raghuwanshi, R. and Singh, H.B. 2015. *Trichoderma* species mediated differential tolerance against biotic stress of phytopathogens in *Cicer arietinum* L. *J. Basic Microbiol.,* 55: 195–206.

Schindler, D.W., Hecky, R.E., Findlay, D.L., Stainton, M.P., Parker, B.R., Paterson, M.J., Beaty, K.G., Lyng, M. and Kasian, S.E.M. 2008. Eutrophication of lakes cannot be controlled by reducing nitrogen input: results of a 37-year whole-ecosystem experiment. *Proc. Nat. Acad. Sci. U.S.A.,* 105: 11254–11258.

Schnepf, A., Leitner, D., Klepsch, S., Pellerin, S. and Mollier, A. 2011. Modelling phosphorusdynamics in the soil-plant system. In: Bünemann, E.K., Obserson, A., Frossard, E. (Eds). Phosphorus in Action: Biological Processes in Soil Phosphorus Cycling. Springer, Heidelberg, pp. 113–133.

Schwartz, M.W., Hoeksema, J.D., Gehring, C.A., Johnson, N.C., Klironomos, J.N., Abbott, L.K. and Pringle, A. 2006. The promise and the potential consequences of the global transport of mycorrhizal fungal inoculum. *Ecol. Lett.,* 9: 501–515.

Sharma, S.B., Sayyed, R.Z., Trivedi, M.H. and Gobi, T. A. 2013. Phosphate solubilizing microbes: Sustainable approach for managing phosphorus deficiency in agricultural soils. *Springer Plus,* 2: 587.

Simpson, R.J., Oberson, A., Culvenor, R.A., Ryan, M.H. and Veneklaas, E.J. 2011. Strategies and agronomic interventions to improve the phosphorus-use efficiency of farming systems. *Plant Soil*, 349: 89–120.

Singh, H.B. 2006. Achievements in biological control of diseases with antagonistic organisms at National Botanical Research Institute, Lucknow. In: Ramanujanand, B. and Rabindra, R.J. (Eds.). Current status of Biological Control of Plant Diseases using Antagonistic Organisms in India. Technical Document, Project Directorate of Biological Control, Bangalore, India. pp. 329-340.

Smith, S.E., Dickson, S., Smith, F.A. 2001. Nutrient transfer in arbuscular mycorrhizas: how are fungal and plant processes integrated? *Aust. J. Plant Physiol.*, 28: 685–696.

Smith, S.E., Jakobsen, I., Grønlund, M. and Smith, F.A. 2011. Roles of arbuscular mycorrhizas in plant phosphorus nutrition: interactions between pathways of phosphorus uptake in arbuscular mycorrhizal roots have important implications for understanding and manipulating plant phosphorus acquisition. *Plant Physiol.*, 156: 1050–1057.

Smith, S.E. and Read, D.J. 2008. Mycorrhizal Symbiosis, Edn 3. Academic Press, New York.

Smith, S.E., Robson, A.D. and Abbott, L.K. 1992. The involvement of mycorrhizas in assessment of genetically dependent efficiency of nutrient uptake and use. *Plant Soil*, 146: 169–179.

Smith, S.E., Smith, F.A., Jakobsen, I. 2004. Functional diversity in arbuscular mycorrhizal (AM) symbioses: the contribution of the mycorrhizal P uptake pathway is not correlated with mycorrhizal responses in growth or total P uptake. *New Phytol.*, 162: 511–524.

Stewart, J.W.B. and Sharpley, A.N. (1987). Controls on dynamics of soil and fertilizer phosphorus and sulfur. In: Mortdvedt, J.J. and Buxton, D.R. (Eds.). Soil Fertility and organic matter as critical components of production systems. Soil Science Society of America and American Society of Agronomy Publ, #19, Madison, WI, pp. 101–112.

Stewart, J.W.B. and Tiessen, H. 1987. Dynamics of soil organic phosphorus. *Biogeochem.*, 4: 41–60.

Sturmer, S.L. 2012. A history of the taxonomy and systematics of arbuscular mycorrhizal fungi belonging to the phylum Glomeromycota. *Mycorrhiza*, 22: 247–258.

Tarafdar, J.C. and Marschner, H. 1994. Phosphatase activity in the rhizosphere and hyphosphere of VA-mycorrhizal wheat supplied with inorganic and organic phosphorus. *Soil Biol. Biochem.*, 26: 387–395.

Tejesvi, M.V., Ruotsalainen, A.L., Markkola, A.M. and Pirttilä, A.M. 2010. Root endophytes along a primary succession gradient in northern Finland. *Fungal Divers.*, 41: 125–134.

Thomson, B.D., Clarkson, D.T. and Brain, P. 1990. "Kinetics of Phosphorus Uptake by the Germ-tubes of the Vesicular- Arbuscular Mycorrhizal Fungus, *Gigaspora magarita"*. *New Phytol.*, 116: 647–653.

Tilak, K.V.B.R., Ranganayaki, N., Pal, K.K., De, R., Saxena, A.K., Nautiyal, C.S., Mittal, S., Tripathi, A.K. and Johri, B.N. 2005. Diversity of plant growth and soil health supporting bacteria. *Cur. Sci.*, 89: 136–150.

Tilman, D., Fargione, J., Wolff, B., D'Antonio, C., Dobson, A., Howarth, R., Schindler, D., Schlesinger, W.H., Simberloff, D. and Wackhamer, D. 2001. Forecasting agriculturally driven global environmental change. *Sci.*, 292: 281–284.

Van Kauwenbergh, S.J. 2010. World phosphate rock reserves and resources, IFDC, Muscle Shoals, AL.

Vassilev, N., Vassileva, M., Azcon, R. and Medina, A. 2001. Preparation of gel-entrapped mycorrhizal inoculum in the presence or absence of *Yarrowia lypolytica*. *Biotechnol. Lett.* 23: 907–909.

Venkateswarlu, B., Rao, A.V., Raina, P. and Ahmad, N. 1984. Evaluation of phosphorus solubilization by microorganisms isolated from arid soil. *J. Indian Soc. Soil Sci.*, 32: 273–277.

Walder, F., Brule, D., Koegel, S., Wiemken, A., Boller, T. and Courty, P.E. 2015. Plant phosphorus acquisition in a common mycorrhizal network: regulation of phosphate transporter genes of the Pht1 family in sorghum and flax. *New Phytol.*, 205: 1632–1645.

Wang, B., Yeun, L.H., Xue, J.Y., Liu, Y. and Ane Qiu, Y.L. 2010. Presence of three mycorrhizal genes in the common ancestor of land plants suggests a key role of mycorrhizas in the colonization of land by plants. *New Phytol.*, 186: 514-525.

Wegmueller, S., Svistoonoff, S., Reinhardt, D., Stuurman, J., Amrhein, N. and Bucher, M. 2008. A transgenic dTph1 insertional mutagenesis system for forward genetics in mycorrhizal phosphate transport of Petunia. *Plant J.*, 54: 1115-1127.

Widada, J., Damarjaya, D.I. and Kabirun, S. 2007. The interactive effects of arbuscular mycorrhizal fungi and rhizobacteria on the growth and nutrients uptake of sorghum in acid soil. In: Rodriguez-Barrueco C. and Velazquez, E. (Eds.). First International Meeting on Microbial Phosphate Solubilization, Springer, pp. 173–177.

Yang, S.Y. and Paszkowski, U. 2011. Phosphate import at the arbuscule: Just a nutrient? *Mol. Plant Microb.*, 24: 1296–1299.

Yang, S-Y., Gronlund, M., Jakobsen, I., Grotemeyer, M.S., Rentsch, D., Miyao, A., Hirochika, H., Kumar, C.S., Sundaresan, V. and Galanihi, N. 2012. Non-redundant regulation of rice arbuscular mycorrhizal symbiosis by two members of the PHOSPHATE TRANSPORTER1 gene family. *Plant Cell*, 24: 4236–4251.

Zhang, Q., Blaylock, L.A. and Harrison, M. J. 2010. Two *Medicago truncatula* half- ABC transporters are essential for arbuscules development in arbuscular mycorrhizal symbiosis. *Plant Cell*, 22: 1483–1497.

Zhou, K., Binkley, D. and Doxtader, K.G. 1992. A new method for estimating gross phosphorus mineralization and immobilization rates in soils. *Plant Soil,* 147: 243–250.

Zhu, F., Qu, L., Hong, X. and Sun, X. 2011. Isolation and characterization of a phosphate-solubilizing halophilic bacterium *Kushneria* sp. YCWA18 from Daqiao Saltern on the coast of Yellow Sea of China. *Evid. Based Complement Alternat. Med.,* 615032.

Zhu, J., Zhang, C. and Lynch, J.P. 2010. The utility of phenotypic plasticity for root hair length for phosphorus acquisition. *Funct. Plant Biol.,* 37: 313–322.

— *Part IV* —
Arbuscular Mycorrhizal Fungi and Stress Tolerance

2017, Mycorrhizal Fungi
Editors: **Ashok Aggarwal and Kuldeep Yadav**
Published by: **ASTRAL INTERNATIONAL PVT. LTD., NEW DELHI**

Pages **247–273**

11

Arbuscular Mycorrhizal Symbiosis: A Boon for Sustainable Legume Production under Salinity and Heavy Metal Stress

Neera Garg*, Purnima Bhandari, Lakita Kashyap and Sandeep Singh

*Department of Botany,
Panjab University, Chandigarh – 160 014
Corresponding Author: garg_neera@yahoo.com; gargneera@gmail.com

ABSTRACT

The problem of soil degradation and loss of arable lands due to heavy metal (HM) toxicity and/or by salinity have recently attracted attention of the scientists worldwide. Moreover, production of legumes, which are the major contributors of nitrogen input in sustainable agro-ecosystem, have been reported to be seriously affected due to the presence of these contaminants in the soils. Various approaches have been adopted by scientists to reclamate disturbed soils, among which mycorrhiza-assisted remediation (MAR) has gained interest in the last decade. Arbuscular mycorrhizal (AM) fungi establish mutualistic symbiosis with the majority of higher plants and have been reported to contribute to the plant growth in disturbed soils by restricting the movement of toxicants within plant tissues and increasing the plant access to various other immobile nutrients. Legumes have the ability to form effective symbioses with both AM fungi and Rhizobium thereby impart functional complementarity for sustainable legume production. This review unfolds recent advances in mitigating abiotic constraints at biochemical, physiological and molecular levels.

Keywords: AM fungi, Legume, Salinity, Heavy metal stress.

1. Introduction

Legumes, belonging to family *Fabaceae*, constitute a major portion of human food throughout the world and are highly suitable for agroforestry ecosystem. In addition, by associating with *Rhizobium*, they are major contributors of fixed atmospheric nitrogen (N_2), consequently leading to an increase in crop growth and yield especially in conventional or derelict soils (Perveen *et al.*, 2015). Nevertheless, legumes are constantly exposed to various abiotic and biotic stresses, including drought stress, salinity stress, metal stress, ionic toxicity *etc.* which have been reported to have devastating impacts on plant growth and productivity besides causing deterioration in the quality of plant product (Wani *et al.*, 2016). According to Acquaah (2007), approximately 70 per cent of worldwide reduction in agricultural yield is due to the direct effect of abiotic stresses. Moreover, combinations of two or more abiotic stresses can occur concurrently which aggravates the severity of the problem (Wani *et al.*, 2016). Among various stresses, two common environmental changes that pose challenges for land managers in both agricultural as well as industrialized areas are soil salinization and the contamination of soils with heavy metals (HMs; Emamverdian *et al.*, 2015; Moray *et al.*, 2016).

Salinity is a result of long-term natural build-up of salts in the soil or in surface water or from irrigation and is widely responsible for increasing the concentration of dissolved salts in the soil profile to a level that impairs plant growth and result in abandoning agricultural land (Munns, 2005; Egamberdiyeva *et al.*, 2007; Manchanda and Garg, 2008). Of all the salts, sodium chloride (NaCl) is the most soluble and abundant salt released making it an imperative contributor to soil salinity (Munns and Tester, 2008). Over 800 million hectare (ha) of land throughout the world are salt affected, either by salinity (397 million ha) or by sodicity (434 million ha) (FAO, 2005; Manchanda and Garg, 2008; Munns and Tester, 2008). In Asia alone, approximately 21.5 million ha of land area is salt affected, with India having 6.7 million ha of saline area (Mandal *et al.*, 2009; Nazar *et al.*, 2011). Salinity induces an initial *osmotic stress* that occurs due to the reduction in the osmotic potential of the soil solution which reduces the amount of water available to the plant, causing *'physiological drought'* (Jahromi *et al.*, 2008); second is the *ion specific effect* that occurs due to the accumulation of toxic ions, such as Na^+ and Cl^- which negatively affects cellular metabolism (Munns *et al.*, 2006); third effect which is the outcome of two effects includes *oxidative stress* that occurs due to the excessive generation of reactive oxygen species (ROSs). This disrupts the structure of enzymes, alters the rate of photosynthesis and respiration, damages cell organelles and plasma membrane and cause significant disruption of metabolic processes including protein synthesis (Porcel *et al.*, 2012). In addition, under stressed conditions, saline plants take up excessive amounts of Na^+ at the cost of K^+ and Ca^{2+} that disturbs ionic equilibrium (Na^+/Ca^{2+} and Na^+/K^+ ratios) and cause *nutrient imbalance* thereby decreasing nutrient uptake and/or transport from root to the shoot (Munns and Tester, 2008; Hussain *et al.*, 2010; Hasanuzzaman *et al.*, 2011a, b; Fageria *et al.*, 2011; Plaut *et al.*, 2013; Garg and Pandey, 2015).

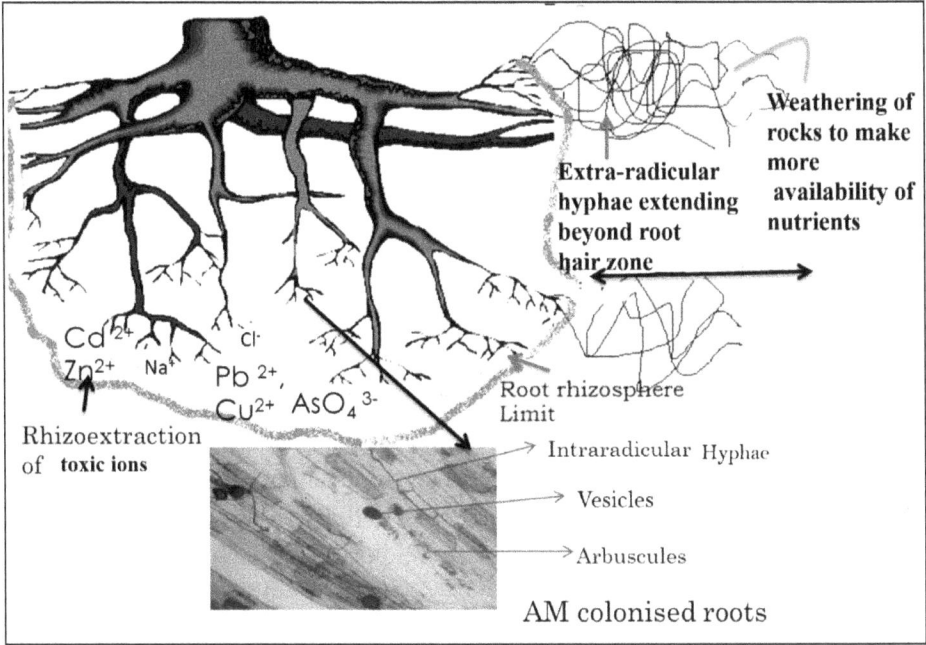

Figure 11.1: Extra Radicular Hyphae Extend Beyond the Zone of Root Hair and Improve the Availability of Mineral Nutrient by Weathering of Rocks or by making Unavailable Forms to available Form. Further HMs are extracted and sequestered in fungal parts or are stabilized in soil by forming complexes with organic aids released by the plants.

In addition, soils are frequently contaminated with HMs such as (cadmium - Cd, copper - Cu, nickel - Ni and zinc - Zn) or with metalloids (such as aluminium - Al, arsenic - As and selenium - Se) that occur due to the rapid development of various industries related to mining, fertilizer, pesticide and leather, which discharge wastes containing HMs directly or indirectly into the soil (Wang and Chen, 2006; Singh *et al.*, 2016). The term "heavy metals" is a general collective term, which applies to the group of metals and metalloids with atomic density greater than 4 g/cm^3, or five times or more, are greater than water (Hawkes, 1997; Kalaivanan and Ganeshamurthy, 2016). At high concentrations, HMs have been reported to cause many morphological, physiological, biochemical and structural changes in plants including growth inhibition, disturbance in cellular homeostasis and cause inhibition of seed germination (Benavides *et al.*, 2005; Goncalves *et al.*, 2007). They interfere with essential enzymatic activities by modifying protein structure or by replacing a vital element, resulting in deficiency symptoms (Muleta and Woyessa,

2012). In addition, HMs have been reported to generate ROSs, thereby causing deterioration of bio-membranes (Yadav, 2010). Some of HMs including lead (Pb), mercury (Hg), silver (Ag) are bio-accumulative in nature which neither break down in the environment nor get easily metabolized, but get accumulated in the ecological food chain through uptake at primary producer level and then through consumption at consumer levels (Kalaivanan and Ganeshamurthy, 2016). When taken in excessive amount, these toxicants (Na^+, Cl^-, Cd, Zn, As) in the soil causes undeniable damage to legume-rhizobial symbiosis and have been reported to affect the survival and ability of rhizobia to form N_2 fixing nodules (Garg and Manchanda, 2008; Arora *et al.*, 2009; Corticerio et al.,2012; Garg and Bhandari, 2012; Garg and Kaur, 2012). For instance, increasing concentration of As delayed nodulation time and decreased number of nodules per plant (Reichman, 2007). In addition, authors also reported that As treatment led to the poor development of root hairs in the inoculated plants and substantially decreased dry matter contents in roots and shoots. In another study, Pajuelo *et al.* (2008) demonstrated that HMs caused a significant decrement in the length of root and decreased number and size of root hairs in *Vigna radiata* and *Medicago sativa* respectively, which are the essential steps for nodulation process. Similar results were revealed by Garg and Kaur (2012) who noted that build-up of Cd and Zn in nodules of *Cajanus cajan* genotypes resulted in sharp reduction in nodule number, nodule dry mass as well as N_2 fixation potential (in terms of leghemoglobin and nitrogenase), with sensitive genotype displaying higher negative effect of metal toxicity when compared with that of tolerant genotype.

Improving plant stress tolerance and upholding crop productivity against such abiotic stresses is a major challenge for sustainable agriculture (Kapoor *et al.*, 2013). Besides the intrinsic capacity of plants to tolerate abiotic stresses, extensive research has been carried out worldwide to evolve efficient, low cost, easily adaptable methods for the abiotic stress management which includes development of tolerant plant species, resource management practices, *etc.* (Grover *et al.*, 2011). Although, most of these approaches are cost-intensive, recent studies have indicated the usage of soil-borne microorganisms in alleviating these stresses. In general, microhabitat of the rhizosphere is a specialised ecosystem, where microbial populations are highly favoured, with whom plants are able to establish beneficial associations thus, acquiring tolerance and/or resistance to stress factors (Singh *et al.*, 2011; Folli-Pereira *et al.*, 2013). Amongst these, arbuscular mycorrhizal (AM) fungi have received considerable attention due to their extensive benefits to host plant. AM fungi, belonging to the phylum Glomeromycota (Schüβler *et al.*, 2001, 2014) are able to establish a symbiotic association with the roots of 80 per cent of terrestrial plants (van der Heijden *et al.*, 1998; Smith and Read, 2008), wherein the symbionts gain all of their carbon from a host plant, whilst delivering a range of benefits to the plant, such as improved nutrient acquisition and resistance to pathogens and abiotic stress (Smith and Read, 2008; Smith *et al.*, 2009). In addition, AM fungi are able to establish "tripartite symbiosis" with *Rhizobium*-legume roots (Lodwig *et al.*, 2003; Zaidi *et al.*, 2003; Garg and Manchanda, 2008), thereby strengthening the nodulation frequency, N_2 fixation efficiency of host plant that ultimately results in better nutrient uptake and plant yield in mycorrhizal plants compared to plants which are associated with either of the two (Barea *et al.*, 2002; van der Heijden *et al.*, 1998, 2006). Under abiotic stress conditions, AM fungi enhances plant growth, productivity and nutrient acquisition by influencing several physiological as well as metabolic plant processes including plant–water relation, rate of photosynthesis, ionic balance, antioxidant production, accumulation of compatible solutes, thus improve plant's capacity to withstand abiotic stresses (Garg and Chandel, 2010; Ruiz-Lozano *et al.*, 2012; Evelin *et al.*, 2012; Kapoor *et al.*, 2013). The purpose of this chapter is to outline briefly the current state of knowledge about the mode of AM establishment in nature, potential of AM fungi on sustainable crop production under contaminated areas *via* which AM fungi imparts stress tolerance towards salinity and HM toxicity, especially in legumes. Besides, the effects of metal contamination and soil salinity on the viability and inoculation capability of mycorrhizal fungi and the effect of different combinations of fungi species or isolates on the beneficial effect of mycorrhizal symbiosis under stressful conditions are discussed. Finally, some testimonials on the changes in the expression pattern of genes involve in stress response of plants are presented under AM fertilizations.

2. AM Symbiosis: Establishment

The development of a functional AM symbiosis is a multi-step process which involves a sequence of recognition events leading to the morphological and physiological integration of two symbionts (Gianinazzi *et al.*, 1995; Giovannetti and Sbrana, 1998; Garg and Chandel, 2010; Kapoor *et al.*, 2013). AM–host symbiotic association can functionally be divided into asymbiotic, presymbiotic and symbiotic stages (Smith and Read, 2008). In the asymbiotic phase, the establishment of AM symbiosis begins with the colonization of a compatible root by the hyphae produced by AM fungal soil propagules, asexual spores or mycorrhizal roots (Requena *et al.*, 1996). Under favourable conditions, plant host roots release strigolactones (SLs) which act as a "rhizospheric plant signals" to the developing spores and hyphal branching and are able to trigger activities at molecular levels including the expression of symbiotic genes (Badri *et al.*, 2009; Garg and Chandel, 2010). Conversely, when the AM fungus starts to branch in the vicinity of the root, plants perceive diffusible fungal signals, called "Myc factors", that induce symbiosis-specific responses in the host root, even in the absence of any physical contact (Parniske, 2008; Genre and Bonfante, 2010; Barea *et al.*, 2014). Pre-symbiotic stage includes the formation of appressorium which is an enlarged extra-radical hyphal tip that attaches to the surface of the host plant root (Hajiboland, 2013). During the last stage *i.e. symbiotic stage*, AM fungi form extra-radical and intraradical structures, among which tree-shape structures – *arbuscules* are formed within the cortical cells which increases the contact area between plant and fungus and are considered to be the primary area of exchange (Calvo-Polanco *et al.*, 2013). In addition, by forming an extensive hyphal network of extra-radical hyphae, AM fungi increase the total absorptive surface area of root and help in acquiring nutrients even beyond the depletion zone which develops around plant roots (George *et al.*, 1992; Smith and Read, 2008; Kapoor *et al.*, 2013). In some case, vesicles are also formed during mycorrhizal symbiosis which serve as carbon (C) storage compartments for the fungi (Smith and Read, 2008; Hajiboland, 2013) and help the host plant under different conditions including stress (Harrison *et al.*, 1999). The extra-radical mycelium grows out of the root exploring the soil in search of mineral nutrients and displays the ability to colonize other susceptible roots (Figure 11.1).

3. Contribution of AM Fungi

Several studies have demonstrated that AM contributes to plant growth *via* enhancing the acquisition of immobile soil nutrients including P, Zn, Cu, Mn and Fe particularly in nutrient depleted soils (Smith and Read, 2008). However, AM-derived benefits accorded by plants are not only confined to the improvement in nutrient uptake but also includes adaptability of plants under adverse environmental conditions such as drought (Augé, 2001; Miransari, 2010), salinity (Evelin *et al.*, 2009; Porcel *et al.*, 2012; Ruiz-Lozano *et al.*, 2012) and HM stress (Garg and Chandel, 2012; Garg and Bhandari, 2012; Garg and Aggarwal, 2012; Garg and Singla, 2012) (Figure 11.1).

4. Tolerance of AM Fungi to Salt and Metal Stress

AM fungi are an integral component of soil microflora and their presence have been reported in severe disturbed areas including saline soils as well as metal-contaminated sites (Gaur and Adholeya, 2004; Vallino *et al.*, 2006; Leung *et al.*, 2007; Yamato *et al.*, 2008; Gamalero *et al.*, 2009; Hammer *et al.*, 2011; Amir *et al.*, 2014), thus indicating that these fungi have evolved characteristics of stress tolerance and has a potential role in bioremediation (Gildon and Tinker, 1983). However, despite the widespread distribution of AM fungi in several ecosystems, less than 250 species have been reported till date (Öpik *et al.*, 2013; Lenoir *et al.*, 2016). Moreover, in case of disturbed soils, exposed to various abiotic stresses, diversity of mycorrhizal fungi is generally found to be lower when compared with those in undisturbed soils. Among the different AM species, variations have been reported by several authors demonstrating the differential response of mycorrhizal fungi to salt and metal toxicity (Hildebrandt *et al.*, 2007; Muleta and Woyessa, 2012; Cicatelli *et al.*, 2013) Various studies have validated that contaminated sites in different parts of the world are dominated by *Glomeraceae* and most commonly occurring AM fungi in saline and metal stressed soils are *Glomus* spp. (Allen and Cunningham, 1983; Pond *et al.*, 1984; Ho, 1987; Krishnamoorthy *et al.*, 2015; Lenoir *et al.*, 2016). Among various *Glomus* spp., *G. fasciculatum, Rhizophagus irregularis* (formely *G. intraradices*), *G. etunicatum and Funneliformis mosseae* (formerly *G. mosseae*) are the most reported species that occur in contaminated soils along with some other mycorrhizal species including *Acaulospora and Gigaspora* at low frequencies (Khade and Adholeya, 2007). The occurrence of spores of *Glomus* spp. in the saline and metal stressed soils suggests that these fungi are the dominant root colonizers in such disturbed soils (Landwehr *et al.*, 2002; Khade and Adholeya, 2007; Grover *et al.*, 2011; Krishnamoorthy *et al.*, 2015).

AM fungal tolerance to HM and salt stress depends upon type of AM isolate, the physico-chemical properties, concentration of metal and amount of bioavailability/extractability of metals (Leyval *et al.*, 1997). Some workers have demonstrated that exotic fungal species have more tolerance to salt and metal stress than native species. Recently Garg and Pandey (2016) documented that exotic AM fungi (*R. irregularis*) displayed higher tolerance, colonization and efficiency than the native AM fungal species under stress conditions. On the other hand, few studies have showed greater benefits of native AM fungi than the non-native/exotic AM fungal isolates under metal stressed conditions (Gaur and Adholeya, 2004).The reason appeared to be physiologically and genetically adaptation of these species to the stress conditions (Querejeta *et al.*, 2006; Pellegrino *et al.*, 2011; Estrada *et al.*, 2013b; Pellegrino and Bedini, 2014). However, several researchers have documented that soil salinity and metal toxicity reduce spore germination, colonization capacity and growth of fungal hyphae which they attributed to be due to the direct effect on the fungi or due to indirect effect on host root inhibition (Repetto *et al.*, 2003; Cittero *et al.*, 2005; Lingua *et al.*, 2008; Garg and Manchanda, 2008; Evelin *et al.*, 2012, Hajiboland 2013). The available data on the effects of salt stress on germination pattern of AM spores indicate that majority of stress effects occur due to the inhibition of spore germination by increasing concentrations of toxic ions (Hirrel, 1981; Estaun, 1989, 1991; Juniper and Abbott, 1993, 2006; as reviewed by Hajiboland, 2013). However, the dependency of host plant on AM symbiosis increases once the association is established, indicating the ecological significance of AM under abiotic stresses (Kumar *et al.*, 2010; Miransari, 2010). The responses of AM fungi depend on both host and fungal species under stress conditions. Several salt and metal-tolerant AM fungal species have been isolated from polluted soils (Turnau *et al.*, 2001) and are known to show better resistance to these stresses than those from uncontaminated soils (Aliasgharzadeh *et al.*, 2001; Gonzalez-Chavez *et al.*, 2002; Malcova *et al.*, 2003; Wu *et al.*, 2010). Reduction in AM colonization with increasing salt concentrations have been reported in many plant species including *Lotus glaber* (Sannazzaro *et al.*, 2006), *Acacia nilotica* (Giri *et al.*, 2003, 2007) and *Lycopersicon esculentum* (Zhi *et al.*, 2010; Hajiboland *et al.*, 2010). Similar results of toxic metal ions were witnessed in case *Zea mays* (Lins *et al.*, 2006; Massoud, *et al.*, 2012), *Glycine max* (Spagnoletti and Lavado, 2015), *Medicago sativa* (Kanwal *et al.*, 2015), *Solanum photeinocarpum* (Tan *et al.*, 2015) and *Cajanus cajan* and *Pisum sativum* (Garg *et al.*, 2015). Despite the negative correlations obtained between salt or metal concentrations in soil and mycorrhizal symbiosis, AM fungal propagules never disappear completely from the soils (Biro *et al.*, 2009; Gamalero *et al.*, 2009), thereby suggesting that these fungi are extremely adapted and are tolerant to different abiotic stresses (Aliasgharzadeh *et al.*, 2001; Khade and Adholeya, 2007; Zarei *et al.*, 2008; Wu *et al.*, 2010).

5. AM Fungi: Stress Ameliorators

The role of AM fungi in mitigating stress-induced effects of metal toxicity and salinity in plants is widely recognized. During the last two decades, large number of studies have focused on the mechanisms through which influence of mycorrhiza on mitigation of HM toxicity as well as salinity in plants could be explained (Garg and Chandel, 2012; Garg and Kaur, 2012, 2013a, b; Ruiz-Lozano *et al.*, 2012; Hajiboland, 2013; Garg and Pandey, 2015; Garg and Bhandari, 2016; Garg and Singla, 2016). AM mediates alleviation of stress effects and improves tolerance of host plants by combining nutritional (enhancing/selective uptake of nutrients and prevention of nutritional disorder), biochemical (accumulation of osmolytes, detoxifies ROSs *via* up-regulating enhanced activities of antioxidant enzymes and molecules), physiological (photosynthetic efficiency) and structural adaptations (Kapoor *et al.*, 2013) which is being discussed briefly under the following sub-sections (Figure 11.2):

5.1. Role of AM Fungi under HM Stress

Although considerable variability in plant responses to mycorrhizal inoculation has been observed in contaminated soils, the potential of AM fungi to buffer metal toxicity have been thoroughly investigated and demonstrated by several authors (Hildebrandt *et al.*, 2007; Upadhyaya *et al.*, 2010; Garg and Aggarwal, 2011; Garg and Chandel, 2012; Muleta and Woyessa, 2012; Garg and Bhandari, 2013; Garg and Kaur, 2013a, b). In one of the green house experimental study, Lin *et al.* (2007) reported that colonization with *F. mosseae* increased the growth of three leguminous plants (*Sesbania rostrata, S. cannabina, Medicago sativa*) indicating that mycorrhization confer plant's resistance to HM stress. Similar reports were documented by Garg and Chandel (2011) and Garg and Kaur (2012) in *Cajanus cajan* genotypes where *F. mosseae* alleviated the toxic effects of Cd and Zn and enhanced plant productivity under stressful environments. Collectively, AM fungi are able to maintain an efficient symbiosis with leguminous plants in soil containing high metal concentrations (Garg and Bhandari, 2012; Garg and Kaur, 2013b). In one of the study, Garg and Kaur (2013b) reported that *F. mosseae* substantially reduced the transport of Cd and Zn in both shoots and seeds of differentially tolerant *Cajanus cajan* genotypes, leading to nominal build-up of metal ions in seeds.

One of the major mechanisms by which mycorrhizal fungi are able to alleviate metal stress includes sequestration of HM in soil and/or in roots which occurs *via* several mechanism including extracellular chelation, cell wall binding and HM

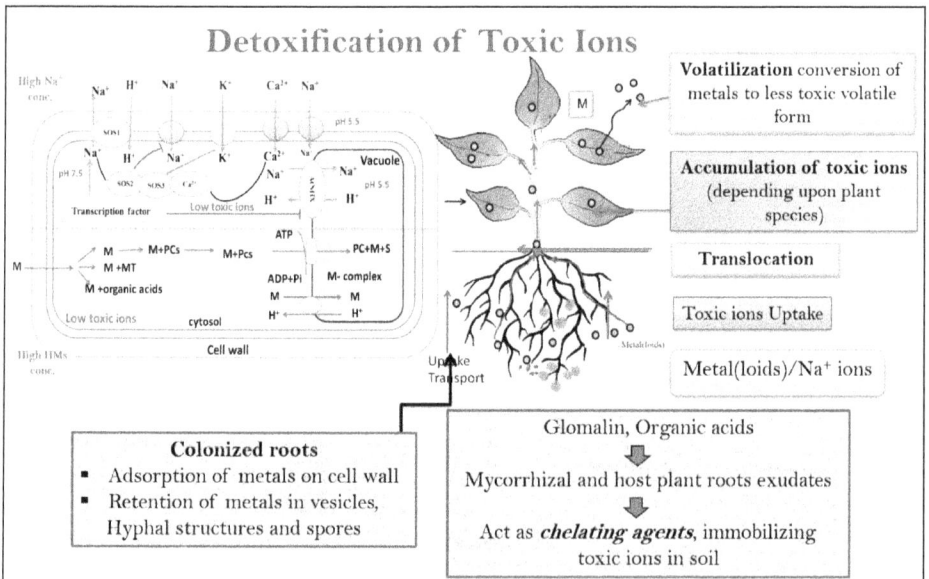

Figure 11.2: Various Detoxification Mechanisms Adopted by AM Fungi in Alleviating HM Toxicity and Salinity Stress.

accumulation in extraradical mycelium (Colpaert *et al.*, 2011; Amir *et al.*, 2014). In order to maintain physiological concentrations of essential metal ions and minimize the exposure to toxic ions, first line of defense adopted by AM includes accumulation of metal ion on wall of spores and hyphae and in the vacuoles, thus, maintaining low cytoplasmic concentrations (González-Guerrero *et al.*, 2007). Fungal cell wall which is composed of chitin containing negatively charged hydroxyls, carboxyls and amino group has been suggested to interact with metals present in soil solution (Strandberg *et al.*, 1981), thereby assist in arresting toxic metals in the soil matrix. Approximately 50 per cent of metal has been suggested to be adsorbed on fungal cell wall (Joner *et al.*, 2000). In one of the study, Christie *et al.* (2004) validated that more than 3 per cent of Zn (on dry weight basis) was found to be bound on the extracellular mycelium of *F. mosseae*, thereby suggesting its potential to alleviate Zn phytotoxicity at contaminated sites. In another study, using EDXS analyses, González-Guerrero *et al.* (2008) validated that higher concentrations of Cu, Zn and Cd were partly localized in the fungal cell wall of *R. irregularis* (Amir *et al.*, 2014). In addition, many filamentous fungi are able to sorb trace elements and are used as commercial bio-sorbants (Morley and Gadd, 1995; Nasim, 2010). Recently, *glomalin*, a glycoprotein produced by mycorrhizal fungi in soil has been suggested to act as metal chelator in influencing metal availability for plants (Wright and Upadhyay, 1998; Gonzalez-Chavez *et al.*, 2004; Ferrol *et al.*, 2009; Gamalero *et al.*, 2009; Kapoor *et al.*, 2013). In one of the study, Gonzalez-Chavez *et al.* (2004) demonstrated that one gram of glomalin was able to extract up to 0.08 mg Cd, 1.12 mg Pb and 4.3 mg Cu from the polluted sites. Several authors have also suggested reduced transfer, as indicated by enhanced root/shoot Cd ratios in AM plants (Joner *et al.*, 2000; Tullio *et al.*, 2003) which may be ascribed to intracellular precipitation of metallic cations with phosphates (Kapoor *et al.*, 2013). Moreover, AM hyphae displayed the ability to bind metals present in roots or in the rhizosphere which in turn resulted in decreased metal translocation from roots to aerial organ (Wasserman *et al.*, 1987; Muleta and Woyessa, 2012). Similar results were observed by Garg and Kaur (2013b) who disclosed that symbiosis with *F. mosseae* caused substantial immobilization of Cd and Zn contents in roots, thereby mediating lower translocation of toxic ions in both shoots and seeds of AM-colonized plants. Recently, Yang *et al.* (2016) documented that AM restricted large amount of Pb in roots, thus preventing future damage to above ground parts.

Once HM has surpassed fungal wall, other defense mechanisms which comes into play include alteration of HM influx transporter processes and an increase in metal efflux through the cell membrane (Meharg, 2003; Ouziad *et al.*, 2005; Amir *et al.*, 2014). In order to inactivate absorbed HMs, strategy of intracellular compartmentalization of metals have been documented (Amir *et al.*, 2014) in which toxic elements are translocated into fungal vacuoles where they are stored away from the cytosol (Göhre and Paszkowski, 2006). In one of the study, Ferrol *et al.* (2009) observed that mycorrhizal spores of AM fungi, grown in a Cu-enriched medium contained a high concentration of Cu. Besides, vesicles of the intraradical mycelium have also been suggested to serve for the storage of HMs (Orlowska *et al.*, 2008; Amir *et al.*, 2014). Thus, immobilization of metals on both extra- and intraradical fungal tissues provides a plausible explanation for enhancement of the barrier for metal transfer from soil to roots and finally translocation to shoots of inoculated plants (Upadhyaya *et al.*, 2010). Several metal chelators such as phytochelatins (PCs, cysteine rich enzymatically synthesized compounds), metallothioneins (MTs, gene encoded cysteine rich compounds) and glutathione (GSH, thiol rich) have been reported to be involved in metal detoxification mechanism which may binds with the metal ions so that free metal concentration is kept at lower level in order to avoid their interference with plant metabolic reactions (Javaid, 2011; Saraswat and Rai, 2011). Three glomeromycotan MTs have been identified in *Gigaspora* and *Glomus* species (Stommel *et al.*, 2001; Lanfranco *et al.*, 2002; González-Guerrero *et al.*, 2007; Amir *et al.*, 2014). GinZnT1 was identified as the first HM transporter in *R. irregularis* by González-Guerrero *et al.* (2005) which was observed to decrease the cytosolic Zn level in yeast. In addition, a putative ABC transporter has been reported by González-Guerrero *et al.* (2006) which is upregulated in the extra radical mycelium of *R. irregularis* under Cd and Cu stress. In contrast, AM colonization resulted in the down regulation of *Lemt2* and *LeNramp1* (encoding MTs) genes, presumably because the content of metals was lower in mycorrhizal plants than non-mycorrhizal ones (Gamalero *et al.*, 2009; Garg and Bhandari, 2014). In their respective study, Garg and Aggarwal (2011) and Garg and Chandel (2011) evaluated the role of *F. mosseae* in the alleviation of Cd and Pb toxicities in *Cajanus cajan* genotypes and concluded that AM fungi enhanced PCs and GSH content in stressed plants that resulted in sequestering of metal ions, thus helped legume species to grow in multi metal-contaminated soils. Moreover, the positive effects of AM fungi were genotype dependent, with tolerant one gaining more benefit than the sensitive counterpart which authors correlated with the higher rate of mycorrhizal colonization in the former than the later. Recently, *R. irregularis* has been validated to increase the expression of the phytochelatin synthase gene (PCS1) in legume plant (*Sophora viciifolia*) exposed to Pb stress, thereby, suggesting another mechanism of Pb immobilization in the roots of mycorrhizal stressed plants (Yang *et al.*, 2016). Additionally, studies have revealed that the fungus struggles to reduce the free radicals produced under metal toxicity by activating its enzymatic and non-enzymatic antioxidant systems (Ferrol *et al.*, 2009). Till date, few genes involved in oxidative stress homeostasis have been identified by several researchers in AM fungi which includes three SOD, ten genes encoding glutathione S-transferases, a glutaredoxin, a gene encoding a protein involved in vitamin B6 biosynthesis and a MT (Benabdellah *et al.*, 2009; Ferrol *et al.*, 2009; Amir *et al.*, 2014; Lenoir *et al.*, 2016). Many studies have documented that AM fungi enhance the ROS-scavenging capacity of plants by up-regulating the antioxidant defense response under metal stress conditions. For instance, in case of *Cajanus cajan* genotypes, Garg and Aggarwal (2012) demonstrated that colonization with *F. mosseae* amplified the activities of antioxidants enzymes – SOD, CAT, POX and GR, as well as maintained GSH/GSSG ratios in mycorrhizal plants, thereby reducing the synthesis of stress metabolites, more in tolerant genotype than the sensitive counterpart under Cd and Pb stress. Similar results were demonstrated by Garg and Kaur (2013a) who concluded that symbiosis with *F. mosseae* attenuated the phytotoxic effects of Cd and Zn in nodules of *Cajanus cajan* in a genotype dependent manner by decreasing metal uptake, oxidative stress and by enhancing defense antioxidant responses that improved N_2 fixing potential of nodules under stressed conditions.

Recently, Jiang *et al.* (2016) observed that colonization with *F. mosseae* ameliorated the phytotoxic effects of Cd by enhancing P acquisition, up-regulating antioxidant activities and ultimately improved growth and phytoremediation efficiency of AM-inoculated *Solanum nigrum* plants. Indirectly, mycorrhizal symbiosis improves plant establishment and growth in metal contaminated soils by influencing pH (Li *et al.*, 1991b), enhancing water availability (Auge, 2001), better P nutrition (Feng *et al.*, 2003) and soil aggregation properties (Rillig and Steinberg, 2002) which ultimately contributes to the dilution of metal concentrations in plant tissues.

5.2. Role of AM Fungi under Salinity Stress

Various studies have demonstrated that mycorrhizal colonization alleviates NaCl-induced retardation of plant growth and fitness (Giri *et al.*, 2007; Evelin *et al.*, 2012; Garg and Pandey, 2015; Garg and Bhandari, 2016). The increased dependency of host plants on mycorrhizal fungi under salinity is an evidence indicating that AM fungi enhances plant growth by improving nutrient uptake, especially phosphorus (P), influences water absorption and increases photosynthetic system (Giri *et al.*, 2003; Garg and Manchanda, 2009; Evelin *et al.*, 2012; Kapoor *et al.*, 2013). Several studies have validated that the main mechanism for enhanced salt tolerance in mycorrhizal plants seems to be AM-mediated improvement in nutrient uptake and translocation under both stressed as well unstressed conditions (Rabie and Almadini, 2005; Garg and Manchanda, 2009; Ruiz-Lozano *et al.*, 2012; Estrada *et al.*, 2013a, b; Garg and Pandey, 2015). As discussed previously, the extended network of fungal hyphae allows mycorrhizal roots to explore more soil volume beyond the depletion zones around roots, thus assists in acquisition of nutrients (Smith and Read, 2008). Among the various nutrients, enhancement of plant P uptake by AM fungi has been frequently reported as one of the main reasons for amelioration of plant growth in salinized soils colonized by AM fungi (Ruiz-Lozano and Azcón, 2000; Garg and Manchanda, 2008; Huang *et al.*, 2013; Latef and Miransari, 2014). According to Evelin *et al.* (2009), AM fungi induced uptake in P in stressed plant plants has been hypothesized to reduce the negative effects of Na^+ and Cl^- ions by maintaining membrane integrity, thus facilitating selective ion uptake as well as compartmentalization within vacuoles, thereby averting ions from interference with different plant metabolic pathways. In addition, several authors have also correlated enhanced P nutrition with increment in antioxidant production and enhancement in nodulation as well as N_2 fixing potential of mycorrhizal legumes (Manchanda and Garg, 2011; Kapoor *et al.*, 2013; Garg and Singla, 2015). In addition, studies have revealed that mycorrhizal colonization counteracts the adverse effects of salinity on nodule senescence and enhances nitrogenase activity due to uptake of some essential micro-nutrients which results both in improved growth of plants (Founoune *et al.*, 2002) or vice versa (Rabie and Almadini, 2005; Garg and Manchanda, 2008). Furthermore, Abd-Alla *et al.* (2014) credited the enhancement of N_2 fixation in *Vicia faba* to the mycorrhiza-mediated mobilization of elements including P, Fe, K and other minerals that are involved in synthesis of nitrogenase and leghemoglobin. Similarly, Garg and Pandey (2015) reported that mycorrhizal fungi improved growth of *Cajanus cajan* genotypes by enhancing the uptake of N and P under salinity stress, with tolerant genotype getting more benefits from mycorrhization than the sensitive one. In addition, improved nutritional status (such as K^+, Ca^{2+}, P and Mg^{2+}) helps in avoiding the specific effects of salt stress on chlorophyll degradation and leaf senescence (Evelin *et al.*, 2012), thus, stimulate photosynthetic capacity over stomatal conductance, thereby, positively influencing plant water status in mycorrhizal plants (Querejeta *et al.*, 2003). Recent studies have demonstrated that AM symbiosis regulates root hydraulic properties including root hydraulic conductivity, which have been linked to regulate the functioning of plant aquaporins (Ruiz-Lozano and Aroca, 2010). In 2009, Aroca *et al.* cloned first aquaporin from *R. irregularis* (*GintAQP1*) and on the basis of evidences, suggested that fungal aquaporins could compensate the downregulation of host plant aquaporins caused by osmotic stress.

Another mechanism which has been proposed to improve growth under salinity includes maintenance of cellular homeostasis. Under saline stress, high levels of Na^+ competes with K^+ for the binding sites, thus alters K^+/Na^+ ratio under stressed conditions (Tester and Davenport, 2003) which not only disrupts the ionic balance in the cytoplasm, but also several other metabolic pathways (Giri *et al.*, 2007). Several studies have suggested that AM fungi exclude Na^+ during its transfer to the plants or by discriminating its uptake from the soil (Hammer *et al.*, 2011) thus, maintaining a fine balance of K^+/Na^+ and Ca^{2+}/Na^+ ratios (Giri *et al.*, 2007; Garg and Manchanda, 2009; Porras-Soriano *et al.*, 2009; Evelin *et al.*, 2012; Garg and Pandey, 2015; Garg and Bhandari, 2016). In one of the study, Estrada *et al.* (2013a) demonstrated differential expression levels for Na^+, K^+ transporters of *Z. mays* putatively involved in $Na^+/$ K^+ homeostasis (phloem expressed K^+ transporter -*ZmAKT2*, plasma membrane localized Na^+/H^+ antiporter - *ZmSOS1* - and *ZmSKOR* - xylem K^+ transporter) in roots during AM colonization. Very recently, Porcel *et al.* (2016) reported that mycorrhiza up-regulated the expression of genes encoding transporters involved in Na^+/K^+ homeostasis in rice under saline conditions, favoured Na^+ extrusion from the cytoplasm and its sequestration into the vacuole, thus enhanced salt tolerance. Moreover, the presence of mycorrhizal fungi can reverse the effect of salinity on K^+ by enhancing K^+ absorption (Giri *et al.*, 2007; Sharifi *et al.*, 2007; Zuccarini and Okurowska, 2008; Calvo-Polanco *et al.*, 2013). However, only few studies have documented the possible contribution of AM fungi in K^+ acquisition (Benito and González-Guerrero, 2014; Garcia and Zimmermann, 2014) which needs further investigations. Although Na^+ has been considered as a toxic ion by various researchers under salinity stress, several studies have suggested that Cl^- could be more toxic than Na^+ and the main factor contributing to membrane integrity loss (Calvo-Polanco *et al.*, 2009, 2013; Tavakkoli *et al.*, 2010). Consequently, by decreasing the uptake of toxic ions, AM fungi has been reported to maintain ionic homeostasis, thus functioning of plant cells under stressful conditions.

However, when toxic ions are not controlled by the above described mechanisms, they become redox active, induces higher levels of ROSs including superoxide radical (O_2^-), hydroxyl radical (OH^-) and hydrogen peroxide (H_2O_2), thus cause *oxidative stress*, thereby altering cellular reactions. Various studies have demonstrated that colonization with mycorrhizal fungi restricts the excessive generation of ROSs by up-regulating defense response in terms of the activities of antioxidant enzymes which includes both enzymatic (Superoxide Dismutase

-SOD, Catalase - CAT, Peroxidase - POX, Ascorbate Peroxidase - APX) as well as non-enzymatic antioxidants (Glutathione -GSH, carotenoids, Ascorbic acid (ASA), tocopherols) (Evelin *et al.*, 2009; Garg and Manchanda, 2009; Hajiboland *et al.*, 2010; Abdel Latef and Chaoxing, 2011; Kapoor *et al.*, 2013; Hameed *et al.*, 2014; Garg and Singla, 2015) that work in concert manner and control the cascades of uncontrolled oxidation, thereby improves plant growth under saline conditions. While working on *Cicer arietinum* genotypes, Garg and Singla (2015) and Garg and Bhandari (2016) documented that *F. mosseae* inoculations enhanced plant salt tolerance by mitigating NaCl-oxidative burden by up-regulating antioxidant defense response, thereby shifting the redox homeostasis (in terms of ASA/DHA and GSH/GSSG ratios) toward a more reduced form in mycorrhizal stressed plants. However, few studies have revealed down-regulation in the activities of antioxidant enzymes in mycorrhizal colonized stressed plants indicating existence of lower oxidative stress in + AM inoculated plants (Kohler *et al.*, 2009).

In addition to the defense response, AM fungi may modify the production and accumulation of different compatible solutes within the plant, mainly sugars and amino acids (especially Proline - Pro). Several studies have reported higher plant Pro content in +AM inoculated plants compared to uninoculated plants under salt (Sharifi *et al.*, 2007; Garg and Manchanda, 2009; Talaat and Shawky, 2011; Abdel-Fattah *et al.*, 2012; Ruiz-Lozano *et al.*, 2012). Besides acting as an osmoregulator, Pro has been validated to play an important role in scavenging free radicals, stabilizing subcellular structures and buffering cellular redox potential under several abiotic stresses (Calvo-Polanco *et al.*, 2013; Hajiboland, 2013; Kumar *et al.*, 2015). In one of the study, while working on *Cicer arietinum* L., Garg and Baher (2013) accredited higher accumulation of Pro in mycorrhizal plants to the significant enhancement observed in Pro anabolic enzymes - pyrroline-5-carboxylate synthetase (P-5-CS) and glutamate dehydrogenase (GDH) activities, with a concomitant decline in Pro catabolic enzyme – Pro dehydrogenase (ProDH) activity under varying degree of salt stress. In contrast, several authors have demonstrated lower amounts of Pro in mycorrhizal stressed plants (Rabie and Almadini, 2005; Jahromi *et al.*, 2008; Sheng *et al.*, 2011) which indicates that mycorrhizal fungi confers lesser enhanced stress resistance *i.e.* reflects less cellular damage (Porcel and Ruiz-Lozano, 2004; Calvo-Polanco *et al.*, 2013). In addition to Pro, mycorrhization alleviates osmotic stress by accumulating sugars including Trehalose which not only act as an osmoprotectant (Garg and Chandel, 2011; Garg and Singla, 2016; Garg and Pandey, 2016) but also helps in improving N_2 fixation and plant productivity. Sugars have been ascribed to directly involved in the synthesis of other compounds, production of energy as well as membrane stabilization (Hoekstra *et al.*, 2001) and regulation of gene expression (Koch, 1996) and have been reported to act as signal molecules (Evelin *et al.*, 2009; Calvo Polanco *et al.*, 2013). Additionally, researchers have validated that by enhancing root biomass and shoot growth, AM fungi protects plant from toxic effects of Na+, thereby causing *salt dilution* effect (Garg and Pandey, 2015; Garg and Bhandari, 2016).

6. Conclusion and Future Perspectives

AM fungi are ubiquitous and their diversity allows them to adapt to various stressful conditions including metal toxicity and salinity. From the large number of studies extensively reviewed in this chapter, it is evident that mycorrhizal symbiosis improves tolerance of legumes towards both the stresses. However, the mitigating effects of myco-symbiont as discussed under stressful conditions are complex and vary with the diversity of physiological and molecular mechanisms involved in these processes. Moreover, such effects also display variation when discuss in relation to the diversity of factors affecting the interaction between metal/soil/plant/fungal symbioses. However, comprehensive studies are needed at molecular, biochemical and physiological levels in order to unfold the actual mechanisms adopted by AM fungi in imparting abiotic stress resilience. An increased understanding of such plant-microbial interactions will not only provide tool for efficient remediation practices, but will also enhance crop production in degraded soils.

References

Abd-Alla, M.H., Issa, A.A. and Ohyama, T. 2014. Impact of harsh environmental conditions on nodule formation and dinitrogen fixation of legumes. *In*: Ohyama, T. (Ed.) Advances in biology and ecology of nitrogen fixation, ISBN: 978-953-51-1216-7, InTech, DOI: 10.5772/56997.

Abdel Latef, A.A.H. and Chaoxing, H. 2011. Effect of arbuscular mycorrhizal fungi on growth, mineral nutrition, antioxidant enzymes activity and fruit yield of tomato grown under salinity stress. *Sci. Hortic.*, 127: 228-233.

Abdel-Fattah, G.M. and Asrar, A.A. 2012. Arbuscular mycorrhizal fungal application to improve growth and tolerance of wheat (*Triticum aestivum* L.) plants grown in saline soil. *Acta Physiol. Plant.*, 34: 267-277.

Acquaah, G. 2007. Principles of Plant Genetics and Breeding. Blackwell, Oxford, UK.

Aliasgharzadeh, N., Rastin, N.S., Towfighi, H. and Alizadeh A. 2001. Occurrence of arbuscular mycorrhizal fungi in saline soils of the Tabriz plain of Iran in relation to some physical and chemical properties of soil. *Mycorrhiza*, 11: 119-122.

Allen, E.B. and Cunningham, G.L. 1983. Effects of vesicular arbuscular mycorrhizae on *Distichlis spicata* under three salinity levels. *New Phytol.*, 93: 227-236.

Amir, H., Jourand, P., Cavaloc, Y. and Ducousso, M. 2014. Role of mycorrhizal fungi in the alleviation of heavy metal toxicity in plants. *In*: Zakaria M. Solaiman, Z.M., Abbott, L.K. and Varma, A. (Eds.) Mycorrhizal Fungi: Use in Sustainable Agriculture and Land Restoration, Soil Biology, Springer-Verlag Berlin Heidelberg, 41: 241-258.

Aroca, R., Bago, A., Sutka, M., Paz, J.A., Cano, C., Amodeo, G. and RuizLozano J.M. 2009. Expression analysis of the first arbuscular mycorrhizal fungi aquaporin described reveals concerted gene expression between salt-stressed and nonstressed mycelium. *Mol. Plant-Microbe Interact.*, 22: 1169-1178.

Augé, R.M. 2001. Water relations, drought and vesicular-arbuscular mycorrhizal symbiosis. *Mycorrhiza*, 11: 3-42.

Badri, D.V., Quintana, N., El Kassis, E.G., Kim, H.K., Choi, Y.H., Sugiyama, A., Verpoorte, R., Martinoia, E., Manter, D.K. and Vivanco, J.M. 2009. An ABC transporter mutation alters root exudation of phytochemicals that provoke an overhaul of natural soil microbiota. *Plant Physiol.*, 151: 2006-2017.

Barea, J.M., Toro, M., Orozco, M.O., Campos, E. and Azcón, R. 2002. The application of isotopic (32P and 15N) dilution techniques to evaluate the interactive effect of phosphate-solubilizing rhizobacteria, mycorrhizal fungi and Rhizobium to improve the agronomic efficiency of rock phosphate for legume crops. Nutr. Cycl. Agroecosyst (in press).

Benabdellah, K., Merlos, M.A., Azcón-Aguilar, C. and Ferrol, N. 2009. GintGRX1, the first characterized glomeromycotan glutaredoxin, is a multifunctional enzyme that responds to oxidative stress. *Fungal Gen. Biol.*, 46: 94-103.

Benavides, M.P., Gallego, S.M. and Tomaro, M.L. 2005. Cadmium toxicity in plants. *Braz. J. Plant Physiol.*, 17: 21-34.

Benito, B. and González-Guerrero, M. 2014. Unravelling potassium nutrition in ectomycorrhizal associations. *New Phytol.*, 201: 707–709.

Calvo-Polanco, M., Sa´nchez-Romera, B. and Aroca, R. 2013. Arbuscular Mycorrhizal Fungi and the Tolerance of Plants to Drought and Salinity. *In:* Aroca, R. (Ed.) Symbiotic Endophytes, Soil Biology 37, Springer-Verlag Berlin, Heidelberg.

Calvo-Polanco, M., Sa´nchez-Romera, B. and Aroca, R. 2014. Mild Salt Stress Conditions Induce Different Responses in Root Hydraulic Conductivity of *Phaseolus vulgaris* Over-Time. *PLOS ONE*, 9: e90631.

Calvo-Polanco, M., Zwiazek, J.J., Jones, M.D. and MacKinnon, M.D. 2009. Effects of NaCl on responses of ectomycorrhizal black spruce (*Picea mariana*), white spruce (*Picea glauca*) and jack pine (*Pinus banksiana*) to fluoride. *Physiol. Plant.*, 135: 51-61.

Cicatelli, A., Lingua, G., Todeschini, V., Biondi, S., Torrigiani, P. and Castiglione, S. 2010. Arbuscular mycorrhizal fungi restore normal growth in a white poplar clone grown on heavy metal-contaminated soil, and this is associated with upregulation of foliar metallothionein and polyamine biosynthetic gene expression. *Ann. Bot.*, 106: 791–802.

da Silva Folli-Pereira, M., Meira-Haddad, L.A., da Cruz Houghton, C.M.N.S. V.and Kasuya, M.C.M. 2016. Plant-Microorganisminteractions:Effectsonthetoleranceo fplantstobioticand abioticstresses. *In*: Hakeem, K.R., Ahmad, P. and Ozturk, M. (Eds.) Crop Improvement: New approaches and modern techniques, Springer Science+Business Media, LLC, US, pp. 209-238.

Egamberdiyeva, D. 2007. The effect of plant growth promoting bacteria on growth and nutrient uptake of maize in two different soils. *Appl. Soil Ecol.*, 36: 184-189.

Emamverdian, A., Ding, Y., Mokhberdoran, F. and Xie, Y. 2015. Heavy metal stress and some mechanisms of plant defense response. *Scientific. World. J.* ID: 756120.

Estaun, M.V. 1989. Effect of sodium chloride and mannitol on germination and hyphal growth of the vesicular-arbuscular mycorrhizal fungus *Glomus mosseae*. *Agric. Ecosyst. Environ.*, 29: 123-129.

Estrada, B., Aroca, R., Maathuis, F.J.M., Barea, J.M. and Ruiz-Lozano, J.M. 2013a. Arbuscular mycorrhizal fungi native from a Mediterranean saline area enhance maize tolerance to salinity through improved ion homeostasis. *Plant Cell Environ.*, 36: 1771-1782.

Estrada, B., Barea, J.M., Aroca, R. and Ruiz-Lozano, J.M. 2013b. A native *Glomus intraradices* strain from a Mediterranean saline area exhibits salt tolerance and enhanced symbiotic efficiency with maize plants under salt stress conditions. *Plant Soil*, 366: 333-349.

Evelin, H., Giri, B. and Kapoor, R. 2012. Contribution of *Glomus intraradices* inoculation to nutrient acquisition and mitigat ion of ionic imbalance in NaCl-stressed *Trigonella foenum*-graecum. *Mycorrhiza*, 22: 203-217.

Evelin, H., Kapoor, R. and Giri, B. 2009. Arbuscular mycorrhizal fungi in alleviation of salt stress: A review. *Ann. Bot.*, 104: 1263-1280.

Fageria, N.K., Gheyi, H.R. and Moreir, A. 2011. Nutrient bioavailability in salt affected soils. *J. Plant Nutrition.*, 3: 945-962.

FAO (2005) Salt-affected soils from sea water intrusion: Strategies for rehabilitation and management. *Food and Agriculture Organization of the UN.*

Feng, G., Song, Y.C., Li, X.L. and Christi, P. 2003. Contribution of arbuscular mycorrhizal fungi to utilization of organic sources of phosphorus by red clover in a calcareous soil. *Appl. Soil Ecol.* 22: 139–148.

Ferrol, N., González-Guerrero, M., Valderas, A., Benabdellah, K. and Azcón-Aguilar, C. 2009. Survival strategies of arbuscular mycorrhizal fungi in Cu-polluted environments. *Phytochem. Rev.*, 8: 551–559.

Founoune, H., Duponnois, R., Bâ, A.M., Sall, S., Branget, I., Lorquin, J., Neyra, M. and Chotte, J.L. 2002. Mycorrhiza helper bacteria stimulate ectomycorrhizal symbiosis of *Acacia holosericea* with *Pisolithus albus*. *New Phytol.*, 153: 81–89.

Gamalero, E., Lingua, G., Berta, G. and Glick, B.R. 2009. Beneficial role of plant growth promoting bacteria and arbuscular mycorrhizal fungi on plant responses to heavy metal stress. *Can. J. Microbiol.*, 55: 501-514.

Garcia, K. and Zimmermann, S.D. 2014. The role of mycorrhizal associations in plant potassium nutrition. *Front. Plant Sci.*, 5: 337.

Garg, N. and Kaur, H. 2013a. Response of antioxidant enzymes, phytochelatins and glutathione production towards Cd and Zn stresses in *Cajanus cajan* (L.) Millsp. genotypes colonized by arbuscular mycorrhizal fungi. *J. Agron. Crop Sci.*, 199: 118–133.

Garg, N. and Singla, P. 2012. The role of *Glomus mosseae* on key physiological and biochemical parameters of pea plants grown in arsenic contaminated soil. *Sci. Horti.*, 143: 92-101.

Garg, N. and Aggarwal, N. 2011. Effects of interactions between cadmium and lead on growth, nitrogen fixation, phytochelatin, and glutathione production in mycorrhizal *Cajanus cajan* (L.) Millsp. *J. Plant Growth Regul.*, 30: 286-300.

Garg, N. and Aggarwal, N. 2012. Effect of mycorrhizal inoculations on heavy metal uptake and stress alleviation of *Cajanus cajan* (L.) Millsp. Genotypes grown in cadmium and lead contaminated soils. *Plant Growth Regul.*, 66: 9–26.

Garg, N. and Baher, N. 2013. Role of arbuscular mycorrhizal symbiosis in proline biosynthesis and metabolism of *Cicer arietinum* L. (Chickpea) genotypes under salt stress. *J. Plant Growth Regul.*, 32: 767-778.

Garg, N. and Bhandari, P. 2012. Influence of cadmium stress and arbuscular mycorrhizal fungi on nodule senescence in *Cajanus cajan* (L.) Millsp. *Int. J. Phyto.*, 14: 62-74.

Garg, N. and Bhandari, P. 2014. Cadmium toxicity in crop plants and its alleviation by arbuscular mycorrhizal (AM) fungi: An overview. *Plant Biosyst.*, 148: 609-621.

Garg, N. and Bhandari, P. 2015. Interactive effects of silicon and arbuscular mycorrhiza in modulating ascorbate-glutathione cycle and antioxidant scavenging capacity in differentially salt-tolerant *Cicer arietinum* L. genotypes subjected to long-term salinity. *Protoplasma,* DOI 10.1007/s00709-015-0892-4.

Garg, N. and Bhandari, P. 2016. Silicon nutrition and mycorrhizal inoculations improve growth, nutrient status, K^+/Na^+ ratio and yield of *Cicer arietinum* L. genotypes under salinity stress. *Plant Growth Regul.*, 78: 371-387.

Garg, N. and Chandel, S. 2010. Arbuscular mycorrhizal networks: process and functions. A review. *Agron. Sustain. Dev.*, 30: 581–599.

Garg, N. and Chandel, S. 2011. Effect of mycorrhizal inoculation on growth, nitrogen fixation, and nutrient uptake in *Cicer arietinum* (L.) under salt stress. *Turk. J. Agric. For.*, 35: 205-214.

Garg, N. and Chandel, S. 2011. The effects of salinity on nitrogen fixation and trehalose metabolism in mycorrhizal *Cajanus cajan* (L.) Millsp. plants. *J. Plant Growth Regul.*, 30: 490-503.

Garg, N. and Kaur, H. 2013b. Impact of cadmium-zinc interactions on metal uptake, translocation and yield in pigeonpea genotypes colonized by arbuscular mycorrhizal fungi. *J. Plant Nutr.* 36(1): 67-90.

Garg, N. and Manchanda, G. 2008. Effect of arbuscular mycorrhizal inoculation on salt-induced nodule senescence in *Cajanus cajan* (pigeonpea). *J. Plant Growth Regul.*, 27: 115-124.

Garg, N. and Manchanda, G. 2009. Role of arbuscular mycorrhizae in the alleviation of ionic, osmotic and oxidative stresses induced by salinity in *Cajanus cajan* (L.) Millsp. (pigeonpea). *J. Agron. Crop Sci.* 195: 110-123.

Garg, N. and Pandey, R. 2015. Effectiveness of native and exotic arbuscular mycorrhizal fungi on nutrient uptake and ion homeostasis in salt stressed *Cajanus cajan* L. (Millsp.) genotypes. *Mycorrhiza*, 25: 165-80.

Garg, N. and Pandey, R. 2016. High effectiveness of exotic arbuscular mycorrhizal fungi is reflected in improved rhizobial symbiosis and trehalose turnover in *Cajanus cajan* genotypes grown under salinity stress. *Fungal Ecol.*, 21: 57-67.

Garg, N. and Singla, P. 2015. Naringenin-and Funneliformis mosseae-mediated alterations in redox state synchronize antioxidant network to alleviate oxidative stress in *Cicer arietinum* L. Genotypes under salt stress. *J. Plant Growth Regul.*, 34: 595-610.

Garg, N. and Singla, P. 2016. Stimulation of nitrogen fixation and trehalose biosynthesis by naringenin (Nar) and arbuscular mycorrhiza (AM) in chickpea under salinity stress. *Plant Growth Regul.*, DOI 10.1007/s10725-016-0146-2.

Genre, A. and Bonfante, P. 2010. The making of symbiotic cells in arbuscular mycorrhizal roots. *In:* Koltai, H. and Kapulnik, Y. (Eds.) Arbuscular mycorrhizas: Physiology and function, Springer Science+Business Media B.V., DOI 10.1007/978-90-481-9489-6_3, pp. 57-71.

Gianinazzi, S., Gianinazzi-Pearson, V., Franken, P., Dumas-Gaudot, E., van Tuinen, D., Samra, A., Martin-Laurent, F., and Dassi, B. 1995. Molecules and genes involved in mycorrhizal functioning. *In:* Stoechi, V., Bonfante, P. and Nuti, M. (Eds) Biotechnologies of Ectomycorrhizae, New York, pp 67-76.

Giasson, P., Karam, A. and Jaouich, A. 2008. Arbuscular mycorrhizae and alleviation of soil stresses on plant growth. *In:* Zaki Anwar Siddiqui, Z.A., Akhtar, M.S. and Futai, K. (Eds.) Mycorrhizae: Sustainable agriculture and forestry, Springer Science + Business Media B.V, pp. 99-134.

Gildon, A. and Tinker, P.B. 1983. Interactions of vesicular-arbuscular mycorrhiza infections and heavy metals in plants. II. The effects of infection on uptake of copper. *Trans. Br. Mycol Soc.*, 77: 648–649.

Giovannetti, M and Sbrana, C. 1998. Meeting a non-host: the behaviour of AM fungi. *Mycorrhiza*, 8: 123-130.

Giri, B., Kapoor, R. and Mukerji, K.G. 2003. Influence of arbuscular mycorrhizal fungi and salinity on growth, biomass, and mineral nutrition of *Acacia auriculiformis*. *Biol. Fertil. Soils*, 38: 170-175.

Giri, B., Kapoor, R. and Mukerji, K.G. 2007. Improved tolerance of *Acacia nilotica* to salt stress by arbuscular mycorrhiza, *Glomus fasciculatum* may be partly related to elevated K/Na ratios in root and shoot tissues. *Microb. Ecol.*, 54: 753-760.

Gohre, V. and Paszkowski, U. 2006. Contribution of the arbuscular mycorrhizal symbiosis to heavy metal phytoremediation. *Planta*, 223: 1115-1122.

Goncalves, J.F., Becker, A.G., Cargnelutti, D., Tabaldi, L.A., Pereira, L.B., Battisti, V., Spanevello, R.M., Morsch, V.M., Nicoloso, F.T. and Schetinger, M.R.C. 2007. Cadmium toxicity causes oxidative stress and induces response of the antioxidant system in cucumber seedlings. *Braz. J. Plant Physiol.*, 19: 223-232.

Gonzalez-Chavez, C., Harris, P.J., Dodd, J. and Meharg, A.A. 2002. Arbuscular mycorrhizal fungi confer enhanced arsenate resistance on *Holcus lanatus*. *New Phytol.*, 155: 163-171.

González-Chavez, M.C., Carrillo-Gonzalez, R., Wright, S.F. and Nichols, K.A. 2004. The role of glomalin, a protein produced by arbuscular mycorrhizal fungi, in sequestering potentially toxic elements. *Environ. Pollut.*, 130: 317–323.

González-Guerrero, M., Azcon-Aguilar, C. and Ferrol, N. 2006. GintABC1 and GintMT1 are involved in Cu and Cd homeostasis in *Glomus intraradices*. Abstracts of the 5th International conference on mycorrhiza, 23–27 July, Granada, Spain.

González-Guerrero, M., Azcón-Aguilar, C., Mooney, M., Valderas, A., MacDiarmid, C.W., Eide, D.J., Ferrol, N. 2005. Characterization of a *Glomus intraradices* gene encoding a putative Zn transporter of the cation diffusion facilitator family. *Fungal Genet. Biol.*, 42: 130-140.

González-Guerrero, M., Cano, C., Azcón-Aguilar, C. and Ferrol, N. 2007. GintMT1 encodes a functional metallothionein in *Glomus intraradices* that responds to oxidative stress. *Mycorrhiza*, 17: 327-335.

González-Guerrero, M., Melville, L.H., Ferrol, N., Lott, J.N.A., Azcon-Aguilar, C. and Peterson, R.L. 2008. Ultrastructural localization of heavy metals in the extraradical mycelium and spores of the arbuscular mycorrhizal fungus *Glomus intraradices*. *Can. J. Microbiol.*, 54: 103–110.

Grover, M., Ali, S., Sandhya V., Rasul, A. and Venkateswarlu, B. 2011. Role of microorganisms in adaptation of agriculture crops to abiotic stresses. *World. J. Microbiol. Biotechnol.*, 27:1231–1240.

Hajiboland, R. 2013. Role of arbuscular mycorrhiza in amelioration of salinity. *In:* Ahmad, P., Azooz, M.M. and Prasad, M.N.V. (Eds.). Salt stress in plants: Signalling, omics and adaptations, Springer, New York, pp. 301-354.

Hajiboland, R., Aliasgharzadeh, N., Laiegh, S.F. and Poschenrieder, C. 2010. Colonization with arbuscular mycorrhizal fungi improves salinity tolerance of tomato (*Solanum lycopersicum* L.) plants. *Plant Soil*, 331: 313-327.

Hameed, A., Dilfuza, E., Abd-Allah, E.F., Hashem, A., Kumar, A. and Ahmad, P. 2014. Salinity stress and arbuscular mycorrhizal symbiosis in Plants. *In*: Miransari, M. (Ed.) Use of microbes for the alleviation of soil stresses, Vol 1. Springer, New York, pp. 139-159.

Hammer, E., Nasr, H., Pallon, J., Olsson, P. and Wallander, H. 2011. Elemental composition of arbuscular mycorrhizal fungi at high salinity. *Mycorrhiza*, 21: 117-129.

Harrison, M.J. 1999. Molecular and cellular aspects of the arbuscular mycorrhizal symbiosis. *Annu. Rev. Plant Physiol. Plant Mol. Biol.*, 50: 361-389.

Hasanuzzaman, M., Hossain, M.A. and Fujita, M. 2011a. Nitric oxide modulates antioxidant defense and the methylglyoxal detoxification system and reduces salinity-induced damage of wheat seedlings. *Plant Biotechnol. Rep.*, 5: 353–365.

Hasanuzzaman, M., Hossain, M.A. and Fujita, M. 2011b. Selenium-induced up-regulation of the antioxidant defense and methylglyoxal detoxification system reduces salinity-induced damage in rapeseed seedlings. *Biol. Trace Elem. Res.*, 143: 1704–1721.

Hawkes, J.S. 1997. Heavy metals. *J. Chem. Edu.*, 74: 1369-1374.

Hirrel, M.C. 1981. The effect of sodium and chloride salts on the germination of *Gigaspora margarita*. *Mycologia*, 73: 610-617.

Ho, I. and Trappe, J. M. 1987 Enzymes and growth substances of *Rhizopagon* species in relation to mycorrhizal hosts and infrageneric taxonomy. *Mycologia*, 79: 553-558.

Hoekstra, F.A., Golovina, E.A. and Buitink, J. 2001. Mechanisms of plant desiccation tolerance. *Trends Plant Sci.*, 6: 431-8.

Huang, J., Gu, M., Lai, Z., Fan, B., Shi, K., Zhou, Y.-H., Yu, J.-Q. and Chen, Z. 2010. Functional analysis of the Arabidopsis *pal* gene family in plant growth, development, and response to environmental stress. *Plant Physiol.*, 153: 1526-1538.

Huang, J.C., Lai, W.A., Singh, S., Hameed, A. and Young, C.C. 2013. Response of mycorrhizal hybrid tomato cultivars under saline stress. *J. Soil Sci. Plant Nutr.*, 13: 469-484.

Hussain, K., Nisar, M.F., Majeed, A., Nawaz, K., Bhatti, K.H., Afgan, S., Shafzac, A. and Zia-ul-Hussnian, S. 2010. What molecular mechanism is adapted by plants during salt stress tolerance? *Afr. J. Biotechnol.*, 9: 416-422.

Jahromi, F., Aroca, R., Porcel, R. and Ruiz-Lozano, J.M. 2008. Influence of salinity on the in vitro development of *Glomus intraradices* and on the *in vivo* physiological and molecular responses of mycorrhizal lettuce plants. *Microb. Ecol.*, 55: 45-53.

Javaid, A. 2011. Importance of Arbuscular Mycorrhizal fungi in phytoremediation of heavy metal contaminated soils. *In*: Khan, M.S., Zaidi, A., Goel, R. and Mussarrat, J. (Eds), Biomanagement of metal-contaminated soils, The Netherlands: Springer Science. pp. 125-141.

Jiang, Q-Y., Tan, S-Y., Zhuoa, F., Yang, D-J., Yec, Z-H. and Jing, Y-X. 2016. Effect of *Funneliformis mosseae* on the growth, cadmium accumulation and antioxidant activities of *Solanum nigrum*. *Appl. Soil Ecol.*, 98: 112–120.

Joner, E.J., Briones, R. and Leyval, C. 2000. Metal-binding capacity of arbuscular mycorrhizal mycelium. *Plant Soil*, 226: 227–234.

Juniper, S. and Abbott, L.K. 1993. Vesicular-arbuscular mycorrhizas and soil salinity. *Mycorrhiza*, 4: 45-57.

Juniper, S. and Abbott, L.K. 2006. Soil delays germination and limits growth of hyphae from propagules of arbuscular mycorrhizal fungi. *Mycorrhiza*, 16: 371-379.

Kalaivanan, D. and Ganeshamurthy, A.N. 2016. Mechanisms of heavy metal toxicity in plants. *In:* Rao, N.K.S., Shivashankara, K.S. and Laxman, R.H. (Eds.) Abiotic stress physiology of horticultural crops, Springer, India, pp. 85-102.

Kapoor, R., Evelin, H., Mathur, P. and Giri, B. 2013. Arbuscular mycorrhiza approaches for abiotic stress tolerance in crop plants for sustainable agriculture, *In*: Tuteja, N. and Gill, S.S. (Eds.) Plant acclimation to environmental stress, Springer Science+Business Media, New York, DOI 10.1007/978-1-4614-5001-6_14, pp. 359-401.

Khade, S. W. and Adholeya, A. 2007. Feasible bioremediation through arbuscular mycorrhizal fungi imparting heavy metal tolerance: A retrospective. *Bioremed. J.*, 11: 33-43.

Koch, K.E. 1996. Carbohydrate-modulates gene expression in plants. *Ann. Rev. Plant Physiol. Plant Mol. Biol.*, 47: 509-540.

Kohler, J., Hernández, J.A., Caravaca, F. and Roldán, A. 2009. Induction of antioxidant enzymes is involved in the greater effectiveness of a PGPR versus AM fungi with respect to increasing the tolerance of lettuce to severe salt stress. *Environ. Exp. Bot.*, 65: 245-252.

Krishnamoorthy, R., Kim, C-G., Subramanian, P., Kim, K-Y., Selvakumar, G. and Sa, T-M. 2015. Arbuscular mycorrhizal fungi community structure, abundance and species richness changes in soil by different levels of heavy metal and metalloid concentration. *PLoS ONE,* 10(6): e0128784. doi:10.1371/journal.pone.0128784.

Kumar, A., Dames, J.F., Gupta, A., Sharma, S., Gilbert, J.A. and Ahmad, P. 2015. Current developments in arbuscular mycorrhizal fungi research and its role in salinity stress alleviation: a biotechnological perspective. *Crit. Rev. Biotechnol.*, 35: 461-74.

Kumar, A., Sharma, S. and Mishra, S. 2010. Influence of arbuscular mycorrhizal (AM) fungi and salinity on seedling growth, soluteaccumulation and mycorrhizal dependency of *Jatropha curcas* L. *J. Plant Growth Regul.*, 29: 297-306.

Lanfranco, L., Bolchi, A., Ros, E.C., Ottonello, S. and Bonfante, P. 2002. Differential expression of a metallothionein gene during the presymbiotic versus the symbiotic phase of an arbuscular mycorrhizal fungus. *Plant Physiol.*, 130: 58–67.

Latef, A.A.H.A and Miransari, M. 2014. The role of arbuscular mycorrhizal fungi in alleviation of salt stress. *In:* Miransari, M. (Ed.) Use of microbes for the alleviation of soil stresses, Springer Science+Business Media, New York.

Lenoir, I., Fontaine, J. and Sahraoui, L-H. 2016. Arbuscular mycorrhizal fungal responses to abiotic stresses: A review. *Phytochemis.,*123: 4-15.

Leung, H.M., Ye, Z.H. and Wong, M.H. 2007. Survival strategies of plants associated with arbuscular mycorrhizal fungi on toxic mine tailings. *Chemosphere*, 66: 905–915.

Leyval, C., Turnau, K. and Haselwandter, K. 1997. Effect of heavy metal pollution on mycorrhizal colonization and function: physiological, ecological and applied aspects. *Mycorrhiza*, 7: 139-153.

Li, X.L., George, E., Marschner, H. 1991. Phosphorus depletion and pH decrease at the root-soil and hyphae-soil interfaces of VA mycorrhizal white clover fertilized with ammonium. *New Phytol.*, 119: 397–404.

Lins, C.E.L., Cavalcante, U.M.T., Sampaio, E.V.S.B., Messias, A.S. and Maia, L.C. 2006. Growth of mycorrhized seedlings of *Leucaena leucocephala* (Lam.) de Wit. in a copper contaminated soil. *App. Soil Ecol.,* 31: 181-185.

Lodwig, E.M., Hosie, A.H.F., Bourdes, A., Findlay, K., Allaway, D., Karunakaran, R., Downie, J.A. and Poole, P.S. 2003. Amino-acid cycling drives nitrogen fixation in the legume-Rhizobium symbiosis. *Nature,* 422: 722–726.

Malcova, R., Vosatka, M. and Gryndler, M. 2003. Effects of inoculation with *Glomus intraradices* on lead uptake by *Zea mays* L. and *Agrostis capillaris* L. *Appl. Soil Ecol.,* 23: 255-267.

Manchanda, G. and Garg N. 2011. Alleviation of salt-induced ionic, osmotic and oxidative stresses in *Cajanus cajan* nodules by AM inoculation. *Plant Biosyst.,* 145: 88-97.

Manchanda, G. and Garg, N. 2008. Salinity and its effects on the functional biology of legumes. *Acta Physiol. Plant.,* 30: 595-618.

Mandal, S., Yadav, S. and Nema, R.K. 2009. Antioxidants: A Review. *J. Chem. Pharm. Res.,* 1: 102-104.

Massoud, O.N., El-Sabagh, S.M., Morsy, E.M. and Megahed, M.K.M. 2012. Alleviation of Certain Heavy Metals Toxicity on *Zea mays* by Arbuscular Mycorrhiza. *J. App. Sci. Res.,* 8: 3491-3502.

Meharg, A. A. 2003. The mechanistic basis of interactions between mycorrhizal associations and toxic metal cations. *Mycol. Res.,* 107: 1253–1265.

Miransari, M. 2010. Contribution of arbuscular mycorrhizal symbiosis to plant growth under different types of soil stress. *Plant Biol.,* 12: 563-569.

Moray, C., Goolsby, E.W. and Bromham, L. 2015. The phylogenetic association between salt tolerance and heavy metal hyperaccumulation in angiosperms. *Evol. Biol.,* DOI 10.1007/s11692-015-9355-2.

Morley, G.F. and Gadd, G.M. 1995. Sorption of toxic metals by fungi and clay minerals. *Mycol. Res.,* 99: 1429–1438.

Muleta, D. and Woyessa, D. 2012. Importance of arbuscular mycorrhizal fungi in legume production under heavy metal-contaminated soils. *In*: Zaidi, A., Wani, P.A. and Khan, M.S. (Eds.) Toxicity of heavy metals to legumes and bioremediation,London, UK: Springer, pp. 219–241.

Munns, R. 2005. Genes and salt tolerance: bringing them together. *New Phytol.* 167: 645-663.

Munns, R. and Tester M. 2008. Mechanisms of salinity tolerance. *Annu. Rev. Plant. Biol.* 59: 651-681.

Munns, R., James, R.A. and Lauchli, A. 2006. Approaches to increasing the salt tolerance of wheat and other cereals. *J. Exp. Bot.* 57: 1025-1043.

Nasim, G. 2010. The role of arbuscular mycorrhizae in inducing resistance to drought and salinity stress in crops. *In*: Ashraf, M., Ozturk,M. and Ahmad,

M.S.A. (Eds.) Plant Adaptation and Phytoremediation, Springer-Verlag, Berlin Heidelberg, pp. 119-141.

Nazar, R., Iqbal, N., Syeed, S. and Khan, N.A. 2011. Salicylic acid alleviates decreases in photosynthesis under salt stress by enhancing nitrogen and sulfur assimilation and antioxidant metabolism differentially in two mungbean cultivars. *J. Plant Physiol.*, 168: 807–815.

Öpik, M., Zobel, M., Cantero, J.J., Davison, J., Facelli, J.M., Hiiesalu, I., Jairus, T., Kalwij, J.M., Koorem, K., Leal, M.E., Liira, J., Metsis, M., Neshataeva, V., Paal, J., Phosri, C., Põlme, S., Reier, Ü., Saks, Ü., Schimann, H., Thiéry, O., Vasar, M. and Moora, M. 2013. Global sampling of plant roots expands the described molecular diversity of arbuscular mycorrhizal fungi. *Mycorrhiza*, 23: 411-430.

Orlowska, E., Mesjasz-Przybylowicz, J., Przybylowicz, W., Turnau, K. 2008. Nuclear macroprobe studies of elemental distribution in mycorrhizal and non-mycorrhizal roots of Ni-hyperaccumulator Berkheya coddii. *X Ray Spectrom.*, 37: 129–132.

Ouziad, F., Hildebrandt, U., Schmelzer, E. and Bothe, H. 2005. Differential gene expressions in arbuscular mycorrhizal-colonized tomato grown under heavy metal stress. *J. Plant Physiol.*, 162: 634-649.

Pajuelo, E., Rodriguez-Llorente, I.D., Dary, M. and Palomares, A.J. 2008. Toxic effects of arsenic on *Sinorhizobium–Medicago sativa* symbiotic interaction. *Environ. Pollut.* 154: 203–211.

Parniske, M. 2008. Arbuscular mycorrhiza: The mother of plant root endosymbioses. *Nat. Rev. Microbiol.*, 6: 763-775.

Pellegrino, E. and Bedini, S. 2014. Enhancing ecosystem services in sustainable agriculture: biofertilization and biofortification of chickpea (*Cicer arietinum* L.) by arbuscular mycorrhizal fungi. *Soil Biol. Biochem.*, 68: 429-439.

Pellegrino, E., Bedini, S., Avio, L., Bonari, E. and Giovannetti, M. 2011. Field inoculation effectiveness of native and exotic arbuscular mycorrhizal fungi in a Mediterranean agricultural soil. *Soil Biol. Biochem.*, 43: 367-376.

Perveen, R., Faizan, S. and Ansari, A.A. 2015. Phytoremediation using leguminous plants: Managing cadmium stress with applications of arbuscular mycorrhiza (AM) fungi. *In:* Ansari, A.A., Gill, S.S., Gill, R., Lanza, G.R. and Newman, L. (Eds.) Phytoremediation: Management of Environmental Contaminants, 2: 131-142.

Plaut, Z., Edelstein, M. and Ben-Hur, M. 2013. Overcoming salinity barriers to crop production using traditional methods *Crit. Rev. Plant Sci.*, 32. 250-291.

Pond, E.C., Merge, J.A. and Jarrell, W.M. 1984. Improved growth of tomato in salinized soil by vesicular-arbuscular mycorrhizal fungi collected from saline soils. *Mycologia*, 76: 74-84.

Porcel, R. and Ruíz-Lozano, J.M. 2004. Arbuscular mycorrhizal influence on leaf water potential, solute accumulation, and oxidative stress in soybean plants subjected to drought stress. *J. Exp. Bot.*, 55: 1743-1750.

Porcel, R., Aroca, R. and Ruiz-Lozano, J.M. 2012. Salinity stress alleviation using arbuscular mycorrhizal fungi. A review. *Agron. Sustain. Dev.*, 32: 181-200.

Porcel, R., Aroca, R., Azcon, R. and Ruiz-Lozano, J.M. 2016. Regulation of cation transporter genes by the arbuscular mycorrhizal symbiosis in rice plants subjected to salinity suggests improved salt tolerance due to reduced Na^+ root-to-shoot distribution. *Mycorrhizae*, DOI 10.1007/s00572-016-0704-5.

Porras-Soriano, A., Soriano-Martin, M.S., Porras-Piedra, A. and Azcon, R. 2009. Arbuscular mycorrhizal fungi increased growth, nutrient uptake and tolerance to salinity in olive trees under nursery conditions. *J. Plant Physiol.*, 166: 1350-1359.

Querejeta J. I., Allen M. F., Caravaca F. A. and Roldan A. 2006. Differential modulation of host plant $\delta^{13}C$ and $\delta^{18}O$ by native and nonnative arbuscular mycorrhizal fungi in a semiarid environment. *New Phytol.*, 169: 379-387.

Querejeta, J.I., Egerton-Warburton, L.M. and Allen, M.F. 2003. Direct nocturnal water transfer from oaks to their mycorrhizal symbionts during severe soil drying. *Oecologia*. 134: 55-64.

Rabie, G.H. and Almadini, A.M. 2005. Role of bioinoculants in development of salt-tolerance of *Vicia faba* plants under salinity stress. *Afr. J. Biotechnol.*, 4: 210-222.

Reichman, S.M.Ã. 2007. The potential use of the legume–Rhizobium symbiosis for the remediation of arsenic contaminated sites. *Soil Biol.*, 39: 2587–2593

Requena, N., Jeffries, P. and Barea, J.M., 1996. Assessment of natural mycorrhizal potential in a desertified semiarid ecosystem. *Appl. Environ. Microbiol.*, 62: 842-847.

Rillig, M. C. and Steinberg, P. D. 2002. Glomalin production by an arbuscular mycorrhizal fungus: A mechanism of habitat modification. *Soil Biol. Biochem.*, 34: 1371–1374.

Ruiz-Lozano, J.M. and Aroca, R. 2010. Modulation of aquaporin genes by the arbuscular mycorrhizal symbiosis in relation to osmotic stress tolerance. Aquaporin in AM Plants Under Osmotic Stress. *In*: Sechback, J. and Grube, M. (Eds.) Symbiosis and Stress: Joint ventures in biology, cellular origin, life in extreme habitats and astrobiology, Springer Science+Business Media B.V. DOI 10.1007/978-90-481-9449-0_17. pp. 357-374.

Ruiz-Lozano, J.M. and Azcón, R. 2000. Symbiotic efficiency and infectivity of an autochthonous arbuscular mycorrhizal *Glomus* sp from saline soils and *Glomus deserticola* under salinity. *Mycorrhiza*, 10: 137-143.

Ruiz-Lozano, J.M., Porcel, R., Azcón, C. and Aroca, R. 2012. Regulation by arbuscular mycorrhizae of the integrated physiological response to salinity in plants: New challenges in physiological and molecular studies. *J. Exp. Bot.*, 63: 4033-4044.

Sannazzaro, A.I., Ruiz, O.A., Albetró, E.O. and Menéndez, A.B. 2006. Alleviation of salt stress in *Lotus glaber* by *Glomus intraradices*. *Plant Soil*, 285: 279-287.

Saraswat, S. and Rai, J.P.N. 2011. Mechanism of metal tolerance and detoxification in mycorrhizal fungi. *In*: Khan, M.S., Zaidi,A., Goel,R. and Mussarrat, J. (Eds.)

Biomanagement of metal-contaminated soils, The Netherlands: Springer Science. pp. 225–240.

Schüßler A, Schwarzott D, and Walker C. 2001. A new fungal phylum, the Glomeromycota: phylogeny and evolution. *Mycological Research,* 105: 1413-1421.

Schüßler, A. 2014. Glomeromycota: Species list. [WWW document] URL http://schuessler.userweb.mwn.de/amphylo [accessed 9 November 2013].

Sharifi, M., Ghorbanli, M. and Ebrahimzadeh, H. 2007. Improved growth of salinity-stressed soybean after inoculation with salt pre-treated mycorrhizal fungi. *J. Plant Physiol.,* 164: 1144-1151.

Sheng, M., Tang, M., Zhang, F. and Huang, Y. 2011. Influence of arbuscular mycorrhiza on organic solutes in maize leaves under salt stress. *Mycorrhiza,* 21: 423-430.

Singh, B.R., Singh, A., Mishra, S., Naqvi, A.H. and Singh, H.B. 2016. Remediation of heavy metal- contaminated agricultural soils using microbes. *In:* Singh, D.P., Singh, H.B. and Prabha, R. (Eds.), Microbial inoculants in sustainable agricultural productivity, Springer, India, pp. 115-132.

Singh, J.S., Pandey, V.C. and Singh, D.P. 2011. Efficient soil microorganisms: A new dimension for sustainable agriculture and environmental development. Agric, *Ecosys Environ.,* 140: 339-353.

Smith, S.E. and Read, D.J. 2008. Mycorrhizal symbiosis, third edn. Academic press, San Diego, pp. 787.

Spagnoletti, F. and Lavado, R.S. 2015. The arbuscular mycorrhiza *Rhizophagus intraradices* reduces the negative effects of arsenic on soybean plants. *Agronomy,* 5: 188-199.

Stommel, M., Mann, P. and Franken, P. 2001. EST-library construction using spore RNA of the arbuscular mycorrhizal fungus *Gigaspora rosea. Mycorrhiza,* 10: 281–285.

Talaat, N.B. and Shawky, B.T. 2011. Influence of arbuscular mycorrhizae on yield, nutrients, organic solutes, and antioxidant enzymes of two wheat cultivars under salt stress. *J. Plant Nutr. Soil Sci.,* 174: 283-291.

Tavakkoli, E., Rengasamy, P. and McDonald, G.K. 2010. High concentrations of Na+ and Cl– ions in soil solution have simultaneous detrimental effects on growth of faba bean under salinity stress. *J. Exp. Bot.,* 61: 4449–4459.

Tester, M. and Davenport, R. 2003. Na⁺ tolerance and Na⁺ transport in higher plants. *Ann. Bot.,* 91: 503-527.

Tulllu, M., Pierandrei, F., Salerno, A., Rea, E. 2003. Tolerance to cadmium of vesicular arbuscular mycorrhizae spores isolated from cadmium-polluted and unpolluted soil. *Biol. Fertil. Soil,* 37: 211–214.

Turnau, K. and Mesjasz-Przybylowicz, J. 2003. Arbuscular mycorrhizal of Berkheya codii and other Ni-hyperaccumulating members of Asteraceae from ultramafic soils in South Africa. *Mycorrhiza,* 13: 185–190.

Upadhyaya, H., Panda, S.K., Bhattacharjee, M.K. and Dutta S. 2010. Role of arbuscular mycorrhizal in heavy metal tolerance in plants: prospects for phytoremediation. *J. Phytol.*, 2: 16-27.

Vallino, M., Massa, N., Lumini, E., Bianciotto, V., Berta, G. and Bonfante, P. 2006. Assessment of arbuscular mycorrhizal fungal diversity in roots of *Solidago gigantea* growing in a polluted soil in Northern Italy. *Environ. Microbiol*, 8: 971-983.

van der Heijden, M.G., Klironomos, J.N., Ursic, M., Moutoglis, P., StreitwolfEngel, R., Boller, T., Wiemken, A. and Sanders, I.R. 1998. Mycorrhizal fungal diversity determines plant biodiversity, ecosystem variability and productivity. *Nature*, 396: 69-72.

van der Heijden, M.G.A., Streitwolf-Engel, R., Riedl, R., Siegrist, S., Neudecker, A., Ineichen, K., Boller, T., Wiemken, A. and Sanders, I.R. 2006. The mycorrhizal contribution to plant productivity, plant nutrition and soil structure in experimental grassland. *New Phytol.*, 172: 739–752.

Wang, J. and Chen, C. 2006. Biosorption of heavy metals by *Saccharomyces cerevisiae*: A review. *Biotechnol. Adv.*, 24: 427–451.

Wani, S. H., Kumar, V., Shriram, V. and Sah S. K. 2016. Phytohormones and their metabolic engineering for abiotic stress tolerance in crop plants. *The Crop Journal*, doi:10.1016/j.cj.2016.01.010.

Wassermann, J.L., Mineo, L., Majumdar, S.K. and Vantyne, C. 1987. Detection of heavy metals in oak mycorrhizae of northeastern Pennsylvania forests, using X- ray microanalysis. *Can. J. Bot.*, 65: 2622-2627.

Wu, Q.S., Zou, Y.N. and He, X.H. 2010. Contributions of arbuscular mycorrhizal fungi to growth, photosynthesis, root morphology and ionic balance of citrus seedlings under salt stress. *Acta Physiol. Plant.*, 32: 297-304.

Yadav, S.K. 2010. Heavy metals toxicity in plants: An overview on the role of glutathione and phytochelatins in heavy metal stress tolerance of plants. *S. Afr. J. Bot.*, 76: 167–179.

Yamato, M., Ikeda, S. and Iwase, K. 2008. Community of arbuscular mycorrhizal fungi in coastal vegetation on Okinawa Island and effect of the isolated fungi on growth of sorghum under salt treated conditions. *Mycorrhiza*, 18: 241-249.

Yang, Y., Liang, Y., Han, X., Chiu, T-Y., Ghosh, A., Chen, H. and Tang, M. 2016. The roles of arbuscular mycorrhizal fungi (AMF) in phytoremediation and tree-herb interactions in Pb contaminated soil. *Sci. Rep.*, 6:20469 | DOI: 10.1038/srep20469.

Zaidi, A., Khan, M.S. and Amil M. 2003. Interactive effect of rhizotrophic microorganisms on yield and nutrient uptake of chickpea (*Cicer arietinum* L.). *Eur. J. Agron.*, 19: 15-21.

Zarei, M., Saleh-Rastin, N., Jouzani, G.S., Savaghebi, G. and Buscot, F. 2008. Arbuscular mycorrhizal abundance in contaminated soils around a zinc and lead deposit. *Eur. J. Soil Biol.*, 44: 381-391.

Zhi, H., Chao-Xing, H., Zhong-Qun, H., Zhi-Rong, Z. and Zhi-Bin, Z. 2010. The effects of arbuscular mycorrhizal fungi on reactive oxyradical scavenging system of tomato under salt tolerance. *Agric. Sci. China,* 9: 1150-1159.

Zuccarini, P. and Okurowska, P. 2008. Effects of mycorrhizal colonization and fertilization on growth and photosynthesis of sweet basil under salt stress. *J. Plant Nutr.,* 31: 497-513

2017, Mycorrhizal Fungi   *Pages 275–290*
Editors: Ashok Aggarwal and Kuldeep Yadav
Published by: ASTRAL INTERNATIONAL PVT. LTD., NEW DELHI

12

Potential Use of AM Fungi for Better Utilization of Fly Ash in Agroecosystem

Harbans Kaur Kehri*, Rani Mishra, Pallavi Rai
and Ovaid Akhtar

Sadasivan Mycopathology Laboratory,
Department of Botany, University of Allahabad,
Allahabad – 211 002, Uttar Pradesh
*Corresponding Author: kehrihk@rediffmail.com

ABSTRACT

More than 70 per cent of the world's energy is generated by coal-based thermal power plants. The coal-based thermal power plants are powered by combustion of bituminous and sub-bituminous coal. During this combustion huge amount of mineral residue is produced known as fly ash. The prevalent practice of disposal of such huge amounts of fly ash in India is to dump it in lagoons, ponds, semi-agriculture lands and wastelands, which not only requires more than 30,000 hectares of agricultural and forest lands, but also becomes a potential source of metal contamination in surface and groundwater, threatening human health. To minimize the hazardous effects of fly ash, large scale utilization of fly ash in different areas is gaining global attention. Efforts are being made to reutilize fly ash for more useful and profitable exploitations and commercial purposes.

Because of their ability to provide essential macro and micronutrients to plants for nutrition, fly ash is being considered for amending agricultural soils to improve both chemical and physical properties. A large number of demonstrative trials have been executed by different technological institutes and laboratories at various sites in dispersed locations across the country under varied agro-climatic conditions on a spread of crops, forestry and horticulture species. But according to a number of workers fly ash being rich in trace/heavy metals, long time repeated applications to the soil may result in hyper accumulation of the heavy metals, which may impart toxicity in the soils and hence, in plants. According to them there is a need to take care while applying fly ash in agriculture in view of the fear of much harm in the minds.

The AM fungi played most fascinating and key role in the amelioration of toxic effects of heavy metals for hosts in heavy metal contaminated soils. Application of AM fungi as bioremediation agents can be exploited suitably for the management of fly ash. In the present communication researches carried-out till date for fly ash management in agriculture through AM fungi have been reviewed.

Keywords: Fly ash, Agro-ecosystem, Heavy metal toxicity and AM fungi.

1. Introduction

More than 70 per cent of the world's energy is generated by coal-based thermal power plants. The coal-based thermal power plants are powered by combustion of bituminous and sub-bituminous coal. During this combustion huge amount of mineral residue is produced known as fly ash. In India, there are about forty major coal-based thermal power plants, which are working as the major commercial energy source. A typical modern coal-based thermal power plant burns 0.7 tons of coal per megawatt per hour, consequently generating fly ash at the rate of 0.28 tons per megawatt per hour. As per available estimates, the production of fly ash in India is likely to touch 175 million tons per year by 2020.

The prevalent practice of disposal of such huge amounts of fly ash in India is to dump it in lagoons, ponds, semi-agriculture lands and wastelands, which not only requires more than 30,000 hectares of agricultural and forest lands, but also becomes a potential source of metal contamination in surface and groundwater, threatening human health (Nawaz, 2013). Due to the seepage associated with dumped fly ashes, toxic heavy metals enter into the natural draining system, which result in siltation and clogs the system and reduces the pH and potability of water making it turbid.

To minimize the hazardous effects of fly ash, large scale utilization of fly ash in different areas is gaining global attention. Since, national and international regulations are being formulated aiming at environmental protection and eco-system conservation, the disposal of fly ash has become more and more costly. Efforts are being made to reutilize fly ash for more useful and profitable exploitations and commercial purposes.

In India, major areas for fly ash utilization are construction and biomass production. Construction area includes cement production, brick manufacturing, road embankments, sea-port fill, *etc.*, while biomass production covers agriculture, forestry, and floriculture. Because of their ability to provide essential macro and micronutrients to plants for nutrition, fly ash is being considered for amending agricultural soils to improve both chemical and physical properties (Yeledhalli *et al.*, 2007; Kene *et al.*, 1991; Pichtel and Hayes, 1990; Jala and Goyal, 2006).

2. Fly Ash in Soil Improvement

Fly ash consists of fine, glasslike particles, which range in particle size from 0.01 to 100 μm (Davison *et al.*, 1974) that are predominantly spherical in shape, has a low bulk density, high surface area and light texture (Asokan *et al.*, 2005; Jala and Goyal, 2006). It shows a wide variation in their physico-chemical and mineralogical properties depending on the nature of parent coal, conditions of combustion, type of emission control devices, storage and handling methods (Jala and Goyal, 2006). The pH of fly ash may vary from 4.5 to 12.0 depending largely on the sulphur content of the parent coal (Elseewi *et al.*, 1978). According to Ainsworth and Rai (1987) fly ashes with Ca/S ratios of less than about 2.5 generate acid extracts, whereas, fly ashes with Ca/S ratios higher than 2.5 produce alkaline extracts. Fly ash contains essential macro-nutrients like P, K, Ca, Mg and S and micro-nutrients including Fe, Mn, Zn, Cu, B and Mo. It is also substantially rich in trace elements like Hg, Co and Cr. Because of these characteristics fly ash amendment improves the physico-chemical properties of the soil to a greater extent. Lime in fly ash readily reacts with acidic components in soil and releases nutrients such as S, B and Mo in the form and amount beneficial to crop plants. Fly ash improves the physical properties and nutrient status of soil (Rautaray *et al.*, 2003). Fly ash has been used for correction of S and B deficiency in acidic soils (Chang *et al.*, 1977). Fail and Wochok (1977) reported that addition of appropriate quantities of fly ash alters the texture of sandy and clayey soil to loamy. Chang *et al.* (1977) and Page *et al.* (1979) reported modification in bulk density of soil on fly ash application. Because of the dominance of silt-size particles in fly ash, this material may often be substituted for topsoil in surface mine lands, thereby enhancing physical conditions of soil, especially water holding capacity.

3. Fly Ash in Agriculture

From extensive studies carried out by various research and development agencies on varied agro climatic conditions and soil crop combinations with broad objectives of building confidence towards safe disposal and scientific utilization of fly ash, it has been shown that fly ash has a vast potential for being utilized in terrestrial and agroecosystems (Kalra *et al.*, 2003; Singh *et al.*, 2000). A large number of demonstrative trials have been executed by different technological institutes and laboratories at various sites in dispersed locations across the country under varied agro-climatic conditions on a spread of crops, forestry and horticulture species (Cervelli *et al.*, 1987; Chapman, 1984; Dhankhar, 2003; Furr *et al.*, 1977; Jha *et al.*, 2000; Mandal and Saxena, 1998). They have concluded that, the fly ash is an important resource material for agriculture.

Application of weathered coal fly ash at 5 per cent resulted in higher seed germination rate and root length of lettuce (*Lactuca sativa*) (Lau and Wong, 2001). Many researchers have demonstrated that fly ash amendment increased the biomass production/crop yield of wheat (*Tritiucm aestivum*), alfalfa (*Medicago sativa*), barley (*Hordeum vulgare*), bermuda grass (*Cynodon dactylon*), sabai grass (*Eulaiopsis binata*), mung (*Vigna unguiculata*) and white clover (*Trifolium repens*) (Garg *et al.*, 2005; Page *et al.*, 1979; Grewal *et al.*, 2001; Hill and Lamp, 1980; Martens, 1971; Elseewi *et al.*,

1980a; Elseewi *et al.*, 1980b; Weinstein *et al.*, 1989; Sridhar *et al.*, 2006; Basu *et al.*, 2006; Thetwar *et al.*, 2006). Kalra *et al.* (2003) reported that the shoot and root growth and yield of wheat (*Triticum aestivum* L.), mustard (*Brassica juncea* L.), lentil (*Lens esculenta* Moench.), rice (*Oryza sativum* L.) and maize (*Zea mays* L.) increased by the addition of varying levels of fly ash at the time of sowing/transplantation. Fly ash amendment also improved the performance of oilseed crops such as sunflower (*Helianthus* sp.), sesame (*Sesamum indicum*), turnip (*Brassica rapa*) and groundnut (*Arachis hypogaea*) (Jala and Goyal, 2006; Inam, 2007, Basu *et al.*, 2007; Thetwar *et al.*, 2006; Sao *et al.*, 2007).

Furr *et al.* (1977) demonstrated that alfalfa (*Medicago sativa*), sorghum (*Sorghum bicolor*), field corn (*Zea mays*), millet (*Echinochloa crusgalli*), carrots (*Daucus carota*), onion (*Allium cepa*), beans (*Phaseolus vulgaris*), cabbage (*Brassica oleracea*), potatoes (*Solanum tuberosum*) and tomatoes (*Lycopersicon esculentum*) grew on a slightly acidic soil (pH 6.0) treated with 125 MT ha of un-weathered fly ash and that these crops showed higher contents of As, B, Mg and Se. Fly ash amendments have corrected plant nutritional deficiencies of Mg (Hill and Lamp, 1980), Mo (Elseewi *et al.*, 1980b), S (Elseewi *et al.*, 1980b; Hill and Lamp, 1980) and Zn (Martens, 1971). This resulted in better growth and yield of plants.

Fly ash has also been tried alone or in combination with materials like farm yard manure, sewage sludge, water hyacinth, microbial cultures, gypsum and lime and has been found to improve the growth, yield and nutrients uptake in various agricultural crops, plantations and vegetables.

4. Limitations of Fly Ash Utilization

Besides the pros of fly ash utilization in agriculture, there are several cons associated with long term use of fly ash. Fly ash is deficient in N because it is volatilized during the combustion. It is also deficient in P and is low in microbial activity. Moreover, being rich in trace/heavy metals, long time repeated applications to the soil may result in hyper accumulation of the heavy metals, which may impart toxicity in the soils and hence, in plants (Asokan *et al.*, 1995; Saxena *et al.*, 1998). Because of these limitations, the sole application of fly ash has been reported to reduce the germination and establishment of transplants.

Though the researches till date have proved that Indian fly ash is safer than those produced in other countries especially on account of lower content of sulphur, toxic/heavy elements and radio nuclides, care must be taken while applying fly ash in agriculture in view of the fear of much harm in the minds.

In order to nullify the adverse effects of fly ash and to improve the N and P status of soils and crops there is a need to explore the potentialities of bio-inoculants, especially the nitrogen fixers, phosphate (P) solubilizers and P-scavengers; the arbuscular mycorrhizal (AM) fungi. The most fascinating and key role of AM fungi played for hosts in fly ash is the amelioration of toxic effects of heavy metals (del Vel *et al.*, 1999; Joner *et al.*, 2000; Jamal *et al.*, 2002). In fact, it is the AM fungi which protect the host plants from heavy metal toxicity in the fly ash.

5. Arbuscular Mycorrhizal (AM) Fungi

Arbuscular mycorrhizal (AM) fungi belong to phylum glomeromycota, form symbiotic mycorrhizal association with the roots of most of terrestrial plants. It has been established during the last few decades that this association is helpful in improving the overall performance of the plants. This association is gaining importance in the reclamation of various types of lands and researches on AM fungi have reached a stage of practical application in many countries for reclaiming the adverse sites. AM fungi have a number of assets to prove it a magic tool *e.g.* AM fungi increase the absorptive surface of roots manifold and improve the uptake of nutrients and water resulting in better performance of host plants. Mycorrhization ensures better supply of water and nutrients especially phosphorus, calcium and zinc to the plants, thereby improved growth and productivity. AM fungi have been reported to increase the activity of nitrogen fixing organisms. AM fungi play a vital role in mineral cycling, energy flow and plant establishment in disturbed ecosystems, arid and semi-arid zones as well as wastelands. They have been proved to be a potential tool for reclamation of alkali, calcareous, saline and other types of problematic soils (Tiwari *et al.*, 2008). Their ability to initiate plant succession in virgin lands, to increase tolerance to heavy metals and drought has been recognized. AM fungi have been shown to decrease transplanting shocks and improve soil structure. Their association with plants has been reported to affect osmotic balance and hydraulic conductivity of roots and improve plant-water relations. It has also been reported to increase lignifications of root tissue, increase chitinolytic activity, higher concentration of phosphorus and amino acids or production of phytoalexins. Mycorrhizal association has been shown to affect the hormone levels, initiate rooting and helps in the root development. AM technology has proved to be effective even under unfavourable conditions related to nutrients or water. The technology has proved to be helpful not only in the establishment but also the growth and productivity of plants.

There are conclusive evidences to show that AM fungi play a vital role in metal tolerance (del Vel *et al.*, 1999) and accumulation (Jamal *et al.*, 2002). External mycelium of AM fungi provides a wider exploration of soil volumes by spreading beyond the root exploration zone thus providing access to greater volume of heavy metals present in the rhizosphere. A greater volume of metals is also stored in the mycorrhizal structures in the root and in spores. For example, concentrations of over 1200 mg kg^{-1} of Zn have been reported in fungal tissues of *Glomus mosseae* and over 600 mg kg^{-1} in *G. versiforme* (Chen *et al.*, 2001). Another important feature of this symbiosis is that AM fungi can increase plant establishment and growth despite high levels of soil heavy metals (Enkhtuya, 2002), due to better nutrition (Chen *et al.*, 2001; Feng *et al.*, 2003), water availability (Auge, 2001) and soil aggregation properties (Rillig and Steinberg, 2002; Kabir and Koide, 2000) associated with this symbiosis.

Several biological and physical mechanisms have been proposed to explain metal tolerance of AM fungi and AM fungal contribution in metal tolerance to host plants. Immobilization of metals in the fungal biomass is one such mechanism involved (Zhu *et al.*, 2001; Li and Christie, 2001). Reduced transfer, as indicated by enhanced root/shoot Cd ratios in AM plants, has been suggested as a barrier in metal transport (Joner *et al.*, 2000; Tullio *et al.*, 2003). This may occur due to intracellular precipitation of metallic cations with PO^{4-}. Turnau *et al.* (1993) have demonstrated greater accumulation of Cd, Ti and Ba in fungal structures than in the host plant cells. Uptake into hyphae may be influenced by absorption on hyphal walls as chitin has an important metal-binding capacity (Zhou, 1999). Thus, AM fungal metal tolerance includes adsorption of heavy metal onto fungal cell walls present on and in plant tissues, or onto or into extraradical mycelium in soil (Joner *et al.*, 2000), chelation by compounds such as siderophores and metallothionens released by fungi or other rhizosphere microbes, and sequestration by plant-derived compounds like phytochelatins or phytates (Joner and Leyval, 1997). These fungi also secrete some Glomalin Related Soil Proteins (GRSPs) from the hyphae into the soils, which form water-stable aggregates (Rillig and Steinberg, 2002; Kabir and Koide, 2000). Other possible metal tolerance mechanisms include dilution by increased root or shoot growth, exclusion by precipitation onto polyphosphate granules, and compartmentalization into plastids or other membrane-rich organelles (Turnau *et al.*, 1993; Kaldorf *et al.*, 1999). Indirect mechanisms include the effect of AM fungi on rhizosphere characteristics such as changes in pH (Li *et al.*, 1991), microbial communities (Olsson *et al.*, 1998) and root-exudation patterns (Laheurte *et al.*, 1990). AM fungi associated with metal-tolerant plants, *e.g.* metallophytes, may contribute to the accumulation of heavy metals in plant roots in a non-toxic form. The accumulation of heavy metals in the fungal structures as suggested by their high heavy metal-binding capacity (Joner *et al.*, 2000) could represent a biological barrier. Reduced Cd translocation from roots to shoots in the presence of AM fungi in roots of *Trifolium subterraneum* has been shown (Schüepp *et al.*, 1987; Joner and Leyval, 1997).

6. AM Fungi vs Fly Ash

AM fungi play key roles for establishment, survival of plant species, and improve soil properties in stressed environments (Ortega-Larrocea *et al.*, 2010) including fly ash treated soil. As plants grow, their roots along with AM fungal mycelium hold the fly ash together, making it less prone to being air-borne. AM fungi provide nutrition to the plants by sequestering the nutrients from the fly ash and translocating them to the plants. This makes the utilization of the nutrients highly efficient, which ultimately enhance the biomass production in associated plants. In addition, as already discussed in the previous sections, AM fungi decrease heavy metal toxicity in fly ash after few years of plantations. The AM fungi through their mycelial network accumulate heavy metals from fly ash and retain them within their cells. Besides, supplying nutrients, it produces acids that combine with some heavy metals to form compounds that are less mobile and less likely to pollute groundwater by surface runoff. Moreover, the contribution of AM fungi in heavy metal rich soils and fly ash may depend upon AM fungi, nature of fly ash, concentration and types of heavy metals in them, associated host plants, pH (Li *et al.*, 1991), microbial communities (Olsson *et al.*, 1998), root-exudation patterns (Laheurte *et al.*, 1990), nutrients status of fly ash and local environmental conditions.

Several AM fungi have been reported from the fly ash deposits and their potential roles have been described for the plants' establishment in the fly ash dumping sites.

Selvam and Mahadevan (2002) surveyed AM fungi in an abandoned lignite fly ash pond, overburden dumps and reclaimed overburden dumps of Neyveli, Tamil Nadu. In ash pond, 15 AM species (*Acaulospora gerdemannii, Gigaspora decipiens, G. gigantea, G. margarita, Glomus citricola, G. fasciculatum, G. formosanum, G. fulvum, G. maculosum, G. magnicaule, G. mosseae, G. tenebrosum, Sclerocystis pachycaulis, Scutellospora erythropa* and *S. fulgida*) were isolated. From the overburden and reclaimed overburden dumps, 4 AM species (*G. fasciculatum, G. mosseae, Sclerocystis microcarpus,* and *Scutellospora verrucosa*) and 13 AM fungal species (*Acaulospora gerdemannii, Entrophospora colombiana, Gigaspora gigantea, G. margarita, Glomus fasciculatum, G. macrocarpum, G. mosseae, G. vesiculiferum, Sclerocystis pachycaulis, S. sinuosa, Scutellospora coralloidea, S. erythropa* and *S. persica*) were isolated, respectively. In all the sites, *Glomus mosseae* was the dominant AM fungus.

Babu and Reddy (2011a) studied root colonization and diversity of AM fungi in plants growing in fly ash pond. Eight species could be separated morphologically, while phylogenetic analyses after PCR amplification of the ITS region followed by RFLP and sequencing revealed seven different AM fungal sequence types. Phylogenetic analysis showed that these sequences cluster into four discrete groups, belonging to the genus *Glomus* and *Archaeospora*.

Kulshreshtha and Khan (1999) studied the effect of *Glomus caledonium* and *Rhizobium* sp. on the roots of *Vigna mungo* grown in fly ash. They have found that AM fungi and root nodulating bacterium protected the plants from the harmful effects caused by fly ash.

Adholeya (2000) used the AM technology at the fly ash dumping sites of two power stations in Badarpur (Delhi) and Korba (Chattisgarh). They have utilized the AM technology for the plantation of *Shorea robusta, Tectona grandis, Dalbergia sissoo, Albizzia procera, Casuarina equisetifolia, Melia azadirach, Acacia nilotica, Dendrocalamus strictus, Populus euphratica, Eucalyptus tereticornis, Bombex ceiba, Gmelina arborea, Mentha arvensis, Vetiver zizanoides, Polianthes tuberose, Helianthus* sp., *Tagetes erecta, Sesbania aculeata and Agave sisalana.* On soil amendment with organic matter (farmyard manure and compost) without any inorganic fertilizer, they have recorded a significant restoration in soil quality with respect to porosity and water holding capacity, percentage nitrogen, available phosphorus, available potassium, organic carbon, dehydrogenase activity and total microbial population after two years of plantation. Further, heavy metal toxicity in fly ash also decreased as a result of reclamation activities. The height of various tree species at Korba showed a tremendous improvement during the growth period.

Medicinal plants such as cornmint (*Mentha arvensis*) and vetiver (*Vetiver zizanoides*) have been successfully planted in fly-ash mixed with 20 per cent farmyard manure and mycorrhiza (Adholeya *et al.*, 1997; Sharma *et al.*, 2001a; Sharma *et al.*, 2001b). Amendment of different fly ash soil combinations resulted in high yield of aromatic grasses, particularly palmarosa (*Cymbopogon martini*) and citronella (*Cymbopogon nardus*), which was due to increased availability of major plant nutrients.

Reddy and Garampalli (2002) studied the effect of fly ash amendment at three concentrations (10, 20 and 30 per cent) on the infectivity of *Glomus aggregatum* in low fertile soil using pigeon pea (*Cajanus cajan*) cv. *Maruti* and chickpea (*Cicer arietinum*) cv. *Annigeri*. They have found negative impact of fly ash concentration on AM colonization. They further reported positive impact of fly ash alone on both the plants. In terms of growth response, *G. aggregatum* was found more effective in chickpea than in pigeon pea.

Bi *et al.* (2003) in an experiment with AM fungi and fly ash, demonstrated that, root colonization with AM fungi led to the successful growth of maize in soil overlying fly ash. According to them, AM fungi can assist plants in the exploitation of nutrients from fly ash and may help them to resist excessive salt (Na) accumulation in the shoots. AM fungi may also contribute to the re-establishment of a general microflora, and of a sustainable agricultural system combined with the appropriate use of fertilizers. They have reported that, external hyphae of the AM fungi contribute to P uptake from the substrate, possibly including the fly ash. AM fungi alter the soil microbial communities in rhizosphere directly or indirectly through changes in root exudation patterns (Barea *et al.*, 2005) and enhance the soil enzyme activities (Wang *et al.*, 2006). AM fungi form hyphal network in outside the roots and along with dense root, assist in binding of fly ash particles.

Singh *et al.* (2006) reported that AM inoculation in *Jatropha curcas* plants growing in fly ash showed a significant increase in the shoot and root dry mass in comparison to non-inoculated plants. This was mainly due to the increased P concentration in the root and shoot, which was higher in AM inoculated plant than non-inoculated plant.

Inoculation of plants with spores of AM fungal consortia (*Glomus etunicatum, Glomus heterogama, Glomus maculosum, Glomus magnicaule, Glomus multicaule, Glomus rosea, Scutellospora heterogama*, and *Scutellospora nigra*) along with colonized root pieces increased the growth (84.9 per cent), chlorophyll (54 per cent), and total P content (44.3 per cent) of *Eucalyptus tereticornis* seedlings grown on fly ash compared to non-inoculated seedlings (Babu and Reddy, 2011a). The growth improvement was due to increased P nutrition and decreased Al, Fe, Zn, and Cu accumulations.

Rollie Verma (2008) while working on the salt affected soils of Handia, Allahabad found that these soils were not supportive for the establishment and growth of plants due to high pH, high EC and poor nutrient status. Even these soils were not conducive for the growth of microbial population *viz.* Arbuscular mycorrhizae, PSB and N_2 fixer. She utilized the capacity of fly ash to buffer soil pH and to decrease the EC of the salt affected soil. Addition of 45 per cent fly ash series showed maximum reduction in pH and EC. This fly ash when supplemented with organic matter (*Cynodon*/compost) a synergistic effect on the growth performance and microbial population was recorded. Better growth performance of *Cicer arietinum* and *Phaseolus mungo* was recorded with increasing concentrations of fly ash *i.e.* 15, 30 and 45 per cent. Best growth performance for both the crops was recorded at 45 per cent fly ash concentration. *Cynodon* was better organic input in comparison to compost for improving the nutrient status of the soil and growth of the crops. Triple inoculation (AM fungi, PSB and N_2 fixer) gave better results in comparison to non-inoculated control and single as well as dual inoculations. Triple inoculation (AM fungi, PSB and N_2 fixer) along with 45 per cent fly ash and *Cynodon* gave best performance in terms of establishment, survival, root/shoot biomass and yield.

In another experiment, Babu and Reddy (2011b) studied the effect of fly ash-adapted AM fungi and phosphate solubilizing fungus *Aspergillus tubingensis* on the growth, nutrient, and metal uptake of bamboo (*Dendrocalamus strictus*) plants grown in fly ash. Dual inoculation of these fungi significantly increased the P (150 per cent), K (67 per cent), Ca (106 per cent), and Mg (180 per cent) in shoot tissues compared to control plants. It was also found that, Al and Fe content were significantly reduced (50 per cent and 60 per cent, respectively) due to the presence of AM fungi and *A. tubingensis*. The physicochemical and biochemical properties of fly ash were also improved compared to those of individual inoculation and control. The results clearly showed that combination of AM fungi and *A. tubingensis* elicited a synergistic effect by increasing plant growth and uptake of nutrients with reducing heavy metal translocation.

Pallavi Rai (2014) reported a high rate of mortality in fly ash added series at the time of emergence. In comparison to 10 per cent fly ash added series, 20 per cent and 30 per cent fly ash series showed more mortality of the plants of pea and cowpea. However, addition of organic matter (*Cynodon*) and other bio inoculants *viz.*, PSF (Phosphate Solubilizing Fungi), *Rhizobium* and AM fungi caused a significant reduction in rate of mortality while dual inoculation (AM+ *Rhizobium*) and triple inoculation (PSF + AM + *Rhizobium*) series completely nullified the ill effects of fly ash. Her findings clearly showed that the microbial inoculants improved not only the mycorrhizal status of these two crops, but also their biomass and yield, however, the degree of improvement varied with the type of soil treatment and the microbial inoculants. Photosynthetic pigments (Chlorophyll a, Chlorophyll b) content increased with addition of 10 per cent fly ash, while decreased with addition of 20 per cent fly ash and 30 per cent fly ash, without any amendment while carotenoids content increased with addition of 30 per cent fly ash, while decreased with addition of 10 per cent fly ash and 20 per cent fly ash, without any amended. Proline accumulation was more in 20 per cent and 30 per cent fly ash series as well as in combination with organic matter in comparison to control. All other series showed a significant decrease in the proline content. However the minimum proline content was recorded in a series where soil was amended with organic matter, 10 per cent fly ash and inoculated with AM, PSF and nitrogen fixer. Addition of lower dose of fly ash *i.e.*, 10 per cent fly ash in a soil gave the best performance of both the crops. In her investigation fly ash was found beneficial for the cultivation of cowpea and pea and showed immense potential to improve their growth, nodulation and yield especially in combination with organic matter and microbial inoculants.

Singh (2014) recorded *Acaulospora scrobiculata, A. denticulata, Glomus mosseae, G. clarum, G. deserticola, G. fasciculatum, G. tortuosum, Glomus intraradicies* and *Gigaspora* sp. as the most dominant species present in the rhizosheric soils of the plants growing in the vicinity of fly ash dumping site (IFFCO, Phulpur). An adverse effect of high amount of fly ash (30 per cent and 45 per cent) in the agricultural soil was recorded in terms of reduced root/shoot biomass and quality and quantity of flowers in *Gladiolus grandiflorus* cvs. Snow Princess, Picardy and Pricilla, while 15 per cent fly ash addition enhanced the root/shoot biomass and quality and quantity of flowers. Slight increase in the root/shoot biomass and quality and quantity of flowers in all the three cultivars of *Gladiolus grandiflorus* was recorded when organic matter (*Cynodon*) was added alone or in combination of fly ash. Addition of consortium of AM fungi native to fly ash site nullified the adverse effects of heavy metals present in fly ash to a greater extent and improved the overall performance of all the three cultivars of *Gladiolus grandiflorus* by providing nutritional as well as non-nutritional benefits. Inoculation with the consortium of AM fungi native to fly ash site also altered the physiology of the plants. In comparison to non-mycorrhizal plants, mycorrhizal plants showed increase in photosynthetic pigments and protein content and reduction in proline, ascorbic acid content and catalase activity. The best performance of all the three cultivars of *Gladiolus grandiflorus* in terms of growth and physiological parameters was recorded when the soil was amended with *Cynodon* and 45 per cent fly ash and inoculated with the consortium of AM fungi and PSF (*Aspergillus awamori*). Amongst all the three cultivars of *Gladiolus grandiflorus,* cv. Pricilla showed the maximum improvement in all the parameters.

It is now established that, AM fungi do much more for host plants in fly ash, however, spontaneous selection of infective and effective AM fungi for fly ash management is a long process. It has been demonstrated that the use of adapted AM fungal strains, in restoration and bioremediation studies is more effective than applying non-adapted strains (Vivas *et al.,* 2005).

7. Conclusions and Future Prospective

Application of AM fungi as bioremediation agents can be exploited suitably for the management of fly ash. The regulated use of fly ash in agronomy not only improve the soil nutrient status, as it contains almost all the essential micro and macronutrients, but also solve the disposal problem. Fly ash may be used as substitute for lime in acidic soils or as fertilizers amendments. By improvement in physical and nutritional properties, fly ash increase the productivity of plants in terms of yield and biomass. The major limitation of long term fly ash amendment in the agricultural and terrestrial eco-system is the toxic effect caused by the heavy metals found in them. AM fungi provide tremendous opportunity in reducing this toxic effect of heavy metals. Application of AM fungi in fly ash amended soils or in fly ash over burden dumps greatly increase the remediation of substrates in zero time and provide a biological means of assuring plant production at a low cost.

Despite of many researches carried-out till date, the fly ash management through AM fungi is still a thirst area of research. AM fungal technology have wide vistas and opportunities to solve the problems associated with fly ash use in soil improvement and biomass production, but there is a need of upsurge in this field of research. The AM fungi studied till date in fly ash dumping sites do not cover even a fraction of actual diversity, because they are more diverse than we see with our eyes. There is a need for exploration of AM fungi from such sites. They must be carrying some very important and fascinating features, which could be utilized in the fly ash management and this will surely enhance the productivity and sustainability of terrestrial and agroecosystem.

References

Adholeya, A. 2000. Utilization of fly ash for commercial plant production and environmental protection using microbes. In: Second International Conference on Fly Ash Disposal and Utilization, New Delhi, Vol. II, pp. 32-35.

Adholeya, A., Sharma, M.P. and Bhatia, N.P. 1997. Mycorrhiza biofertilzer: a tool for reclamation of wasteland and bioremediation. In: National symposium on microbial technologies for environmental management and resource recovery. New Delhi, India, TERI.

Ainsworth, C.C. and Rai, D. 1987. *Chemical characterization of fossil fuel combustion wastes: Final report* (No. EPRI-EA-5321). Pacific Northwest Lab., Richland, WA (USA); Electric Power Research Inst., Palo Alto, CA (USA).

Asokan, P., Saxena, M. and Asolekar, S.R. 2005. Coal combustion residues environmental implications and recycling potentials. *Resour. Conserv. Recycl.*, 43: 239-262.

Asokan, P., Saxena, M. and Bose, S.K.Z. 1995. In: Proceedings of the workshop on Fly ash Management in the State of Orissa. Bhubhaneshwar, India, pp. 64-75.

Augé, R.M. 2001. Water relations, drought and vesicular-arbuscular mycorrhizal symbiosis. *Mycorrhiza*, 11: 3-42.

Babu, A.G. and Reddy M.S. 2011a. Diversity of Arbuscular Mycorrhizal Fungi Associated with Plants Growing in Fly Ash Pond and Their Potential Role in Ecological Restoration, *Curr. Microbiol.*, 63: 273-280.

Babu, A. G. and Reddy M. S. 2011b. Dual Inoculation of Arbuscular Mycorrhizal and Phosphate Solubilizing Fungi Contributes in Sustainable Maintenance of Plant Health in Fly Ash Ponds. *Water Air Soil Pollut.*, 219: 3-10.

Barea, J.M., Pozo, M.J., Azcón, R. and AzcoÂn-Aguilar, C. 2005. Microbial co-operation in the rhizosphere. *J. Exp. Bot.*, 56: 176-178.

Basu, M., Bhadoria, P.B.S. and Mahapatra, S.C. 2007. Role of soil amendments in improving groundnut productivity of acid lateritic soils. *Int. J. Agric. Res.*, 2(1):87-91.

Basu, M., Mahapatra, S.C. and Bhadoria, P.B.S. 2006. Exploiting fly ash as soil ameliorant to improve productivity of Sabai grass (*Eulaliopsis binata* (Retz) C.E.) under acid lateritic soil of India. *Asian J. Plant Sci.*, 5. 1027-1030.

Bi, Y. L., Li, X. L., Christie, P., Hu, Z. Q. and Wong, M. H. 2003. Growth and nutrient uptake of arbuscular mycorrhizal maize in different depths of soil overlying coal fly ash. *Chemosphere*, 50: 863-869.

Cervelli, S., Petruzzelli, G. and Perna, A. 1987. Fly ashes as an emendment in cultivated soils. *Water, Air, Soil Pol.*, 33: 331-338.

Chang, A.C., Lund, L.J., Page, A.L. and Warneke, J.E. 1977. Physical properties of fly ash-amended soils. *J. Environ. Qual.,* 6: 267-270.

Chapman, S.L. 1984. Fly ash as a fertilizer and lime source in Arkansas. In *Arkansas Academy Science Proceedings*, 28: 20-22.

Chen, B., Christie, P. and Li, X. 2001. A modified glass bead compartment cultivation system for studies on nutrient and trace metal uptake by arbuscular mycorrhiza. *Chemosphere*, 42: 185-192.

Davison, R.L., Natusch, D.F., Wallace, J.R. and Evans Jr, C.A. 1974. Trace elements in fly ash. Dependence of concentration on particle size. *Envir. Sc. and Technol.,* 8: 1107-1113.

del Val, C., Barea, J.M. and Azcón-Aguilar, C., 1999. Assessing the tolerance to heavy metals of arbuscular mycorrhizal fungi isolated from sewage sludge-contaminated soils. *Appl. soil Ecol.*, 11: 261-269.

Dhankhar, R. 2003. Impact of thermal power plant discharge on crop plant harvested soils. *J. Ind. Env. Protec.,* 23: 519-524.

Elseewi, A.A., Grimm S.R. and Page, A.L. 1980a. Chemical characterization of fly ash aqueous systems. *J. Environ. Qual.*, 9: 424-428.

Elseewi, A.A., Straughan, I.R. and Page, A.L. 1980b. Sequential cropping of fly ash-amended soils: Effects on soil chemical properties and yield and elemental composition of plants. *Sci. Tot. Environ.,* 15: 247-259.

Enkhtuya, B., Rydlová, J. and Vosátka, M. 2002. Effectiveness of indigenous and non-indigenous isolates of arbuscular mycorrhizal fungi in soils from degraded ecosystems and man-made habitats. *Appl. Soil Ecol.,* 14: 201-211.

Fail, Jr. J.L. and Wochok, Z.S. 1977. Soyabean growth on fly ash amended strip mine spoils. *Plant Soil*, 48: 473-484.

Feng, G., Song, Y.C., Li, X.L. and Christie, P. 2003. Contribution of arbuscular mycorrhizal fungi to utilization of organic sources of phosphorus by red clover in a calcareous soil. *Appl. Soil Ecol.*, 22: 139-148.

Furr, A.K., Parkinson, T.F., Hinrichs, R.A., Van Campen, D.R., Bache, C.A., Gutenmann, W.H., St. John Jr, L.E., Pakkala, I.S. and Lisk, D.J. 1977. National survey of elements and radioactivity in fly ashes. Absorption of elements by cabbage grown in fly ash-soil mixtures. *Env. Sci. Technol.,* 11: 1194-1201.

Garg, R.N., Pathak, H. and Das, D.K. 2005. Use of fly ash and biogas slurry for improving wheat yield and physical properties of soil. *Environ. Monit. Assess.,* 107: 1-9.

Grewal, K.S., Yadav, P.S. and Mehta, S.C. 2001. Direct and residual effect of fly ash application to soil on crop yield and soil properties. *Crop Res.,* 21: 60-65.

Hill, M.J. and Lamp, C.A. 1980. Use of pulverized fuel ash from Victorian brown coal as a source of nutrients for pasture species. *Ani. prod. Sci.,* 20: 377-384.

Inam, A. 2007. Use of fly ash in turnip (*Brassica rapa* L.) cultivation. *Pollut. Res.,* 26: 39-42.

Jamal, A., Ayub, N., Usman, M. and Khan, A.G., 2002. Arbuscular mycorrhizal fungi enhance zinc and nickel uptake from contaminated soil by soybean and lentil. *J. I. Phytoremediation*, 4: 205-221.

Jala, S. and Goyal, D. 2006. Fly ash as a soil ameliorant for improving crop production-a review. *Bioresource Technol.*, 97:1136-1147.

Jha, R.K., Anbazhagan, B., Srivastava, N.K., Jha, G.K., Jha, S.K., Das, M.C., Tripathi, R.C., Roy, R.R.P., Gupta, S.K., Tripathi, P.S.M., Singh, G., Manoharan, V. and Sanmugasundaram, R. 2000. Proc. 2nd Intern. Conf. fly ash disposal and utilization edited by Varma C.V.J., Rao S.V.R., Kumar V. and Krishnamurthy R. published by Central Board of Irrigation and Power, New Delhi. Vol. I, pp. 20-31.

Joner, E.J., Briones, R. and Leyval, C. 2000. Metal-binding capacity of arbuscular mycorrhizal mycelium. *Plant Soil*, 226: 227-234.

Joner, E.J. and Leyval, C. 1997. Uptake of 109Cd by roots and hyphae of a *Glomus mosseae/Trifolium subterraneum* mycorrhiza from soil amended with high and low concentrations of cadmium. *New Phytol.*, 135: 353-360.

Kabir, Z. and Koide, R.T. 2000. The effect of dandelion or a cover crop on mycorrhiza inoculum potential, soil aggregation and yield of maize. *Agri., Eco. Env.*, 78: 167-174.

Kaldorf, M., Kuhn, A. J., Schroder, W. H., Hildebrandt, U. and Bothe, H. 1999. Selective element deposits in maize colonized by a heavy metal tolerance conferring arbuscular mycorrhizal fungus. *J. Pl. Physiol.*, 154: 718-728.

Kalra, N., Jain, M.C., Joshi, H.C., Chaudhary, R., Kumar, S., Pathak, H., Sharma, S.K., Kumar, V., Kumar, R., Harit, R.C. and Khan, S.A. 2003. Soil properties and crop productivity as influenced by fly ash incorporation in soil. *Env. Monit. Assess.*, 87: 93-109.

Kene, D.R., Lanjewar, S.A., Ingole, B.M. and Chaphale, S.D. 1991. Effect of application of fly ash on physico-chemical properties of soil. *J. Soils Crops.*, 1: 11-18.

Kulshreshtha, M. and Khan, M.W. 1999. Impact of sulphur dioxide on root colonization by VAM fungi and nodulation on blackgram. *Ind. Phytopath.*, 52: 182-184.

Laheurte, F., Leyval, C. and Berthelin, J. 1990. Root exudates of maize, pine and beech seedlings influenced by mycorrhizal and bacterial inoculation. *Symbiosis*, 9: 111-116.

Lau, S.S.S. and Wong, J.W.C. 2001. Toxicity evaluation of weathered coal fly ash amended manure compost. *Water Air Soil Pollut.*, 128: 243-254.

Li, X. and Christie, P. 2001. Changes in soil solution Zn and pH and uptake of Zn by arbuscular mycorrhizal red clover in Zn-contaminated soil. *Chemosphere*, 42: 201-207.

Li, X.L., George, E. and Marschner, H. 1991. Phosphorus depletion and pH decrease at the root–soil and hyphae–soil interfaces of VA mycorrhizal white clover fertilized with ammonium. *New Phytol.*, 119: 397-404.

Mandal, S. and Saxena, M. 1998. Pp. 1 in Proc. of Regional workshop cum symposium on fly ash disposal and utilization KSTPS, Kota, Rajasthan, September. pp 15-16.

Martens, D.C. 1971. Availability of plant nutrients in fly ash. *Compos. Sci.*, 12: 15-19.

Nawaz I. 2013. Disposal and Utilization of Fly Ash to Protect the Environment. *I. Jour. Inn. Res. Sci. Eng. Tech.*, 2: 5259-5266.

Olsson, P.A., Francis, R., Read, D.J. and Söderström, B. 1998. Growth of arbuscular mycorrhizal mycelium in calcareous dune sand and its interaction with other soil microorganisms as estimated by measurement of specific fatty acids. *Plant Soil*, 201: 9-16.

Ortega-Larrocea M.D., Xoconostle-Cázares, B., Maldonado-Mendoza, I.E., Carrillo-González, R., Hernández-Hernández, J., Garduño, M.D., López-Meyer, M., Gómez-Flores, L., and González-Chávez, M.C.A. 2010. Plant and fungal biodiversity from metal mine wastes under remediation at Zimapan, Hidalgo, Mexico. *Environ. Pollut.*, 158: 1922-1931.

Page, A.L., Elseewi, A.A. and Straughan, I.R. 1979. Physical and chemical properties of fly ash from coal-fired power plants with special reference to environmental impacts. *Resi. Rev.*, 71: 83-120.

Pichtel, J.R. and Hayes, J.M. 1990. Influence of fly ash on soil microbial activity and populations. *J. Env. Quality*, 19: 593-597.

Rai, Pallavi 2014. Effect of fly ash application on the growth, nodulation and mycorrhization in certain legumes. D. Phil. Thesis, University of Allahabad, Allahabad.

Rautaray, S.K., Ghosh, B.C. and Mittra, B.N. 2003. Effect of fly ash, organic wastes and chemical fertilizers on yield, nutrient uptake, heavy metal content and residual fertility in a rice–mustard cropping sequence under acid lateritic soils. *Bioresource Technol.*, 90: 275-283.

Reddy, C.N. and Garampalli, H.R. 2002. Effect of fly ash on VAM formation and growth response of pulse crops infested with *Glomus aggregatum* in sterile soil. In: Manohrachary, C., Purohit, D.K., Ram Reddy, S., Singra Charya, M.A. and Girishm, S. (Eds.), *Frontiers in Microbial Biotechnology and Plant Pathology*, pp. 205-212.

Rillig, M.C. and Steinberg, P.D. 2002. Glomalin production by an arbuscular mycorrhizal fungus: a mechanism of habitat modification? *Soil Biol. Biochem.*, 34: 1371-1374.

Sao, S., Gothalwal, R. and Thetwar, L.K. 2007. Effects of fly ash and plant hormones treated soil on the increased protein and amino acid contents in the seeds of ground nut (*Arachis hypogaea*). *Asian J. Chem.*, 19: 1023-1026.

Saxena, M., Chauhan, A. and Asokan, P. 1998. Fly ash vermicompost form non-ecofroendly organic wastes. *Polln. Res.*, 17: 5-11.

Selvam, A. and Mahadevan, A. 2002. Distribution of mycorrhizas in an abandoned fly ash pond and mined sites of Neyveli Lignite Corporation, Tamil Nadu, India. *Basic Appl. Ecol.*, 3: 277-284.

Sharma, M.P. Tanu, U. and Reddy, G. 2001a. Herbage yield of Mentha arvensis DC. Holms as influenced by AM fungi inoculation and farmyard manure application grown in fly ash over burdens amended with organic matter. In: International ash utilization symposium, KY, USA.

Sharma, M.P., Tanu, U. and Adholeya, A. 2001b. Growth and yield of Cymbopogon martini as influenced by fly ash, AM fungi inoculation and farmyard manure application. In: Proceedings of the 7th international symposium on soil and plant analysis, Edmonton, AB, Canada.

Singh, A.P. 2014. Impact of Fly Ash and AM Fungi on the Flower Quality of Gladioli. D. Phil. Thesis, University of Allahabad, Allahabad.

Singh, R., Adholeya, A. and Mukerji, K.G. 2000. Mycorrhiza in Control of Soil- Borne Pathogens. In: Mukerji, K.G., Chamola, B.P. and Singh, J. (Eds.). Mycorrhizal Biology, Academic Plenum Publishers, New York, pp. 173-196.

Singh, R., Sharma, M.P. and Adholeya, A. 2006. Screening of *Jatropha curcas* germplasm from different provenances for cultivation in fly ash overburdens using arbuscular mycorrhiza fungi. *Mycorrhiza News*, 18: 24-27.

Sridhar, R., Duraisamy, A. and Kannan, L. 2006. Utilization of thermal power station waste (fly ash) for betterment of plant growth and productivity. *Pollut. Res.,* 25: 119-122.

Thetwar, L.K., Sahu, D.P. and Vaishnava, M.M. 2006. Analysis of *Sesamum indicum* seed oil from plants grown in fly ash amended acidic soil. *Asian J. Chem.,* 18: 481-484.

Tiwari, S., Kumari, B. and Singh, S.N. 2008. Evaluation of metal mobility/immobility in fly ash induced by bacterial strains isolated from the rhizospheric zone of *Typha latifolia* growing on fly ash dumps. *Biores. tech.,* 99: 1305-1310.

Tullio, M., Pierandrei, F., Salerno, A. and Rea, E. 2003. Tolerance to cadmium of vesicular arbuscular mycorrhizae spores isolated from a cadmium-polluted and unpolluted soil. *Biol. Fert. Soils,* 37: 211-214.

Turnau, K., Kottke, I. and Oberwinkler, F. 1993. Element localization in mycorrhizal roots of *Pteridium aquilinum* (L.) Kuhn collected from experimental plots treated with cadmium dust. *New Phytol.,* 123: 313-324.

Verma Rollie, 2008. Management of Fly Ash and Wasteland Soils through VAM Technology. D. Phil. Thesis, University of Allahabad, Allahabad.

Vivas, A., Barea, J.M. and Azco´n, R. 2005. Interactive effect of *Brevibacillus brevis* and *Glomus mosseae*, both isolated from Cd contaminated soil, on plant growth, physiological mycorrhizal fungal characteristics and soil enzymatic activities in Cd polluted soil. *Env. Pollut.,* 134: 257-266.

Wang, F.Y., Lin, X.G., Yin, R. and Wu, L.H. 2006. Effects of arbuscular mycorrhizal inoculation on the growth of *Elsholtzia splendens* and *Zea mays* and the activities of phosphatase and urease in a multi-metal-contaminated soil under unsterilized conditions. *Appl. Soil Ecol.,* 31: 110-119.

Yeledhalli, N.A., Prakash, S.S., Gurumurthy, S.B. and Ravi, M.V. 2007. Coal fly ash as modifier of physico-chemical and biological properties of soil. *J. Karnataka Agri. Sci.*, 20(3): 531-534.

Zhou, J.L. 1999. Zn biosorption by *Rhizopus arrhizus* and other fungi. *Appl. Microb. Biotechnol.*, 51: 686-693.

Zhu, Y., Christie, P. and Laidlaw, A.S. 2001. Uptake of Zn by arbuscular mycorrhizal white clover from Zn-contaminated soil. *Chemosphere*, 42: 193-199.

2017, Mycorrhizal Fungi

Editors: Ashok Aggarwal and Kuldeep Yadav

Published by: ASTRAL INTERNATIONAL PVT. LTD., NEW DELHI

Pages 291–309

13

Arbuscular Mycorrhizal Fungi: An Eco-Friendly Bio-Resource for Enhancing Nutrient Use Efficiency and Drought Tolerance in Agricultural Crops

V.K. Suri[1], Anil Kumar[2] and Anil K. Choudhary[3]*

[1]*Former Vice-Chancellor,*
CSA University of Agriculture and Technology, Kanpur, U.P.
[2]*Agriculture Science Centre, GADVASU, Tarn Taran, Punjab*
[3]*Division of Agronomy,*
ICAR-Indian Agricultural Research Institute, New Delhi
*Corresponding Author: anilhpau2010@gmail.com

ABSTRACT

Adoption of Arbuscular mycorrhizal fungi (AMF) as a biofertilizer in crop production might have proved as an effective tool in mitigating nutrient and water stresses, improving crop quality and soil health under different conditions. This is possible due to the fact that arbuscular mycorrhizal fungi extend root system into the soil through ramifying hyphae thereby increasing its exploratory area for harnessing nutrients and water for plant growth. Studies conducted under various conditions involving arbuscular mycorrhizal fungi resulted in better crop stand following increased nutrient acquisition mediated by fungal hyphae in the soil, improved soil-plant-water relations and plant tolerance to a variety of abiotic stresses, etc. AMF inoculated plants usually maintained higher tissue water content imparting greater drought resistance to plants during soil moisture stress. Fungal hyphae penetrate the soil pores, which are inaccessible to root hairs thereby absorbing water which is unavailable to non-mycorrhizal plants. Studies further shows that improvement in soil structure following arbuscular mycorrhizal fungi inclusion in different cropping system. Actually AMF produce certain glycoprotein which is capable of binding soil particles leading to aggregate formation. Fungal hyphae further bind soil aggregates leading to improved soil structure, which in turn exert a significant impact on water holding capacity. Overall, AMF have ability to reduce the fertility constraints and meet out moisture requirement of host. Further, AM fungi revealed a tremendous potential in enhancing water use efficiency, improved plant succulence and imparting drought resistance to plants in drying soils.

Keywords: *AM fungi, Nutrient acquisition, Soil structure, Water use efficiency.*

1. Introduction

Many microorganisms form symbiosis with plants that ranges from parasitic to mutualistic. Amongst various associations, arbuscular mycorrhizal mutualistic symbiosis is most widespread. Arbuscular mycorrhizal fungi are obligate symbionts which grow in association with living tissues (Al-Raddad, 1995). Arbuscular mycorrhizal symbiosis occurs between fungi and the majority of terrestrial plants (Schubler *et al.*, 2001) and involves an intimate relationship between plant roots and fungal hyphae. The AM fungi receive carbon compounds/nutritional requirements from host plant roots and in turn, supply nutrients *viz.* phosphorus (P), zinc (Zn), nitrogen (N), potassium (K), calcium (Ca), copper (Cu), *etc.* to plants and also enhance water uptake (Barea and Jeffries, 1995; Bai et. al., 2016a). The arbuscular mycorrhizal fungi are capable of enhancing nutrient availability and water use efficiency of crops (Yadav *et al.*, 2015a; Kumar *et al.*, 2016a; Bai *et al.*, 2016b). This is possible due to the fact that it extends root system into the soil through ramifying hyphae thereby increasing its exploratory area for harnessing nutrients and water (Harrier and Watson, 2003; Choudhary, 2011).

The AM fungi have very broad specificity towards plants including various agricultural, horticultural and forest species (Mohammad *et al.*, 1995). However, ability to form AM symbiosis has been lost in about 10 per cent of plants and is completely absent in members of *Brassicaceae* and *Chenopodiaceae* families (Tester *et al.*, 1987). Besides plant nutrition, AM symbiosis has ability to alter plant water relations and responses to drought significantly (Auge, 2001; Kumar *et al.*, 2015a). However, mycorrhizal effects on plant water relations are not as dramatic and consistent as those on nutrient acquisition and host growth. AMF symbiosis has ability to protect host plants against the detrimental effects of drought (Auge, 2001; Ruiz- Lozano, 2003). AMF inoculated plants usually maintain higher tissue water content imparting greater drought resistance to plants during soil moisture stress (Kumar *et al.*, 2015a, 2016a). Fungal hyphae penetrate the soil pores, which are inaccessible to root hairs thereby absorbing water which is unavailable to non-mycorrhizal plants, thereby enhance water use efficiency of the plants.

AM fungi also play a crucial role in soil health improvement (Choudhary, 2011). The AM fungi develop an extensive extra-radical hyphal network that grows away from the root, through the rhizosphere and into the surrounding bulk soil matrix. Above network makes a significant contribution to improvement of soil texture and water relations (Tisdall and Oades, 1982; Miller and Jastrow, 1990; Tisdall, 1991). The extra-radical hyphae of above fungi have been shown to be more important than root length and/or bacterial populations in stabilizing soil aggregates (Schreiner *et al.*, 1997). Several studies have demonstrated the contribution of the AM symbiosis to plant drought tolerance resulting from a combination of physical, nutritional, physiological and cellular effects (Ruiz-Lozano, 2003). There are sharp indications of fertilizer P economy by about 25 per cent (Yadav *et al.*, 2015b; Suri and Choudhary, 2012; Suri and Choudhary, 2013c; Kumar *et al.*, 2016c) and enhancement in water-use efficiency of crops to the extent of 15-30 per cent following AMF inoculation (Kumar *et al.*, 2015a; Yadav *et al.*, 2015a). Further, improvement in plant-water relations vis-à-vis drought resistance, crop productivity, soil aggregation and water holding capacity in different cropping system has been reported by number of researchers (Kumar *et al.*, 2015a, 2016a; Bai, 2014; Kumar, 2010; Yadav *et al.*, 2015b; Suri *et al.*, 2010).

Thus, AM fungi have proved as an effective and eco-friendly bio-resource for enhancing nutrient and water use efficiency of agricultural crops. This is a low-cost farm input for resource-poor farmers, who ill afford expensive external nutrient inputs especially P. AMF also revealed a tremendous potential in enhancing water use efficiency, improved fruit succulence and imparting drought resistance to plants at moisture stress.

2. Mechanism of Nutrient and Water Absorption

The AM fungi expand surface area of plant root system by 10 to 1000 fold into the soil through ramifying hyphae, thereby increasing its exploratory area for harnessing nutrients and water (Marshner and Dell, 1994). Further, AM fungi are capable of mineralizing organic-P and solublizing inorganic-P in soils by releasing various enzymes (chitinase, peroxidase, cellulase, protease, phosphatase, *etc.*) and organic acids (oxalic, malic acids, *etc.*) (Zou *et al.*, 1995; Chen *et al.*, 2007). On the other hand, AMF symbiosis probably affect the water relations of plants indirectly by improving P nutrition at initial stage (Safir *et al.*, 1971). The main absorption apparatus of mycorrhizal fungi is extension hyphae with a diameter of 2-5 μm which penetrate soil pores inaccessible to root hairs (10- 20 μm) and hence, absorb water that is not available to non-mycorrhizal plants (Gong *et al.*, 2000). Because the number of extension hyphae is far more than that of root hairs, the area of surface where AMF plant and soil interacted increased greatly resulting into more water absorption.

In addition, colonization of plant roots with AM fungi might change the architecture of roots, which may enhance the interaction of root and soil (Atkinson, 1994). Further, AMF plant explore larger soil volume through extension of root system into soil profile by way of development of higher order laterals through ramification of fungal hyphae associated with it (Song, 2005; Suri and Choudhary, 2013b; Kumar *et al.*, 2016b).

2.1. AM Fungi and Water Absorption

Auge (2006) considered a relationship between plant water balances or drought resistance with indirect effects associated with changes in plant size and phenology, as size of a plant can affect its water relations. The AM symbiosis often affects the plant size. Larger plants have deeper or larger root system that has ability to extract water reserve more extensively, thereby affecting plant water relations (Fitter, 1985; Koide, 1993). The high dry matter production, following AM inoculation might partially explain why mycorrhizal plants gave higher Water Use Efficiency (WUE) than non-mycorrhizal ones (Duan *et al.*, 1996; Al-Karaki and Al-Radded, 1997). Mycorrhizal plants have ability to maintain stomatal conductance and hence, maintain leaf water potential or water content (Allen and Allen, 1986; Auge *et al.*, 1986; Auge *et al.*, 1987; Osonubi, 1994; Duan *et al.*, 1996; Kumar *et al.*, 2016a).

Main absorption apparatus of mycorrhizal fungi is extension hyphae, which penetrate soil pores inaccessible to root hairs and hence, absorb water that is not available to non-mycorrhizal plants (Gong *et al.*, 2000; Koide, 1993; Farahani *et al.*, 2008). Further, AMF inoculated plants have also shown increased root growth in soybean (Suri *et. al.*, 2010), maize (Suri and Choudhary, 2013a), okra and pea (Kumar *et al.*, 2016c) resulting from extraction of water from deeper layer. AM hyphae generally absorb the phosphorus from the dense soil location, where the host roots could not access (Li *et al.*, 1994). Further, under drought conditions the uptake of highly mobile nutrients such as NO_3^- can also be enhanced by mycorrhizal associations (Azcon *et al.*, 1996; Subramanian and Charest, 1999).

Farahani *et al.* (2008) further attributed higher WUE to a greater absorption of water and P by plants thereby, boosting biological yield. Actually, each factor which promotes biological yield would naturally enhance WUE also. The AM symbiosis often results in altered rates of water movement into, through and out of host plants, with consequent effects on tissue hydration and leaf physiology (Auge, 2001). Mycorrhizal plants develop more roots and extract more water to sustain high growth rate, thereby resulting into higher water use (Nagarathna *et al.*, 2007).

Enhanced plant water status and water-use-efficiency following AMF inoculation have also been reported by Yusnaini *et al.* (1999), Farahani *et al.* (2008) and Yadav *et al.* (2015a). AMF inoculation led to higher consumptive use of water during crop growth as well as higher water use efficiency than non mycorrhizal plants (Singh and Idani, 2007). Kumar *et al.* (2016a) registered significantly higher relative leaf water content in okra-pea cropping system with AMF inoculation. Mycrorrhizal plants enhance WUE to the extent of 5–17 per cent and 12–35 per cent in okra and pea, respectively over non mycorrhizal ones. Moreover, mycorrhizal plants also maintain higher tissue water content imparting greater drought resistance to plants over non–mycorrhizal plants at moisture stress conditions (Kumar *et al.*, 2016a). Many workers have reported better growth of AM inoculated plants in drying soils owing to improved exploitation of bound water as fungal hyphae provide access to soil water below the permanent wilting point (Dakessian *et al.,* 1986; Bethlenfalvay *et al.,* 1988b; Franson *et al.,* 1991).

2.2. AM Fungi and Nutrient Acquisition

Many nutrients are directly and indirectly involved in improving and maintaining water relations of host plants. While fungi obtain carbon compounds/ nutritional requirements from host plant roots, in turn, they supply plant nutrients such as N, P, K, Ca, Cu, Zn, etc, which are absorbed by them from the soil (Barea and Jeffries, 1995). Many workers have reported an increase in plant growth resulting from above AMF association and the same has been attributed to increased mineral element uptake mediated by fungal hyphae in the soil, improved soil-plant-water relations and plant tolerance to a variety of abiotic stresses, *etc.* (Harrier and Watson, 2003; Smith and Read, 1998; Kumar *et al.,* 2015b). Kumar *et al.* (2017) reported increased acquisition of N, P, K, Ca, B and Mo uptake in legume based okra-pea cropping system. Similarly, Yadav (2012) reported increases of 16.3, 18.2 and 6.0 per cent in N, P and K acquisition following AMF inoculation over non-AMF plants. Similar, results have been demonstrated by Bai *et al.* (2016c), Suri *et al.* (2011b) and Suri *et al.* (2006) under acid soil environment. Enhancement of N, K, Ca, Mg, Fe and Cu uptake by AMF have also been reported by several researchers (Smith and Read, 1997; Clark and Zeto, 2000).

Enhanced acquisition of P in AMF inoculated plants to the extent of 39 and 26 per cent has been reported by Bryla and Duniway (1997); Sorial (2001) under P deficient and moisture stress conditions. Similarly, Clark and Zeto (2000) reported significant enhancement in N, K, Ca, Mg, Zn, and Fe and Cu uptake under AMF inoculation, both under acid and alkaline soil conditions. The AMF inclusion in legume based cropping system enhanced N, P and K uptake by pea to the tune of 16, 70 and 41 per cent. Further, increase in uptake of above nutrients was to the extent of 62, 43 and 28 per cent in okra (Singh *et al.,* 2004). Marschner and Dell (1994) have reported increased acquisition of boron (B) in AM colonized plants. Enhanced 'B' and 'Zn and Cu' acquisition following AM fungi inoculation has also been reported by Clark and Zeto (2000) and Marschner and Dell (1994), respectively. Similarly, increase in uptake of Chlorine (Cl), bromine (Br), caesium (Cs), cobalt (Co), molybdenum (Mo), Nickel (Ni), cadmium (Cd) and lead (Pb) has also been registered by different workers (Buwalda *et al.,* 1983; Ellis *et al.,* 1995; Rogerrs and Williams, 1986; Raju *et al.,* 1987; Killham and Firestone, 1983; Cooper and Tinker, 1978; Guo *et al.,* 1996; Diaz *et al.,*1996).

Inoculation of *arhar* (legume crop) with AM fungi have shown to mobilize a good amount of P from both soluble and insoluble sources, thereby increasing yield and uptake of P by crop (Misra and Pattanayak, 1997). Similarly, George *et al.* (1992) reported significantly higher concentration of nutrients (Fe, Zn and Cu) in mycorrhizal plants as compared to non-mycorrhizal ones. Further, the concentration of Mn in stover of AMF inoculated plants was lower than that in non-mycorrhizal ones. In an experiment covering P deficient soils, Li *et al.* (1991a) concluded that AMF hyphae absorbed P from deeper soil layers, where host roots failed to reach. In addition, AMF root colonization change root architecture, which in turn, increases root-soil interaction vis-à-vis nutrient exploration (Atkinson, 1994). AM fungi also play an important role in transport of inorganic N (George *et al.,* 1992; Hawkins and George, 1999; Hawkins *et al.,* 2000; Johansen *et al.,*1993, 1994). Hyphae of AM fungi have been shown to take up amino acids and transport N to the plant roots (Nasholm *et al.,* 1998). Improved acquisition of K, Ca and Mg in AMF inoculated plants in acidic conditions has been reported by several authors and suggested that acquisition of above nutrients depends on some factors such as soil pH, host plant and isolate of AM fungi (Medeiras *et al.,* 1994; Clark and Zeto, 1996; He *et al.,* 1997).

3. AM Fungi and Soil Physico-chemical Properties

AM fungi have been shown to improve productivity of crops in soils even with low fertility (Jeffries, 1987). AM fungi are especially important for increasing the acquisition of slowly diffusing ions such as PO_4^{3-} (Jacobsen *et al.,* 1992), immobile nutrients *viz.* P, Zn and Cu (Lambert *et al.,* 1979; George *et al.,* 1992, 1996; Ortas *et al.,* 1996; Liu *et al.,* 2002) and other nutrients such as Cadmium (Guo *et al.,* 1996). Improved P nutrition has been shown to increase in infertile and P fixing soils of the tropics (Dodd, 2000). Similarly, improved P and micronutrient nutrition following AM fungal inoculation has also been in medium P acid Alfisol (Kumar *et al.,* 2014, 2017; Yadav, 2012; Bai, 2014). Further, mycorrhizal fungi can also improve absorption of N from NH^{4+}-N mineral fertilizers and transporting the same to host plant (Ames *et al.,* 1983; Johansen *et al.,* 1993), thus increase biomass production in soils with low K, Ca and Mg (Liu *et al.,* 2002).

Improvement in soil structure following AMF inoculation has been registered by several researchers. AMF improve soil structure by way of binding of soil aggregates, involving their hyphal network, thus enhancing soil moisture retention capacity (Tisdall, 1991; Staddon *et al.,* 2003; Hamblin, 1985; Rillig, 2004). Hyphae of AM fungi, develop an extensive extra-radical hyphal network that grows into the soil matrix and holds primary soil particles together via physical entanglement. Above network makes a significant contribution to improvement of soil texture and water relations (Tisdall and Oades, 1982; Miller and Jastrow, 1990; Tisdall, 1991). Further, hyphae of mycorrhizal fungi produce a glycoprotein called glomalin which is capable of binding soil particles leading to aggregate formation and in turn, improve soil structure/soil moisture retention capacity (Wright and Upadhyaya, 1998; Wright *et al.,* 1998; Rillig, 2004).

AM fungi also improve soil structure indirectly, by enhancing root biomass, root length, root surface area and root volume density (Auge 2001; Kumar *et al.,* 2016a). Above factors in turn, provide comparatively more water availability to mycorrhizal plants as compared to non-mycorrhizal ones. AM fungi also improve water holding capacity of soil particles significantly (Kumar *et al.,* 2016a). The extra-radical hyphae of above fungi have been shown to be more important than root length and/or bacterial populations in stabilizing soil aggregates (Schreiner *et al.,* 1997). Actually, AMF chemically enmesh and stabilize micro-aggregates and smaller macro-aggregates into macro-aggregate structures (Miller and Jastrow, 2000). As AM fungi affect the soil structure, it might affect moisture retention properties and which in turn, affect the behaviour of plants growing in the soil, especially when the soils are relatively dry.

There are indications of soil organic carbon build-up in AM fungi imbedded soils, which might improve a lot in the long-term (Kumar *et al.*, 2015b). Similarly, there are reports in literature, which suggest a substantial increase in soil organic carbon following continuous and long-term application of AMF biofertilizer/culture (Suri and Choudhary, 2014; Paul *et al.*, 2011). The AMF enhance soil organic carbon level by way of addition of extra roots *i.e.* mycorrhizal roots as well as improved plant root biomass due to extra P availability for root growth resulting from AMF application.

4. Impact of Agricultural Practices on AMF Population

4.1. Soil Conditions

Propagules of mycorrhizal fungi may be absent from soils where severe soil disturbance has resulted in topsoil loss, or where host plants are limited by adverse soil or site factors such as salinity, aridity, waterlogging or climatic extremes (Brundrett, 1991). Studies have found reduced levels of mycorrhizal propagules in highly disturbed habitats (Danielson, 1985, Jasper *et al.*, 1992, Pfleger *et al.*, 1994, Brundrett *et al.*, 1996). Similarly, soil disturbance resulting from agricultural tillage, soil animal activities, fire and erosion can also reduce levels of mycorrhizal fungus propagules (Habte *et al.*, 1988, O'Halloran *et al.*, 1986, Read and Birch, 1988, Vilarino and Arines, 1991). Observations in natural ecosystems have shown that plants with mycorrhizal associations are often less common than non-mycorrhizal species in soils which are waterlogged or saline, but that some mycorrhizal plants are normally present in even the worst soils (Brundrett, 1991). Excessive salt levels in soil inhibit mycorrhizal formation and restrict their activity. Some species however, can tolerate above conditions (Malajczuk *et al.*, 1981; Juniper and Abbott, 1993).

Soils under low-input management particularly organic ones show higher AM fungus spore populations than soils under conventional management (Galvez *et al.*, 2001; Douds *et al.*, 1993, 1995; Suri and Choudhary, 2014). Survival of AMF in soil may also be affected by the presence or absence of crops and the crop being grown (Troeh *et al.*, 2003). The activity of AMF in soil may be greatly limited by soil fumigation, non-responsive plant varieties or crop rotations with low AMF dependency and non-mycorrhizal crops. Further, there are evidences of reduced mycorrhization in plants with salicylic acid contents, suggesting that enhanced salicylic acid levels in plants delay AMF root colonization (Medina *et al.*, 2003).

4.2. Tillage

With tillage there is disruption of hyphal network, which act as inoculum to field grown crops. Disturbance of these hyphae by tillage lead to decreased colonization. Kabir *et al.* (1997) reported more fungal hypha density in no till as compared to reduced tillage, further it was minimum in conventional tillage system. Reduced/no till system has potential advantages in maintaining fungal hyphal network in soil as it provide greater ground cover, less erosion, more organic matter and better soil structure compared to conventional agricultural practices. Moreover, soil disturbances have a long-term impact on aggregate stability (Jastrow, 1987) as when fungal hyphal network are disrupted following tillage, it exerted a negative influence on aggregate stability (Wright and Upadhaya, 1998).

4.3. Fertilizer Use

Nitrogen and phosphatic fertilizers seem to have a negative impact on AMF formation. There is considerable reduction in spore numbers with the application of large quantities on nitrogen and phosphorus, some workers however, reported increased AMF infection with increasing N levels, but at higher levels of P, infection decreased with increasing N. Under low P conditions AM inoculation increased, however, under high P conditions AM fungi inoculation depressed the colonization potential. The soil having higher P status or heavy application of P fertilizer reduces the colonization of roots by AM fungi (Suri *et al.*, 2011a; Suri *et al.*, 2013; Kumar *et al.*, 2016c; Bai, 2016a; Kumar *et al.*, 2014). The AM fungal contribution to plant nutrient utilization has been shown to be reduced at high levels of readily available P (Thingstrup *et al.*, 2000). Similarly, higher organic matter status in soil likely to mask the effect of AM fungi application in field crops (Kumar, 2012). Joner (2000) has also reported negative influence of organic matter on AM fungi. Sewage sludge containing P, N and heavy metals can reduce mycorrhizal activity.

4.4. Crop Rotation

Pre-cropping with non host crops does not affect the AMF population in the soil. However, the pre-cropping practices with host crops results in increased AMF population in soil. Some workers further, reported decreased AMF population when pre-cropping with crops like kale and mustard was practised. So, proper crop rotation and intercropping enhance natural AMF. Legume based cropping systems, generally increase the AMF population. Bagayoko *et al.* (2000) observed that AM colonization rate was higher in early season within cereals grown in rotation than in a monoculture. The inclusion of non-mycorrhizal crops (crops that don't form AMF association) within crop rotations has been shown to decease the early growth, P uptake (Arihawa and Karasawa, 2000), AM fungal colonization (Douds *et al.*, 1997) and yield (Arihawa and Karasawa, 2000; Gavito and Miller, 1998a) of subsequent crops population in soil. The intercropping has the potential to improve resource use. Zhu *et al.* (2001) have demonstrated that modern cultivars are less responsive to AM fungal colonization than older ones.

4.5. Use of Fungicides

Fungicides used to control diseases, also affect the AMF natural population in soils. Thus, reduces the nutrient uptake. However, there are certain reports in literature indicating that, it is difficult to generalize the impact of pesticides on AM fungi and their symbiotic interactions since both beneficial and detrimental effects have been reported (Trappe *et al.*, 1984; Johnson and Pfleger, 1992). Certain alteration in agricultural practices, legume based cropping systems, use of organic inputs has tremendous potential to maintained AMF levels in the soil (Kumar *et al.*,

2016d). Further, manipulation of agricultural systems that favour AMF colonization must be practiced.

5. Conclusion and Future Perspective

Studies have clearly indicated alteration in water relation of host plants following AM fungal inoculation. Simply, AMF symbiosis enlarges absorption areas of host plant and improves nutritional status. However, AM fungal effects on plant water relations are not as dramatic and consistent as those on acquisition and host growth. The use of mycorrhizal fungi as biofertilizer significantly enhance relative leaf water content, xylem water potential and water use efficiencies of crops under different conditions. AM fungi increase the drought resistance of the crops especially under phosphorus and water stressed situations. Overall, AMF reveal a tremendous potential in enhancing water use efficiency, improved plant succulence and imparting drought resistance to plants at moisture stress. Moreover, AM fungi have ability to reduce some of the fertility constraints.

From above presentation it is apparent that arbuscular mycorrhizal symbiosis play fundamental roles in shaping plant communities and terrestrial ecosystems. However, despite being aware of potentiality of arbuscular mycorrhizal fungi in agricultural research, it has not been possible to use on commercial scale due to various bottlenecks involving biotrophic nature of AMF, lack of location specific efficient strains, *etc*. To overcome above bottleneck, it will be necessary to undertake short and long-term research involving multidisciplinary scientists. Major research focus should be on the production of efficient and sustainable AM culture for crop plants to reduce inorganic fertilizer especially P application rates. Efforts should be made to propagate the fungi in the field itself and to explore the possibility of developing location specific efficient strains. Further, manipulation of agricultural systems to favour AMF colonization must occur only if people may understand that arbuscular mycorrhizal fungi make a positive contribution to yield and are vital for maintenance of ecosystem health and sustainability.

References

Al-Karaki, G.N. and Al-Raddad, A. 1997. Effects of arbuscular mycorrhizal fungi and drought stress on growth and nutrient uptake of two wheat genotypes differing in drought resistance. *Mycorrhiza,* 7: 83–88.

Allen, E.B. and Allen, M.F. 1986. Water relations of xeric grasses in the field: interactions of mycorrhizas and competition. *New Phytol.*, 104: 559–571.

Al-Raddad, A. 1995. Mass production of *Glomus mosseae* spores. *Mycorrhiza,* 5: 229 231.

Ames, R.N., Reid, C.P.P., Porter, L. and Cambardella, C. 1983. Hyphal uptake and transport of nitrogen from two 15N-labelled sources by *Glomus mosseae*, a vesicular-arbuscular mycorrhizal fungus. *New Phytol.*, 95: 381-396.

Arihawa, J. and Karasawa, T. 2000. Feecet of previous crops on arbuscular mycorrhizal formation and growth of succeeding maize. *Soil Sci. Plant Nutr.*, 46: 43-51.

Atkinson, D. 1994. Impact of mycorrhizal colonization on root architecture, root longevity and the formation of growth regulators. In: Gianinazzi S *et al.* (Eds). *Impact of arbuscular mycorrhizas on sustainable agriculture and natural ecosystem*, pp. 89-99.

Auge, R.M. 2006. Water relations, drought and VA mycorrhizal symbiosis. *Mycorrhiza*, 11: 3–42.

Auge, R.M., Schekel, K.A. and Wample, R.L. 1986. Greater leaf conductance of well-watered VA mycorrhizal rose plants is not related to phosphorus nutrition. *New Phytol.*, 103:107-116.

Auge, R.M., Schekel, K.A. and Wanple, R.L. 1987. Rose leaf elasticity changes in response to mycorrhizal colonization and drought acclimation. *Physiol. Plant.*, 70: 175-182.

Auge, R.M., Stodola, J.W. Tims, J.E. and Saxton, A.M. 2001. Moisture retention properties of a mycorrhizal soil. *Plant Soil*, 230: 87-97.

Auge, RM. 2001. Water relations, drought and VA mycorrhizal symbiosis. *Mycorrhiza*, 11: 3-42.

Azcon, R., Gomes, M. and Tobart, R. 1996. Physiological and nutritional responses by *Lactuca sativa* L. to nitrogen sources and mycorrhizal fungi under drought stress conditions. *Biol. Fert. Soils*, 22:156-161.

Bagayoko, M., Buerkert, A., Lung, G., Bationo, A. and Romheld, V. 2000. Cereal-legume rotation effects on cereal growth in Sudano-Sahelian West Africa: soil mineral nitrogen, mycorrhizae and nematodes. *Plant Soil*, 218: 103-116.

Bai, B. 2014. Nitrogen and phosphorus economy through inoculation with arbuscular mycorrhizal Fungi (AMF) and *Rhizobium* in garden pea (*Pisum sativum* L.), M.Sc. Thesis, CSK HPKV, Palampur (HP), India.

Bai, B., Suri, V.K., Kumar, A. and Choudhary, A.K. 2016a. Influence of dual-inoculation of AM fungi and *Rhizobium* on growth indices, production economics and nutrient use efficiencies in garden pea (*Pisum sativum* L.). *Commun. Soil Sci. Plant Anal.*, 47: 941-954.

Bai, B., Suri, V.K., Kumar, A. and Choudhary, A.K. 2016b. Influence of *Glomus–Rhizobium* symbiosis on productivity, root morphology and soil fertility in garden pea in Himalayan acid Alfisol. *Commun. Soil Sci. Plant Anal.*, 47: 787-798.

Bai, B., Suri, V.K., Kumar, A. and Choudhary, A.K. 2016c. Tripartite symbiosis of *Pisum–Glomus–Rhizobium* lead to enhanced productivity, nitrogen and phosphorus economy, quality and biofortification in garden pea in a Himalayan acid Alfisol. *J. Plant Nutr.*, Published, DOI: 10.1080/01904167.2016.1263320.

Barea, J.M. and Jeffries, P. 1995. Arbuscular mycorrhizas in sustainable soil-plant systems. In: Varma A, Hock B (Eds) *Mycorrhiza: Structure, Function, Molecular Biology and Biotechnology.* Springer, Berlin Heidelberg New York, pp. 521-561.

Bethlenfalvay, G.J., Brown, M.S., Ames, R.N. and Thomas, R.E. 1988. Effect of drought on host and endophyte development in mycorrhizal soybeans in relation to water use and phosphate uptake. *J. Plant Physiol.*, 72: 565-571.

Brundrett, M. 1991. Mycorrhizas in natural ecosystem. *In*: Macfayden, Begon, M and Fitter AH (Eds). *Advances in Ecological Research*, Academi Press London, pp.171-173.

Brundrett, M., Beegher, N., Dell, B., Groove, T. and Malajczuk, N. 1996. Working with mycorrhizas in Forestry and Agriculture. *ACIAR Monograph* 32: 374p. ISBN186320 181 5.

Bryla, D.R. and Duniway, J.M. 1997. Effects of mycorrhizal infection on drought tolerance and recovery in safflower and wheat. *Plant Soil*, 197: 95- 103.

Buwalda, J.G., Stribley, D.P. and Tinker, P.B. 1983. Increased uptake of anions by plants with vesicular-arbuscular mycorrhizal. *Plant Soil*, 71: 463-467.

Chen CR, Condron LM and Xu ZH. 2007. Impacts of grassland afforestation with coniferous trees on soil phosphorus dynamics and associated microbial processes: A review. *Forest Ecol. Manag.*, 255: 396–409.

Choudhary, A.K. 2011. Resource conservation technologies under changing climate in north-western Himalayas. *In:* "Summer School on climate variability and its impact on crop production - Physiological perspective towards mitigation strategies" at AAU, Jorhat (Assam) held w.e.f. August 23 to September 12, 2011. *Summer School Compendium, AAU, Jorhat*, pp: 199-206.

Clark, R.B. and Zeto, S.K. 1996. Iron acquisition by mycorrhizal maize grown on alkaline soil. *J. Plant Nutr.*, 19: 247–264.

Clark, R.B. and Zeto, S.K. 2000. Mineral acquisition by arbuscular mycorrhizal plants. *J. Plant Nutr.*, 23: 867-902.

Cooper, K.M. and Tinker, P.B. 1978. Translocation and transfer of nutrients in vesicular-arbuscular mycorrhizas. II. Uptake and translocation of phosphorus, zinc and sulphur. *New Phytol.*, 81: 43-52.

Dakessian, S., Brown, M.S. and Bethlenfalvay, G.J. 1986. Relationship of mycorrhizal growth enhancement and plant growth with soil water and texture. *Plant Soil*, 94: 439–443.

Danielson, R.M. 1985. Mycorrhizae and reclamation of stressed terrestrial environments. In: Tate RL, Klein DA (eds) *Soil Reclamation Processes - microorganisms, Analyses and Applications*. Marcel Dekker, New York. pp. 173-201.

Diaz, G., AzconAguilar, C. and Honrubia, M. 1996. Influence of arbuscular mycorrhizae on heavy metal (Zn and Pb) uptake and growth of Lygeum spartum and Anthyllis cytisoides. *Plant Soil*, 180: 241–249.

Dodd, J.C. 2000. The role of arbuscular mycorrhizal fungi in agro-and natural ecossystems. *Outlook on Agriculture*, 29: 55-62.

Douds, D.D., Galvez, L., Franke-Snyder, M., Reider, C. and Drinkwater, L.E. 1997. Effect of compost addition and crop rotation point upon VAM fungi. *Agriculture Ecosystem Environment*, 65: 257-266.

Douds, D.D., Galvez, L., Janke, R. and Wagoner, P. 1995. Effect of tillage and farming system upon populations and distribution of veisculararbuscular mycorrhizal fungi. *Agric. Ecosys. Environ.*, 52: 111-118.

Douds, D.D., Janke, R.R. and Peters. 1993. VAM fungus spore populations and colonization of roots of maize and soybean under conventional and low-input sustainable agriculture. *Agric. Ecosys. Environ.*, 43: 325- 335.

Duan, X., Neuman, D.S., Reiber, J.M., Green, C.D., Saxton, A.M. and Auge, R.M. 1996. Mycorrhizal influence on hydraulic and hormonal factors implicated in the control of stomatal conductance during drought. *J. Exp. Bot.*, 47: 1541–1550.

Ellis, J.R., Larsen, H.J., Boosalis, M.G. 1985. Drought resistance of wheat plants inoculated with vesicular-arbuscular mycorrhizae. *Plant Soil*, 86: 369–378.

Farahani, A., Lebaschi, H., Hussein, M., Hussein, S.A., Reza, V.A. and Jahanfar, D. 2008. Effects of arbuscular mycorrhizal fungi, different levels of phosphorus and drought stress on water-use-efficiency (WUE), relative water content and proline accumulation rate in coriander (*Coriandrum sativum* L.). *J. Med. Plant Res.*, 2: 125-131.

Fitter, A.H. 1985. Functioning of vesicular-arbuscular mycorrhizas under field conditions. *New Phytol.*, 99: 257–265.

Franson, R.L., Milford, S.B. and Bethlenfalvay, G.J. 1991. The *Glycine-Glomus-Bradyrhizobium* symbiosis. XI. Nodule gas exchange and efficiency as a function of soil and root water status in mycorrhizal soybean. *Physiol. Plant.*, 83: 476-482.

Galvez, L., Douds, D.D., Drinkwate, L.E. and Wagoner, P. 2001. Effect of tillage and farming system upon VAM fungus populations and mycorrhizas and nutrient uptake of maize. *Plant Soil*, 228: 299-308.

Gavito, M.E. and Miller, M.H. 1998. Early phosphorus nutrition, mycorrhizae development, dry matter partitioning and yield of maize. *Plant Soil*, 199: 177-186.

George, E., Gorgus, E., Schmeisser, A. and Marschner, H. 1996. A method to measure nutrient uptake from soil by mycorrhizal hyphae. *In:* Mycorrhizas in Integrated System from Genes to plant Development (Eds.) Azcon-Aguilar and JM Barea), Luxembourg European Community.

George, E., Haussler, K., Vetterlein, D., Gorgus, E. and Marschner, H. 1992. Water nutrient translocation by hyphae of Glomus mosseae. *Can. J. Bot.*, 70: 2130–2137.

George, E., Romheld, V. and Marschner, H. 1994. Contribution of mycorrhizal fungi to micronutrient uptake by plants. In: J.A. Monthey, D.E Crowley and D.G. Luster (Eds.), *Biochemistry of Metal Micronutrients in the Rhizosphere*, CRC Press. pp. 93-109.

Gong, Q., Xu, D. and Zhong, C. 2000. Study on biodiversity of mycorrhizae and its application. *Beijing: Chinese forest press*, pp. 51-61.

Guo, Y., George, E. and Marschner, H. 1996. Contribution of an arbuscular mycorrhizal fungus to uptake of Cadmnium and Nickel in bean by maize plants. *Plant Soil*, 184: 195-205.

Habate, M. and Manjunath, A. 1991. Categories of vesicular-arbuscular mycorrhizal dependency of host species. *Mycorrhiza*, 1: 3-12.

Hamblin, A.P. 1985. The influence of soil structure on water movement, crop root growth, and water uptake. *Advances in Agronomy*, 38: 95–158.

Harrier, L.A. and Watson, C.A. 2003. The role of arbuscular mycorrhizal fungi in sustainable cropping systems. *Advances in Agronomy*, 42: 185-225.

Hawkins, H. and George, E. 1999. Effect of nitrogen status on the contribution of arbuscular mycorrhizal hyphae to plant nitrogen uptake. *Physio. Planta*, 105: 694-700.

Hawkins, H., Johansen, A. and George, E. 2000. Uptake and transport of organic and inorganic nitrogen by arbuscular mycorrhizal fungi. *Plant Soil*, 226: 275-285.

He, S., Fang, D., Wang, S., Wu, G., Wang, D. and Bei, Z. 1997. Effects of VA mycorrhizal fungus on phosphorus and potassium uptake in tea seedlings. *Acta Agric. Nuc. Sin.*, 11: 45-48.

Jacobsen, I., Abbott, L.K., Robson, A. 1992. External hyphae of vesiculararbuscular mycorrhizal fungi associated with *Trofoluim subterraneum* L. I. Spread of hyphae and phosphorus inflow into roots. *New Phytol.*, 120:371-380.

Jasper, D.A., Robson, A.D., Abbott, L.K. 1987. The effect of surface mining on the infectivity of vesicular-arbuscular mycorrhizal fungi. *Aust. J. Bot.*, 35: 641-652.

Jastrow, J.D. 1987. Changes in soil aggregation associated with tallgrass restoration. *Am. J. Bot.*, 74: 1656-1664.

Jeffries, P. 1987. Use of mycorrhiza in agriculture. *Crit. Rev. Biotechnol.*, 5: 319-357

Johanssen, A., Jakobsen, I. and Jessen, E.S. 1994. Hyphal N transport by a vesiculararbuscular mycorrhizal fungus associated with cucumber grown at three nitrogen levels. *Plant Soil*, 160: 1-9.

Johanssen, A., Jakobsen, I., Jessen, E.S. 1993. External hyphae of vesicular-arbuscular mycorrhizal fungi associated with Trifolium subterraneum L. 3. Hyphal transport of ^{32}P and ^{15}N. *New Phytol.*, 124: 61-68.

Johnson, N.C., Pfleger, F.L. 1992. VA mycorrhizas and cultural stresses. In: Betlhenfalvay, G.J. (Ed) *Mycorrhizas in Sustainable Agriculture*, Wisconsis, Madison, USA.

Joner, E.J. and Levylal, C. 1997. Uptake of roots by roots and hyphae of Glomus mosseae/Trifolium subterraneum mycorrhiza from soil amenede with high and low concentrations of cadmium. *New Phytol.*, 135: 53-360.

Juniper, S. and Abbott, L. 1993. Vesicular-arbuscular mycorrhizas and soil salinity. *Mycorrhiza*, 4: 45-57.

Kabir, Z., O'Halloran, I.P., Fyles, J.W. and Hamel, C. 1997. Seasonal changes of arbuscular mycorrhizal fungi as affected by tillage practices and fertilization: hyphal density and mycorrhizal root colonization. *Plant Soil*, 192: 285-293.

Killham, K. and Firestone, M.K. 1983. Vesicular-arbuscular mycorrhizal mediation of grass response to acid and heavy metal deposition. *Plant Soil,* 72:39-48.

Koide, R. 1993. Physiology of the mycorrhizal plant. *Advances in Plant Pathology,* 9: 33–54.

Kumar, A. 2012. Phosphorus and rain-harvested water economy through vesicular arbuscular mycorrhizae (VAM) in okra-pea sequence. Ph.D. Thesis, CSK HPKV, Palampur (HP), India

Kumar, A., Choudhary, A.K and Suri, V.K. 2016a. Influence of AM fungi, inorganic phosphorus and irrigation regimes on plant water relations and soil physical properties in okra (*Abelmoschus esculentus* l.)–pea (*Pisum sativum* l.) cropping system in Himalayan acid Alfisol. *J. Plant Nutr.,* 39: 666-682.

Kumar, A., Choudhary, A.K. and Suri, V.K. 2015a. Influence of AM–fungi and applied phosphorus on growth indices, production efficiency, phosphorus–use efficiency and fruit–succulence in okra (*Abelmoschus esculentus*)–pea (*Pisum sativum*) cropping system in an acid Alfisol. *Indian J. Agric Sci.,* 85: 1030-1037.

Kumar, A., Choudhary, A.K. and Suri, V.K. 2017. Agronomic bio–fortification and quality enhancement in okra–pea cropping system through arbuscular mycorrhizal fungi at varying phosphorus and irrigation regimes in Himalayan acid Alfisol. *J. Plant Nutr.,* Published Online, DOI: 10.1080/01904167.2016.1267208.

Kumar, A., Choudhary, A.K. and Suri, V.K. 2016c. Influence of AM fungi and inorganic phosphorus on fruit characteristics, root morphology, mycorrhizal colonization and soil phosphorus in okra–pea production system in Himalayan acid Alfisol. *Indian J. Hort.* 73 (2): 213-218.

Kumar, A., Choudhary, A.K., Suri, V.K. and Rana, K.S. 2016c. AM fungi lead to fertilizer phosphorus economy and enhanced system productivity and profitability in okra (*Abelmoschus esculentus*) – pea (*Pisum sativum*) cropping system in Himalayan acid Alfisol. *J. Plant Nutr.,* 39 (10): 1380-1390.

Kumar, A., Choudhary, A.K., Pooniya, V., Suri, V.K. and Singh, U. 2016d. Soil factors associated with micronutrient acquisition in crops – Biofortification perspective. In: Singh *et al.* (Eds). Biofortification of Food Crops, Springer, India, pp. 159-176.

Kumar, A., Suri, V.K., and Choudhary, A.K. 2014. Influence of inorganic phosphorus, VAM fungi, and irrigation regimes on crop productivity and phosphorus transformations in okra (*Abelmoschus esculentus* L.) – pea (*Pisum sativum* L.) cropping system in an Acid Alfisol. *Commun. Soil Sci. Plant Anal.,* 45: 953-967.

Kumar, A., Suri, V.K., Choudhary, A.K., Yadav, A., Kapoor, R., Sandal, S. and Dass, A. 2015b. Growth behavior, nutrient harvest index and soil fertility in okra–pea cropping system as influenced by AM fungi, applied phosphorus and irrigation regimes in Himalayan acid Alfisol. *Commun Soil Sci Plant Anal.,* 46: 2212-2233.

Kumar, S. 2010. Studies on Vesicular Arbuscular Mycorrhizae (VAM) in meeting phosphorus needs of okra-wheat sequence in Typic Hapludalf. Ph.D. Thesis, CSK HPKV, Palampur (HP), India.

Lambert, D.H., Baker, D.E., Cole, H. Jr. 1979. The role of mycorrhizae in the interactions of phosphorus with zinc, copper and other elements. *Soil Sci. Soc. Am. J.,* 43: 976-980.

Li, X., Marschner, H. and George. 1991. Acquisition of phosphorus and copper by VA mycorrhizal haphae and root-to-shoot transport in white clover. *Plant Soil,* 136: 49-57.

Li, X., Zhou, W. and Cao, Y. 1994. Acquisition of phosphorus by VA-mycorrhizal hyhpaefrom the dense soil. *Plant Nutr. Fert. Sci.,* 1: 55-60.

Liu, A., Hamel, C., Elmi, A., Costa, C., Ma, B. and Smith, D.L. 2002. Concentrations of K, Ca and Mg in maize colonised by arbuscular mycorrhizal fungi under field conditions. *Can. J. Soil Sci.,* 82: 271-278.

Malajczuk, N., Linderman, R.G., Kough, J. and Trappe, J.M. 1981. Presence of vesicular-arbuscular mycorrhizae in *Eucalyptus* spp. and *Acacia* sp. and their absence in *Banksia* sp. after inoculation with *Glomus fasciculatus*. *New Phytol.,* 78: 567-572.

Marschner, H. and Dell, B. 1994. Nutrient uptake in mycorrhizal symbiosis. *Plant Soil,* 159: 89-102.

Medeiros, C.A.B., Clark, R.B. and Ellis, J.R. 1994. Growth and nutrient uptake of sorghum cultivated with vesicular-arbuscular mycorrhizal isolates at varying pH. *Mycorrhiza,* 4: 185-191.

Medina, M.J.H., Gagnon, H., Piche, Y., Ocampo, J.A., Garrido, J.M.G. and Vierheilig, H. 2003. Root colonization by arbuscular mycorrhizal fungi is affected by the salicilic acid content of the plant. *Plant Sci.,* 164: 993-998.

Mehrotra, R.S. and Aneja, K.R. 2003. An Introduction to Mycology. New Age Pvt. Ltd.

Miller, R.M. and Jastrow, J.D. 1990. Hierarchy of root and mycorrhizal fungal interactions with soil aggregation. *Soil Bio. Biochem.,* 22: 579-584.

Miller, R.M. and Jastrow, J.D. 2000. Mycorrhizal fungi influence soil structure. *In:* Kapulnik, Y. and Douds, D.D. (Eds.) *Arbuscular mycorrhizas: physiology and function*. Kluwer Academic Publishers, Dordrecht, Netherlands, pp. 3–18.

Misra, U.K. and Pattanayak, S.K. 1997. Characterization of rock phosphates for direct use in different cropping sequences. Technical report of the US- India fund project, number: In- AES-708, GRANT Number: FG-IN-744, 1991-1995.

Mohammad, M.J., Pan, W.L. and Kennedy, A.C. 1995. Wheat responses to vesicular arbuscular mycorrhizal fungal inoculation of soils from eroded toposequence. *Soil Sci. Soc. Am. J.,* 59: 1086-1090

Nagarathna, T.K., Prasad, T.G., Bagyaraj, D.J. and Shadakshari, Y.G. 2007. Effect of arbuscular mycorrhiza and phosphorus levels on growth and water use efficiency in sunflower at different soil moisture status. *J. Agric. Technol.,* 3(2): 221-229.

Nasholm, T., Ekbald, A., Nordin, A., Giesler, R., Hogberg, M. and Hogberg, P. 1998. Boreal forest plants take up organic nitrogen. *Nature*, 392: 914-916

O'Holloran, I.P., Miller, M.H., Arnold, G. 1986. Absorption of P by corn (*Zea mays* L.) as influenced by soil disturbance. *Can. J. Soil Sci.*, 66: 287-302.

Ortas, I., Harries, P.J. and Rowell, D.I. 1996. Enhanced uptake of phosphorus by mycorrhizal sorghum plants as influenced by form of nitrogen. *Plant Soil*, 184: 255-264.

Osonubi, O. 1994. Comparative effects of vesicular-arbuscular mycorrhizal inoculation and phosphorus fertilization on growth and phosphorus uptake of maize (*Zea mays* L.) and sorghum (*Sorghum bicolor* L.) plants under drought-stressed conditions. *Biol. Fertil. Soil.*, 18: 55-59.

Paul, J., Suri, V.K., Sandal, S.K. and Choudhary, A.K. 2011. Evaluation of targeted yield precision model for soybean and toria crops on farmers' fields under sub-humid sub-tropical northwestern Himalayas. *Commun. Soil Sci. Plant Anal.*, 42: 2452–2460.

Pfleger, F.L., Stewart, E.L., Noyd, R.K. 1994. Role of VAM fungi in mine revegetation. In: *A Reappraisal of Mycorrhizae in Plant Health*. Ed. by: Pfleger, F., Linderman, B. (Eds.) The American Phytopathological Society, St. Paul, MN. pp. 47-81.

Raju, P.S., Clark, R.B., Ellis, J.R. and Maranville, J.W. 1987. Vesicular-arbuscular mycorrhizal infection effects on sorghum growth, phosphorus efficiency and mineral nutrient uptake. *J. Plant Nutr.*, 10:1331-1339

Read, D.J. and Birch, C.P.D. 1988. The effects and implications of disturbance of mycorrhizal mycelial systems. *Proc. Royal Soc. Edinburgh*, 94: 13-24.

Rillig, M.C. 2004. Arbuscular mycorrhizae, glomalin and soil aggregation. *Can. J. Soil Sci.*, 84: 355–363.

Rogers, R.D. and Williams, S.E. 1986. Vesicular-arbuscular mycorrhizae: influence on plant uptake of caesium and cobalt. *Soil Biol. Biochem.*, 18: 371-376

Ruiz-Lozano, J.M. 2003. Arbuscular mycorrhizal symbiosis and alleviation of osmotic stress: new perspectives for molecular studies. *Mycorrhiza*, 13: 309–317.

Safir, G.R., Boyer, J.S. and Gerdemann, J.W. 1971. Mycorrhizal enhancement of water transport in soybean. *Sci.*, 172: 581–583.

Schreiner, R.P., Mihara, K.L., McDaniel, H. and Bethlenfalvay, G.J. 1997. Mycorrhizal fungi influence plant and soil functions and interactions. *Plant Soil*, 188: 199-209.

Schubler, A., Schawarzott, D. and Walker, C. 2001. A new fungal phylum, the *Glomeromycota*: phylogeny and evolution. *Mycorrhizal Res.*, 105: 1414-1421

Singh, R.J. and Idani, L.K. 2007. Field response of summer mungbean to VAM fungus and water management. *J. Food Legumes*, 20: 62-64.

Singh, T.R., Singh, S., Singh, S.K., Singh, M.P. and Srivastva, B.K. 2004. Effect of integrated nutrient management on crop nutrient uptake and yield under okra-pea-tomato cropping system in a Mollisol. *Ind. J. Hort.*, 61: 312-314.

Smith, S.E. and Read, D.J. 1997. Mycorrhizal symbiosis. 2nd ed. 605 pp. Academic Press, London, UK.

Smith, S.E. and Read, D.J. 1998. Mycorrhizal symbiosis. *J. Ecol.*, 85: 925-926.

Song, H. 2005. Effects of VAM on host plant in the condition of drought stress and its mechanisms. *Electronic J. Biol.* 1: 44-48.

Sorial, M.E. 2001. Growth, phosphorus uptake and water relations of wheat infected with an arbuscular mycorrhizal fungus under water stress. *Ann. Agric. Sci.,* 39: 909-931.

Staddon, P.L., Ramsey, C.B., Ostle, N., Ineson, P. and Fitter, A.H. 2003. Rapid turnover of hyphae of mycorrhizal fungi determined by AMS microanalysis of C-14. *Science,* 300: 1138–1140.

Subramanian, K.S. and Charest, C. 1998. Arbuscular mycorrhizae and nitrogen assimilation in maize after drought and recovery. *J. Plant Physiol.,* 102: 285–296.

Subramanian, K.S., Charest, C., Dwyer, L.M. and Hamilton, R.I. 1997. Effects of arbuscular mycorrhizae on leaf water potential, sugar content and P content during drought and recovery of maize. *Can. J. Bot.,* 75: 1582–1591.

Suri, V.K. and Choudhary, A.K. 2012. Fertilizer economy through vesicular arbuscular mycorrhizal fungi under soil-test crop response targeted yield model in maize–wheat–maize crop sequence in Himalayan acid Alfisol. *Commun. Soil Sci. Plant Anal.,* 43: 2735-2743.

Suri, V.K. and Choudhary, A.K. 2013a. Effect of vesicular arbuscular mycorrhizal fungi and phosphorus application through soil-test crop response precision model on crop productivity, nutrient dynamics, and soil fertility in soybean–wheat–soybean crop sequence in an acidic Alfisol. *Commun. Soil Sci. Plant Anal.,* 44: 2032-2041.

Suri, V.K. and Choudhary, A.K. 2013b. Effect of vesicular arbuscular mycorrhizae and applied phosphorus through targeted yield precision model on root morphology, productivity and nutrient dynamics in soybean in an acid Alfisol. *Commun. Soil Sci. Plant Anal.,* 44: 2587-2604.

Suri, V.K. and Choudhary, A.K. 2013c. *Glycine-Glomus-Phosphate Solubilizing Bacteria* interactions lead to fertilizer phosphorus economy in soybean in a Himalayan Acid Alfisol. *Commun. Soil Sci. Plant Anal.,* 44: 3020-3029.

Suri, V.K. and Choudhary, A.K. 2014. Comparative performance of geographical isolates of *Glomus mosseae* in field crops under low input intensive P–deficient acid Alfisol. *Commun. Soil Sci. Plant Anal.,* 45: 101-110.

Suri, V.K., Chander, G., Choudhary, A.K. and Verma, T.S. 2006. Co-inoculation of VA-mycorrhizae (VAM) and phosphate solubilizing bacteria (PSB) in enhancing phosphorus supply to wheat in Typic Hapludalf. *Crop Res.,* 31: 357-361.

Suri, V.K., Choudhary, A.K. and Chander, G. 2010. Effect of VAM fungi and applied phosphorus through STCR precision model on growth, yield and nutrient dynamics in maize in an acid Alfisol. *Prog. Agri.,* 10: 12-18.

Suri, V.K., Choudhary, A.K. and Kumar, A. 2013. VAM fungi spore populations in different farming situations and their effect on productivity and nutrient dynamics in maize and soybean in Himalayan acid Alfisol. *Commun. Soil Sci. Plant Anal.*, 44: 3327-3339.

Suri, V.K., Choudhary, A.K., Chander, G. and Verma, T.S. 2011a. Influence of vesicular arbuscular mycorrhizal fungi and applied phosphorus on root colonization in wheat and plant nutrient dynamics in a phosphorus-deficient acid Alfisol of western Himalayas. *Commun. Soil Sci. Plant Anal.*, 42: 1177-1186.

Suri, V.K., Choudhary, A.K., Chander, G., Gupta, M.K. and Dutt, N. 2011b. Improving phosphorus use through co-inoculation of vesicular arbuscular mycorrhizal (VAM) fungi and phosphate solubilizing bacteria (PSB) in maize in an acid Alfisol. *Commun. Soil Sci. Plant Anal.*, 42: 2265-2273.

Tester, M., Smith, S.E. and Smith, F.A. 1987. The phenomenon of non-mycorrhizal plants. *Can. J. Bot.*, 65: 419-431.

Thingstrup, I., Kahiluouto, H. and Jakobsen, I. 2000. Phosphate transport by hypahe of field communities of arbuscular mycorrhizal fungi at two levels of P fertilization. *Plant Soil*, 221: 181-187.

Tisdall, J.M. 1991. Fungal hyphae and structural stability of soil. *Austr. J. Soil Res.* 29: 729-743.

Tisdall, J.M. and Oades, J.M. 1982. Organic matter and water stable aggregates in soils. *J. Soil Sci.*, 33: 141-163.

Trappe, J.M., Molina, R. and Castellano, M. 1984. Reactions of mycorrhizal fungi and mycorrhizal formation to pesticides. *Ann. Rev. Phytopathol.*, 22: 331-359.

Troeh, Z.I. and Loynachan, T.E. 2003. Endomycorrhizal fungal survival in continuous corn, soybean and fallow. *Agron. J.*, 95: 224-230.

Vilarino, A. and Arines, J. 1991. Numbers and viability of vesicular-arbuscular fungal propagules in field soil samples after wildfire. *Soil Biol. Biochem.*, 23: 1083-1087.

Wright, S.F. and Upadhyaya, A. 1998. A survey of soils for aggregate stability and glomalin, a glycoprotein produced by hyphae of arbuscular mycorrhizal fungi. *Plant Soil*, 198: 97–107.

Wright, S.F., Upadhyaya, A. and Buyer, J.S. 1998. Comparison of n-linked oligosaccharides of glomin from arbuscular mycorrhizal fungi and soils by capillary electrophoresis. *Soil Biol. Biochem.*, 30: 1853-1857.

Yadav, A. 2012. Phosphorus and rain-harvested water economy through vesicular arbuscular mycorrhizae (VAM) in garden pea. M.Sc. Thesis, CSK HPKV, Palampur (HP), India.

Yadav, A., Suri, V.K., Kumar, A. and Choudhary, A.K. 2015b. Influence of AM fungi and inorganic phosphorus on growth, green pod yield and profitability of pea (*Pisum sativum* L.) in Himalayan acid Alfisol. *Ind. J. Agron.* 60: 163-167.

Yadav, A., Suri, V.K., Kumar, A. Choudhary, A.K. and Meena, A.L. 2015a. Enhancing plant water relations, quality and productivity of pea (*Pisum sativum* L.) through

AM fungi, inorganic phosphorus and irrigation regimes in a Himalayan acid Alfisol. *Commun. Soil Sci. Plant Anal.*, 46: 80-93.

Yusnaini, S., Niswati, A., Nugroho, S.G., Muludi, K. and Irawati, A. 1999. The effect of vesicular arbuscular mycorrhizae inoculation on the yield of corn treated with temporary water stress during vegetative and generative phases. *J. Tanah Tropika (Indonesia)*, 5: 1-6.

Zhu, Y.G., Smith, S.E., Barritt, A.R. and Smith, F.A. 2001. Phosphorus efficiencies and mycorrhizal responsiveness of old and modern wheat cultivars. *Plant Soil*, 237: 249-255.

2017, Mycorrhizal Fungi
Editors: Ashok Aggarwal and Kuldeep Yadav
Published by: ASTRAL INTERNATIONAL PVT. LTD., NEW DELHI

Pages 311–333

14

Saline Soils and Possible Ways for Reclamation using Arbuscular Mycorrhizal Fungi

Romana M. Mirdhe, H.C. Lakshman and
Jyoti Puttaradder*

Post Graduate Department of Studies in Botany, Microbiology Laboratory,
Karnatak University, Dharwad – 580 003, Karnataka
*Corresponding Author: j.puttaradder@gmail.com

ABSTRACT

Saline soils are deficient in organic matter and their fertility is very low. Arbuscular Mycorrhizal colonized plants are able to grow and survive better in stress conditions through an increase in uptake of nutrients specially 'P', Zn', 'Cu' and H_2O in addition to the increased tolerance of plants to adverse conditions treated by unfavorable factors related to soil salinity. The present paper deals with causes of salinity, effect of salinity on plants, methods of Amelioration, AM Fungi symbiosis and its role and effect of salinity stress on AM fungal colonization, effect of salinity on germination of spores and plants grown in saline treatment with mycorrhizae. AM fungi have role in improving water relations of plants in sandy areas. However, it is not clear whether this effect is a direct result of fungal invasion and can be attributed improving water flow through hyphae or secondary response due to improved nutrition or physiological alterations of the host. Studies conducted showed that AM fungal technology may have great deals to contribute in improving the productivity and profitability of saline lands.

However, there is need to select promising AM fungal species or strains suitable to saline lands. Possibilities of using a combination of symbiotic/non- symbiotic nitrogen fixtures, phosphate solubilizers and AM fungi to give maximum benefit to the plants growing under saline stress conditions will have to be explored.

Keywords: *Arbuscular mycorrhizal fungi (AMF), Saline soils, Strategies, Reclamation, Nutrients, Absorption, Coastal sand dunes.*

1. Introduction

Salinity stress is a world wide phenomenon which leads to the poor development of crop. Several national and international laboratories/institutions are involved in the production of crops in saline soils, which cover about one billion hectares of the world's land area. In India salinity is a major problem, nearly seven million hectare of the land is not in use due to saline/stressed conditions. Salinization of soil is a serious problem and is increasing steadily in many parts of the world, in particular in arid and semiarid areas (Giri *et al.*, 2003; Al-Karaki, 2006). Saline soils occupy 7 per cent of the earth's land surface (Ruiz-Lozano *et al.*, 2001) and increased salinization of arable land will result in to 50 per cent land loss by the middle of the 21st century (Wang *et al.*, 2003).

Plants growing in saline conditions mainly in arid and semi arid zones may suffer from high salt stress and temperature conditions. Absorption of constituent ions of saline habitat lead to the detrimental accumulations of ions and decreased absorption of essential nutrients which result in the imbalanced nutrient level in plants. The saline soils contain toxic concentration of soluble salts in the root zone. Soluble salts consist of chlorides and sulphates of sodium, calcium, magnesium. Because of the white encrustation formed due to salts, the saline soils are also called white alkali soils (Lakshman *et al.*, 2000)

2. Causes of Salinity

In arid and semi arid areas salts formed during weathering are not fully leached. During the periods of heavy rainfall the soluble salts are leached from the more permeable high areas to low lying areas, and where-ever the drainage is restricted, salts accumulate on the soil surface, as water evaporates. The excessive irrigation of uplands containing salts resulting in the accumulation of salts in the valleys. In areas having salt layer at lower depths in the profile, seasonal irrigation may favour the upward movement of salts. Salinity is also caused if the soils are irrigated with saline water. In coastal areas the ingress of sea water induces salinity in the soil (Table 14.1).

2.1. Effect of Salinity on Plants

High osmotic pressure decreases the water availability to plants resulting in retardation of growth rate. As a result of retarded growth rate, leaves and stems of affected plants are stunted. Development of thicker layer of surface wax imparts bluish green tinge on leaves. Due to high EC germination per cent of seeds is reduced (Figure 14.1).

Table 14.1: Showing the Characteristics of Saline Soil

Parameters	Details
pH	Less than 8.3
EC	More than 4.0 m.mhos/cm
ESP (exchangeable sodium per cent)	Less than 15
Chemistry of soil solution	Dominated by sulphate and chloride ions and low in exchangeable sodium
Effect of electrolytes on soil particles	Flocculation due to excess soluble salts.
Main effect on plant	High osmotic pressure of soil solution
Geographic distribution	Arid, semi arid and coastal areas.
Diagnosis under field condition	Presence of white crust, Presence of *Chloris barborata* (weed), Patchy growth of plants.

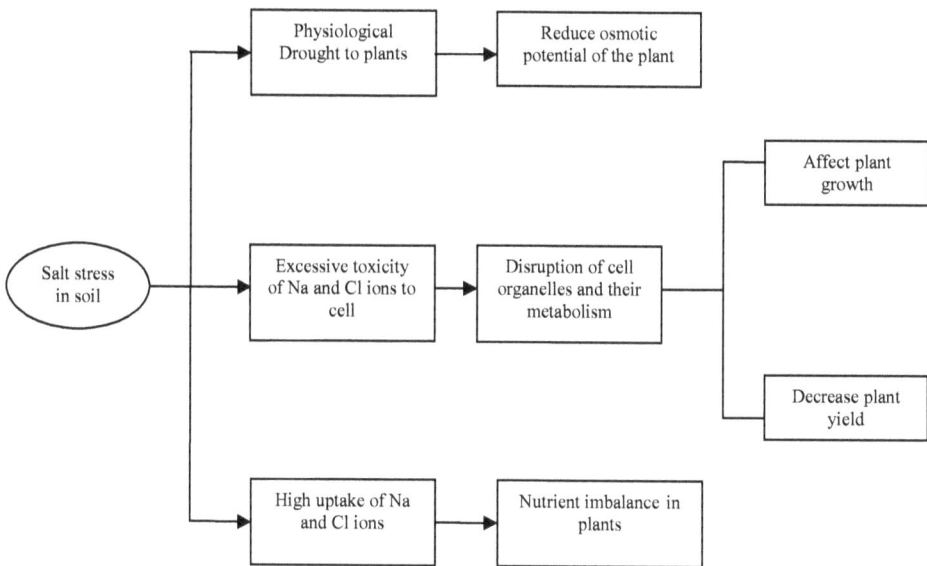

Figure 14.1: Showing the Effect of Salt Stress on Metabolism and Growth of Plants (Source: Evelin *et al.*, 2009).

2.2. Crops Suitable for Cultivation in Saline Soils

Barley, Sugar beet, Cotton, Sugarcane, Mustard, Rice, Maize, Red gram, Green gram, Sunflower, Linseed, Sesame, Bajra, Sorghum, Tomato, Cabbage, Cauliflower, Cucumber, Pumpkin, Bitter gourd, Beetroot, Guava, Asparagus, Banana, Spinach, Coconut, Grape, Date palm, Pomegranate.

3. Methods of Amelioration

☆ The salts are to be leached below the root zone and not allowed to come up. However, this practice is some-what difficult in deep and fine textured soils containing more salts in the lower layers. Under these conditions, a provision of some kind of sub-surface drains becomes important.

☆ The required area is to be made into smaller plots and each plot should be bounded to hold irrigation water.

☆ Separate irrigation and drainage channels are to be provided for each plot.

☆ Plots are to be flooded with good quality water up to 15-20 cms and puddled. Thus, soluble salts will be dissolved in the water.

☆ The excess water with dissolved salts is to be removed into the drainage channels.

☆ Flooding and drainage are to be repeated 5 or 6 times till the soluble salts are leached from the soil to a safer limit.

☆ Green manure crops like *Daincha* can be grown up to flowering stage and incorporated into the soil. Paddy straw can also be used.

☆ Super phosphate, Ammonium sulphate or Urea can be applied in the last puddle. Ammonium chlorides should not be used.

☆ Scrape the salt layer on the surface of the soil with spade.

☆ Grow salt tolerant crops like Sugar beet, Tomato, Beet root, Barley *etc.*

Before sowing, the seeds are to be treated by soaking the seeds in 0.1 per cent salt solution for 2 to 3 hours.

During last two decades much attention has been paid towards the role of Saline stress on the growth and development of plants. Scientists have developed their confidence to protect the crop from salinity by various methods such as:

i. The use of micro-propagated plants.

ii. The use of Cell and tissue culture and evolving plantlets growing *in vitro/ in vivo* conditions.

iii. Inoculation with appropriate phosphate solubilizing/saline tolerant bacteria and other microorganisms, Arbuscular Mycorrhizal Fungi under salinity and drought conditions.

iv. Inoculation of plants with appropriate indigenous/efficient strains of mycorrhizal fungi and transplanted to saline soils.

4. Symbiosis of AM Fungi

Arbuscular mycorrhizal fungi are common in Coastal Sand Dunes (CSDs) throughout the world (Sturmer and Bellei, 1994). Functions of AM fungi on CSDs include nutrient uptake, improvement of soil structure and formation of sand aggregates, which supports plant succession and dune stability (Koske and Poison, 1984). In Rhode Island, *Acaulospora breviligulata* shows high AM spore density (Koske and Halvorson, 1981). Members of Asteraceae, Papilionaceae and Poaceae on the Italian CSDs were heavily colonized by AM fungi (Giovannetti, 1985). In CSDs of US Atlantic Coast. *Gigaspora gigantea* was most dominant (Lee and Koske,

1994). Several new species of AM fungi have been reported from Poland CSDs (Arun, 2002). Subtropical CSDs (Florida, Mexico, Baja California, Brazil, Japan, Australia and Pakistan) have been surveyed for AM fungi. Maximum plant species (97 per cent) in mobile and stable dunes of the Gulf of Mexico colonized by AM Fungi (Corkidi and Rincon, 1997). *Scutellospora erythropa* (Bahamas), *Acaulospora scrobiculata* and *Gigaspora albida* (North America) (Koske and Walker, 1984) and *Glomus* spp. (Japanese dunes) (Abe *et al.*, 1994) were most dominant. Roots of 23 out of 31 vascular plant species of Hawaiian Islands consist of AM fungi (dominant: *Glomus microaggregatum* and *Sclerocystis sinuosa*). Most of the seedlings grown on the drift line area in Hawaiian Islands consist of AM fungi (Koske and Gerruna, 1990).

A few recent surveys are available from the sand dunes of the Indian subcontinent. Along the 36 km coast of Tamil Nadu, out of 56 plant species, 35 were colonized by AM fungi (Mohankumar *et al.*, 1988). Fourteen AM fungi were recovered on examination of 31 plant species on the CSDs of Tuticorin, Tamil Nadu (Ragupathy *et al.*, 1998). Twelve plant species of the west coast of Karnataka yielded 16 AM fungi, among them, *Scutellospora gregaria* and *Glomus albidum* were most frequent (Kulkarni *et al.*, 1997). Among five AM fungal spores recovered on non-vegetated dunes, *Glomus albidum* and *G. lacteum* were dominant. Twenty-eight CSD plant species of the west coast of India yielded 30 AM fungi (Beena *et al.*, 2001b). The major plant species of southwest CSDs, *Ipomoea pes-caprae* and *Launaea sarmentosa* showed 41 and 28 AM fungi respectively (Beena *et al.*, 1997, 2000a). Root colonization was highest during post-monsoon in both plant species, while spore density during summer (*I. pes-caprae*) or monsoon (*L. sarmentosa*). Species richness and diversity of AM fungi associated with *I. pes-caprae* of 100 km stretch of the southwest coast in moderately disturbed dunes (MDD) and severely disturbed dunes (SDD) have been studied by Beena *et al.* (2000b). The AM fungal colonization, richness and diversity were coincided with vegetation cover and rhizosphere nitrogen. *Glomus mosseae*, *G. dimorphicum* and *Gigaspora gigantea* were most common irrespective of MDD and SDD. *Polycarpaea corymbosa* of southwest CSDs yielded 12 species of AM fungi (Beena *et al.*, 2001a). One hundred eighty AM spores belonging to 29 species were recovered from 900 g soil samples collected from vegetation-free sites of mid dunes of southwest coast of India (Bhagya *et al.*, 2005), Spore recovery increased with increasing number of samples, while species richness saturated at about 20 samples. *Acaulospora denticulata* and *Scutellospora calospora* were dominant, while *Acaulospora denticulata*, *Glomus intraradices* and *Scutellospora calospora* showed wide distribution. Significant difference was seen between AM spore density with legume richness and density in mid dunes. The AM spore density was positively correlated with soil organic carbon, total nitrogen and available phosphorus. Surveys carried out on the CSD vegetation of Goa revealed 17 AM fungi (Jaiswal and Rodriguos, 2001; Rodrigues and Jaiswal, 2001). Although a few surveys were undertaken, AM fungal richness and diversity of the Indian CSDs are higher than temperate regions. Hiremath and Lakshman, (2007), have screened in three coastal sand dunes one estuary of Kumata (South Western India) for AM fungal association. A total of 30 AM fungal spores were recovered from coastal sand dunes. Genus *Scutellospora* was most predominant with 12 species followed by *Glomus* 9 species, *Gigaspora* 4 species, *Acaulospora* 2 species, *Sclerocystis* 2 species and *Enterophosphora* one. The highest number of AM spores was recovered during summer, followed by winter and rainy season.

Successful growth of dune plant species depends on AM fungi in soils (Gemma and Koske, 1997). Artificial inoculation of AM fungi (*e.g. Gigaspora gigantea*) enhanced the transplanting success of *Ammophila* on the CSDs (Maun and Baye, 1989). The CSDs with high vegetation cover by *Ammophila* possess elevated AM fungal species diversity as well as spore density (Koske and Halvorson, 1981). Inoculation of *Glomus macrocarpum* and *G. fasciculatum* stimulated the growth of beach grass in unstable foredunes of Scotland (Sylvia, 1986). Association of *U. paniculata* (sea oats) with AM fungi decreased the environmental stresses and resulted in dune stability, Sea oats planted in Florida beach showed high AM fungal colonization, fungi and bacteria than in vacant or recently planted beach sites (Sylvia and Will, 1988). Greenhouse inoculation of *Glomus deserticola* and *G. etunicatum* to *U. paniculata* significantly increased the height, dry mass and plant cover (Sylvia and Burks, 1988). Seedlings colonized by *G. deserticola* and *G. macrocarpum* showed high shoot dry mass (219 per cent), root length (81 per cent), plant height (64 per cent) and number of tillers (53 per cent) at beaches of Miami (Sylvia, 1989). Plant species of embryo dunes, foredunes and stable dunes of Gulf of Mexico were colonized by AM fungi (Corkidi and Rincon, 1997). Rhizomatous perennial grass, *Spartina ciliata* in fixed CSDs of Brazil dominated due to high AM fungal diversity and tropical plants and beach ponds in India (Sturmer and Bellei, 1994; Lakshman 1994, 1999; Lakshman *et al.*, 2000). Strand vegetation of Galapagos Islands and Island of Great Barrier Reef consists of high AM fungal colonization (Schmidt and Scow, 1986; Peterson *et al.*, 1985).

5. Role of Arbuscular Mycorrhizal Fungi (AMF) in Saline Soils

Arbuscular mycorrhizal fungi make symbiotic association with the roots of higher plants, utilize carbohydrates produced by plant, in turn plant is benefited by the increased uptake of mineral nutrients especially P and N. The role of AM fungi in saline and drought conditions is very important for increasing fertility. The potential of mycorrhizal inoculum can be utilized for the growth and development of plants in the stressed conditions through various ways such as:

  i. Enhancing establishment and survival of transplanted seedlings in adverse conditions including pH stress, tempersature stress, heavy metals and toxin stress. (Gardner and Malajczuk, 1988; osonuki *et al.*, 1991; Dixon *et al.*, 1994)

 ii. Improving plant growth rate by increasing nutrient uptake with production of hormones like auxins, cytokines, gibberellins and B vitamins (Ho, 1987; Kraigher *et al.*, 1991).

iii. Improving ability of host plant to compete root/soil borne plant pathogens through strategies such as the use of surplus carbohydrates and secretion of antibiotics (Garrido *et al.*, 1982; Channabasava and Lakshman, 2012).

 iv. Boosting capacity of host plant against stresses by increasing nutrient and water absorption (Boyd, 1987).

v. Increasing efficiency of nutrient recycling with mobilization through biological weathering (Remy *et al.,* 1994).

vi. Stabilization and aggregation of the soil (Fries and Allen, 1991; Lakshman, 1999).

Mycorrhizal technology therefore, has assumed greater relevance in crop production in stressed conditions. They play significant role in the establishment and growth of plant seedlings. Several studies investigating the role of AMF in protection against salt stress have demonstrated that the symbiosis often results in increased nutrient uptake, accumulation of osmoregulators, increased photosynthetic rate and water-use efficiency, suggesting that salt-stress alleviation by AMF results from a combination of nutritional, biochemical and physiological effects. Studies carried out so far have suggested several mechanisms by which AM symbiosis can alleviate salt stress in host plants.

6. Effect of Salinity Stress on AM Fungal Colonization in different Plants

Juniper and Abbott (1993) suggested that activity of AM fungi may be affected by soil salinity. They found that environmental factors which affect the physiology of host plants are likely to affect their mycosymbionts also. It is well known that mycorrhizal associations are dependent on carbohydrate nutrition for the existence but photosynthetic activity of host plants also affects the carbohydrate status and root colonization in saline soils (Jasper *et al.,* 1979; Subramanian and Charest, 1995). It has been found that effect of soil salinity differs in different plant species. The toxic effect of specific ions such as sodium, calcium and chloride, prevalent in saline soil affects AMF colonization (Gildon and Tinker, 1981) and alter enzyme status and absorption of several other macro and micro elements therefore, their induction reduced photosynthesis, respiration and protein synthesis (Epstin, 1972). Salinity of soil is also known to interfere with the uptake of cations such as calcium and magnesium causing physiological drought which reduces AM colonization (Bernstein, 1975; Juniper and Abott, 1993).

AM fungi are reported to reduce the detrimental effects of soil-associated plant stresses, such as lack of nutrients, organic matter, high salinity, or high pH (Sylvia and Williams, 1992; Entry *et al.,* 2002). Most of the studies show that, native AM fungi grow and function better especially in stressed conditions and degraded soils as they are quite adapted to their stressed microhabitat (Caravaca *et al.,* 2003; Calvente *et al.,* 2004). Several researchers have investigated the distribution of AM fungi in different ecological regions and their relation to soil (Aliasgharzadeh *et al.,* 2001; Wang *et al.,* 2004; Shi *et al.,* 2007; Hiremath and Lakshman, 2007; Jyoti *et al.,* 2015). These species and isolates of AM fungi differ in their tolerance to adverse physical and chemical conditions of soil (Aliasgharzadeh *et al.,* 2001). All the plants, except Brassicaceae and Cyperaceae, surveyed in alkaline/sodic soils of the selected site were colonized by the AM fungi but the colonization percentages were significantly low. The results were in accordance to the earlier reports of Brown and Bledsoe (1996), Udaiyan *et al.* (1996), and Wang *et al.* (2004). Mycorrhizal intensity showed a significant correlation with the plant species and magnitude of stress. Lowest mycorrhizal intensity at Zone 1 of all the three sites may be attributed to the fact that high soil pH, low nutrients, poor plant diversity, and vegetation cover severely restrict the colonization of AM fungi (Aliasgharzadeh *et al.*, 2001; Wang *et al.*, 2004). However, AM fungi have the ability to form a zone of altered pH in the adjacent soil (Li, *et al.*, 1971). Mycorrhization in chenopods under different kind of stresses, such as drought, salinity, *etc.*, was earlier reported by Hirrel *et al.* (1978). These results may also support the conclusion of Carvalho *et al.* (2001) that the colonization by Vesicular Arbuscular Mycorrhizae (VAM) under stress conditions depend on more on host plant species than environmental stresses.

7. Effect of Salinity on Germination of Spores

Besides lowering arbuscular mycorrhiza formation, soil salinity also inhibits germination of AM fungal spores. Many workers have reported restricted germination of AM fungal spores in saline soils (Gildon and Tinker, 1981; Hirrel, 1981; Juniper and Abott, 1991; Juniper and Abott, 1992; Juniper and Abott, 1993). Estaun (1989) reported inhibitory effect of NaCI on germination of AM fungal spores. *Gigaspora gigantea* and *Glomus epigaeus* showed reduced spore germination when spores were treated to polyethylene glycol (PEG) (Koske, 1981). Sylvia and Schenck (1983), have reported that soil matric potential and fungal contamination affects germination of spores of three *Glomus* species. Application of NaCl reduced hyphal growth of *Acaulospora trappei*, *Gigaspora decipiens* and *Scutellospora calospora*. However, diameter of hyphae of *Scutellospora calospora* was not affected by the increasing concentration of NaCl.

Arbuscular mycorrhizal hyphae have the ability to extract nitrogen and transport it from soil to plant. AM fungi contain some enzymes which influence nitrogen fixation rates by reducing stress imposed on plants (Raman and Mahadevan, 1996). Nitrogen exists in many forms *viz.*, free nitrogen, nitrate, nitrite ammonium ions and organic nitrogen. Ammonium is less mobile in soil. AM fungal hyphae transport such immobile ammonium to plant roots. AM fungi have shown increased nitrate reductase activity which confers the importance to AM fungi in absorption of nitrate (Ho and Trappe, 1975; Lakshman, 1999). Increased concentration of potassium in mycorrhizal tomato plants was reported. Bethlenfalvay and Franson (1989) found that growth response of soybean to AM fungal inoculation was more related to improve rather than phosphorus nutrition of host plant. Enhanced uptake of Zn in mycorrhizal *Zea mays* growing under calcareous soil was reported which ranged from 16 to 25 per cent. Similar results were also obtained in mycorrhizal *Linum usitatissimum* plants by Wellings and Thompson (1991).

8. Role of AM Fungi in Sand Aggregation

Aggregation of sand is one of the major contributions of sand dune microbes (bacteria, algae and fungi) to stabilize the Coastal sand Dunes (CSDs). Sand aggregates possess a wide range of microbes (*Bacillus* spp., *Pseudomonas* spp., *Nocardia* spp., *Streptomyces* spp. *Aspergillus fumigatus*, *Penicillium* spp. and *Glomus fasciculatum*). Smaller sand aggregates (>1 to <1.4 mm) had a few fungal species than the larger ones (>2 min). The AM fungi in CSDs involve in sand aggregation in three phases: a) entangling the soil particles by hyphae; b) roots with hyphae create conditions suitable to form microaggregates; c) roots and hyphae enmesh and bind the microaggregates to large macroaggregates (Miller and Jastrow, 1990). Besides fungal hyphae, polysaccharides produced by the hyphae firmly bind the microaggregates (Tisdall, 1991). One-gram stable macroaggregate is known to have about 50 m hyphae (Tisdall, 1991). *Phaseolus vulgaris* showed 10 g/kg aggregates without AM fungal colonization, while colonization of *Glomus* resulted in 54 g/kg (Sutton and Sheppard, 1976). Aggregates of sand dunes are known to remain intact even after the death of roots or hyphae (Koske and Poison, 1984). The AM fungi (Acaulosporaceae and Glomaceae) are known to produce extracellular polysaccharide, β1-3 glucan, which serves the soil aggregation (Lemoine *et al.*, 1995). A stable glycoprotein, glomalin produced by *Gigaspora gigantea* is responsible for soil aggregation (Wright and Upadhyaya, 1998). Insoluble hydrophobic glomalin coating is known to protect minerals, microbes and organic matter in aggregates. Interestingly, *G. gigantean* is the most common AM fungus in severely disturbed dunes of southwest coast of India (Beena *et al.*, 2000b). Mycorrhizal roots are known to stimulate specific groups of bacteria, including those producing extracellular polysaccharides (Bagyaraj and Menge, 1978; Meyer and Linderman, 1986). Mineral aggregates in soil are stable in the presence of humic substances (Emerson *et al.*, 1978). Stable aggregates maintain aeration and moisture in soil, which is essential for optimal growth of plants (Oades, 1984).

9. Dune Restoration in different Plants Growing in Coastal Regions

The CSDs of the world are under the threat of human interference (*e.g.* encroachment, agriculture, waste disposal and roads). Large dune areas reclaimed for forest and farmland in New Zealand (Gadgil and Ede, 1998) suggested the implementation of traditional knowledge to restore dunes for future needs. The coastal vulnerability in SE Portugal to SW Spain increased with human disturbance (Garcia-Mora *et al.*, 2000). Building seawalls cannot replace vegetation adjacent to CSDs. There is ample scope to vegetate tropical CSDs and adjacent locations employing a variety of plant species (Table 14.2).

Cropping in multistoried pattern using mat-forming creepers, grasses, sedges, xerophytes, scrubs, herbs and tree species might effectively withstand wind and wave action and restore landscapes. To derive maximum benefits from the CSDs, it is necessary to conserve the belowground biota including AM fungi.

In summary, CSDs constitute major biomes of considerable ecological and economic importance. Members of plant family Poaceae are dominant plant species in temperate dunes, while Asteraceae, Convolvulaceae, Fabaceae and Poaceae in tropical dunes. Stress-tolerant plant species and associated microbes of CSDs need special attention for stabilization. Below ground diversity of mycorrhiza is one of the major factors governing the plant distribution, diversity and functioning of CSD ecosystem. Elimination of mycorrhizal fungi decreases the plant equilibrium as well as productivity, which results in destabilization. Future studies should concentrate

Table 14.2: Vegetation on the Coastal Sand Dunes of the Indian Subcontinent

Mat-forming creepers
Aeluropus lagopoides, Canavalia maritima, Canavalia cathartica, Indigofera aspalathoides, Ipomoea pes-caprae, Launaea sarmentosa, Paspalum vaginatum, Perotis indica, Sesuvium portulacstrum, Sporobolous virginicus, Trachys muricata and Zoysia matrella.

Prostrate/erect herbs and sedges
Allmania nodiflora, Anotis carnosa, Atriplex repens, Atriplex stocksii, Borreria articularis, Borreria stricta, Crotalaria nana, Enicostema hyssopifolium, Euphorbia atoto, Euphorbia rosea, Geniosporum tenuiflorum, hydrophylax maritime, Polycarpaea corymbosa, Polycarpaea spicata and Scaevola plumieri.

Climbers
Dalbergia spinosa, Derris triflorum, Flagellaria indica, Ipomoea macrantha and Parsonsia helicandra.

Plants with perennating organs
Asparagus dumosus, Scilla hyacinthine and Urginea indica.

Scrubs
Acrostichium aureum, Clerodendrum inerme, Dimorphocalyx glabellus, Halopyrum mucronatum, Myriostachya wightiana, Scaevola taccada, Syzygium ruscifolium and Tamarix articulate.

Trees
Acacia planifrons, Ardisia littoralis, Copparis cartilaginea, Calophyllum inophyllum, Euphorbia caducifolia, Hyphaene dichotoma, Messerschmidia argentea, Morinda citrifolia, Pandanus tectorius, Pemphis acidula, Premna serratifolia and Salvadora persica.

Source: Rao and Meher-homji, 1985; Rao and Sherieff, 2002; Sridhar, 2006.

on the importance of mycorrhizae on CSD stabilization and application of stress-tolerant mycorrhizae in agriculture.

11. Effect of AM Fungi on Plant Growth under Saline Conditions

Several workers have paid attention towards the role of AM fungi in improved growth and nutrition of plants under different habitats (Hirrel and Gerdemann, 1980; Ojala *et al.*, 1983). This may be due to increased uptake of phosphorus and other mineral elements which lead to increased growth of plant. AM fungi reduce salinity stress and enhance resistance against plant pathogens and reduce transplantation shock in the multipurpose tree species (Giri, 1997). AM fungi have increased water absorption capacity of the plants, elevated rate of photosynthesis and improved growth rate under stressed conditions. Many workers (Gupta and Krishnamurthy, 1996) pointed out that AM inoculated plants increased the number of adventitious roots, their branching and also suggested that these modifications may be relevant to salt tolerance. Many workers also reported improved growth in mycorrhizal plants like *Parthenium argentatum* (Levy and Krikun, 1980), *Allium cepa, Capsicum annum* and Finger millet (Giri, 1997; Patil and Lakshman, 2003) *Arachis hypogaea* and *Triticum astivum* (Gupta and Krishnamurthy, 1996; Bheemareddy and Lakshman 2010) under salinity stressed conditions.

Pond *et al.* (1984), have reported distinct improved growth of *Lycopersicum esculantum* in saline soil inoculated with *Glomus fasciculatum*. Poss *et al.* (1985) summarized that AM fungi inoculated plant increased growth over uninoculated control plants, under low phosphorus level. However, no response of AM inoculation was reported when 0.8 and 1.6 nmol p^{k-1} were added. *Parthenium argentatum* plant inoculated with *Glomus intraradices* showed increased growth in highly saline soil and decreased amount of arbuscules and vesicles in mycorrhizal roots. Romana and Lakshman (2010), recorded the influence of *Rhizophagus fasciculatus* at lower concentration of saline water on two legumes.

12. Greenhouse Experiment to Understand the Effect of AM Fungi in Saline Stress

AM fungi had significant role in promoting seedling growth and establishment of two legumes crops in saline conditions under green house conditions also (Romana and Lakshman, 2011). Improved nutrient uptake by mycorrhizal plants as compared to non-mycorrhizal ones is probably the major mechanism involved in increasing seedling salinity tolerance. These findings are in consistent with early workers (Marschner and Dell, 1994; Al-Karaki and Al-Raddad, 1997; Al-Karaki, 2000; Bheemareddy and Lakshman 2010; Lakshman, 2014). These observations may be taken in highlighting the potential of using AM pre-inoculated seedling of mycorrhizal responsive plants to re-vegetate saline lands. Non-Mycorrhizal and Mycorrhizal plants treated with varied level of salinity showed significant difference in their dry weight of shoot and root. Non-inoculated control plants have not showed root colonization. However, in mycorrhizal inoculated plants good per cent of root colonization was recorded. Percentage of root colonization was less in different levels of salinity in early stages, but at later stage percentage of root colonization was affected much significantly. In saline treated non-mycorrhizal and mycorrhizal plants, with increasing level of salinity there was decrease in growth parameters like shoot length, fresh weight of shoot, dry weight of shoot, root length, fresh weight of root, dry weight of root, per cent root colonization, number of leaves and per cent of 'P' content in shoot was recorded (Romana and Lakshman, 2011).

In *Phaseolus vulgaris,* the shoot length was found to be highest in AMF inoculated plants, where as lowest found in plants treated with 6ds/m salinity, other parameters like, dry weight of shoot, root length dry weight of root, per cent root colonization P' uptake in shoot N uptake were also found to have similar results. No root colonization was recorded in non-inoculated control plants. But the percentage of root colonization was recorded decreasing with increase in salinity level in mycorrhizal plants (Romana and Lakshman, 2011).

In *Psophocarpus tetragonolobus* plants have responded variedly to different levels of salinity. It was observed that at 90 days percentage of root colonization was adversely affected by increasing soil salinity level though this effect was less significant in early stages. Plants showed significant decrease in all growth parameters with increased salinity level. Overall results showed growth and establishment in most of the experimental plants with *Glomus fasciculatum* (inoculation in saline conditions as compared with non inoculated conditions. Pre inoculation increased mycorrhizal effectiveness and ensured that roots were well colonized under salt stress. Decrease in growth parameters is more pronounced in non- mycorrhizal plants than mycorrhizal plants (Romana and Lakshman, 2011). The non mycorrhizal plants under high salinity have greater concentrations of potassium, sodium and phosphorus (Ojala *et al.*, 1983) where as mycorrhizal plants maintained a steady Na/K balance as non-mycorrhizal plants.

AMF under experimental conditions are affected by many edaphic factors, of which pH plays a very important role (Beena *et al.*, 2001a). Sand movement is another important factor affecting the distribution and composition of coastal plant communities (Martinez and Moreno-Casasola, 1996). These studies suggested that addition of these legumes could be successfully used at places of heavy accretion. But it has been emphasized that, much more needs to be known about the micro-morphological, physiological and genetic variability in AM fungi at different levels of their organization. Conservation and efficient utilization of their bio diversity are of crucial importance for sustainable production system for the equilibrium of natural plant community structure (Walker, 1995).

Mycorrhizal efficiency, the net benefit to the host plant of infection by mycorrhizal fungi, is a function of a particular set of soil and environmental conditions. Competitive ability and fitness are attributes assessed only by the benefit. Although they may be affected by the same environmental and host factors as efficiency. Relative aggressiveness, reproductive ability, persistence, *etc.* are also important. Re-vegetation is one of the most important practices that can help in the reclamation of degraded lands (Lakshman, 2005). Use of mycorrhiza could form the basis of a sustainable method to support the survival of transplanted plants in re-vegetation by increasing the ability of plant root systems to take up nutrients (especially phosphorus) and water.

13. Effect of AM Fungi on Nutrient uptake in Saline Soils

The phosphorus concentration in plant tissues rapidly becomes low under salt stress because phosphate ions precipitate with Ca^{2+} ions in salt-stressed soil and become unavailable to plants (Poss *et al.*, 1985; Munns, 1993). Studies revealed that mycorrhization strongly affects Ca^{2+} in the plant. Cantrell and Linderman (2001) reported increased Ca^{2+} uptake in mycorrhizal lettuce under saline conditions. A higher Ca^{2+} concentration in mycorrhizal than in non-mycorrhizal banana plants was reported by Yano-Melo *et al.* (2003). Moreover, high Ca^{2+} was also found to enhance colonization and sporulation of AMF (Jarstfer *et al.*, 1998). However, in contrast to the reports above, Giri *et al.* (2003) reported that Ca^{2+} concentration remains unchanged in shoot tissues of mycorrhizal and non-mycorrhizal *Acacia auriculiformis* plants. This suggests that AMF may not be so important to the nutrients moving to plant roots by mass flow as compared with nutrients moving by diffusion (Tinker, 1975). Ojala *et al.* (1983) also experimented with different salinity levels and found that AM fungal-inoculated onion had higher concentrations of potassium in shoots and bulbs at all salinity levels. Higher potassium accumulation by mycorrhizal plants in saline soil could be beneficial by maintaining a high K/Na ratio and by influencing the ionic balance of the cytoplasm or Na^+ efflux from plants (Table 14.3). When Na^+ or salt concentration in the soil is high, plants tend to take up more Na^+ resulting in decreased K^+ uptake. Na^+ ions compete with K^+ for binding sites essential for various cellular functions. Potassium plays a key role in plant metabolism.

Table 14.3: Showing the Probable Mechanisms throfugh which AMF Reduce Impacts of High Salinity on Plants

Protects plant from oxidative damage	$\longrightarrow$	Antioxidant production
Photosynthesis	$\longrightarrow$	Increased Mg^{2+}/Na^+ ratio Maintain chlorophyll
Ionic balance Protein synthesis	$\longrightarrow$	High K^+/Na^+ ratio
Water-use efficiency	$\longrightarrow$	Stomatal conductance Transpiration Water absorption

Source: Heikhem *et al.*, 2009.

Microelements are needed by the plant in very small quantities, but play very prominent role in the growth and development of plants. These elements includes Cu, Zn, Mg, Mn, Ca, Co, Cd, *etc.* which are significantly correlated with developmental stages of the plants. Mycorrhizal fungi are known to accumulate greater amount of some microelements especially under stressed conditions (Faber *et al.*, 1990; Smith, 1980). The absorption of microelements is usually limited by the rate of diffusion. Uptake of these elements results in formation of depletion zones around actively growing plant roots. AM fungus prevents plant from Mn toxicity and reduced amount of Mn in red clover (Arines *et al.*, 1989), alfalfa (Shrivastava *et al.*, 1996) and maize (Kothari *et al.*, 1991).

Biosynthesis of chlorophyll is impeded by salt stress, which prevents light harvesting and causes impairment of photosynthesis. Mycorrhiza by improving Mg^{2+} can support a higher chlorophyll concentration (Giri *et al.*, 2003). This suggests that salt interferes less with chlorophyll synthesis in mycorrhizal than non-mycorrhizal plants (Giri and Mukerji, 2004). Effective Mg^{2+}-uptake helps by increasing the chlorophyll concentration and hence improving photosynthetic efficiency and plant growth. Increasing salinity causes a reduction in chlorophyll content (Sheng *et al.*, 2008) due to suppression of specific enzymes that are responsible for the synthesis of photosynthetic pigments (Murkute *et al.*, 2006). A reduction in the uptake of minerals (*e.g.* Mg) needed for chlorophyll biosynthesis also reduces the chlorophyll concentration in the leaf (El-Desouky and Atawia, 1998). A higher chlorophyll content in leaves of mycorrhizal plants under saline conditions has been observed by various authors (Patil and Lakshman, 2003; Giri and Mukerji, 2004; Sannazzaro *et al.*, 2006; Zuccarini, 2007; Colla *et al.*, 2008; Sheng *et al.*, 2008). This suggests that salt interferes less with chlorophyll synthesis in mycorrhizal than in non-mycorrhizal plants (Giri and Mukerji, 2004).

14. Effect on Hydraulic Conductivity

Many workers have studied water transport in terms of hydraulic conductivity of root. Levy and Krikun (1980) determined that Arbuscular mycorrhizal inoculation increased stomatal conductance but did not affect conductivity under well watered condition. It has also been reported that AM association affects stomatal regulation.

Root conductivity of P-red clover was measured by Hardie and Leyton (1981) and found that higher transpiration rate and water demand of arbuscular mycorrhizal plants were not a result of increased plant height, but higher root conductivity. Increased drought tolerance of onion plant was also associated with increased P-nutrition and hydraulic conductivity (Nelson and Safir, 1982). Graham and Syvertson (1984) pointed out that mycorrhizal inoculation of citrus cultivars reduced root-shoot ratio whereas increased root hydraulic conductivity of mycorrhizal plant and was twice that of the non-mycorrhizal control plants. Higher transpiration rates and phosphorus nutrition in mycorrhizal plants were associated to the increased conductivity of roots. Drought stressed root showed lower hydraulic conductivity in comparison to well watered plants (Levy *et al.*, 1983). Higher transpiration rates and root density of AM inoculated plants were responsible for rapid depletion of soil water leading to more severe stress condition during stress period.

15. Conclusion

Arbuscular mycorrhizal fungi alleviate the detrimental effect of salinity through improved nutrient uptake. The reduction in shoot Na uptake and maintaining electrical conductivity of the soil may be significant in helping mycorrhizal plants to survive in saline conditions. The greater effectiveness of combined inoculation may be due to synergistic interaction between both AM fungal species and variation in efficacy among fungal species. AM fungi may bring an improvement in the growth of plants through increased P uptake. Higher rate of photosynthesis and reduction in stomatal conduction and mesophyll resistance resulting from an altered hormonal balance and increase water flow. The role of mycorrhizal fungi in salt stress conditions is still inconclusive. A few studies have demonstrated their effect on osmotic adjustment but all these studies have been conducted under low salt conditions. However, more number of pilot studies are warranted.

References

Abe, J.P., Masuhara, G. and Katsuya, K. 1994. Vesicular arbuscular mycorrhizal fungi in coastal dune plant communities I. Spore formation of *Glomus* sp. predominates under a patch of *Elymus mollis*. *Mycoscience*, 35: 233-238.

Aliasgharzadeh, N., Rastin, N.S., Towfighi, H., and Iizadeh, A. 2001. Occurrence of arbuscular mycorrhizal fungi in saline soils of the Tabriz Plain of Iran in relation to some physical and chemical properties of soil. *Mycorrhiza*, 11: 19-122.

Al-Karaki, G.N. 2000. Growth of mycorrhizal tomato and mineral acquisition under salt stress. Mycorrhiza, 10: 51–54.

Al-Karaki, G.N. 2006. Nursery inoculation of tomato with arbuscular mycorrhizal fungi and subsequent performance under irrigation with saline water. *Sci. Hort.*, 109: 1–7.

Al-Karaki, G.N. and Al-Raddad, A. 1997. Effects of arbuscular fungi and drought stress on growth and nutrient uptake of two wheat genotypes differing in their drought resistance. *Mycorrhiza*, 7: 83-88.

Arines, J., Vilarino, A. and Sainz, M. 1989. Effect of vesicular arbuscular mycorrhizal fungi on Mn uptake by red clover. *Agri. Eco. Environ.*, 29: 1-4.

Arun, A.B. 2002. Studies on Coastal Sand Dune Legumes of Karnataka (India). Ph.D. Thesis, Mangalore University, India. p. 38.

Auge, R.M., Schekel, K.A. and Wample, R.L. 1987. Leaf water and carbohydrate status of VA mycorrhizal rose exposed to drought stress. *Plant Soil*, 99: 291-302.

Bagyaraj, D.J. and Menge, J.A. 1978. Interactions between a VA mycorrhiza and *Azotobacter* and their effects on rhizosphere microflora and plant growth. *New Phytol.* 80: 567-573.

Beena, K.R., Arun, A.B., Raviraja, N.S. and Sridhar, K.R. 2001a. Arbuscular mycorrhizal status of *Polycarpaea corymbosa* (Caryophyllaceae) on sand dunes of west coast of India. *Ecol. Env. Cons.*, 7: 355-363.

Beena, K.R., Arun, A.B., Raviraja, N.S. and Sridhar, K.R. 2001b. Association of arbuscular mycorrhizal fungi with plants of coastal sand dunes of west coast of India. *Trop. Ecol.*, 42: 213-222.

Beena, K.R., Raviraja, N.S. and Sridhar, K.R. 1997. Association of arbuscular mycorrhizal fungi with *Launaea sarmentosa* on maritime sand dunes of west coast of India. *Kavaka*, 25: 53-60.

Beena, K.R., Raviraja, N.S. and Sridhar, K.R. 2000a. Seasonal variations of arbuscular mycorrhizal fungal association with *Ipomoea pes-caprae* of coastal sand dunes, Southern India. *J. Environ. Biol.*, 21: 341-347.

Beena, K.R., Raviraja, N.S., Arun, A.B. and Sridhar, K.R. 2000b. Diversity of arbuscular mycorrhizal fungi on the coastal sand dunes of the west coast of India. *Curr. Sci.*, 79: 1459-1466.

Bernstein, L. 1975. Effects of salinity and sodicity on plant growth. *Annu. Rev. Phytopathol.*, 13: 295-312.

Bethlenfalvay, G.J. and Franson, R.L. 1989. Manganese toxicity alleviated by Mycorrhizae in soybean. *J. Plant. Nutr.*, 12: 952-972.

Bhagya, B., Sridhar, K.R. and Arun, A.B. 2005. Diversity of legumes and arbuscular mycorrhizal fungi in the coastal sand dunes of southwest coast of India. *Int. J. Forest Usufructs Manag.*, 6: 1-18.

Bheemareddy. V.S., Lakshman, H.C. and Mahesh B. Byatanal. (2010). Effect of salt and acid stress on *Triticum aestivum* L. Var. Inoculated with *Glomus fasiculatum*. Libyan Agricultural Research center Journal International. 1 (5): 325-331.

Boyd, R. 1987. The role of ectomycorrhizas in the water relatuions of plants, Ph.D. thooio, University of Sheffield.

Bremmer, J. M. 1960. Determination of nitrogen in soil by the Kjeldahl method. *J Agric. Sci.*, 55: 11-33.

Brown, A.M. and Bledsoe, C. 1996. Spatial and temporal dynamics of mycorrhizas in *Jaumea carnosa*, a tidal saltmarsh halophyte. *J. Ecol.*, 84: 703–715.

Calvente, R, Cano, C., Ferrol, N., Azcon-Aguilar, C., and Barea, J.M. 2004. Analysing natural diversity of arbuscular mycorrhizal fungi in olive tree (*Oleo europaea* L.) plantations and assessment of the effectiveness of native fungal isolates as inoculants for commercial cultivars of olive plantlets. *Appl. Soil Ecol.*, 26: 11-19.

Cantrell, I.C. and Linderman, R.G. 2001. Preinoculation of lettuce and onion with VA mycorrhizal fungi reduces deleterious effects of soil salinity. *Plant Soil*, 233: 269–281.

Caravaca, F., Alguacil, M.M., Figueroa, D., Barea, J.M., and Roldan, A. 2003. Re-establishment of *Retarna sphaerocarpa* as a target species for reclamation of soil physical and biological properties in a semi-arid Mediterranean area. *Forest Ecol. Manage.*, 182: 49-58

Carvalho, L.M., Cacador, I., and Martins-Loucao, M. A. 2001. Temporal and spatial variation of arbuscular mycorrhizas in salt marsh plants of the Tagus estuary (Portugal). Mycorrihiza, 11: 303-309.

Channabasava, A. and Lakshman, H.C. 2012. Mycorrhization effect on biomass of four rare millets of North Karanataka (India). *Int. J. Agr. Sci.*, 2: 235-239.

Charest, C., Dalpe, Y. and Brown, A. 1993. The effect of vesicular arbuscular mycorrhizae and chilling on two maize hybrids. Mycorrhiza, 4: 89-92.

Colla, G., Rouphael, Y., Cardarelli, M., Tullio, M., Rivera, C.M. and Rea, E. 2008. Alleviation of salt stress by arbuscular mycorrhizal in zucchini plants grown at low and high phosphorus concentration. Biol. Fert. Soils, 44: 501–509.

Corkidi, L. and Rincon, E. 1997. Arbuscular mycorrhizae in a tropical sand dune ecosystem on the Gulf of Mexico I. Mycorrhizal status and inoculum potential along a successional gradient. Mycorrhiza, 7: 9- 15.

Dixon, R. K., Rao, M. V. and Garg, V. K. (1994). *In situ* and *in vitro* response of mycorrhizal fungi to salt stress. Mycorrhiza News. 5: 6-8.

El-Desouky, S.A. and Atawia, A.A.R. 1998. Growth performance of citrus rootstocks under saline conditions. *Alexandria J. Agr. Res.*, 43: 231–254.

Entry, J.A., Rygiewicz, P.T., Watrud, L.S., and Donelly, P.K. 2002. Influence of adverse soil conditions on the formation and function of arbuscular mycorrhizas. *Adv. Environ. Res.*, 7: 123-133.

Epstin, E. 1972. Mineral Nutrition of Plants: Principles and Perspectives. Wiley Industrial, New York.

Estaun, M.V. 1989. Effects of Radium chloride and mannitol on germination and hyphal growth the vesicular arbuscular mycorrhizal fungus *Glomus mosseae*. *Agric. Ecosyst. Environ.*, 29: 123-129.

Faber, B.A., Zasoski, R.J., Burau, R.G. and Uriu, K. 1990. Zinc uptake by corn as affected by vesicular-arbuscular mycorrhizae. *Plant Soil.*, 129: 121-130.

Fries, C.F. and Allen, M.F. 1991. Tracking the fates if exotic and local VA mycorrhizal fungi: methods and patterns. *Agric. Ecosys. Environ.*, 34: 87-96.

Gadgil, R.L. and Ede, F.J. 1998. Application of scientific principles to sand dune stabilization in New Zealand: past progress and future needs. *Land Degrad. Devp.*, 9: 131-142.

Garcia-Mora, M.R., Gallego-Fernandez, J.B. and Garcia-Novo, F. 2000. Plant diversity as a suitable tool for coastal dune vulnerability assessment. *J. Coastal Res.* 16: 990-995.

Gardner, J. H. and Malajczuk, N. 1988. Recolonization of rehabilitated bauxite mine sites in Western Australia by mycorrhizal fungi. *For. Ecol. Manag.*, 24: 27-42.

Garrido, N., Becerra, I., Manticorea, C., Oehrens, E., Silvia, M. and Horak, E. 1982. Antibiotis properties of ectomycorrhizas and sporophytic fungi growing on *Pinus radiate*. D. Don. *Mycopathol.*, 77: 93-98.

Gemma, J.N. and Koske, R.E. 1997. Arbuscular mycorrhizae in sand dune plants of the North Atlantic Coast of the US: field and greenhouse inoculation and presence of mycorrhizae in planting stock. *J. Environ. Manag.*, 50: 251-264.

Gerdemann, J.W. and Nicolson T.H. 1963. Spores of mycorrhizal Endogone species extracted from soil by wet sieving and decanting. *Trans. Br. Mycol. Soc*, 46: 235-244.

Gildon, A. and Tinker, P.B. 1981. A heavy metal tolerant strain of mycorrhizal fungus. *Trans. Brit. Myco. Sco.*, 77: 648-649.

Giovannetti, M. 1985. Seasonal variations of vesicular-arbuscular mycorrhizas and endogonaceous spores in a maritime sand dune. *Trans. Br. Mycol. Soc.*, 84: 679-689.

Giri, B. 1997. VAM colonization in MPTS. M. Phil. Thesis, Department of Botany, University of Delhi, Delhi.

Giri, B., Kapoor, R. and Mukerji, K.G. 2003. Influence of arbuscular mycorrhizal fungi and salinity on growth, biomass and mineral nutrition of *Acacia auriculiformis*. *Biol. Fert. Soils*, 38: 170–175.

Giri, B. and Mukerji, K.G. 2004. Mycorrhizal inoculant alleviates salt stress in *Sesbania aegyptiaca* and *Sesbania grandiflora* under field conditions: evidence for reduced sodium and improved magnesium uptake. *Mycorrhiza*, 14: 307–312.

Graham, J.H. and Syversten, J.P. 1984. Influence of vesicular arbuscular mycorrhiza on the hydraulic conductivity of roots of two Citrus rootstocks. *New Phytol.*, 97: 277–284.

Gupta, R. and Krishnamurthy, K.V. 1996. Response of mycorrhizal and non-mycorrhizal *Arachis hypogaea* to NaCl and acid stress. *Mycorrhiza*, 6: 145 149.

Hardie, K. and Leyton, L. 1981. The influence of vesicular-arbuscular mycorrhiza on growth and water relations of red clover. I. In phosphate deficient soil. *New Phytol.*, 89: 599–608.

He, X., Mouratov, S. and Steinberger, Y. 2002. Temporal and spatial dynamics of vesicular-arbuscular mycorrhizal fungi under the canopy of *Zygophyllum dumosum* Boiss. in the Negev Desert. *J. Arid Env.*, 52: 379-387

Hiremath, S.G and Lakshman, H.C. 2007. Survey of AM Fungi in plants growing in coastal beaches and their effect on *Solanum nigrum* L. *Nat. J. Environ. Poll. Tech.* 6: 81-84.

Hirrel, M.C. 1981. The effect of sodium and chloride salts on the germination of *Gigaspora margarita. Mycologia*, 73: 610-617.

Hirrel, M.C. and Gerdemann, J.W. 1980. Improved growth of onion and bell pepper in saline soils by two vesicular mycorrhizal fungi. *Soil Sci. Amer. J.*, 44: 654-655.

Hirrell, M.C, Mehravaran, H. and Crertlemann, J.W. 1978. Vesicular- arbuscular mycorrhizae in Chenopodiaceae anti Cruciferae: do they occur? *Can. J. Bot.* 56: 2813-2817.

Ho, I. 1987. Enzyme activity and phytohormone production of a mycorrhizal fungus *Laccaria laccata. Can. J. For. Res.*, 17: 855-858.

Ho, I. and Trappe, J.M. 1975. Nitrate reducting capacity of two vesicular arbuscular mycorrhizal fungi. *Mycologia*, 67: 886-888.

Huang, R.S., Smith W.K. and Yost, R.S. 1985. Influence of vesicular arbuscular mycorrhiza on growth, water relations and leaf orientation in *Leucaena leucocephala* (Lam.) de Wit. *New Phytol.*, 99: 229-243.

Jackson, M.L. 1973. Soil chemical analysis. Prentice Hall, New Delhi.

Jaiswal, V. and Rodrigues, B.F. 2001. Occurrence and distribution of arbuscular mycorrhizal fungi in coastal sand dune vegetation of Goa. *Curr. Sci.*, 80: 826-827.

Jarstfer, A.G., Farmer-Koppenol, P., and Sylvia, D.M. 1998. Tissue magnesium and calcium affect mycorrhiza development and fungal reproduction. *Mycorrhiza*, 7: 237–242.

Jasper, D.A., Robson, A.D. and Abott, L.K. 1979. Phosphorous and formation of vesicular arbuscular mycorrhiza. *Soil Biol. Biochem.*, 11: 501-505.

Juniper, S. and Abbott, L.K. 1993. Vesicular-arbuscular mycorrhizas and soil salinity. *Mycorrhiza*, 4: 45–57.

Juniper, S. and Abott, L.K. 1991. The effect of salinity on spore germination and hyphal extension of some mycorrhizal fungi. 3rd European Symposium and mycorrhiza diversity of Sheffield (Abstr.) Sheffield, U.K.

Juniper, S. and Abott, L.K. 1992. The effect of change of soil salinity ion growth of hyphae from spores of *Gigaspora decipiens* and *Scutellospora calospora*. (Abstracts), International Symposium on Management of Mycorrhizas in Agriculture, Horticulture and forestry. University of Western Australia.

Juniper, S. and Abott, L.K. 1993. Vesicular arbuscular mycorrhiza and soil salinity. *Mycorrhiza*, 4: 45-57.

Puttaradder, J., Lakshman, H.C. and Shankrappanavar, N.B. 2015. Studies on diversity and selection of suitable AM Fungi on growth and nutrient uptake of *Capsicum annum* L. Var. Pusa Jwala. *Bionano Frontier*, 8: 80-86.

Koide, R. 1985. The effect of VA mycorrhizal infection and phosphorus status on sunflower hydraulic and stomatal properties. *J. Exp. Bot.,* 36: 1087-1098.

Koske, R.E. 1981. *Gigaspora gigantea*: Observation on spore germination of VA-Mycorrhizal fungus. *Mycologia,* 73: 288- 300.

Koske, R.E. and Gemma, J.N. 1990. VA mycorrhizae in vegetation of Hawaiian Coastal strand: evidence for codispersal of fungi and plants. *Amer. J. Bot.,* 77: 466-674.

Koske, R.E. and Gemma, J.N. 1997. Mycorrhizae and succession in plantings of beachgrass in sand dunes. *Amer. J. Bot.,* 84: 118-130.

Koske, R.E. and Halvorson, W.L. 1981. Ecological studies of vesicular arbuscular mycorrhizae in a barrier sand dune. *Can. J. Bot.,* 59: 1413- 1422.

Koske, R.E. and Walker, C. 1984. Gigaspora erythropa, a new species forming arbuscular mycorrhizae. *Mycologia,* 76: 250-255.

Kothari, S.K., Marcherner, H. and Romhald, V. 1991. Effect of vesicular arbuscular mycorrhizal fungus and rhizosphere microorganisms on manganese reduction in rhizosphere and manganese concentration in maize (*Zea mays*). *New Phytol.* 117: 649-655.

Kraigher, H., Grayling, A., Wang, T.L. and Hanke, D.F. 1991. Cytokinin production by two ectomycorrhizal fungi in liquid culture, *Phytochem.,* 30: 2249-2254.

Kulkarni, S.S., Raviraja, N.S. and Sridhar, K.R. 1997. Arbuscular mycorrhizal fungi of tropical sand dunes of west coast of India. *J. Coastal Res.,* 13: 931-936.

Lakshman, H.C. 1999. Dual inoculation with Rhizobium and *G. mosseae* to enhance biomass production and nitrogen fixation in *Pterocarpus marsupium. Nat. Con.,* 97: 105-111.

Lakshman, H. C., Mulla, F. I. and Inchal, R. F. 2000. Some coastal potential plants with AM fungi and their use in land reclamation programme. In: Proce. Coastal zone Management. S.D.M.C.E.T. Dharwad. Spl. pub. 2: 189 – 196.

Lakshman, H.C. 1999. VA-mycorrhizal survey of plant species colonizing beach ponds Karwar in South India. *Asian J. Microbiol. Biotecnol. and Env. Sci.* 1: 15-21.

Lakshman, H.C. 1994. Occurrence of VA-Mycorrhiza in some tropical hydrophytic and xerophytic plants of Dharwad District Krnataka. *J. Nat. Con.* 6: 29-35.

Lakshman, H.C. 1999. Aggregation and sand-dune soil by arbuscular mycorrhizal fungi its use in revegetation practices. *Ind. J. Environ. Ecoplan.,* 2: 247-252.

Lee, P.J. and Koske, R.E. 1994. Gigaspora gigantea: seasonal abundance and ageing of spores in a sand dune. *Mycol. Res.,* 98: 453-457.

Lemoine, M.C., Gollotte, A. and Gianinazzi-Pearson, V. 1995. Localization of β (1-3) Glucan in walls of the endomycorrhizal fungi *Glomus mosseae* (Nicol. and Gerd.) Gerd. and Trappe and *Acaulospora laevis* Gerd. and Trappe during colonization of host roots. *New Phytol.,* 129: 97-105.

Levy, Y., Dodd, J. and Krikun, J. 1983. Effect of irrigation water salinity and rootstock on the vertical distribution of vesicular-arbuscular mycorrhiza in citrus roots. *New Phytol.*, 95: 397–403.

Levy, Y. and Krikum, J. (1980). Effect of vesicular and arbuscular mycorrhiza on *Citrus jambhiri* water relations. *New Phytol.* 85: 25-31.

Levy, Y., Syvertsen, J. P. and Nemec, S. 1983. Effect of drought stress and vesicular arbuscular mycorrhiza on citrus transpiration and hydraulic conductivity of roots. *New Phytol.*, 93: 61-66.

Li, X.I., George F., and Marshhcr, H. 1991. Phosphorus depletion and pH decrease at the root-soil and hyphae-soil interfaces of VA mycorrhizal white clover fertilized by ammonium. *New Phytol.*, 119: 397-404.

Linderman, R.G. and Bethlenfalvay, G.J. 1992. VA mycorrhiza and sustainable agriculture. Soil Science Society of America Madison, WI.

Marschner, H. and Dell, B. 1994. Nutrient uptake in mycorrhizal symbiosis. *Plant Soil,* 159: 89–102.

Martinez, M.L., and Moreno-Casasola, P. 1996. Effects of burial by sand on seedling growth and survival in six tropical sand dune species from the gulf of Mexico. *J. Coastal Res.,* 12: 406-419.

Maun, M.A. and Baye, P.R. 1989. The ecology of *Ammophila breviligulata* Fern. On coastal dune ecosystem. *CRC Crit. Rev. Aquat. Sci.,* 1: 661- 681.

Meyer, J.R. and Linderman, R.G. 1986. Selective influence on populations of rhizosphere or rhizoplane bacteria and actinomycetes by mycorrhizas formed by Glomus fasciculatum. *Soil Biol. Biochem.,* 18: 191-196.

Miller, R.M. and Jastrow, J.D. 1990. Hierarchy of root and mycorrhizal fungal interactions with soil aggregation. *Soil Biol. Biochem.,* 22: 579-584.

Mohankumar, V., Ragupathy, S., Nirmala, C.B. and Mahadevan, A. 1988. Distribution of vesicular arbuscular mycorrhizae (VAM) in the sandy beach soils of Madras Coast. *Current Sci.,* 57: 367-368.

Munns, R. 1993. Physiological responses limiting plant growth in saline soils: some dogmas and hypotheses. *Plant Cell Environ.,* 16:15–24.

Murkute, A.A., Sharma, S., Singh, S.K. 2006. Studies on salt stress tolerance of citrus rootstock genotypes with arbuscular mycorrhizal fungi. *Hort. Sci.* 33: 70–76.

Nelson, C.E. and Safir, G.R. 1982a. The water relations of well-watered mycorrhizal and non-mycorrhizal onion plants. *J. Amer. Soc. Hort. Sci.,* 107: 281-288.

Nelson, C.E. and Safir. G.R. 1982b. Increased drought tolerance of mycorrhizal onion plants caused by improved phosphorus nutrition. *Planta,* 154: 407-413.

Nicolson, T.H. 1960. Mycorrhizae in the Graminae. II. Development in different habitats particularly sand dunes. *Can. J. Soil Sci.,* 43: 132-145.

Oades, J.M. 1984. Soil organic matter and structural stability: mechanisms and implications for management. *Plant Soil,* 76: 319-337

Ojala, J.C., Jarell, W.M., Menge, J.A. and Johnson, E.L.V. 1983. Influence of mycorrhizal fungi on the nutrition and yield of onion in saline soil. *Agron. J.*, 75: 255-259.

Osonubi, O. 1994. Comparative effects of vesicular arbuscular mycorrhizal inoculation and phosphorus fertilization on growth and phosphorus uptake of maize (*Zea mays* L.) and sorghum (*Sorghum bicolor* L.) plants under drought stressed conditions. *J. Plant Nutr.*, 18: 55-59.

Osonuki, O.K., Mulongoy, O.O., Atayesa, M.O. and Okari, D.U.U. 1991. Effects of ectomycorrhizal and VAM fungi on drought tolerance of four leguminous woody seedlings. *Plant Soil*, 136: 131-143.

Patil, G.B. and Lakshman, H.C. 2003. Effect of AMF and saline water with and without additional phosphate on *Elusine coracana* (Finger millet). *Indian J. Env. Ecopl.* 7: 477-482.

Peterson, R.L., Ashford, A.E. and Allaway, W.G. 1985. Vesiculararbuscular mycorrhizal association of vascular plants on Heron Island, a Great Barrier Reef coral cay. *Aust. J. Bot.*, 33: 669-676.

Philips, J.M. and Hayman, D.S. 1970. Improved procedure for clearing roots and staining parasitic and vesicular-arbuscular mycorrhizal fungi for rapid assessment of infection. *Trans. Br. Mycol. Soc.*, 55: 158.

Pond, E.C., Menge, T.A. and Jarrell, W. M. 1984. Improved growth of tomato in salinized soil by vesicular arbuscular mycorrhizal fungi collected from saline soil. *Mycologia*, 76: 74-84.

Poss, J.A., Pond, E.C., Menge, T.A. and Jarrell, W.M. 1985. Effect of salinity on mycorrhizal onion and tomato in soil with and without additional phosphate. *Plant Soil*, 88: 307-319.

Ragupaty, S., Nagarajan, G. and Mahadevan, A. 1998. Mycorrhizae in coastal sand dunes of Tuticorin, Tamil Nadu. *J. Environ. Biol.*, 19: 281-284.

Raman, N. and Mahadevan, A. 1996. Mycorrhizal Research- A priority in Agriculture. In: Mukerji, K.G. (Ed.) Concepts in Mycorrhizal Research, Kluwer Academic Publisher, Netherlands, pp. 41-75.

Rao, T.A. and Meher-Homji, V.M. 1985. Strand plant communities of the Indian sub-continent. *Proc. Indian Acad. Sci. (Plant Sci.)*, 94: 505-523.

Rao, T.A. and Sherieff, A.N. 2002. Coastal Ecosystem of the Karnataka State, India II. Beaches. Karnataka Association for the Advancement of Science, Bangalore, India.

Remy, W., Taylor, T.N., Hass, II. and Kerp, H. 1994. Four hundred million year old vesicular-arbuscular mycorrhizas. *Proc. Nat. Acad. Sci.*, 91: 11841-11849.

Rodrigues, B.F. and Jaiswal, V. 2001. Arbuscular mycorrhizal (AM) fungi from coastal sand dune vegetation of Goa. *Ind. J. For.*, 24: 18-20.

Romana, M.M. and Lakshman, H.C. 2011. Effect of AM fungi salinity on two important legumes. *J. Theor. Exp. Biol.*, 8: 1-10.

Ruiz-Lozano, J.M., Collados, C., Baream, J.M., and Azcón, R. 2001. Arbuscular mycorrhizal symbiosis can alleviate drought induced nodule senescence in soybean plants. *Plant Physiol.*, 82: 346–350.

Sannazzaro, A.I., Ruiz, O.A., Albetro,´ E.O. and Mene´ndez, A.B. 2006. Alleviation of salt stress in *Lotus glaber* by *Glomus intraradies*. Plant Soil, 285: 279–287.

Schmidt, S.K. and Scow, K.N. 1986. Mycorrhizal fungi on the Galapagos Islands. Biotropica, 18: 236-240.

Sheng, M., Tang, M., Chan, H., Yang, B., Zhang, F. and Huang, Y. 2008. Influence of arbuscular mycorrhizae on photosynthesis and water status of maize plants under salt stress. *Mycorrhiza*, 18: 287–296.

Shi, Z.Y., Zhang, L.Y., Li, X.L., Feng, G., Tian, C.Y., and Christie, P. 2007. Diversity of arbuscular mycorrhizal fungi associated with desert ephemerals in plant communities of Junggar Basin, northwest China. *App. Soil Ecol.*, 35: 10-20.

Shrivastava, D., Kapoor, R., Shrivastava, S.K. and Mukerji, K.G. 1996. Vesicular arbuscular Mycorrhiza: an overview. In: Mukerji, K.G. (Ed.) Concepts in Mycorrhizal Research, Kluwer Academic Publisher, Netherlands, pp. 1-39.

Smith, S.E. 1980. Mycorrhiza of autotrophic higher plants. *Biological Reviews*, 55: 475-510.

Sridhar, K.R. 2006. Arbuscular Mycorrhizal Fungi of Coastal Sand Dunes. In: Bukhari, M.J. and Rodrigues, B.F. (Eds.) Techniques in Mycorrhizae, pp. 30-42.

Sturmer, S.L. and Bellei, M.M. (1994). Composition and seasonal variation of spore population of arbuscular mycorrhizal fungi in dune soils on the island of Santa Catarina, Brazil. *Can. J. Bot.*, 72: 359-363.

Subramanian, S. and Charest, C. 1995. Influence of arbuscular mycorrhizae on the metabolism of maize under drought stress. *Mycorrhiza*, 5: 273-278.

Sutton, J.C. and Sheppard, B.R. 1976. Aggregation of sand dune soil by endomycorrhizal fungi. *Can. J. Bot.*, 54: 326-333.

Sylvia, D.M and Schenck, N.C. 1983. Germination of chlamydospores of three *Glomus* species as affected by soil matric potential and fungal contamination. *Mycologia*, 75: 30-35.

Sylvia, D.M. 1986. Spatial and temporal distribution of vesicular arbuscular mycorrhizal fungi associated with *Uniola paniculata* in Florida foredunes. *Mycologia*, 78: 728-734.

Sylvia, D.M. 1989. Nursery inoculation of sea oats with vesicular-arbuscular mycorrhizal fungi and outplanting performance of Florida beaches. *J. Coastal Res.* 5: 747-754.

Sylvia, D.M. and Burks, J.N. 1988. Selection of vesicular-arbuscular mycorrhizal fungus for practical inoculation of *Uniola paniculata*. *Mycologia*, 80: 565-568.

Sylvia, D.M. and Will, M.E. 1988. Establishment of vesicular-arbuscular mycorrhizal fungi and other microorganisms on a beach replenishment site in Florida. *Appl. Environ. Microbiol.*, 54: 348-352.

Sylvia, D.M. and Williams, S.E. 1992. Vesicular arbuscular mycorrhizae and environmental stress. In: Linderman, R.G. and Bethlenfalvay, G.F. (Eds.) Mycorrhizae in Sustainable Agriculture. Special publication No. 54, American Society of Agronomy, Madison, WI, pp. 101-124.

Tinker, P.B. 1975. Soil chemistry of phosphorus and mycorrhizal effects on plant growth. In: Sanders, F.E., Mosse, B. and Tinker, P.B. (Eds.) Endomycorrhizas. Academic Press, London, pp. 353–371.

Tisdall, J.M. 1991. Fungal hyphae and structural stability of soil. *Aust. J. Soil Res.,* 29: 729-743.

Trappe, J.M. and Schenck, N.C. 1982. Taxonomy of the fungi forming endomycorrhizae. A vesicular arbuscular mycorrhizal fungi (Endogonales). In: Schenck, N.C. (Ed.) Methods and Principles of Mycorrhizal Research, American Phytopathological Society, St. Paul. Minn. pp. 1-10

Udaiyan, K, Karthikeyan, A., and Muthukumar, T. 1996. Influence of edaphic and climatic factors on dynamics of root colonization and spore density of vesicular arbuscular mycorrhizal fungi in *Acacia farnesiana* Wild. and *A. planifrons* W. et A. *Tree*, 11: 65-71.

Walker, C. 1995. AM or VAM: What's in a word? *In*: Varma, A. and Hock, B. (Eds.) Mycorrhizal structure, function, molecular biology and biotechnology, Springer Verlag. pp. 25-26.

Wang, W., Vinocur, B., Altman, A. 2003. Plant responses to drought, salinity and extreme temperatures: toward genetic engineering for stress tolerance. *Planta,* 218: 1–14.

Wang, X. F., Auler, A. S., Edwards, R. L., Cheng, H., Cristalli, P. S., Smart, P. L., Richards, D. A. and Shen, C.C. 2004. Wet periods in northeastern Brazil over the past 210 kyr linked to distant climate anomalies, *Nature,* 432: 740-743

Wright, S.F. and Upadhyaya, A. 1998. A survey of soils for aggregate stability and glomalin, a glycoprotein produced by hyphae of arbuscular mycorrhizal fungi. *Plant Soil,* 198: 97-107.

Yano-Melo, A.M., Saggin, O.J. and Maia, L.C. 2003. Tolerance of mycorrhized banana (*Musa* sp. cv. Pacovan) plantlets to saline stress. *Agric. Ecosyst. Environ.,* 95: 343–348

Zuccarini, P. 2007. Mycorrhizal infection ameliorates chlorophyll content and nutrient uptake of lettuce exposed to saline irrigation. *Plant Soil Env.,* 53: 283–289.

— Part V —
Molecular Approaches and Biocontrol

2017, Mycorrhizal Fungi
Editors: Ashok Aggarwal and Kuldeep Yadav
Published by: ASTRAL INTERNATIONAL PVT. LTD., NEW DELHI

Pages 337–355

15

Biocontrol of *Fusarium* Wilt of Tomato with special reference Arbuscular Mycorrhizal Fungi

Anju Tanwar[*1], Ishan Saini[2], Esha Jangra[2],
Anil Gupta[1] and Vipin Panwar[3]*

[1]*Botany Department, University College,*
[2]*Department of Botany,*
Kurukshetra University, Kurukshetra – 136 119, Haryana
[3]*Indian Institute of Wheat and Barley Research,*
Karnal – 132 001, Haryana
Corresponding Author: anjutanwarbotany@gmail.com

ABSTRACT

Diseases of vegetable crops are one of the main limiting factors in the modern agriculture, holding potential for a devastating effect on the growth, development and yield of vegetables. Over the past few years, there has been a strong upsurge in the demand for tomato. Root wilt of tomato caused by Fusarium species is considered to be a serious threat in the tomato fruit production all over the world and biological control is a way out. Management of plant diseases through manipulation of associated microorganisms restores the biological equilibrium. Plant microbe interactions are one of the interesting events that contribute for the sustainable agriculture and among all, Arbuscular mycorrhizal (AM) fungi, constitutes an important biological resource in this respect and could be effectively used in vegetable production. Only recently the biocontrol efficiency of AM fungi have been studied with respect to vegetable plants and their diseases. The purpose of this brief review is to provide comprehensive information about the various disease control methods of Fusarium wilt of tomato with special reference to its biological control using Arbuscular mycorrhizal fungi.

Keywords: *Arbuscular mycorrhizal fungi, Tomato, Fusarium wilt, Biocontrol.*

1. Introduction

Vegetables occupy an important position in the dietary habits of the vast majority of the human population in the Indian subcontinent. Besides being nutritionally important, vegetables also contribute to the maintenance of health and the prevention of diseases. Increased vegetable production is the demand of the present agricultural scenario and the occurrence of environmental problems has forced agricultural and horticultural producers and researchers to critically evaluate disease protection systems and their effects on crop quality and environment.

Tomato (*Lycopersicon esculentum* Mill.) is considered as one of the most important cultivated crops in India and there has been a strong upsurge in the demand for tomato (Kumar and Rai, 2007). The crop is affected by many diseases, causing economic loss by reducing the quality and quantity of produce harvested especially by soil borne fungal pathogens (Srinon *et al.*, 2006). Two formae speciales of *F. oxysporum* are known to affect tomato *i.e. F. oxysporum* f. sp. *lycopersici*, the causal agent of severe wilt disease, whereas *F. oxysporum* f. sp. *radicis-lycopersici* causes crown and root rot (Joner, 1991). Among them, root wilt caused by *F. oxysporum* Schlecht. f.sp. *lycopersici* (Sacc.) W.C. Synder H.N. Hansen (Ignjatov, Miloševiæ, Nikoliæ, Gvozdanoviæ-Varga, Joviæiæ, and Zdjelar 2012) is the most prevalent and important disease of tomato and the damage caused by this disease can bring in major economic losses to tomato growers in India (Chowdhury *et al.,* 2009; Ahmed and Upadhyay, 2009; Ojha, and Chatterjee, 2012).

According to El-Khallal (2007) and Ozbay and Newman (2004), *Fusarium oxysporum* is a soil borne fungal pathogen capable of infecting several host roots worldwide, inducing necrosis and wilting symptoms at all the stages of its growth and thereby cause great economic losses. All strains of *F. oxysporum* are saprophytic in nature and able to grow and survive for long periods on the organic matter in soil (Garrett, 1970). While some strains of *F. oxysporum* are pathogenic to different plant species and responsible for severe damage on many economically important plant species (Fravel *et al.,* 2003). The fungus has numerous specialized forms known as formae speciales (f. sp.) that affects a range of host plants causing diseases such as vascular wilt, crown rot, root rot or damping off (Armstrong and Armstrong, 1981). In India, alone 80-100 per cent yield loss is associated with *F. oxysporum* species (Anjani *et al.,* 2004).

They (*Fusarium* sp.) penetrate into the roots, inducing either root-rots or tracheomycosis when they invade the vascular system (Fravel *et al.,* 2003). The disease causes severe losses in tomato production both in field and polyhouse-grown tomato plants, due to stunted seedlings followed by drooping and yellowing of leaves (Jones *et al.,* 1991; Nurset and Steven, 2004). The disease causes great losses, especially on susceptible varieties and when soil and air temperatures are rather high (Agrios, 2005). Occasionally entire field of tomatoes are killed or damaged severely before a crop can be harvested. Losses due to soil-borne diseases on some green house, nursery vegetables including tomato can amount to thousands of dollars annually (Zinati, 2005). It is a devastating disease causing considerable economic losses ranging from 10-80 per cent yield loss in many tomatoes producing area of the country (Keshwan and Chaudhary, 1977).

2. Disease Management

Managing diseases through different strategies also adds to a grower's input costs, thereby cutting into potential profits. The main problem in the management of *Fusarium* wilt is the endophytic nature of the pathogen and its persistence in the soil (Alström, 2001). Environmental problems have forced agricultural and horticultural producers and researchers to critically evaluate production systems and their effects on crop quality and environment. Numerous strategies have been proposed to control *Fusarium* wilt of Tomato (FWT) like, chemical (Amini and Sidovich, 2010), green house control (Paulitz and Belanger, 2001), use of antagonistic fungi and cyanobacteria (Alwathnani and Perveen, 2012), plant products (Singha *et al.,* 2011), organic amendments (Szczech, 1999) induced resistance (El-Khallal, 2007), biocontrol (Alam *et al.,* 2011), selecting resistant varieties of tomato (Silva and Bettiol, 2005) *etc.*

2.1. Chemical Control

Maximum control of the disease is done by the application of large quantities of chemical fungicides like benomyl, captafol and thiram. Chemical control methods can also have both positive and negative impacts and must be carefully considered. Continuous use of these chemicals (fungicides) in the past two decades has led to several problems of environment such as environmental degradation, decrease in the population of beneficial microorganisms, emergence of fungicide resistant strains which has direct impact on disease resistance and health hazards for human (Hartman and Fletcher, 1991). Application of chemical fungicides has been replaced with biocontrol agents because of the emergence of fungicide resistant strains and public concern regarding the health and environmental impacts of these chemicals (John *et al.,* 2010). During the past few decades, several potential beneficial organisms including AM fungi have been isolated, characterized and commercialized (Shali *et al.,* 2010).

It is inconceivable to believe that we can do away with chemical fungicides for plant disease control. But through minimum use of these hazardous chemicals when other methods of controlling plant diseases are not available remarkable disease control can be achieved (Aggarwal *et al.,* 2006a). The disease control through use of systemic fungicides Prochloraz and Carbendazim in hydroponic culture is also reported (Song *et al.,* 2004). Effectiveness of Prochloraz and bromoconazole among all the six fungicides tested in controlling both *in vivo* and *in vitro* FWT has been reported by Jahanshir and Dzhalilov (2010).

2.2. Physical Method

Employment of soil solarization process for the disinfection of *Fusarium* contaminated soil is also commercially practiced mainly in areas which are characterized by high summer air temperatures for the control of FWT (Barakat and Al-Masri, 2012).

2.3. Biological Control

Biological control is an alternative approach against the chemical control measures in controlling plant pathogens and is currently accepted as a key practice for sustainable agriculture as it is based on the management of natural resources (El-Khallal, 2003). The term 'biological control' was used for the first time by a German Plant Pathologist, G.F. Von Tubeuf in the year 1914. Biological control is defined by Baker and Cook (1974) as 'the reduction of inoculum density or disease producing activities of the pathogen or parasite in its active or dormant state, by one or more organisms, accomplished naturally or through manipulation of the environment, host or antagonist or by mass introduction of one or more antagonist'. Use of antagonistic fungi as biocontrol agent in the nursery soils was first demonstrated by Harley in 1921 in an attempt to control damping off of pine caused by *Pythium debaryanum*. Cook (1993) listed three major strategies to consider for the development of introduced fungal or bacterial microorganisms for biological control of plant diseases.

1. Reduce or limit population of the plant pathogen
2. Control or prevent infection by the pathogen; and
3. Control the spread and development of diseases following infection.

In this present scenario, ecofriendly alternative strategies such as use of antagonistic fungi from rhizosphere and endophytic bacteria are being explored. They have the capacity of producing growth stimulating effect, in addition to controlling rhizospheric pathogens. Besides direct interaction with the plant pathogen, bioagents are also reported to induce systemic resistance in plants (Srivastava *et al.*, 2010; Ojha and Chatterjee, 2012). The antagonists must be ecologically fit to survive and function within the particular conditions of the ecosystem; must be present at adequate population levels and be capable of effectively interacting with the pathogen or host plant to provide acceptable disease control (Larkin and Fravel, 1999). Since soil borne pathogens as well as symbiont share common habitat and show differential influence on the growth of host plant, major interest has recently centered on the release of AM fungi and symbiont bacteria in the control of soil borne pathogens (Dar *et al.*, 1997).

2.3.1 Arbuscular Mycorrhizal (AM) Fungi

AM fungi are obligate biotrophs and despite their obligate nature, AM fungal symbionts remain remarkably successful and integral components of plant root systems, contributing to nutrient uptake in the large majority of land plants and stimulate plant growth (Smith and Smith, 2012). They play an important role in natural ecosystems and influence plant productivity, nutrition and disease resistance

(Demir and Akkopru, 2007). AM fungi limit fungal root diseases by strengthening morphological traits of plants with some physiological and microbial modifications in the mycorrhizosphere and by altering the chemical composition of plant tissues (Linderman, 2000; Barea *et al.*, 2002). It benefits plants by stimulating the production of growth regulating substances, increasing photosynthesis, improving osmotic adjustment under stress conditions and increasing resistance to pests and soil borne diseases (Al-Karaki, 2006). Classically, four major groups of mycorrhizal mode of action mechanisms that mediated bioprotection have been considered as stated by Vierheilig *et al.* (2008): (1) direct competition, (2) mechanism mediated by alteration in plant growth, nutrition and morphology, (3) biochemical and molecular changes in mycorrhizal plants that induce pathogen resistance, and (4) alterations in the soil microbiota and development of pathogen antagonism.

Success of AM fungi as biocontrol agent has been established beyond doubt as there are numerous reports indicating the use of AM fungi for the control of soil borne pathogens such as *Fusarium* wilt of lettuce (*Lactuca sativa*), lentil (*Ervum lens*), pea (*Pisum satium*) (Fracchia *et al.*, 2000), *Verticillium* wilt of bell pepper (Garmendia *et al.*, 2004), strawberry (Tahmatsidou *et al.*, 2006), cotton (Kobra *et al.*, 2009), *Pythium* damping off of cucumber (Li *et al.*, 2007), *Sclerotium rolfsii* stem rot of peanut (Ozgonen *et al.*, 2010), *Macrophomina* root rot of chickpea (Akhtar and Siddique, 2010), Black bundle disease of maize caused by *Cephalosporium acremonium* (Veerabhadraswamy and Garampalli, 2011).

The search for promising microbial antagonists has been increased to control various diseases. Evidence reveals that mycorrhizal symbiont offer increased resistance to certain *Fusarium* wilt and root rot diseases of vegetables, such as *Fusarium* wilt of tomato using *Glomus intraradices* and Rhizobacteria (Akkopru and Demir, 2005), AM fungi and *Trichoderma harzianum* T-22 (Mbuthia *et al.*, 2009), for chickpea using *G. hoi*, *G. fasciculatum* and *Rhizobium leguminosarum* Biovar (Singh *et al.*, 2010), for cucumber using AM fungi (Hu *et al.*, 2010), *G. mosseae* and *G. versiforme* (Wang *et al.*, 2012a), *G. versiforme* and *G. intraradices* (Wang *et al.*, 2012b), for Alfaalfa using *Glomus* sp, *G. fasciculatum* and *G. mosseae* (Hwang, 1992), *Fusarium* root rot of common bean using *G. mosseae* and *Rhizobium leguminosarum* (Dar *et al.*, 1997), *G. mosseae*, *G. intraradices*, *G. clarum*, *Gigaspora gigantean*, *G. margarita* (Al-Askar and Rashad, 2010), for *Asparagus officinalis* using *G. intraradices* and *T. harzianum* (Arriola *et al.*, 2000), *Glomus* sp R10 (Matsubra, 2012).

Biocontrol potential of AM fungi could be explained in terms of its ability to change root architecture, improved nutrient uptake, competition with the pathogen for photosynthates, for infection site, activation of plant defense enzymes (chitinase, chitosanase, β–1,3–glucanase and superoxide dismutase), phenolic and phytoalexin production (Avis *et al.*, 2008). There is also an increase in the lignin content of mycorrhizal root system (Linderman, 1992), which makes root stronger and resistant to attack by pathogens. Several other mechanisms have been proposed for AM induced protection including competitive interactions with pathogenic fungi, anatomical or architectural changes in the root system, microbial community changes in the rhizosphere, activation of plant defense mechanism (Wehner *et al.*, 2009).

2.3.2. Antagonistic Fungi

Alternative methods with emphasis on biological control using several other fungi have been tested against FWT to reduce fungicide application and decrease cost of plant production including *Phytophtohora cryptogea* (Attitalla *et al.*, 2001), *Trichoderma* sp. (Osuinde *et al.*, 2002), *Penicillium oxalicum* (Larena *et al.*, 2003), oil suspension of *Emericella nidulans* filtrate (Sibounnavong *et al.*, 2010), *Streptomyces psammoticus* strain (Kim *et al.*, 2011), *Streptomyces miharaensis* (Kim *et al.*, 2012). According to Anitha and Rabeeth (2009) talc formulation of *Streptomyces griseus* and chitin performed well in the management of FWT when applied to the root systems. Root dipping of tomato seedlings in the conidial suspension of *Penicillium* sp EU1003 before planting resulted in significant (49-84 per cent) disease reduction (Sartaj *et al.*, 2011).

Disease control using some non pathogenic fungal strains is also reported. Muslim *et al.* (2003) demonstrated tomato wilt control with hypovirulent binucleate *Rhizoctonia* in green house conditions. Antagonists recovered from *Fusarium* wilt-suppressive soils, especially nonpathogenic *F. oxysporum*, have been used to reduce *Fusarium* wilt diseases (Fravel *et al.*, 2003). About 50-80 per cent reduction in disease incidence by nonpathogenic *F. oxysporum* and *F. solani* against FWT is also reported (Larkin and Fravel, 2002; Shishido *et al.*, 2005). Horinouchi *et al.* (2011) demonstrated effective biocontrol efficiency of *Fusarium equiseti* GF191 against FWT in both hydroponic rockwool and soil systems. Therefore, a variety of soil microorganisms have demonstrated activity in the control of soil borne plant pathogens of tomato and extensive research has been performed in the control of FWT using one or other species of *Trichoderma* as biocontrol agents.

2.3.3. Trichoderma spp.

Fungal antagonistic *Trichoderma* spp. are among the major group of microorganisms that have shown great potential for biological control of several soil borne pathogens (Kucuk and Kivanc, 2003). *Trichoderma* species have high rate of growth, are of common occurrence in soil and root ecosystems and have been recognized as an effective biocontrol agent against several soil borne fungal pathogens like *Fusarium*, *Pythium*, *Sclerotinia*, *Rhizoctonia*, etc. (Howell, 2003).

Weindling's interesting discovery about the antagonism efficiency of *Trichoderma viride* against the soil borne plant pathogen, *Rhizoctonia solani* during 1931-1941 had directed the attention of scientists towards various strains of *Trichoderma*. Weindling (1937) described in detail the mycoparasitism of fungal pathogen *R. solani* and *Sclerotinia americana* causing damping off disease by the hyphae of *Trichoderma* through coiling around hyphae, penetration and subsequent dissolution of its host cytoplasm and secretion of an antibiotic gliotoxin. *Trichoderma* strains colonize the plant roots and influence the synthesis of chloroplast enzymes that increase rate of photosynthesis (Abo–Ghalia and El–Khallal, 2005) or establishing chemical communication and systemically altering the expression of numerous plant genes (Hermosa *et al.*, 2012; Harman *et al.*, 2004).

Among *Trichoderma* strains, *T. harzianum* has been described as novel strain for maximum disease control (Ramezani, 2010; Alwathnani and Perveen, 2012). In an experiment conducted by Ramezani (2010), *T. harzianum* proved most effective strain under *in vitro* and *in vivo* conditions for controlling *Fusarium* while testing five different *Trichoderma* isolates *i.e.*, *T. harzianum*, *T. konningi*, *T. longiconis*, *T. hamatum* and *T. viride*. Similarly, *T. harzianum* induced tomato wilt control under *in vitro* and *in vivo* has been reported by Alwathnani and Perveen (2012). However, Mehra (2008) reported that application of *T. viride* through seedling dip and soil drench can efficiently control FWT.

Their biocontrol ability is due to mycoparasitism, antibiosis, competition with the pathogen for nutrients, direct parasitism, production of various mycolytic metabolites, induced resistance, production of protease and other fungal cell wall degrading enzymes (Verma *et al.*, 2007; Perelló, *et al.*, 2003). Furthermore, they inhibit or degrade pectinases and other enzymes that are essential for plant-pathogenic fungi and may also inuence disease development by inducing resistance in and D or promoting growth of the plant host (Harman *et al.*, 2004; Verma *et al.*, 2007). *Trichoderma* sp. form avirulent plant symbionts and even some strains establish robust and long-lasting colonizations on root surfaces and penetrate into the epidermis and a few cells below this level (Harman *et al.*, 2004). Successful reduction in *Fusarium* wilt in different crops with application of different *Trichoderma* isolates has been achieved where they are used as bio-pesticides, bio-protectants, bio-stimulants and bio-fertilizers on a wide variety of plants (Harma and Kubicek, 1998; Sivan and Chet, 1986; Biswas and Das, 1999; Ramezani, 2009; Morsey *et al.*, 2009; Singh *et al.*, 2009; John *et al.*, 2010; Perveen and Bokhari, 2012).

Similar encouraging results were reported by Seqarra *et al.* (2010), who tested *T. asperellum* strain T34 for the control of FWT in soilless culture through competition for iron and found significant disease control. The effectiveness of *T. asperellum* strain T34 was examined in hydroponically grown tomato plants under five ammonium/nitrate ratios to control FWT and a decrease in the population of rhizospheric *F. oxysporum* and disease severity by T34 under increasing concentrations of ammonia was observed by Borrero *et al.* (2012). Christopher *et al.* (2010) screened various isolates of *T. virens* in suppressing incidence of FWT and reported significant disease reduction by native isolates of *T. virens*. Recently, Barari (2016) isolated twenty eight isolates of *Trichoderma* and tested them against *Fusarium* under *in vitro* conditions and *in vivo* conditions to control FWT.

2.3.4. Soil Bacteria

A variety of soil bacteria have also been demonstrated activity in the control of FWT such as *Pseudomonas* sp. strain MF30 (Attitalla *et al.*, 2001), *Bacillus subtilis* (Abd-Allah *et al.*, 2007; Adebayo and Ekpo, 2004), *Azospirillum brasilense* and *Bacillus subtilis* (Abo-Elyousr and Kamal, 2009), *Pseudomonas* sp. NJ134 supplemented with mannitol (Kang, 2011). Combined use of the biocontrol bacterium *Pseudomonas fluorescens* strain LRB3W1 with reduced fungicide application (benomyl), sufficiently controls FWT as compared to sole application of fungicide (Someya *et al.*, 2006). Inam-Ul-Haq *et al.* (2007) tested antagonistic potential of *Xenorhabdus nematophila* and *Xenorhabdus* sp. associated with entomopathogenic nematodes against FWT under greenhouse conditions and recorded reduction in the disease incidence. Talc and sodium alginate formulation of *Pseudomonas fluorescens* exhibited synergism in promoting tomato plant growth and yield besides controlling FWT as reported by Asha *et al.* (2011). Kang (2012) investigated the potential of *Pseudomonas* sp. NJ134 isolated from field grown tomatoes against *Fusarium* and recommended 10^8 cfu/g soil culture application for effective control of this disease which is due to the production of an antifungal compound polyketide 2,4-diacetylphloroglucinol (DAPG).

2.4. Plant Products/Botanicals

Exploitation of plant products/botanicals for the control of FWT is also well documented to have significant control under *in vitro* conditions. The crude extracts of neem leaf, neem seed and garlic were tested on myclial growth of *Fusarium oxysporum* f. sp. *lycopersici* by Agbenin and Marley (2006) and 100 per cent inhibition of mycelial growth was recorded by dry neem seed extract. Furthermore, plant extracts and plant essential oils have also been reported to be effectively used against grain storage fungi, foliar pathogens and soil borne pathogens (Bowers and Locke, 2000). Singha *et al.* (2011) reported effective control of *Fusarium* wilt of tomato while using crude chloroform extract of *Piper betle* (1 per cent w/w) than Carbendazim and combination of Carbendazim and *Piper betle* extract.

2.5. Organic Amendments

The use of organic amendments had also been employed by many workers for disease control. Szczech (1999) reported control of FWT with vermicompost while combined use of sewage sludge with *T. asperellum* has been reported by Cotxarrera *et al.* (2002). Management of FWT by soil amendment with compost (banana leaves, bagasse, synthetic mushroom compost, paddy straw and spent mushroom compost) has been reported by Raj and Kapoor (1997). Agbenin *et al.* (2004) suggested use of neem seed powder for FWT. Similarly, control of FWT using neem kernel cake powder, grape marc compost and sheep manure in combination with *T. harzianum* has been recommended by Kimaru *et al.* (2004), Borrero *et al.* (2004) and Barakat and Al-Masri (2009) respectively. Al-Banna *et al.* (2016) investigated antifungal effect of *Bacillus thurigiensis* against *Fusarium oxysporum* and observed their potential inhibitory effect.

2.6. Induced Resistance

Induction of systemic resistance in plant against the pathogen is a new and one of the potential technique to check disease development in plants (Abo-Elyousr *et al.*, 2008). In plants, resistance can be enhanced more by the application of some bioagents or through the application of chemical elicitors. Hormone inducers represent an interesting strategy to stimulate the defense system of plants especially when integrated with bioagents. Reduction in *Fusarium* disease incidence, growth promotion and increased metabolic activity of tomato plants on inoculation with AM fungi and two hormone elicitors (Jasmonic acid and salicylic acid) has been reported by El-Khallal (2007). Likewise, Mandal *et al.* (2009) observed induced resistance in tomato against *F. oxysporum* on exogenous application of salicylic acid through root feeding and foliar spray. Houssien *et al.* (2010) documented *T. harzianum* and salicylic acid induced activation of plant defense system in tomato against *Fusarium* through improved growth and physiological defense.

2.7. Integrated Control

Rising awareness regarding the adverse effects of chemical fungicides and an increasing demand for the vegetables have encouraged our farmers to transit to sustainable mode of production systems (Klonsky, 2004). The use of combination of multiple antagonist microrganisms with synergistic modes of action and a different ecological behaviour may also provide improved disease control over singly used organisms. This could be a promising alternative for future biocontrol strategies, especially to control pathogens with high ecological versatility (Grosch *et al.*, 2006). Mixed consortium of bioinoculants have been applied to seeds, seedlings and planting media in several ways to reduce FWT under field and greenhouse conditions with various degree of success such as *Pseudomonas fluorescens* WCS417r + non pathogenic *Fusarium oxysporum* Fo47 (Duijff *et al.*, 1998), use of Fluorescent *Pseudomonas* + non pathogenic *Fusarium* + *T. harzianum* T-22 (Yigit and Dikilitas, 2007). Likewise, dual inoculation with *G. intraradices* + rhizobacteria strains reduced disease severity of FWT by 58.6 per cent (Akköprü and Demir, 2005). According to Tanwar *et al.* (2013) soil inoculation with *G. mosseae* along with root inoculation with conidial suspension of *T. viride* before transplantation can offer better survival and resistance to tomato seedlings against *Fusarium* wilt. Recently significant effect of combined application of *T. harzianum*, *Bacillus subtilis* and Chitosan as foliar spray on controlling FWT has been observed by Bakeer *et al.* (2016). However both experimental and theoretical studies reviewed by Xu *et al.* (2011) suggests that the combined use of biocontrol are not always better to that achieved single treatments.

3. Research from our Laboratory

Several research papers, reviews and book chapters from our laboratory have highlighted the use of environmentally safe approaches to crop disease control towards sustainable agriculture (Aggarwal *et al.*, 1999; Aggarwal *et al.*, 2006a; Aggarwal *et al.*, 2006b; Sharma *et al.*, 2007).

Although lot of work has been done and several fungal bioagents have been isolated, tested and successfully commercialized for the management of *Fusarium* wilt of tomato in green house as well as in the field conditions. However it is also documented that one specific isolate is not effective in controlling all the strains of *Fusarium* worldwide due to genetic variations, therefore specific isolates are needed to control the native pathogen of particular area (Ramezani, 2010). Bioinoculants do not always show phytoprotection against pathogen, indicating that choice of the host cultivar, isolate of the pathogen and choice of bioinoculant is very important (Singh *et al.*, 2010). Proper understanding of the causes behind inconsistencies observed with biocontrol agents is urgent, especially when they are released under the field conditions and multiplicity of a (biotic) factor dynamically interacting (Cazorla and Mercado-Blanco, 2016). Furthermore, there is scarce information regarding use of specific fungal isolate. Hence, experiments are needed in every *Fusarium* affected area for tomato wilt control to find a suitable combination of these biocontrol agents that may increase the plant growth and resistance to pathogen. However before developing such combinations, it should be confirmed that these biological agents have no antagonistic effect on each other (Martinez-Medina *et al.,* 2009).

References

Abd–Allah, E.F., Ezzat, S.M. and Tohamy, M.R. 2007. Bacillus subtilis as an alternative biologically based strategy for controlling *Fusarium* wilt disease in tomato: A histological study. *Phytoparasitica*, 35(5): 474–478.

Abo–Elyousr A.M.K., Hussein M.A.M., Allam A.D.A. and Hassan A.H.M. 2008. Enhanced onion resistance against *Stemphy-lium* leaf blight disease, caused by *Stemphylium vesicarium*, by di–potassium phosphate and benzothiadiazole treat-ments. *Plant Pathol. J.*, 24 (2): 171–177.

Abo–Elyousr. A.M.K. and Kamal, A.M. 2009. Biological control of *Fusarium* wilt in tomato by plant growth promoting yeast and rhizobacteria. *Plant Pathol. J.*, 25(2): 199–204.

Abo–Ghalia, H. and El–Khallal, S.M. 2005. Alleviation of heavy metal stress by arbuscular mycorrhizal fungi and jasmonic acid in maize plants. *Egypt. J. Bot.*, 45: 55–77.

Adebayo, O.S. and Ekpo, E.J.A. 2004. Efficacy of fungal and bacterial biocontrol organisms for the control of *Fusarium* wilt of Tomato. *Nig. J. Hort. Sci.*, 9: 63–68.

Agbenin, N.O. and Marley, P.S. 2006. *In vitro* assay of some plant extracts against *Fusarium oxysporum* f. sp. *lycopersici* causal agent of tomato wilt. *J. Plant Protec. Res.*, 46(3): 117–121.

Agbenin, N.O., Emechebe, A.M. and Marley, P.S. 2004. Evaluation of neem seed powder for *Fusarium* wilt and *Meloidogyne* control on tomato. *Arch. Phytopathol. Plant Protec.*, 37: 319–326.

Aggarwal, A., Parkash, V. and Mehrotra, R.S. 2006a. VAM fungi as biocontrol agents for soil borne pathogens. In: Plant Protection in New Millenium, (Eds.) Gadewar, A.V. and Singh, B.P., Satish Serial Pub., New Delhi, India, pp. 137–162.

Aggarwal, A., Parkash, V., Sharma, D., Sharma, S. and Mehrotra R.S. 2006b. Some environmental safe approaches to crop diseases control toward sustainable agriculture. In: Advancing Frontiers of Ecological Research in India (Eds.) Kandya, A.K. and Gupta, A., Bisen Singh Mahendera Pal Singh Publ., Dehradun, India, pp. 593–614.

Agrios G.N. 2005. Plant Pathology, 5th Edn, Elsevier/Academic Press, Amsterdam, The Netherlands.

Ahmed, M. and Upadhyay, R.S. 2009. Rhozosphere fungi of tomato: their effect on seed germination, seedling growth and the causal agent of wilt of tomato. *J. Mycopathol. Res.*, 7(2): 99–110.

Akhtar, M.S. and Siddique, Z.A. 2010. Effects of AM fungi on the plant growth and root rot diseases of chickpea. *American–Eurasian J. Agric. Environ. Sci.*, 8(5): 544–549.

Akkopru, A. and Demir, R. 2005. Biological control of *Fusarium* wilt in Tomato caused by *Fusarium oxysporum* f. sp. *lycopersici* by AMF *Glomus intraradices* and some rhizobacteria. *J. Phytopath.*, 153(9): 544–550.

Al Banna, L., Khyami-Horani, H., Sadder, M. and Abu Zahra, S. 2016. Efficacy of some local *Bacillus thuringiensis* isolates against soil borne fungal pathogens. *African Journal of Agricultural Research*, 11(19): 1750-1754.

Alam, S.S., Sakamoto, K. and Inubushi, K. 2011. Biocontrol efficiency of *Fusarium* wilt disease by root colonizing fungus *Penicillium* sp. *Soil Sci. Plant Nutr.*, 57: 204–212.

Al–Askar, A.A. and Rashad, Y.M. 2010. Arbuscular mycorrhizaL fungi: A biocontrol agent lower bean *Fusarium* root rot disease. *Plant Physiol. J.*, 9(1): 10–17.

Al–Karaki, G.N. 2006. Nursery inoculation of tomato with arbuscular mycorrhizal fungi and subsequent performance under irrigation with saline water. *Sci. Hortic.*, 109: 1–7.

Alström, S. 2001. Characteristics of bacteria from oilseed rape in relation of their biocontrol activity against *Verticillium dahliae*. *J. Phytopathol.*, 149: 57–64.

Alwathnani, H.A. and Perveen, K. 2012. Biological control of *Fusarium* wilt of tomato by antagonistic and cyanobacteria. *Afr. J. Biotechnol.*, 11(5): 1100–1105.

Amini, J. and Sidovich, D.F. 2010. The effects of fungicides on *Fusarium oxysporum* f. sp. *lycopersici* associated with *Fusarium* wilt of tomato. *J. Plant Protec.* Res., 50: 172–178.

Anitha, A. and Rabeeth, M. 2009. Control of *Fusarium* wilt of Tomato by bioformulations of *Streptomyces griseus* in green house conditions. *Afr. J. Basic Appl. Sci.*, 1(1–2): 9.14.

Anjani, K., Raoof, M.A., Reddy, P.A.V., Rao, C.H. 2004. Sources of resistance to major castor (*Ricinus communis*) disease. *Plant Genet. Resour. Newslett.*, 137: 46–48.

Armstrong, G.M. and Armstrong, J.K. 1981. Formae speciales and races of Fusarium oxysporum causing wilt diseases. In: Fusarium: Diseases, biology and taxonomy, (Eds.) Nelson, P.E., Toussoun, T.A. and Cook, R.J. (eds.), University Park: The Pennsylvania State University Press, pp. 392–399.

Arriola, L.L., Hausbeck, M.N., Rogers, J. and Safir, G.R. 2000. The effect of *Trichoderma harzianum* and arbuscular mycorrhizae on *Fusarium* root rot in Asparagus. *Hortic. Technol.*, 10: 141–144.

Asha, B.B., Chandra, N.S., Udaya, S.A.C., Srinivas, C. and Niranjana, S.R. 2011. Biological control of *Fusarium oxysporum* f. sp. *lycopersici* causing wilt of tomato by *Pseudomonas fluorescens. Int. J. Microbiol. Res.*, 3(2): 79–84.

Attitalla, I.H., Johnson, P., Brishammar, S., Quinatanila, P. 2001b. Systemic resistance to *Fusarium* wilt of tomato induced by *Phytophthora cryptogea. J. Phytopathol.*, 149: 373–380.

Avis, T.R., Gravel, V., Antoun, H. and Tweddell, R.J. 2008. Multifaceted beneficial effects of rhizospheric microorganisms on plant health and productivity. *Soil Biol. Biochem.*, 40: 1733–1740.

Bakeer, A.R.T., El-Mohamedy, R.S.R., Saied, N.M. and Abd-El-Kareem, F. 2016. Field suppression of *Fusarium* soil borne diseases of Tomato plants by the combine application of bio agents and Chitosan. *British Biotechnology J.* 13(3): 1-10.

Baker, K.F. and Cook, R.J. 1974. Biological Control of Plant Pathogens. American Phytopathological Society, St. Paul, MN, pp. 433.

Barakat, R.M. and Al–Masri, M.I. 2009. *Trichoderma harzianum* in combination with sheep manure amendedment enhances soil suppressiveness of *Fusarium* wilt of Tomato. *Phytopathol. Mediterr.*, 48: 385–395.

Barakat, R.M. and AL–Masri, M.I. 2012. Enhanced Soil Solarization against *Fusarium oxysporum* f. sp. *lycopersici* in the Uplands. *Int. J. Agron.*, doi:10.1155/2012/368654.

Barea, J.M., Gryndler, M., Lemanceau, P., Schüepp, H., Azcon, R., 2002. The rhizosphere of mycorrhizal plants. In: Mycorrhizal Technology in Agriculture, (Eds.) Gianinnazi, S., Schüepp, H., Barea, J.M. and Haselwandter, K., Birkhäuser Verlag, Basel, pp. 1–18.

Biswas, K.K. and Das, N.D. 1999. Biological control of pigeon pea wilt caused by *Fusarium udum* with *Trichoderma* spp. *Ann. Plant Prot. Sci.*, 7:46–50.

Borrero, C., Trillas, M.I., Delgado, A. and Avilés, M. 2012. Effect of ammoniumDnitrate ratio in nutrient solution on control of *Fusarium* wilt of tomato by *Trichoderma asperellum* T34. *Plant Pathol.*, 61: 132–139.

Borrero, C., Trillas, M.I., Ordovás, J., Tello, J.C. and Avilés, M. 2004. Predictive factors for the suppression of *Fusarium* wilt of tomato in plant growth media. *Phytopathology*, 94(10): 1094–1101.

Bowers, J.H. and J.C. Locke. 2000. Effect of botanical extracts on population density of *Fusarium oxysporum* in soil and control of *Fusarium* wilt in the green house. *Plant Dis.*, 88: 300–305.

Cazorla, F.M. and Mercado-Blanco. J. 2016. Biological control of tree and woody plant diseases: an impossible task? *BioControl* 61:233–242.

Chowdhury, S., Chauduri, T.R. and Kundu, S. 2009. Tomato R–genes against wilt: present status and future prospect. *J. Mycopathol. Res.*, 47(2): 175–180.

Christopher, D.J., Raj, T.S., Rani U.S. and Udhayakumar, R. 2010. Role of defense enzymes activity in tomato as induced by *Trichoderma virens* against *Fusarium* wilt caused by *Fusarium oxysporum* f sp. *lycopersici*. *J. Biopesti.*, 3(1): 158–162.

Cook, R.J. 1993. Making greater use of introduced microorganisms for biological control of plant pathogen. *Annu. Rev. Phytopathol.*, 31: 53–80.

Cotxarrera, L., Trillas–Gray, M.I., Steinberg, C. and Alabouvette, C. 2002. Use of sewage sludge compost and *Trichoderma asperellum* isolates to suppress *Fusarium* wilt of tomato. Soil Biol. Biochem., 34: 467–476.

Dar, G., Zargar, M.Y. and Beigh, G.M. 1997. Biocontrol *of Fusarium* root rot in the common bean *(Phaseolus vulgaris* L.) by using *Glomus mosseae* and *Rhizobium leguminosarum. Microbial Ecol.,* 34: 74–80.

Demir, S. and Akkopru, A. 2007. Using of arbuscular mycorrhizal fungi (AMF) for biocontrol of soil born fungal plant pathogens. In: Biological control of plant diseases, (Eds.) Chincholkar, S.B. and Mukerji, K.G., Haworth Press, USA, pp. 17–37.

Duijff, B.J., Pouhair, D., Olivain, C., Alabuvette, C. and Lemanceau, P. 1998. Implication of system induced resistance in the suppression *of Fusarium* wilt of Tomato by *Pseudomonas fluorescens* WCS417r and by non *pathogenic Fusarium oxysporum* 47. *Eur. J. Plant Pathol.,* 104: 903–910.

El–Khallal, S.M. 2007. Induction and modulation of resistance in tomato plants against *Fusarium* wilt disease by bioagent fungi (arbuscular mycorrhiza) and/or hormonal elicitors (jasmonic acid and salicylic acid): 1–changes in growth, some metabolic activities and endogenous hormones related to defense mechanism. *Aust. J. Basic Appl. Sci.,* 1(4): 691–705.

Fracchia, S., Garcia–Romera, I., Godeas, A. and Ocampo, J.A. 2000. Effects of the saprophytic fungi *F. oxysporum* on AM colonization and growth of plants in greenhouse and field trials. *Plant Soil,* 223: 175–184.

Fravel, D., Olivain, C. and Alabouvette, C. 2003. *Fusarium oxysporum* and its biocontrol. *New Phytol.,* 157: 493–502.

Garmendia, I., Goicoechea, N. and Aguireolea, J. 2004. Effectiveness of three *Glomus* species in protecting pepper *(Capsicum annuum* L.) against *verticillium* wilt. *Biol. Control,* 31: 296–305.

Garrett, S.S. 1970. Pathogenic root infecting fungi. Cambridge University Pree, London. pp. 194.

Ghanbarzadeh, B., Safaie, N., Goltapeh, E.M., Danesh, Y.R. and Khelghatibana, F. 2016. Biological control of *Fusarium* basal rot of onion using *Trichoderma harzianum* and *Glomus mosseae J. Crop Prot.,* 5 (3):359-368.

Grosch, R., Scherwinski, K., Lottmann, J. and Berg, G. 2006. Fungal antagonists of the plant pathogen *Rhizoctonia solani*: selection, control efficacy and influence on the indigenous microbial community. *Mycol. Res.,* 110: 1464–1474.

Harley, C. 1921. Damping off in forest nurseries. U.S. Dept. Agri. Bull. 934: 1–99.

Harman, G.E. and Kubicek, C.P. 1998. *Trichoderma* and *Gliocladium,* Vol. 2. Enzymes, biological control and commercial applications. Taylor and Francis, London, pp. 393.

Harman, G.E., Howell, C.R., Viterbo, A., and Chet, I. 2004. *Trichoderma* spp.— Opportunistic avirulent plant symbionts. *Nat. Rev. Microbiol.,* 2: 43–56.

Hartman, J.R. and Fletcher, J.T. 1991. Fusarium crown and root rot of tomato in UK. Plant Pathology. Oxford: Blackwell Scientific Publications, 40: 85–92.

Hermosa, R., Viterbo, A., Chet, I. and Monte, E. 2012. Plant–beneficial effects of *Trichoderma* and of its genes. *Microbiol.*, 158, 17–25.

Horinouchi, H., Watanabe, H., Taguchi, Y., Muslim, A. and Hyakumachi, M. 2011. Biological control of *Fusarium* wilt of tomato with *Fusarium equiseti* GF191 in both rock wool and soil systems. *BioControl*, 56(6): 915–923.

Houssien, A.A., Ahmed, S.A. and Ismail, A.A. 2010. Activation of tomato plant defense response against *Fusarium* wilt disease using *Trichoderma harzianum* and salicylic acid under green house condition. *Res. J. Agric. Biol. Sci.*, 6: 328–338.

Howell, C.R. 2003. Mechanisms employed by *Trichoderma* species in the biological control of plant diseases. The history and evolution of current concepts. *Plant disease*, 87: 4–10.

Hu, J.L., Lin, X.G., Wang, J.H., Shen, W.S., Wu, S., Peng, S.P. and Mao, T.T. 2010. Arbuscular mycorrhizal fungi enhances suppression of cucumber *Fusarium* wilt in green house soils. *Pedosphere*, 20(5): 586–593.

Hwang, S.F. 1992. Effect of Vesicular– Arbuscular Mycorrhizal fungi on the development of *Verticillium* and *Fusarium* wilt of Alfalfa. *Plant Disease*, 239–243.

Ignjatov, M., Miloševi , D., Nikoli , Z., Gvozdanovi –Varga, J., Jovi i , D., and Zdjelar, G. 2012. *Fusarium oxysporum* as causal agent of tomato wilt and fruit rot. J. Pestic. *Phytomed.* (Belgrade), 27(1): 25–31.

Inam–Ul–Haq, M., Gowen, S.R., Javed, N., Shahina, F., Izhar–Ulhaq, M., Humayoon, N. and Pembroke, B. 2007. Antagonistic potential of bacterial isolates associated with entomopathogenic nematodes against tomato wilt caused by Fusarium oxysporum f.sp., lycopersici under greenhouse conditions. *Pak. J. Bot.*, 39(1): 279–283.

Jahanshir, A. and Dzhalilov, S. 2010. The effects of fungicides on *Fusarium oxysporum* f. sp. *lycopersic*i associated with *Fusarium* wilt of tomato. *J. Plant Prot. Res.*, 50(2): 172–178

John R.P., Tyagi R.D., Prévost D., Brar S.K., Pouleur S. and Surampalli R.Y. 2010. Mycoparasitic Trichoderma viride as a biocontrol agent *against Fusarium oxysporum* f. sp. *adzuki* and *Pythium arrhenomanes* and as a growth promoter of soybean. *Crop Prot.*, 29: 1452–1459.

Jones, J.P. 1991. Fusarium wilt. In: Compendium of tomato diseases, (Eds.) Jones, J.B., Stall R.E. and Zitter, T.A., St. Paul, Minnesota: APS.

Kang, B.R. 2011. Mannitol amendment as a carbon source in a bean–based formulation enhances biocontrol efficacy of a 2, 4–diacetylphloroglucinol–producing *Pseudomonas* sp. NJ134 against Tomato *Fusarium* wilt. *Plant Pathol. J.*, 27(4): 390–395.

Kang, B.R. 2012. Biocontrol of Tomato *Fusarium* wilt by a novel genotype of 2,4–Diacetylphloroglucinol producing *Pseudomonas* sp. NJ134. *Plant Pathol. J.*, 28(1): 93–100.

Kesavan, V. and Chaudhary, B. 1977. Screening for resistance to *Fusarium* wilt of tomato. *Sabro J.,* 9: 51–65.

Kim, J.D., Han, J.W., Lee, S.C., Lee, D., Hwang, I.C. and Kim, B.S. 2011. Disease control effect of Streverteness produced by *Streptomyces psammoticus* against tomato *Fusarium* wilt. *J. Agric. Food Chem.,* 59(5): 1893–1899.

Kim, J.D., Han, J.W., Hwang, I.C., Lee, D. and Kim, B.S. 2012. Identification and biocontrol efficacy of *Streptomyces miharaensis* producing filipin III against *Fusarium* wilt. *J. Basic Microbiol.,* 52(2):150–159.

Kimaru, S.K., Waudo, S.W., Monda, E., Seif, A.A. and Birgen, J.K. 2004. Effect of neem kernel cake powder (NCCP) on *Fusarium* wilt of tomato when used as soil amendment. *J. Agric. Rural Dev. Trop. Subtrop.,* 105(1): 63–69.

Klonsky, K. 2004. Organic agricultural production in California. p. 241–256. In: "California Agriculture: Dimensions and Issues" (J. Siebert, ed.). University of California, Giannini Foundation of Agricultural Economics, Division of Agriculture and Natural Resources, Berkeley, CA, USA, 304 pp.

Kobra, N., Jalil, K. and Youbert, G. 2009. Effect of three *Glomus* species as biocontrol agents against *Verticillium* induced wilt in cotton. *J. Plant Prot. Res.,* 49(2): 234–239.

Kucuk, C. and Kivanc, M. 2003. Isolation of *Trichoderma* spp. and determination of their antifungal, biochemical and physiological features. *Turk. J. Biol.,* 27: 247–253.

Kumar, N.R. and Rai, M. 2007. Performance, compititiveness and determinants of tomato to export from India. Agricultureal Economics Research Review. 20(conference issue), 551-562.

Larkin, R.P. and Fravel, D.R. 1999. Mechanisms of action and dose–response relationships governing biological control of *Fusarium* Wilt of Tomato by nonpathogenic *Fusarium* spp. *Phytopathology,* 89(12): 1152–1161.

Larkin, R.P. and Fravel, D.R. 2002. Effects of varying environmental conditions on biological control of *Fusarium* wilt of tomato by nonpathogenic *Fusarium* spp. *Crop Prot.,* 21: 539–543.

Linderman, R.G. 2000. Effects of mycorrhizas on plant tolerance to diseases. In: Arbuscular Mycorrhizas: Physiology and Function, (Eds.) Kapulnik, Y. and Douds, D.D.J., Dordrecht, the Netherlands: Kluwer Academic Publishers, pp. 345–365.

Mandal, S., Mallick, N. and Mitra, A. 2009. Salicylic acid induced resistance to *Fusarium oxysporum* f.sp. *lycopersici* in tomato. *Plant Physiol. Biochem.,* 47(7): 642–649.

Martïnez-Medina, A., Roldán, A., Albacete, A. and Pascual, J.A. 2009. Interactions between arbuscular mycorrhizal fungi and *Trichoderma harzianum* and their effects on *Fusarium* wilt in melon plants grown in seedling nurseries. *J Sci Food Agric.,* 89: 1843–1850.

Matsubara, Y.I. 2012. Tolerance to *Fusarium* root rot and changes in antioxidative ability in mycorrhizal asparagus plants. HortSci., 47(3): 356–360.

Mbuthia, L.W., Alten, H.V. and Grunewaldt–Stoker. 2009. The interactions of Arbuscular mycorrhiza fungi (AMF) with other bio–control agent in the control of *Fusarium oxysporum* f.sp. *lycopersici*. In: Biophysical and Socio–economic frame conditions for the sustainable management of natural resources, Tropentag, Oct 6–8, 2009, Hamburg.

Mehra, R. 2008. Biocontrol of *Fusarium* wilt of tomato caused *by Fusarium oxysporum* f.sp. *lycopersici. Plant Disease Res.*, 23(2): 51–54.

Morsy, E.M., Abdel–Kawi, K.A. and Khalil, M.N.A. 2009. Efficiency of *Trichoderma viride* and *Bacillus subtilis* as Biocontrol Agents against *Fusarium solani* on Tomato Plants. *Egypt. J. Phytopathol.*, 37(1): 47–57.

Muslim, A., Horinouchi, H. and Hyakumachi, M. 2003. Biological control of *Fusarium* wilt of tomato with hypovirulent binucleate Rhizoctonia in green house conditions. *Mycosci.*, 44: 77–84.

Nusret, O. and Steven, E.N. 2004. *Fusarium* crown and root rot of tomato and control methods. Plant Pathol. J., 3: 9–18

Ojha, S. and Chatterjee, N.C. 2012. Induction of resistance in tomato plants against *Fusarium oxysporum* f. sp. *lycopersici* mediated through salicylic acid and *Trichoderma. J. Plant Prot. Sci.*, 52: 220–225.

Osuinde, M.I., Aluya, E.I. and Emoghene, A.O. 2002. Control of *Fusarium* wilt of tomato (*Lycopersicon esculentum* Mill) by *Trichoderma* species. *Acta Phytopathol. Entomol. Hungarica*, 37(1–3): 47–55.

Ozbay, N. and Newman, S. E. 2004. *Fusarium* crown and root rot of tomato and control methods. *Pl. Path. J.*, 3: 9-18.

Ozgonen, H., Akgul, D.S. and Erkilic, A. 2010. The effects of arbuscular mycorrhizal fungi on yield and stem rot caused by *Sclerotium rolfsii* Sacc. in peanut. *Afr. J. Agric. Res.*, 5(2): 128–132.

Paulitz, T.C. and Belanger, R.R. 2001. Biological control in greenhouse systems. *Ann. Rev. Phytopathol.*, 39: 103–33.

Perelló, A., Mónaco, C., Sisterna, M. and Dal Bello, G. 2003. Biocontrol efficacy of *Trichoderma* isolates for tan spot of wheat in Argentina. *Crop Prot.*, 22: 1099–1106.

Perveen, K. and Bokhari, N.A. 2012. Antagonistic activity of *Trichoderma harzianum* and *Trichoderma viride* isolated from soil of date palm field against *Fusarium oxysporum. Afr. J. Microbiol. Res.*, 6(13): 3348–3353.

Raj, H. and Kapoor, I.J. 1997. Management of *Fusarium* wilt of tomato by soil amendments with composts. *Indian Phytopath.*, 50(3): 387–395.

Ramezani, H. 2009. Efficacy of some fungal and bacterial bioagents against *Fusarium oxysporum* f.sp. on chickpea. *Plant Prot. J.*, 1(1): 108– 113.

Rompalli, R., Mehendrakar, S.R. and Venkata, P.K. 2016. Evaluation of potential bio-control agents on root-knot nematode *Meloidogyne incognita* and wilt causing fungus *Fusarium oxysporum* f.sp. *conglutinans in vitro. African J. Agric. Res..,* 15 (19):798-805.

Sartaj, A.S., Sakamoto, K. and Inubushi, K. 2011. Effects of *Penicillium* sp. EU1003 inoculation on tomato growth and Fusarium wilt. *Hort.Res.,* 65: 69–73.

Segarra, G., Casanova, E., Avilés, M. and Trillas, I. 2010. *Trichoderma asperellum* strain T34 controls *Fusarium* wilt disease in tomato plants in soilless culture through competition for iron. *Microb. Ecol.,* 59(1): 141–9.

Shali, A., Ghasemi, S., Ahmadian, G., Ranjbar, G., Dehestani, A., Khalesi, N., Motallebi, E. and Vahed, M. 2010. *Bacillus pumilus* SG2 chitinases induced and regulated by chitin, show inhibitory activity against *Fusarium graminearum* and *Bipolaris sorokiniana. Phytoparasitica,* 38: 141–147.

Shishido, M., Miwa, C., Usami, T., Amemiya, Y. and Johnson, K.B. 2005. Biological control efficacy of *Fusarium* wilt of Tomato by nonpathogenic *Fusarium oxysporum* FoB2 in different environments. *Phytopathology,* 95(9): 1072–1080.

Sibounnavong, P., Keoudone, C., Soytong, K., Divina, C.C. and Kalaw, S.P. 2010. A new mycofungicide from *Emericella nidulans* against tomato wilt caused by *Fusarium oxysporum* f. sp. *lycopersici in vivo. J. Agric. Technol.,* 6(1): 19–30.

Silva, J.C. and Bettol, W. 2005. Potential of non–pathogenic *Fusarium oxysporum* isolate for control of *Fusarium* wilt of tomato. *Fitopatologia Brasileria,* 30: 409–412.

Singh, B.K., Singh, R., Upadhyay, R.S. and Rai, B. 2009. Antagonistic potential of *Trichoderma virens* against *Fusarium* wilt disease of tomato, pigeon pea and lentil for biological control. *J Indian bot. Soc.,* 88 (3 and 4): 37–43.

Singh, P.K., Mishra, M. and Vyas, D. 2010. Interactions of vesicular arbuscular mycorrhizal fungi with *Fusarium* wilt and growth of the tomato. *Indian Phytopath.,* 63(1): 30–34.

Singh, P.K., Singh, M. and Vyas, D. 2010. Biocontrol of *Fusarium* wilt of chickpea using arbuscular mycorrhizal fungi and *Rhizobium leguminosarum* Biovar. Caryologica, 63: 349–353.

Singha, I.M., Kakoty, Y., Unni, B.G., Kalita, M.C., Das, J., Naglot, A., Wann, S.B. and Singh, L. 2011. Control of *Fusarium* wilt of tomato caused by *Fusarium oxysporum* f. sp. *lycopersici* using leaf extract of *Piper betle* L.: a preliminary study. *World J. Microbiol. Biotecnol.,* 27: 2583–2589.

Sivan, A. and Chet, I. 1986. Biological control of *Fusarium* spp. in cotton, wheat and muskmelon by *Trichoderma harzianum. J. Phytopathol.,* 116: 39–47.

Smith, S.E. and Smith, F.A. 2012. Fresh perspectives on the roles of arbuscular mycorrhizal fungi in plant nutrition and growth. *Mycologia,* 104(1):1–13.

Someya, N., Tsuchiya, K., Yoshida, T., Noguchi, M.T. and Sawada, H. 2006. Combined use of the biocontrol bacterium *Pseudomonas fluorescens* strain LRB3W1 with reduced fungicide application for the control of tomato *Fusarium* wilt. *Biocontrol Sci.,* 11(2):75–80.

Song, W., Zhou, L., Yang, C., Cao, X., Zhang, L. and Lio, X. 2004. Tomato *Fusarium* wilt and its chemical control strategies in a hydroponic system. *Crop Prot.*, 23(3): 234–247.

Srinon, W., Chuncheen, K., Jirattiwarutkul, K., Soytong, K., and Kanokmedhakul, S. 2006. Efficacies of antagonistic fungi against *Fusarium* wilt disease of cucumber and tomato and the assay of its enzyme activity. *J. Agric. Technol.*, 2: 191–201.

Srivastava, R., Khalid, A., Singh, U.S., and Sharma, A.K. 2010. Evaluation of arbuscular mycorrhizal fungus, *Pseudomonas fluorescens* and *Trichoderma harzianum* formulation against *Fusarium oxysporum* f. sp. *lycopersici* for the management of tomato wilt. *Bio Control*, 53: 24–31.

Szczech, M.M. 1999. Suppressiveness of vermicompost against *Fusarium* wilt of Tomato. *J. Phytopath.*, 147: 155–161.

Tahmatsidou, V., O'Sullivan, J., Cassells, A.C., Voyiatzis, D. and Paroussi, G. 2006. Comparison of AMF and PGPR inoculants for the suppression of *Verticillium* wilt of strawberry (*Fragaria ananassa* cv. *Selva*). *Appl. Soil Ecol.*, 32: 316–324.

Tanwar, A., Aggrwal, A. and Panwar, V. 2013. Arbuscular mycorrhizal fungi and *Trichoderma viride* medicated *Fusarium* with control in tomato. *Biocontrol Sc. Tech.*, 23(5):485-498.

Veerabhadraswamy, A.L. and Garampalli, R.H. 2011. Effects of arbuscular mycorrhizal fungi in the management of Black Bundle disease of maize caused by *Cephalosporium acremonium*. *Sci. Res. Reporter*, 1(2): 96–100.

Verma, M., Brar, S.K., Tyagi, R.D., Sahai, V., Prévost, D., Valéro, J.R., and Surampalli, R.Y. 2007. Bench–scale fermentation of *Trichoderma viride* on wastewater slidge: rheology, lytic enzymes and biocontrol activity. *Enz. Microb. Technol.*, 41, 764–771.

Verma, M., Brar, S.K., Tyagi, R.D., Sahai, V., Prévost, D., Valéro, J.R., and Surampalli, R.Y. 2007. Bench–scale fermentation of *Trichoderma viride* on wastewater slidge: rheology, lytic enzymes and biocontrol activity. *Enz. Microb. Technol.*, 41, 764–771.

Vierheilig, H., Steinkellner, S., Khaosaad, T. and Garcia–Garrido, J.M. 2008. The biocontrol effect of mycorrhization on soilborne fungal pathogens and the autoregulation of the AM symbiosis: One mechanism, two effects. In: Mycorrhiza, (Ed.) Varma, A., Springer–Verlag Berlin Heidelberg.

Wang, C., Li, X. and Song, F. 2012a. Protecting cucumber from *Fusarium* wilt with Arbuscular mycorrhizal fungi. *Commun. Soil Sci. Plant Anal.*, 43(22): 2851–2864.

Wang, C.X., Li, X.L., Song, F.Q., Wang, G.Q. and Li, B.Q. 2012b. Effects of Arbuscular mycorrhizal fungi on *Fusarium* wilt and disease resistance related enzyme activity in cucumber seedling root. *Chinese J. Eco–Agric.*, 20(1): 53–57.

Wehner, J., Antunes, P.M., Powell, J.R., Mazukatow, J. and Rillig, M.C. 2009. Plant pathogen protection by arbuscular mycorrhizas: A role for fungal diversity? Pedobiologia, doi: 10.1016/j.pedobi.2009.10.002.

Weindling, R. 1932. *Trichoderma lignorum* as a parasite of other soil fungi. *Phytpathology*, 22: 837–845.

Xu, X.M., Jeffries, P., Pautasso, M. and Jeger, M. J. 2011. Combined Use of Biocontrol agents to manage plant diseases in theory and practice. *Phytopath.*, 101 (9): 1024 -1031.

Yigit, F. and Dikilitas, M. 2007. Control of *Fusarium* wilt of tomato by combination of Fluorescnet *Pseudomonas*, non pathogenic Fusarium and *Trichoderma harzianum* T–22 in green house conditions. *Plant Pathol. J.*, 6(2): 159–163.

Zinati, G.M. 2005. Compost in the 20th Century: A tool to control plant diseases in nursery and vegetable crops. *HortTech.*, 15: 61–66.

2017, Mycorrhizal Fungi

Editors: **Ashok Aggarwal and Kuldeep Yadav**

Published by: **ASTRAL INTERNATIONAL PVT. LTD., NEW DELHI**

Pages 357–376

16

Molecular Mechanisms and Events Coupled with Mycorrhizal Symbiosis

Navnita Sharma*, Ashish Kumar, Esha Jangra and Ashok Aggarwal

Department of Botany, Kurukshetra University, Kurukshetra, Haryana
*Corresponding Author: navneetasharma20@gmail.com

ABSTRACT

Roots of the most terrestrial plants engage themselves in a symbiotic relationship with fungi belonging to Glomeromycota under nutrient deficient conditions. Arbuscular mycorrhizal symbiosis is a biotrophic interaction where intracellular accommodation of fungi in plant roots benefits the host with high Phosphorus and Nitrogen uptake in exchange for the photo assimilates of the plant as their only source of carbon. Before the physical contact between the two partners exchange of molecular information occurs. While the release of some soluble factors by plants activates the fungal partner, some compounds are also released by the fungal partner who triggers the signal transduction pathways needed for symbiotic modus of plant cells. Here in this review, an attempt has been made to describe some of the recent discoveries on signaling events during symbiosis, their elicited responses including immune responses, role of plant and fungal proteins, phytohormones, nutrients and transcription factors, during rhizospheric conversation and the way in which they are perceived by their hosts.

Keywords: CSSP, Strigolactones, GRAS-type transcription factors, DELLA proteins, Auxins.

1. Introduction

Soil microbial communities are the fundamental component for agro ecosystem functioning as well as success in organic agriculture (Gosling *et al.*, 2006). In mineral scarce soils, microbial relationships have been adapted by plants for nutrient uptake enhancement. Besides arbuscular mycorrhizal symbiosis, legumes and several other plants develop an association for nitrogen fixation by *Rhizobia* in their root nodules. From various studies on these mechanisms, several genes for Rhizobial accommodation have been found indispensible for arbuscular mycorrhizal symbiosis also (Oldroyd, 2013). Such genes common to both symbiosis form so called Common Symbiosis Signaling Pathway (CSSP). Arbuscular Mycorrhizae (AM), a symbiosis formed between AMF and > 80 per cent of land plants is the most prevalent mutualistic interaction between microbes and plants sustaining their mineral and phosphate nutrition. Even after the absence of sexual reproduction, AMF have existed for at least 460 million years which has been evident from the fossils of AMF with other land plants or fossils of the spores (Redecker, 2000). From these discoveries it could be hypothesized that movement of plants to the harsh environment has been assisted by AMF (Bonfante and Genre, 2008). In this symbiosis spores of fungi belonging to *Glomeromycota* (Schußler *et al.*, 2001) give out extra radical hyphae or hyphae may come out of the nearby roots colonized by AM fungi which through subsequent transcellular infection as well as hypopodial attachment, enter into the rhizodermis resulting into the formation of highly branched structures in the inner cortex called as arbuscules (Parniske, 2008). Arbuscules are regarded as the main sites for exchange of nutrients and are transient structures (Harrison, 2005). Exchange of some diffusible molecules initiate AM formation.

2. Strigolactones Provoke Fungal Activity by Activating Mitochondria

A variety of mechanisms including development of lateral roots, root hair growth for increasing the surface area (Péret *et al.*, 2011) and secretion of organic acids as well as Phosphatase free Pi from complexes (Plaxton and Tran, 2011) in the soil have been employed by plants for nutrient and phosphorus uptake. Alternatively, association of AMF increases the surface area because of extension of extra radical mycelia beyond depletion zone of rhizosphere (Javot *et al.*, 2007). Deficiency of phosphorus is sensed systematically as well as locally and several signaling pathways are involved in plant response, Strigolactones (SLS) being most important of them (Czarnecki *et al.*, 2013). Strigolactones are carotene derived plant hormones which are produced by the roots of the plants and are extremely important for plant development. Their nomenclature is based on genus *Striga* (Cook *et al.*, 1966) as they were discovered originally as germination stimulants of parasitic weeds. Later on after 40 years, their importance as the factors which bring about changes in AMF development and as hormones essential for plant architecture was recognized. Particularly, SLS released by plants bring about germination in members of Glomeraceae and branching as well as hyphal growth in the members belonging to Gigasporaceae. Perception system of AMF for SLS is highly sensitive as very low concentrations of 10 nM of the synthetic model-SL GR24 can bring about mitochondrial enlargement, nuclear division, increase in NADH dehydrogenase activity, NADH content as well as ATP content in *G. rosea* hyphae required for proliferation of hyphae (Besserer *et al.*, 2006, 2008). The conditions of phosphate starvation boost up biosynthesis and exudation of strigolactones, a condition which supports AM colonization (Yoneyama *et al.*, 2012).

3. Specific Function of GRAS Type Transcription Factors

The understanding about plant signaling pathways regulating the recognition and response of AM fungi has been derived from study of nodulation mutants recognized previously in model legume plant species (Charpentier and Oldroyd 2010). Most of these mutants have been proved to be defective for mycorrhizal symbiosis and the genes underlying are thus commonly known as SYM (Symbiosis) genes (Kistener and Parniske 2002). Different microbial symbionts activate a common signaling pathway which raises a problem of understanding the ability of plants to achieve specificity in activation of two racially different genetic programs. Identification of genetic components with specific function in either AM pathway or nodulation is critical. Different transcription factors acting downstream of SYM pathway within the nodulation pathway have been identified among which NSP1 and NSP2 belong to GRAS family (Kalo *et al.*, 2005). Nodulation signaling pathway (NSP) 1 and 2 and two GRAS-type transcription factors which are important for nodulation in legumes (Smit *et al.*, 2005) are also concerned with biosynthesis of strigolactones under phosphate shortage conditions. They bring about proper regulation of *DWARF27* (*D27*) that encodes β-carotene isomerase an enzyme helpful in catalyzing first step of strigolactone synthesis (Liu *et al.*, 2011). At root surface formation of hyphopodia at the tip of growing hypha is the first step to mark the onset of colonization. Two genes RAM1 and RAM2 have been identified as the novel components particular to the AM pathway (Wang *et al.*, 2012). It has been seen that hyphopodia were formed by *Gigaspora gigantea* on cell wall fragments of roots isolated from the carrot host plant but for non host *Beta vulgaris*, they were lacking (Nagahashi and Douds, 1996). Capability to bear nodulation was retained in the plants having mutation in either gene but a decrease in mycorrhization with decrease in number of hyphopodia has been shown. Glycerol-3-Phosphate Acyl Transferase (GPAT) is the encoded RAM2 which brings about the synthesis of cutin monomers (Wang *et al.*, 2012). Expression of RAM2 is regulated by a GRAS transcription factor which is being encoded by RAM1 (Gobbato *et al.*, 2012). RAM2 plants have shown a severe defect in arbuscule formation in addition to hyphopodia, suggesting their role in arbuscule development (Wang *et al.*, 2012). Also another role of RAM2 in stimulation of appresorium formation in case of *Phytophtora palmivora* an oomycete has also been suggested (Wang *et al.*, 2012). On germination, signal molecules are released by AMF which provoke the expression of many plant genes, accumulation of starch in roots, spiking of calcium in rhizodermal cells, and formation of lateral roots before the colonization (Mukherjee and An'e, 2011). A mixture of different active molecules including tetra- or pentachitooligosaccharides, N-acetylglucosamine oligosaccharides, chitooligosaccharides (Genre *et al.*, 2013) along with lipochitoligosaccharides same as that of NOD factor released by *Rhizobia* is present in germinating spore exudates (Maillet *et al.*, 2011). Chitooligosaccharide induced signaling depends upon *DMI1* (doesnot make infection), *DMI2*, and *DMI3*, genes which are essential for AMF and Rhizobial symbiosis. In anticipation of the fungal infection a tube like sub cellular structure called as prepenetration apparatus (PPA) is formed (Genre *et al.,* 2005). PPA formation involves the movement of nucleus to transverse the cell vacuole away from the place of hyphopodium formation. Initially PPA was discovered in the epidermal cells in direct contact with fungal hyphopodia. PPA is just similar to the preinfection thread seen during root nodule symbiosis (Fournier *et al.,* 2008) as in either of cases way of microbial entry is determined by transvacoular structure in a fully differentiated cortical cell (Parniske 2008). In cells where PPT and PPA formation is not accompanied by microbial infection their formation appears to be reversible (Sieberer *et al.,* 2009).

4. Protein Pattern Associated with AM Fungi

The host plants perceive Myc factors produced by arbuscular mycorrhizal fungi (AMF) which induces the symbiotic programme through the SYM pathway (Does not make infection DMI1, DMI2, DMI3) leading to onset of NSP2/RAM1/RAM2-mediated signal via the production of cutin monomers at the cell surface. The cellular remodeling events at fungal partner involve protein GinSTE12 and in host plants MSP1, Vapyrin, D3, Expansin are involved. AMF as potential colonizers are recognized by host plants with the help of pattern recognition receptors which perceive microbe-associated molecular patterns (MAMPs). Thus signaling cascade is induced which probably entails reactive oxygen species (ROS) production (Kiirika *et al.,* 2012) mediated by Rac1 resulting into production of defense related compounds leading to MAMP-triggered immunity (MTI). In reaction to this SP7 protein effectors is secreted by AMF into the cytosol of plants which when targets the nucleus, block the ERF19-mediated transcriptional programme upon interaction with nucleus. Suppression of MTI is possibly brought about by the SYM pathway for instance DMI3-dependent down regulation of ACRE264 which is a defence related protein (Siciliano *et al.,* 2007a, b). A flow chart for protein pattern associated with mycorrhizal symbiosis during early stages has been shown in the Figure 16.1.

The proteins which by partial or total loss of their function play role in sustaining intra- radical or arbuscule development have been shown in Figure 16.1.

5. Genetic Control over the Development of Arbuscules

Arbuscule and prepenetration apparatus formation is coupled with cytoskeleton restructuring (Genre and Bonfante, 1998). Around the arbuscule, a basket like structure is formed by microtubules and actin filaments, nucleus migrates to the middle of the branched arbuscule and cell mitochondria as well as plastids proliferate which supply amino and fatty acids to meet the demands of periarbuscular membrane synthesis and deformation of vacuole also occurs (Pumplin and Harrison, 2009). Mechanisms underlying arbuscule development proceeds normally in many cases although, limited and patchy distribution has also been reported (Parniske, 2008). Reverse genetics have identified certain genes which affect the development of arbuscules. Vapyrin gene which codes for a cytoplasmic protein of unknown function brings about a total block down of

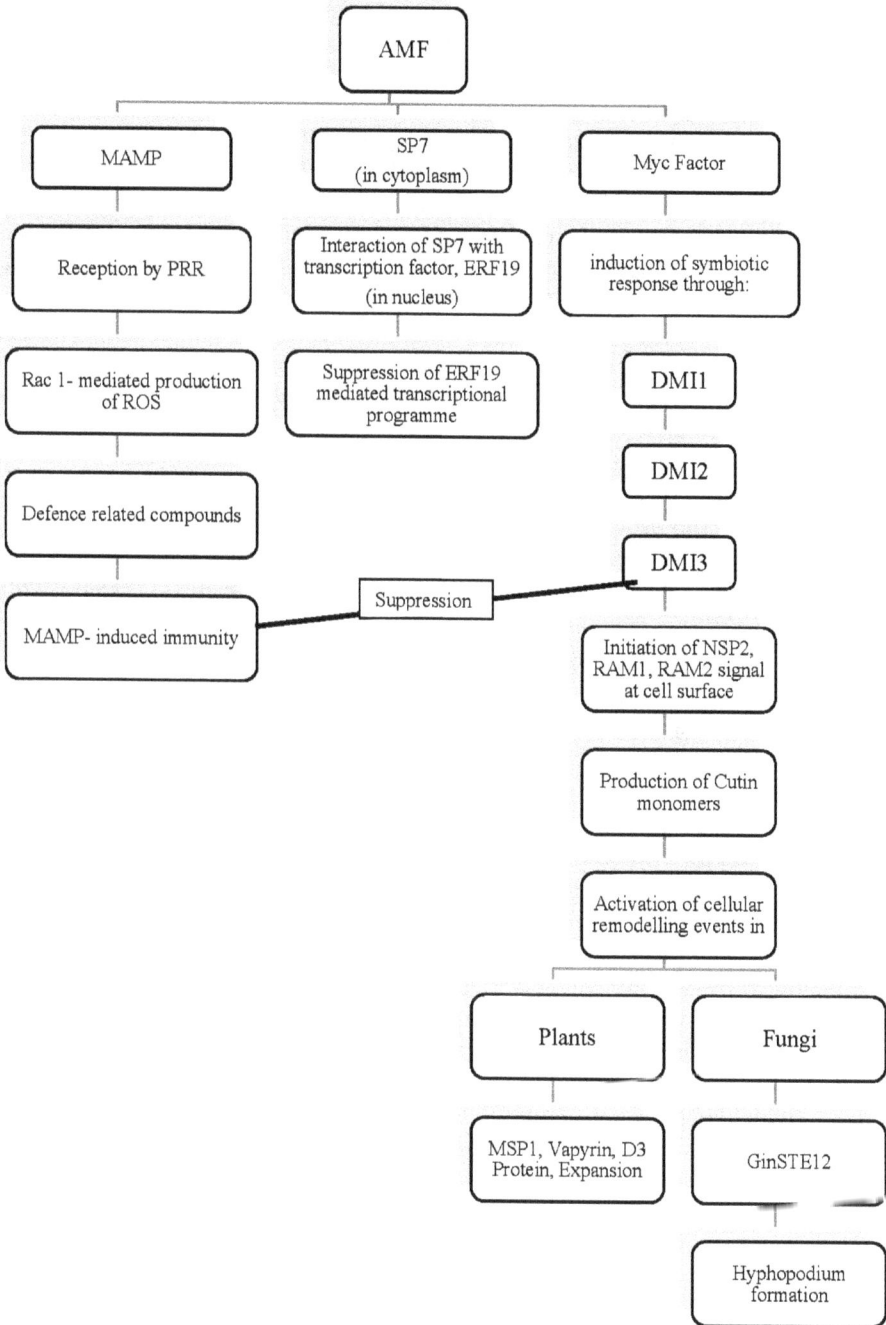

Figure 16.1: Protein Pattern Associated with AM Fungal Entry into Host Root Cells during Initial Stages of Infection.

arbuscule formation and epidermal prepenetration by RNA interference knockdown (Pumplin, 2010). Between the cortical cells only intercellular hyphae develop in such cases. Correspondingly, arbuscular morphological defects arise due to silencing of a phosphate transporter (Javot *et al.*, 2007), subtilin protease (Takeda *et al.*, 2009) or two ABC transporters (Zhang *et al.*, 2010). Findings reveal that genes important for development of arbuscules may not be dependent on SYM signaling pathway. Like any other intracellular hypha, perifungal membrane surrounds all the thin arbuscule branches. The host plasma membrane moulds into an arbuscule by following the surface of every branch. In an arbusculated cell the maximum spatial complexity is achieved by the interface having same composition like the primary cell wall. PPA like structures prearranged both in advance of penetration along the hypha and through the cell wall as little aggregates predicting the fungal branch formation are involved in gathering of membrane and cell wall material in extensive amount as well as proliferation of Golgi apparatus, *trans* -Golgi network and secretory vesicles (Pumplin, 2009). Endoplasmic reticulum and subtilases induced by AM secretion at the interface of symbiosis occur as a result of periarbuscular interface assembly. Plastids interconnect with each other by stromules and multiply and organize around the fungus while, Strigolactones are secreted by the roots which stimulate fungal metabolism and hyphal branching. Presymbiosis responses in the root tissues are elicited by signals from extra radical hyphae and spores (Bonfante and Raquena, 2011).

In rhizodermal cells nuclear and perinuclear Ca^{2+} oscillation occurs due to perception of fungal signals within minutes (Chabaud *et al.*, 2011). Many components of common symbiosis signaling pathway mediate Ca^{+2} signal transduction. Receptor kinase (DMI2, SYMRK), LRR (leucine-rich repeat), and a potassium channel of the nuclear membrane (DMI1, Castor/Pollux) are needed for production of specific Ca^{2+} spiking patterns. Ca^{2+} signals are transduced and decoded by interacting transcription factor (IPD3, Cyclops) and a Ca^{2+}/calmodulin-dependent protein kinase (DMI3,CCaMK) at CSSP terminus (Singh *et al.*, 2014). Not all the CSSP components are involved in plant responses, alternative pathways possibly exist (Nadal and Paszkowski, 2013). After pre-symbiosis communication, AM fungi get into the root cells by forming hyphopodia through apoplastic transcellular compartment after prepenetration apparatus formation (Genre *et al.*, 2008). Substantial transcriptional reprogramming occurs as a result of root colonization by AM fungi (Hohnjec *et al.*, 2005) leading to the generation of fungal and plant proteins collectively defining a functional mycorrhizal root. Presymbiotic responses in the host are also elicited by fungal short-chain chitin oligomers as well as sulfated and non sulfated lipochitooligosaccharides (s/nsMyc-LCOs) which are structurally similar to that of rhizobial Nod-factor LCOs r (Maillet *et al.*, 2011). Genome-wide expression studies have shown that Nod, sMyc- and nsMyc-LCOs, induce a particular set of genes, pointing towards LCO-specific perception in the pre-symbiosis phase.

6. Control of AM Symbiosis by Nutrient Signals

Nutritional status is an important factor for determining the degree of mycorrhizal colonization as depression in the colonization has been observed repeatedly under high Pi supply. Starvation of nitrogen and to some extent potassium, calcium or iron starvation overrules the suppressive effect of increased concentration of phosphorus on the AM root colonization (Nouri *et al.*, 2014) suggesting the control of symbiosis by the plants in function of their requirements for nutrients as per the Liebig's law of the minimum. Several nutrients have shown to influence the development of AMF especially (Pi) phosphorus. An inverse correlation between hypopodium number on maize roots and phosphorus status of shoot has been noticed which shows systematic suppression of AM by high (Pi) phosphorus concentration (Braunberger *et al.*, 1991). By fertilizing one half of the root system with high concentration of Pi and another half maintained at low Pi, an inhibition of AM colonization in entire root system has been noticed in Petunia and Pea (Breuillin *et al.*, 2010). Long distance signal travelling from root to shoot causes the regulation of suppression or promotion of AM colonization by different conditions of phosphorus. Members of miR399 family could be responsible for long distance signaling due to their role in signaling of Pi starvation (Gu *et al.*, 2011). An increase in the expression of miR399 members on AM colonization in tomato and *Medicago* leaves has been noticed. Time and again low levels of miR399 target PHO_2 gene which arbitrate protein degradation and has proved to be important in phosphate starvation response (Branscheid *et al.*, 2010). It was hypothesized that in order to allow constant colonization despite of elevated phosphate content of shoot due to AM symbiosis, higher expression of miR399 members may cause high phosphate starvation responses.

7. Roles of Phytohormones

Different workers have demonstrated in detail the role of abscissc acid and jasmonates in AM symbiosis. Formation of lateral roots is stimulated by AM fungi (Ola´h *et al.*, 2005) which are colonized by fungus preferentially (Gutjahr *et al.*, 2009). Auxin, a phytohormone having important role in root initiation and growth (Overvoorde *et al.*, 2010) proves to be a good candidate for involvement of mycorrhizae. Auxins are known to be produced by root endophytic fungi and ectomycorrhizal fungi (Sirrenberg *et al.*, 2007). With host species and time after inoculation, the response of root axuin concentrations to AM colonization varies. For instance, in case of tobacco and leek, concentration of indole-3-acetic acid (IAA) is unchanged (Shaul-Keinan *et al.*, 2002) but in colonized roots of soyabean the concentration is higher (Meixner *et al.*, 2005). Elevation in concentration of root derived auxins (indole-3-butyric acid) has been noticed in maize and nasturtium roots infected with mycorrhizae (Jentschel *et al.*, 2007). During early stages, auxin signaling is important for various processes including pre-symbiosis signaling (Hanlon and Coenen, 2011). The fact that auxin controls the strigolactone concentrations and thus regulates the early interaction has been confirmed by Foo, (2013) while working with auxin-deficient bushy (bsh) mutant of pea (*Pisum sativum*). On the other hand, normal appearing arbuscules were developed when the fungus entered into bsh-roots which show that auxins are not needed for arbuscule formation. During the formation of hypo podium and the consequent root colonization, phytohormones are involved in cross talk between AM fungi and plant roots. An indirect evidence showing the involvement of auxin in formation of arbuscular mycorrhizae comes from the finding which reveals the involvement of auxin in controlling the expression of Arabidopsis genes taking part in the synthesis of strigolactones (Bainbridge *et al.*, 2005; Ongaro and Leyser, 2008; Hayward *et al.*, 2009) which are carotenoid derivatives emitted by plants inducing responses in fungus (Koltai *et al.*, 2010). Nevertheless, there is lack of direct genetic evidence for requirement of auxin in establishment of mycorrhizae though, a number of auxin related mutants are existing in *Arabidopsis thaliana*, like most of the brassicaseous members, this species shows resistance to mycorrhizal colonization. Consequently, in *Arabidopsis* mutants AM formation can not be assessed directly. In *Tropaeolum majus* it has been seen that during early stages of colonization arbuscular mycorrhizal fungi altered auxin biosynthesis and enhanced auxin levels (Kerstin *et al.*, 2007).

Beside Auxins, other hormones also play an important role in AM symbiosis like, ET and strigolactones control the beginning steps while others control maintenance of symbiosis or arbuscule formation. During initiation of symbiosis and on early AM specific gene expression ET is supposed to have an adverse effect (Mukherjee and Ane, 2011). Zso¨go¨n *et al.*, (2008) has shown Inhibitory effect of ethylene on AM colonization by using the ethylene-insensitive mutant never ripe and ethylene-overproducing mutant epinastic in tomato (Geil *et al.*, 2001). Reduced intensity of mycorrhizal root colonization has been noticed in epinastic plants (Torres de Los Santos *et al.*, 2011). Reverse of the effect of never ripe has been seen by Torres de Los Santos *et al.*, (2011) in ripening inhibitor (RIN) mutant which showed enhanced colonization suggesting the role of RIN pathway in mycorrhizal colonization. RIN obstructs the ripening including climacteric ethylene production which is a MADS-box transcription factor (Vrebalov *et al.*, 2002). So, the effects of RIN are consistent with negative effect of ethylene on mycorrhizal colonization.

The negative effect of gibberellins on arbuscule formation has been confirmed many times by application of GA_3 to mycorrhizal roots (Foo *et al.*, 2013). Especially, Gibberellins control the formation of arbuscules through DELLA proteins which themselves regulate arbuscule formation positively. More arbuscule formation has been noticed in mycorrhizal roots of pea mutant which was defective in early step of GA biosynthesis than the corresponding wild types (Foo *et al.*, 2013). Highly reduced number of arbuscules in rice and *M. truncatula* (Yu *et al.*, 2014) has been noticed due to absence of DELLA proteins which negatively regulate the GA signaling. Substantial transcriptional and post-transcriptional reprogramming of host roots accompanies the establishment of arbuscules, which eventually define the protein composition of the cells containing arbuscules. Candidate genes which encode regulatory micro RNAs and key check points of AM development based on cellular expression profiles were identified. The role of ABA (Abscisic Acid) in development of mycorrhizae has been investigated by Herrera-Medina *et al.*, (2007) who used ABA deficient mutant sitiens in tomato to examine for the first time the effect of endogenous changes in level of ABA on *Glomus intraradices* colonization. In the mutant roots reduction in frequency and intensity of colonization has been noticed. Moreover, arbuscular morphology was less developed in a mutant which was revealed by reduced alkaline Phosphatase activity. Restoration of intensity and frequency of colonization to the wild type levels has been noticed by application

Figure 16.2: Genes Involved in Symbiosis (Redrawn from Mohanta and Bae, 2015).

GENE	FUNCTION	PHENOTYPE	REFERENCS
EARLY NODULIN (ENOD12)	SYM8 pathway signal transduction	Not studied	(Albrecht & Lapeyrie 1998)
DMI2	Myc-LCOs signaling	Reduced colonization	(Maillet et al., 2011)
LNP Lectin nucleotide	Nod factor binding protein	Nodulation	(Roberts et al., 2013)
NUP85	Nuclear trafficking	Reduced colonization	(Kanamori et al., 2006)
NUP133	Nuclear trafficking	Reduced colonization	(Saito et al., 2007)
NENA	Nuclear trafficking	Reduced colonization	(Groth et al., 2010)
NSP1	Transcription factor	Reduced colonization	(Maillet et al., 2011)
RAM1	Transcription factor	No hypopodium	(Wang et al., 2012)
RAM2	Cutin monomer biosynthesis	No hypopodium	(Gobbato et al., 2012)
	Unknown	No arbuscules	(Pumplin et al., 2010)
	Periarbuscular membrane formation	Stunted arbuscules	(Ivanov et al., 2012)
	Unknown	Less colonization and arbuscules	(Takeda et al., 2009)
STR1	Unknown	Less colonization and arbuscules	(Gutjahr et al., 2012)
DELLA	Transcriptional Regulator	Arbuscle formation	(Floss et al., 2013)

Genes controlling symbiosis.

DMI1	Calcium spiking, PPA formation	Less colonization, Fails to assemble PPA	(Ané et al., 2004; Genre et al., 2005)
DMI3	Calcium spiking	No colonization	(Catoira et al., 2000; Capoen et al., 2011)
IPD3	Interacts with DMI3	No arbuscules	(Horváth et al., 2011)
	Calcium spiking	Reduced colonization	(Capoen et al., 2011)
	Calcium spiking, Interacts with CCaMK in nucleus	Impaired in calcium spiking	(Charpentier et al., 2008, Singh et al., 2014)

Genes controlling calcium signaling.

Figure 16.2–*Contd...*

Gene	Function	Mutant phenotype	Reference
CCD7/CCD8	Strigolactone synthesis	Less colonization	(Gomez-Roldan et al., 2008)
D27	Strigolactone synthesis	Not tested	(Alder et al., 2012)
PDR1	Strigolactone transport	Reduced colonization	(Kretzschmar et al., 2012)
NCED	Strigolactone synthesis	Not studied	(López-Ráez et al., 2010)
D14/DAD2/HTD2	Perceive strigolactone signaling	Not studied	(Zhou et al., 2013)
DWARF53 (D53)	Repressor of strigolactone synthesis	Insensitive to strigolactone signaling	(Zhou et al., 2013)
DWARF3 (D3)	Strigolactone response	Fungal colonization defects	(Yoshida et al., 2012)
DWARF14	Strigolactone response	Higher AM fungal colonization	(Yoshida et al., 2012)

Genes controlling strigolactone biosynthesis and signaling.

Gene	Function	Mutant phenotype	Reference
PCT	Cotyledon formation	Presymbiotic root branching	(Hanlon &Coenen, 2011)
EBR	Auxin signaling	Less mycorrhizal colonization	(Foo, 2013)
DELLA	Gibberellin signaling	Less mycorrhizal colonization	(Foo et al., 2013)
	Jasmonic acid biosynthesis	Less mycorrhizal colonization	(Li et al., 2002)

Genes involved in hormonal regulation.

Gene	Function	Mutant phenotype	Reference
PT4	Symbiotic phosphate uptake	Less symbiotic phosphate uptake	(Javot et al., 2007)
PT10	Symbiotic phosphate uptake	Not studied	(Tamura et al., 2014)
PT11	Symbiotic phosphate uptake	Less symbiotic phosphate uptake	(Paszkowski et al., 2002)
AMT2	Symbiotic nitrate uptake	Not studied	(Guether et al., 2009)

Genes involved In nutrient iptake.

of ABA to the sitiens plants which support the indispensability of ABA for root functionality and full AM colonization. Yet a part of this outcome of ABA deficiency on colonization may be due to higher production of ethylene in the mutants. The fact that deficiency of ABA increases the ethylene levels which down regulates mycorrhizal intensity has been supported by the use of sitiens as well as another ABA deficient mutant notabilis, transgenic plants and inhibitors of ethylene as well as ABA biosynthesis. Though ABA deficiency inhibits the arbuscule formation directly (Mart´n-Rodr´guez *et al.*, 2011) probably due to cell wall modification and up regulation of genes related to defense (Garc´a-Garrido *et al.*, 2010).

8. Role of Plastid Derived Metabolites

In plastids two types of apocarotenoids (carotene cleavage compounds) are manufactured upon AM symbiosis which causes a typical yellow coloration of mycorrhizal roots (Stark and Fester, 2006). These compounds which accumulate in cytosol/vacuoles originate from a common carotenoid precursor and are of unknown function. It is evident from data that DXS2 (1-deoxy-D-xylulose 5-phosphate synthase) dependent MEP (methylerythritol phosphate) pathway based products are required to sustain functionality of mycorrhizae during later symbiosis satges. In tomato a knockdown approach performed on carotenoid cleavage dioxygenase genes as gene 7 (Cdd7) located downstream in the pathway of apcarotenoid synthesis led to reduction in concentration of AM induced apocarotenoids in Cdd7 antisense lines coupled to decreased abundance of arbuscules. It has also been seen that high supply of Pi exogenously to mycorrhizal roots repressed the genes involved in carotenoid biosynthesis in addition to those which are involved in intracellular accommodation and P transport (Beruillin *et al.*, 2010). The data indicates direct or indirect role of apocarotenoids in functioning and maintainance of arbuscules. The role of plant derived phytohormone Jasmonic acid located in plastids as a regulator of AM symbiosis has been reviewed by Hause and Schaarschmidt (2009).

9. Discussion

Although, past few years have brought exhilarating discoveries in the area of signaling events and molecular mechanisms coupled with AM symbiosis still, further refinement in this field is required. Molecular events related to signaling and nutrient acquisition process have been well elucidated by different workers. It is expected that future insights as well as methodological refinements will definitely improve our knowledge on identification and characterization of plant genes which are related to AM symbiosis. Few questions regarding their conservation among the terrestrial plants and their relevance for other beneficial microbial interactions need to be further addressed.

References

Akiyama, K., Matsuzaki, K. and Hayashi, H. 2005. Plant sesquiterpenes induce hyphal branching in arbuscular mycorrhizal fungi. *Nature*, 435: 824–827.

Albrecht, C., Geurts, R., Lapeyrie, F. and Bisseling, T. 1998. Endomycorrhizae and rhizobial Nod factors both require SYM8 to induce the expression of the early nodulin genes PsENOD5 and PsENOD12A. *Plant J.*, 15: 605–614.

Alder, A., Jamil, M., Marzorati, M., Bruno, M., Vermathen, M., Bigler, P., Ghisla, S., Bouwmeester, H., Beyer, P. and Al–Babili, S. 2012. The path from β–carotene to carlactone, a strigolactone– like plant hormone. *Science*, 335: 1348–1351.

Ané, J.M., Kiss, G.B., Riely, B.K., Penmetsa, R.V., Oldroyd, G.E.D., Ayax, C., Lévy, J., Debellé, F., Baek, J.M. and Kalo, P. 2004. DMI1 required for bacterial and fungal symbioses in legumes. *Science*, 303: 1364–1367.

Azcon–Aguilar, C., Rodriguez–Navarro, D.N. and Barea, J.M. 1981. Effects of ethrel on the formation and responses to VA mycorrhiza in *Medicago* and *Triticum*. *Plant Soil*, 60: 461–468.

Baier, M.C., Barsch, A., Kuster, H. and Hohnjec, N. 2007. Antisense repression of the nodule enhanced sucrose synthase leads to a handicapped nitrogen fixation mirrored by specific alterations in the symbiotic transcriptome and metabolome. *Plant Physiol.*, 145:1600–1618.

Bainbridge, K., Sorefan, K., Ward, S. and Leyser, O. 2005. Hormonally controlled expression of the *Arabidopsis* MAX4 shoot branching regulatory gene. *Plant J.*, 44: 569–580.

Besserer, A., Bécard, G., Jauneau, A., Roux, C. and Séjalon Delmas, N. 2008. GR24, a synthetic analog of strigolactones, stimulates the mitosis and growth of the arbuscular mycorrhizal fungus *Gigaspora rosea* by boosting its energy metabolism. *Plant Physiol.*, 148: 402–413.

Besserer, A., Puech Pagès, V., Kiefer, P., Gomez Roldan, V., Jauneau, A., Roy, S., Portais, J.C., Roux, C., Bécard, G. and Séjalon Delmas, N. 2006. Strigolactones stimulate arbuscular mycorrhizal fungi by activating mitochondria. *PLoS Biol.*, 4: e226.

Bonfante, P. and Genre, A. 2008. Plants and arbuscular mycorrhizal fungi: An evolutionary developmental perspective. *Trends Plant Sci.*, 13: 492–498.

Bonfante, P. and Genre, A. 2010. Mechanisms underlying beneficial plant–fungus interactions in mycorrhizal symbiosis. *Nat. Commun.*, 1: 48.

Bonfante, P. and Requena, N. 2011. Dating in the dark: how roots respond to fungal signals to establish arbuscular mycorrhizal symbiosis. *Curr. Opin. Plant Biol.*, 4: 451–7.

Branscheid, A., Sieh, D., Pant, B.D., May, P., Devers, E.A. and Elkrog, A. 2010. Expression pattern suggests a role of MiR399 in the regulation of the cellular response to local Pi increase during arbuscular mycorrhizal symbiosis. *Mol. Plant Microbe Interac.*, 23: 915–926.

Braunberger, P.G., Miller, M.H. and Peterson, R.L. 1991. Effect of phosphorous nutrition on morphological characteristics of vesicular–arbuscular mycorrhizal colonization of maize. *New Phytol.*, 119: 107–113.

Breuillin, F., Schramm, J., Hajirezaei, M., Ahkami, A., Favre, P., Druege, U., Hause, B., Bucher, M., Kretzschmar, T., Bossolini, E., Kuhlemeier, C., Martinoia, E., Franken, P., Scholz, U. and Reinhardt, D. 2010. Phosphate systemically inhibits development of arbuscular mycorrhiza in *Petunia hybrida* and represses genes involved in mycorrhizal functioning. *Plant J.*, 64: 1002–1017.

Capoen, W., Sun, J., Wysham, D., Otegui, M.S., Venkateshwaran, M., Hirsch, S., Miwa, H., Downie, J.A., Morris, R.J., Ané, J.M. and Oldroyd, G.E.D. 2011. Nuclear membranes control symbiotic calcium signaling of legumes. *Proc. Natl. Acad. Sci. USA.*, 108: 14348–14353.

Catoira, R., Galera, C., de Billy, F., Penmetsa, R.V., Journet, E.P., Maillet, F., Rosenberg, C., Cook, D., Gough, C. and Dénarié, J. 2000. Four genes of – controlling components of a nod factor transduction pathway. *Plant Cell*, 12: 1647–1666.

Chabaud, M., Genre, A., Sieberer, B.J., Faccio, A., Fournier, J., Novero, M., Barker, D.G. and Bonfante, P. 2011. Arbuscular mycorrhizal hyphopodia and germinated spore exudates trigger Ca^{2+} spiking in the legume and nonlegume root epidermis. *New Phytol.*, 189: 347–355.

Charpentier, M., Bredemeier, R., Wanner, G., Takeda, N., Schleiff, E. and Parniske, M. 2008. *Lotus japonicus* CASTOR and POLLUX are ion channels essential for perinuclear calcium spiking in legume root endosymbiosis. *Plant Cell*, 20: 3467–3479.

Charpentier, M., and Oldroyd, G. (2010). How close we are to N fixing cereals? *Curr.Opin. Plant Biol.*, 13: 556–64.

Cook, C.E., Whichard, L.P., Turner, B., Wall, M.E. and Egley, G.H. 1966. Germination of witchweed (Striga Lutea Lour.): Isolation and properties of a potent stimulant. *Science*, 154: 1189–1190.

Czarnecki, O., Yang, J., Weston, D.J., Tuskan, G.A. and Chen, J.G. 2013. A dual role of strigolactones in phosphate acquisition and utilization in plants. *Int. J.Mol. Sci.*, 14: 7681–7701.

El Ghachtouli, N., Paynot, M., Martin–Tanguy, J., Morandi, D. and Gianinazzi, S. 1996. Effect of polyamines and polyamine biosynthesis inhibitors on spore germination and hyphal growth of *Glomus mosseae*. *Mycol. Res.*, 100: 597–600.

Feddermann, N., Muni, R.R., Zeier, T., Stuurman, J., Ercolin, F., Schorderet, M. and Reinhardt, D. 2010. The PAM1 gene of petunia, required for intracellular accommodation and morphogenesis of arbuscular mycorrhizal fungi, encodes a homologue of VAPYRIN. *Plant J.*, 64: 470–481.

Floss, D.S., Levy, J.G., Levesque–Tremblay, V., Pumplin, N. and Harrison, M.J. 2013. DELLA proteins regulate arbuscule formation in arbuscular mycorrhizal symbiosis. *Proc. Natl. Acad. Sci., U S A.* 110: E5025–5034.

Floss, D.S., Schliemann, W., Schmidt, J., Strack, D. and Walter, M.H. 2008. RNA interference–mediated repression of MtCCD1 in mycorrhizal roots of –causes accumulation of C27 apocarotenoids, shedding light on the functional role of CCD1. *Plant Physiol.*, 148: 1267–1282.

Foo, E. 2013. Auxin influences strigolactones in pea mycorrhizal symbiosis. *J. Plant Physiol.*, 170 :523–528.

Foo, E., Ross, J.J., Jones, W.T. and Reid, J.B. 2013. Plant hormones in arbuscular mycorrhizal symbioses: an emerging role for gibberellins. *Ann.Bot.*, 111: 769–779.

Fournier, J., Timmers, A.C.J., Sieberer, B.J., Jaunea, A., Chabaud, M. and Barker, D.G. 2008. Mechanism of infection thread elongation in root hairs of –and dynamic interplay with associated rhizobial colonization. *Plant Physiol.*, 148: 1985–1995.

García–Garrido, J.M., Morcillo, R.J., Rodríguez, J.A. and Bote, J.A. 2010. Variations in the mycorrhization characteristics in roots of wild–type and ABA–deficient tomato are accompanied by specific transcriptomic alterations. *Mol. Plant Microbe Interac.*, 23: 651–664.

Geil, R.D., Peterson, R.L. and Guinel, F.C. 2001. Morphological alterations of pea (*Pisum sativum* cv. Sparkle) arbuscular mycorrhizas as a result of exogenous ethylene treatment. *Mycorrhiza* 11: 137–143.

Genre, A. and Bonfante, P. 1998. Actin vs tubulin configuration in arbuscule-containing cells from mycorrhizal tobacco roots. *New Phytol.*, 140 (4): 745–752.

Genre, A., Chabaud, M., Balzergue, C., Puech Pagès, V., Novero, M., Rey, T., Fournier, J., Rochange, S., Bécard, G., Bonfante, P. and Barker, D.G. 2013. Short chain chitin oligomers from arbuscular mycorrhizal fungi trigger nuclear Ca^{2+} spiking in –roots and their production is enhanced by strigolactone. *New Phytol.*, 198: 190–202.

Genre, A., Chabaud, M., Faccio, A., Barker, D.G. and Bonfante, P. 2008. Prepenetration apparatus assembly precedes and predicts the colonization patterns of arbuscular mycorrhizal fungi within the root cortex of both and *Daucus carota*. *Plant Cell*, 20: 1407–1420.

Genre, A., Chabaud, M., Timmers, T., Bonfante, P. and Barker, D.G. 2005. Arbuscular mycorrhizal fungi elicit a novel intracellular apparatus in –root epidermal cells before infection. *Plant Cell*, 17: 3489–3499.

Gobbato, E., Marsh, J.F., Vernié, T., Wang, E., Maillet, F., Kim, J., Miller, J.B., Sun, J., Bano, S.A. and Ratet, P. 2012. A GRAS type transcription factor with a specific function in mycorrhizal signaling. *Curr. Biol.*, 22: 2236–2241.

Gobbato, E., Wang, E., Higgins, G., Bano, S., Henry, C., Schultze, M. and Oldroyd, G. 2013. RAM1 and RAM2 function and expression during Arbuscular Mycorrhizal Symbiosis and Aphanomyces euteiches colonization. *Plant Signal. Behav.*, 8: e26049.

Gomez–Roldan, V., Fermas, S., Brewer, P.B., Puech–Pages, V., Dun, E.A., Pillot, J.P., Letisse, F., Matusova, R., Danoun, S. and Portais, J.C. 2008. Strigolactone inhibition of shoot branching. *Nature*, 455: 189–194.

Gosling, P.A., Hodge, G., Goodlass, G.D. and Bending, 2006. Arbuscular mycorrhizal fungi and organic farming. *Agric. Ecosyst. Environ.*, 113:17–35.

Groth, M., Takeda, N., Perry, J., Uchida, H., Dräxl, S., Brachmann, A., Sato, S., Tabata, S., Kawaguchi, M., Wang, T.L. and Parniske, M. 2010. NENA, a *Lotus japonicus* homolog of Sec13, is required for rhizodermal infection by arbuscular mycorrhiza fungi and rhizobia but dispensable for cortical endosymbiotic development. *Plant Cell*, 22: 2509–2526.

Gu, M., Chen, A., Dai, X., Liu, W. and Xu, G. 2011. How does phosphate status influence the development of the arbuscular mycorrhizal symbiosis? *Plant Signal Behav.*, 6: 1300–1304.

Guether, M., Neuhäuser, B., Balestrini, R., Dynowski, M., Ludewig, U. and Bonfante, P. 2009. A mycorrhizal–specific ammonium transporter from *Lotus japonicus* acquires nitrogen released by arbuscular mycorrhizal fungi. *Plant Physiol.*, 150: 73–83.

Gutjahr, C., Casieri, L. and Paszkowski, U. 2009. *Glomus intraradices* induces changes in root system architecture of rice independently of common symbiosis signaling. *New Phytol.*, 182: 829–837.

Gutjahr, C., Radovanovic, D., Geoffroy, J., Zhang, Q., Siegler, H., Chiapello, M., Casieri, L., An, K., An, G. and Guiderdoni, E. 2012. The half–size ABC transporters STR1 and STR2 are indispensable for mycorrhizal arbuscule formation in rice. *Plant J.*, 69: 906–920.

Hanlon, M.T. and Coenen, C. 2011. Genetic evidence for auxin involvement in arbuscular mycorrhiza initiation. *New Phytol.*, 189: 701–709.

Harrison, M.J. (2005). Signaling in the arbuscular mycorrhizal symbiosis. *Annu. Rev. Microbiol.*, 59: 19–42.

Hause, B. and Schaarschmidt, S. 2009. The role of jasmonates in mutualistic symbiosis between plants and soil–born microorganisms. *Phytochemistry*, 70: 1589–1599.

Hayward, A., Stirnberg, P., Beveridge, C. and Leyser, O. 2009. Interactions between auxin and strigolactone in shoot branching control. *Plant Physiol.*, 151: 400–412.

Helber, N., Wippel, K., Sauer, N., Schaarschmidt, S., Hause, B. and Requena, N. 2011. A versatile monosaccharide transporter that operates in the arbuscular mycorrhizal fungus *Glomus* sp. is crucial for the symbiotic relationship with plants. *Plant Cell*, 23: 3812–3823.

Herrera–Medina, M.J., Steinkellner, S., Vierheilig, H., Ocampo, J.A. and García–Garrido, J.M. 2007. Abscisic acid determines arbuscule development and functionality in the tomato arbuscular mycorrhiza. *New Phytol.*, 175: 554–564.

Hohnjec, N., Vieweg, M.F., Puhler, A., Becker, A. and K uster, H. 2005. Overlaps in the transcriptional profiles of –roots inoculated with two different *Glomus* fungi provide insights into the genetic program activated during arbuscular mycorrhiza. *Plant Physiol.*, 137: 1283–1301.

Horváth, B., Yeun, L.H. Domonkos, A., Halász, G., Gobbato, E., Ayaydin, F., Miró, K., Hirsch, S., Sun, J. and Tadege, M. 2011. IPD3 is a member of the common symbiotic signaling pathway required for rhizobial and mycorrhizal symbiosis. *Mol. Plant Microbe Interac.*, 24: 1345–1358.

Isayenkov, S., Mrosk, C., Stenzel, I., Strack, D. and Hause, B. 2005. Suppression of allene oxide cyclase in hairy roots of –reduces jasmonate levels and the degree of mycorrhization with *Glomus intraradices*. *Plant Physiol.*, 139:1401–1410.

Ishii, T., Shrestha, Y.H., Matsumoto, I. and Kadoya, K. 1996. Effect of ethylene on the growth of vesicular–arbuscular mycorrhizal fungi and on the mycorrhizal formation of trifoliate orange roots. *J. Jpn. Soc. Hortic. Sci.*, 65: 525–529.

Ivanov, S., Fedorova, E.E., Limpens, E., De Mita, S., Genre, A., Bonfante, P. and Bisseling, T. 2012. Rhizobium–legume symbiosis shares an exocytotic pathway required for arbuscule formation. *Proc. Natl. Acad. Sci., USA.* 109: 8316–8321.

Javot, H., Penmetsa, R.V., Terzaghi, N., Cook, D.R. and Harrison, M.J. 2007. A phosphate transporter indispensable for the arbuscular mycorrhizal symbiosis. *Proc. Natl. Acad. Sci., USA.* 104: 1720–1725.

Javot, H., Pumplin, N. and Harrison, M.J. 2007. Phosphate in the arbuscular mycorrhizal symbiosis: Transport properties and regulatory roles. *Plant Cell Environ.*, 30: 310–322.

Jentschel, K., Thiel, D., Rehn, F. and Ludwig–Mu¨ller, J. 2007. Arbuscular mycorrhiza enhances auxin levels and alters auxin biosynthesis in *Tropaeolum majus* during early stages of colonization. *Physiol Plant.*, 129: 320–333.

Kalo, P., Gleason, C., Edwards, A., Marsh, J., Mitra, R.M., Hirsch, S., Jabob, J., Sims, S., Long, S.R. and Kaldorf,Rogers, J. 2005. Nodulation signaling in legumes requires NSP2A member off GRAS family of transcriptional regulators. *Science*, 308: 1786–89.

Kanamori, N., Madsen, L.H., Radutoiu, S., Frantescu, M., Quistgaard, E.M.H., Miwa, H., Downie, J.A., James, E.K., Felle, H.H. and Haaning, L.L. 2006. A nucleoporin is required for induction of Ca2+ spiking in legume nodule development and essential for rhizobial and fungal symbiosis. *Proc. Natl. Acad. Sci. USA.*, 103: 359–364.

Kiirika, L.M., Bergmann, H.F., Schikowsky, C., Wimmer, D., Korte, J., Schmitz, U., Niehaus, K. and Colditz, F. 2012. Silencing of the Rac1 GTPase MtROP9 in *Medicago truncatula* stimulates early mycorrhizal and oomycete root colonizations but negatively affects rhizobial infection. *Plant Physiol.*, 159: 501–516.

Kistener, C. and Parniske M. 2002. Evolution of signal transduction in intracellular symbiosis. *Trends Plant Sci.*, 7: 511–18.

Koltai, H., LekKala, S.P., Bhattacharya, C., Mayzlish–Gati, E., Resnick, N., Wininger, S., Dor, E., Yoneyama, K., Hershenhorn, J. and Joel, D.M. 2010. A tomato strigolactone–impaired mutant displays aberrant shoot morphology and plant interactions. *J. Exp. Bot.*, 61: 1739–1749.

Kretzschmar, T., Kohlen, W., Sasse, J., Borghi, L., Schlegel, M., Bachelier, J.B., Reinhardt, D., Bours, R., Bouwmeester, H.J. and Martinoia, E. 2012. A petunia ABC protein controls strigolactone–dependent symbiotic signalling and branching. *Nature*, 483: 341–344.

Lin, S.I., Chiang, S.F., Lin, W.Y., Chen, J.W., Tseng, C.Y. and Wu, P.C. 2008. Regulatory network of microRNA399 and PHO2 by systemic signaling. *Plant Physiol.*, 147: 732–746.

Liu, J., Elmore, J.M., Fuglsang, A.T., Palmgren, M.G., Staskawicz, B.J. and Coaker, G. 2009. RIN4 functions with plasma membrane H+–ATPases to regulate stomatal apertures during pathogen attack. *PLoS Biol.,* 7: e1000139.

Liu, T.Y., Huang, T.K., Tseng, C.Y., Lai, Y.S., Lin, S.I. and Lin, W.Y. 2012. PHO2–dependent degradation of PHO1 modulates phosphate homeostasis in Arabidopsis. *Plant Cell,* 24: 2168–2183.

Liu, W., Kohlen, W., Lillo, A., Op den Camp, R., Ivanov, S., Hartog, M., Limpens, E., Jamil, M., Smaczniak, C. and Kaufmann, K. 2011. Strigolactone biosynthesis in –and rice requires the symbiotic GRAS–type transcription factors NSP1 and NSP2. *Plant Cell,* 23: 3853–3865.

López–Ráez, J., Kohlen, W., Charnikhova, T., Mulder, P., Undas, A.K., Sergeant, M.J., Verstappen, F., Bugg, T.D.H., Thompson, A.J., Ruyter–Spira, C. and Bouwmeester, H. 2010. Does abscisic acid affect strigolactone biosynthesis? *New Phytol.,* 187: 343–354.

Maillet, F., Poinsot, V., André, O., Puech Pagès, V., Haouy, A., Gueunier, M., Cromer, L., Giraudet, D., Formey, D., Niebel, A., Martinez, E.A., Driguez, H., Bécard, G. and Dénarié, J. (2011). Fungal lipochitooligosaccharide symbiotic signals in arbuscular mycorrhiza. *Nature,* 469: 58–63.

Martín–Rodríguez, J.A., León–Mocillo, R., Vierheilig, H., Ocampo, J.A., Ludwig–Müller, J. and García–Gardio, J.M. 2011. Ethylene–dependent/ethylene–independent ABA regulation of tomato plants colonized by arbuscular mycorrhizal fungi. *New Phytol.,* 190: 193–205.

Martín–Rodriguez, J..A, León–Morcillo, R., Vierheilig, H., Ocampo Bote, J.A., Ludwig–Müller, J. and García–Garrido, J.M. 2010. Mycorrhization of the *notabilis* and *sitiens* tomato mutants in relation to abscisic acid and ethylene contents. *J.Plant Physiol.,* 167: 606–613.

Meixner, C., Ludwig–Mu'ller, J., Miersch, O., Gresshoff, P., Staehelin, C. and Vierheilig, H. 2005. Lack of mycorrhizal autoregulation and phytohormonal changes in the supernodulating soybean mutants. *Planta,* 222: 709–715.

Miller, J.B., Pratap, A., Miyahara, A., Zhou, L., Bornemann, S., Morris, R.J. and Oldroyd, G.E. 2013. Calcium/calmodulin–dependent protein kinase is negatively and Positively regulated by calcium, providing a mechanism for decoding calcium responses during symbiosis signaling. *Plant Cell,* 25: 5053–5066.

Mohanta, T. K. and Bae, H. 2015. Functional genomics and signaling events in mycorrhizal symbiosis, *J.Plant Interac.,* 10:1, 21-40.

Nadal, M. and Paszkowski, U. 2013. Polyphony in the rhizosphere: presymbiotic communication in arbuscular mycorrhizal symbiosis. *Curr. Opin. Plant Biol.,*16: 473–479.

Nagahashi, G., Douds, D.D. and Jr Abney, G.D. 1996. Phosphorus amendment inhibits hyphal branching of the VAM fungus *Gigaspora margarita* directly and indirectly through its effect on root exudation. *Mycorrhiza,* 6: 403–408.

Nouri, E., Breuillin–Sessoms, F., Feller, U. and Reinhardt, D. 2014. Phosphorus and nitrogen regulate arbuscular mycorrhizal symbiosis in *Petunia hybrida*. *PLos ONE*, 9: e90841.

Oldroyd, G.E.D. 2013. Speak, friend, and enter: Signalling systems that promote beneficial symbiotic associations in plants. *Nat. Rev. Microbiol.*, 11: 252–263.

Ongaro, V. and Leyser, O. 2008. Hormonal control of shoot branching. *J. Exp.Bot.*, 59: 67–74.

Overvoorde, P., Fukakai, H. and Beekman, T. 2010. Auxin control of root development. *Cold Spring Harbor Perspectives in Biology*, 2: a001537.

Parniske, M. 2008. Arbuscular mycorrhiza: The mother of plant root endosymbiosis. *Nat.Rev. Microbiol.*, 6: 763–775.

Paszkowski, U., Kroken, S., Roux, C. and Briggs, S.P. 2002. Rice phosphate transporters include an evolutionarily divergent gene specifically activated in arbuscular mycorrhizal symbiosis. *Proc. Natl. Acad. Sci. USA.*, 99: 13324–13329.

Péret, B., Clément, M., Nussaume, L. and Desnos, T. 2011. Root developmental adaptation to phosphate starvation: Better safe than sorry. *Trends Plant Sci.*, 16: 442–450.

Plaxton, W.C. and Tran, H.T. 2011. Metabolic adaptations of phosphate starved plants. *Plant Physiol.*, 156: 1006–1015.

Pumplin, N. and Harrison, M.J. 2009. Live–cell imaging reveals periarbuscular membrane domains and organelle location in –roots during arbuscular mycorrhizal symbiosis. *Plant Physiol.*, 151: 809–819.

Pumplin, N., Mondo, S.J., Topp, S., Starker, C.G., Gantt, J.S. and Harrison, M.J. 2010. Vapyrin is a novel protein required for arbuscular mycorrhizal symbiosis. *Plant J.*, 61: 482–494.

Recorbet, G., Abdallah, C., Renaut, J., Wipf1, D., and Gaudot, E. D. 2013. Protein actors sustaining arbuscular mycorrhizal symbiosis: underground artists break the silence. *New phytol.*, 199: 26–40.

Redecker, D. 2000. Glomalean fungi from the Ordovician. *Science*, 289: 1920–1921.

Roberts, N.J., Morieri, G., Kalsi, G., Rose, A., Stiller, J., Edwards, A., Xie, F., Gresshoff, P.M., Oldroyd, G.E.D., Downie, J.A. and Etzler, M.E. 2013. Rhizobial and mycorrhizal symbioses in *Lotus japonicus* require lectin nucleotide phosphohydrolase, which acts upstream of calcium signaling. *Plant Physiol.*, 161: 556–567.

Schaarschmidt, S., Gonzalez, M.C., Roitsch, T., Strack, D., Sonnewald, U. and Hause, C. 2007. Regulation of arbuscular mycorrhization by carbon. The symbiotic interaction cannot be improved by increased carbon availability accomplished by root–specifically enhanced invertase activity. *Plant Physiol.*, 143: 1827–1840.

Schüßler, A., Schwarzott, D. and Walker, C. 2001. A new fungal phylum, the Glomeromycota: phylogeny and evolution. *Mycol. Res.*,105: 1413–1421.

Shaul–Keinan, O., Gadkar, V., Ginzberg, I., Gru"nzweig, J.M., Chet, I., Elad, Y., Wininger, S., Belausov, E., Eshed, Y. and Atzmon, N. 2002. Hormone concentrations in tobacco roots change during arbuscular mycorrhizal colonization with *Glomus intraradices. New Phytol.*, 154: 501–507.

Siciliano, V., Genre, A., Balestrini, R., Cappellazzo, G., deWit, P.J., Bonfante, P. 2007 (a). Transcriptome analysis of arbuscular mycorrhizal roots during development of the prepenetration apparatus. *Plant Physiol.*, 144: 1455–1466.

Siciliano, V., Genre, A., Balestrini, R., Dewit, P.J. and Bonfante, P. 2007(b). Prepenetration apparatus formation during AM infection is associated with a specific transcriptome response in epidermal cells. *Plant Signal. Behav.*, 2: 533–535.

Sieberer, B.J., Chabaud, M., Timmers, C.A., Monin, A., Fournire, J. and Barker, D.G. 2009. A nuclear–targeted cameleon demonstrates intranuclear Ca^{2+} spiking in –root hairs in response to rhizobial nodulation factors. *Plant Physiol.*, 151: 1197–1206.

Singh, S., Katzer, K., Lambert, J., Cerri, M. and Parniske, M. 2014. CYCLOPS, A DNA–binding transcriptional activator, orchestrates symbiotic root nodule development. *Cell Host Microbe*, 15: 139–152.

Singh, S. and Parniske, M. 2012. Activation of calcium– and calmodulin–dependent protein kinase (CCaMK), the central regulator of plant root endosymbiosis. *Curr. Opin. Plant Biol.*, 15: 444–453.

Sirrenberg, A., Go"bel, C., Grond, S., Czempinski, N., Ratzinger, A., Karlovsky, P., Santos, P., Feussner, I. and Pawlowski, K. 2007. *Piriformospora indica* affects plant growth by auxin production. *Physiol Planta.*,131: 581–589.

Smit, P., Raedts, J., Portyanko, V., Debelle,´ F., Gough, C., Bisseling, T. and Geurts, R. 2005. NSP1 of the GRAS protein family is essential for rhizobial Nod factor-induced transcription. *Science*, 308: 1789–1791.

Smith, S.E. and Read, D.J. 1997. Mycorrhizal symbiosis. Academic Press, London, etc

Strack, D. and Fester, T. 2006. Isoprenoid metabolism and plastid reorganization in arbuscular mycorrhizal roots. *New Phytol.*, 172: 22–34.

Takeda, N., Sato, S., Asamizu, E., Tabata, S. and Parniske, M. 2009. Apoplastic plant subtilases support arbuscular mycorrhiza development in *Lotus japonicus. Plant J.*, 58: 766–777.

Tamura, Y., Kobae, Y., Mizuno, T. and Hata, S. 2014. Identification and expression analysis of arbuscular mycorrhiza–inducible phosphate transporter genes of soybean. *Biosci. Biotechnol. Biochem.*, 76: 309–313

Torres de Los Santos, R., Vierheilig, H., Ocampo, J.A. and García–Garrido, J.M. (2011). Altered pattern of arbuscular mycorrhizal formation in tomato ethylene mutants. *Plant Signal. Behavi.*, 6: 755–758.

Vogel, J.T., Walter, M.H., Giavalisco, P., Lytovchenko, A., Kohlen, W., Charnikhova, T., Simkin, A.J., Goulet, C., Strack, D. and Bouwmeester, H.J. 2010. SlCCD7 controls strigolactone biosynthesis, shoot branching and mycorrhiza–induced apocarotenoid formation in tomato. *Plant J.*, 61: 300–311.

Vrebalov, J., Ruezinskly, D., Padamnabhan, V., White, R., Medrano, D., Schuch, W. and Giovannoni, J. 2002. A MADS–box gene necessary for fruit ripening in tomato *ripening inhibitor rin locus. Science,* 296.

Wang, E., Schornack, S., Marsh, J.F., Gobbato, E., Schwessinger, B., Eastmond, P., Schultze, M., Kamoun, S. and Oldroyd, G.E.D. 2012. A common signaling process that promotes mycorrhizal and oomycete colonization of plants. *Curr. Biol.,* 4: 2242–6.

Yoneyama, K., Xie, X., Kusumoto, D., Sekimoto, H., Sugimoto, Y., Takeuchi, Y. and Yoneyama, K. 2012. Nitrogen deficiency as well as phosphorus deficiency in sorghum promotes the production and exudation of 5 deoxystrigol, the host recognition signal for arbuscular mycorrhizal fungi and root parasites. *Planta,* 227: 125–132.

Yoshida, S., Kameoka, H., Tempo, M., Akiyama, K., Umehara, M., Yamaguchi, S., Hayashi, H., Kyozuka, J. and Shirasu, K. 2012. The D3 F–box protein is a key component in host strigolactone responses essential for arbuscular mycorrhizal symbiosis. *New Phytol.,* 196: 1208–1216.

Yu, N., Luo, D., Zhang, X., Liu, J., Wang, W., Jin, Y., Dong, W., Liu, J., Liu, H. and Yang, W. 2014. A DELLA protein complex controls the arbuscular mycorrhizal symbiosis in plants. *Cell Res.,* 24: 130–133.

Zhang, Q., Blaylock, L.A. and Harrison, M.J. 2010. Two half ABC transporters are essential for arbuscule development in arbuscular mycorrhizal symbiosis. *Plant Cell,* 22: 1483–1497.

Zhou, F., Lin, Q., Zhu, L., Ren, Y., Zhou, K., Shabek, N., Wu, F., Mao, H., Dong, W. and Gan, L. 2013. D14–SCFD3–dependent degradation of D53 regulates strigolactone signalling. *Nature,* 504: 406–410.

Zso¨go¨n, A., Lambais, M.R., Benedito, V.A., Figueira, A.V.O. and Peres, L.E.P. 2008. Reduced arbuscular mycorrhizal colonization in tomato ethylene mutants. *Sci. Agric.,* 65: 259–267.

Index